Operational Amplifiers

Operational Amplifiers

Fifth edition

George Clayton
and
Steve Winder

OXFORD AMSTERDAM BOSTON LONDON NEW YORK PARIS SAN DIEGO
SAN FRANCISCO SINGAPORE SYDNEY TOKYO

Newnes
An imprint of Elsevier Science
Linacre House, Jordan Hill, Oxford OX2 8DP
200 Wheeler Road, Burlington, MA01803

First published by Newnes-Butterworth 1971
Second edition 1979
Reprinted by Butterworths 1981, 1982, 1983, 1985, 1986
Third edition 1992
Fourth edition 2000
Fifth edition 2003

British Library Cataloguing in Publication Data
A catalogue record for this book is available from the British Library

ISBN 07506 5914 9

Typeset by Newgen Imaging Systems (P) Ltd., Chennai, India
Printed and bound in Great Britain by Biddles Ltd
www.biddles.co.uk

For information on all Newnes publications visit our website at www.newnespress.com

Contents

Preface

Operational amplifiers have been in use for many years. Originally they were built using discrete transistor circuits, but the development of the integrated circuit (IC) has revolutionized analogue circuit design. The operational amplifier was one of the first analogue integrated circuits, because of its usefulness as a building block in many circuit designs. The popularity of the operational amplifier has resulted in a shortened name 'op-amp' to be commonplace. The term op-amp will be used extensively in this book.

The op-amp's popularity stems from its versatility. It is a high-gain DC amplifier that has differential inputs; the output voltage is the voltage difference between the two inputs multiplied by the gain. Passive components can be used to provide feedback, and this controls the gain and function of the op-amp circuit overall. Passive negative feedback components result in a linear response, i.e. the output is proportional to the input. Passive positive feedback results in switching or oscillation. Sometimes active components such as transistors and diodes are used in the feedback loop to give a non-linear response; typical applications are logarithmic amplifiers or precision rectifiers.

My interest in op-amp circuits began while I was an apprentice technician. One of the first books that I bought was Clayton's *Operational Amplifiers* (first edition). It is therefore fitting that I should be asked by the publisher to edit the fifth edition. In my previous employment as a circuit design engineer for British Telecom, and now as a field applications engineer for Supertex Inc., I have used op-amps in hundreds of circuits. For me, one valuable application is in active filter circuits (refer to Chapter 9 and to my book, *Analog and Digital Filter Design,* ISBN 0–7506–7547–0).

In this fifth edition of *Operational Amplifiers* I have added more on active filters, especially gyrator and frequency-dependent negative resistance circuits. Throughout the book I have updated and added material, where appropriate. This includes the important practical guidelines about passive components used in op-amp circuits. Although placed near the end of the book, in Chapter 10, this information is important and should not be overlooked.

Steve Winder, 2002

Acknowledgements

Figures 2.1, 2.3, 2.30, 3.4–3.9, 4.22–4.25, 7.25, 7.29, 9.30, 9.31, 9.36, 9.37, 10.26 and 10.28 were obtained from IMSI's MasterClips® and MasterPhotos™ Premium Image Collection, 1895 Francisco Blvd East, San Rafael, CA 94901-5506, USA.

Appendix 5 is reproduced with the permission of Maxim Integrated Products Inc. Maxim is not responsible for any errors or omissions in the reproduction of these data sheets. Before using any information in these data sheets for design purposes, please contact Maxim to ensure you have the current version.

1 Fundamentals

1.1 Introduction

The term 'operational amplifier' describes an important amplifier circuit that can form the basis of audio and video amplifiers, filters, buffers, line drivers, instrumentation amplifiers, comparators, oscillators, and many other analogue circuits. The operational amplifier is commonly referred to as an op-amp. Although the op-amp circuit can be designed from discrete components, it is almost always used in integrated circuit (IC) form.

The op-amp is a simple building block. It has two inputs, one is called the inverting input (often labelled −) and the other is called the non-inverting input (often labelled +). Usually op-amps have a single output, but special op-amps used in radio frequency circuits have two outputs. Only single output devices will be described in detail, and the symbol used in circuit diagrams is shown in Figure 1.1.

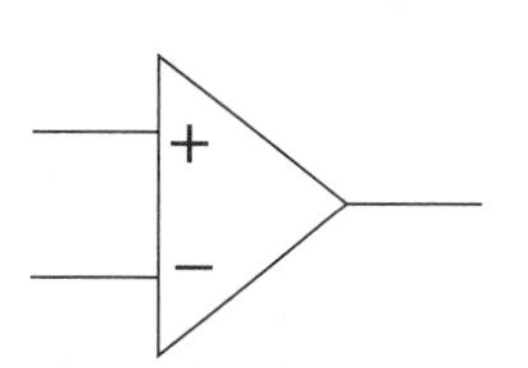

Figure 1.1 *Op-amp symbol*

The op-amp also has two power supply connections, one for the positive rail and one for the negative rail. Many op-amp circuits have a mid-rail supply connected to earth, although the op-amp itself has no specific mid-rail supply connection. Some op-amps are specifically designed for single supply operation, and more details of these are provided later.

The op-amp is a high gain DC amplifier (the DC gain is usually >100 000; or >100 dB). With suitable capacitive coupling, the op-amp is used in many AC amplifier circuits. The output voltage is simply the difference in voltage between the inverting and non-inverting inputs, multiplied by the gain. Thus, the op-amp is a differential amplifier. If the inverting (−) input has the higher potential, the output voltage will become more negative. If the non-inverting (+) input has the higher potential, the output will become more positive. Since the gain is very high, the differential voltage between the input terminals is usually very small.

The op-amp must have feedback in order to perform useful functions. Most designs use negative feedback to control the gain and to provide linear operation. Negative feedback is provided by components, such as resistors, connected between the op-amp's output and its inverting (−) input. Non-linear circuits, such as comparators and oscillators, use positive feedback by having components connected between the op-amp's output and its non-inverting (+) input.

It is not essential that the user of op-amps is familiar with the details of their internal circuits. However, a little knowledge of the internal circuits does help understanding, particularly the input and output circuits. The user should understand the function of the external terminals provided by the manufacturer. In order to be able to select the best amplifier for a particular application, the user should be familiar with the terms used to specify the op-amp's performance.

1.2 The ideal op-amp

When analysing feedback circuits, it is convenient to assume that the amplifier has certain ideal characteristics.

- The output of the ideal differential input amplifier depends only on the difference between the voltages applied to the two input terminals.
- The performance is entirely dependent on input and feedback networks.
- No current flows into the amplifier input terminals.
- The frequency response extends from zero to infinity, ensuring a response to all DC and AC signals, with zero response time and no phase change with frequency.
- The amplifier is unaffected by the load.
- When the input signal voltage is zero, the output signal will also be zero – regardless of the input source resistance.

1.3 Feedback and the ideal op-amp

There are two basic ways of applying feedback to an op-amp: Figure 1.2(a) shows the inverting configuration, the non-inverting configuration being illustrated in Figure 1.2(b). In both circuits, the signal fed back from the output to the input is proportional to the output voltage. Feedback takes place via the resistor R_2 connected between the output and the inverting input terminal of the amplifier. Phase inversion through the amplifier ensures that the feedback is negative.

The action of both circuits may be understood if a small positive voltage e_ε is assumed to exist between the differential input terminals of the amplifier. The op-amp's output voltage will be equal to the negative supply rail, because of the infinite gain. The signal fed back will be in opposition to e_ε, so forcing the differential input voltage towards zero.

Now suppose that e_ε is a small negative voltage. The op-amp's output voltage will be equal to the positive supply rail and feedback is in opposition to e_ε. Again, this forces the differential input voltage towards zero. Thus, negative feedback always forces the differential input voltage to be zero.

This is an extremely important point and is worth restating in an alternative form. When the op-amp's output is fed back to the inverting input terminal, the output voltage will always take on that value required to drive

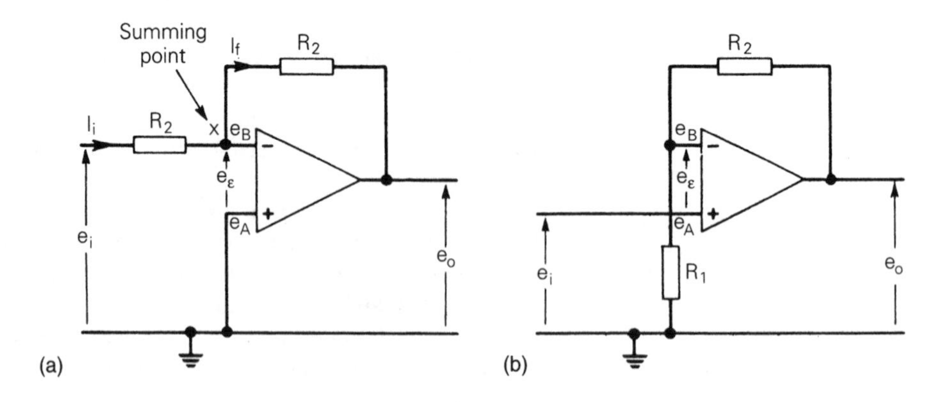

Figure 1.2 *Two basic feedback circuits*

the differential input voltage to zero. For an ideal op-amp having infinite gain, the error voltage e_ε is zero.

In the case of a practical op-amp having large but finite gain, the error voltage e_ε is small but non-zero. The effect of this error voltage will be discussed in the next chapter.

A second basic aspect of the ideal circuit follows from the assumed infinite input impedance of the amplifier. In the circuit of Figure 1.2(a), no current can flow into the op-amp so that any current arriving at the point X, as a result of an applied input signal, must flow through the feedback path R_2.

If instead of the single resistor R_1 connected to the inverting input terminal there are several alternative signal paths, the sum of these several currents arriving at point X must flow through the feedback path. It is for this reason that the phase-inverting input terminal of an operational amplifier (point X) is sometimes referred to as the amplifier summing point.

The two basic aspects of ideal performance are called the summing point restraints; they are so important that they are repeated again.

1. When negative feedback is applied to the ideal amplifier, the differential input voltage is zero.
2. No current flows into either input terminal of the ideal amplifier.

The two statements form the basis of all simplified analyses of operational feedback circuits; we use them to derive closed-loop gain expressions for the circuits of Figure 1.2.

In Figure 1.2(a), the non-inverting input is connected to earth. But with negative feedback, the inverting input has the same potential as point X, so this is known as a 'virtual earth'. Thus the current I_i flowing through R_1 is found simply by dividing the input voltage by the resistance of R_1. An alternative expression is to say the input voltage is I_i times the value of R_1. Since no current flows into the op-amp input, the currents through R_1 and R_2 are equal. The output voltage is the negative product of the I_i times the value of R_2. The gain (amplification, or A) is given by dividing the output voltage by the input voltage. This is

$$A = -\frac{I_i R_2}{I_i R_1}$$

The current I_i can be cancelled to give

$$A = -\frac{R_2}{R_1}$$

If R_2 is less than R_1, fractional gains are possible.

In the case of the non-inverting amplifier of Figure 1.2(b), the voltage at both inputs must be equal. No current flows into either of the op-amp's inputs, so potential divider R_1 and R_2 determine the voltage at the inverting input. So voltage $e_i = e_A$ applied to the non-inverting input causes the output voltage to become positive until the fraction at the inverting input is equal. The fraction is given by the expression:

$$e_B = \frac{e_o R_1}{(R_1 + R_2)}$$

Feedback forces the two inputs to have an equal potential, so $e_A = e_B$

$$e_A = e_i = \frac{e_o R_1}{(R_1 + R_2)}$$

The gain is e_o/e_i, so transposing the equation we get:

$$A = \frac{e_o}{e_i} = \frac{R_1 + R_2}{R_1}$$

$$A = 1 + \frac{R_2}{R_1}$$

Notice that the gain can never be less than 1. A short circuit between the output and the inverting input creates a buffer with unity gain. In theory, this buffer has infinite input impedance and zero output impedance.

The inverting and non-inverting amplifiers have two main differences. The first difference is the sign of the closed-loop gain. More important is the difference in effective input resistance that they present to the signal source e_i.

The effective input resistance of the ideal inverter measured at the amplifier summing point is zero. Feedback prevents the voltage at this point from changing; the point acts as a virtual earth. Note that any current supplied to this point does not actually flow to earth but flows through the feedback path R_2. The resistor R_1 thus determines the input current, I_i, in Figure 1.2(a). The input resistance presented to the signal source is equal to the value of R_1.

Consider the non-inverting circuit Figure 1.2(b) where the only connection to the non-inverting pin is the signal source. An ideal op-amp in this circuit takes no current from the signal source and thus has infinite input impedance.

The simple closed-loop expressions show that, in the ideal case, the gain depends only on the values of series and feedback components, not on the amplifier itself. Real amplifiers introduce departures from the ideal, and these are conveniently treated as errors. Errors can be made very small and one of the main features of the op-amp approach to analogue circuit design is the accuracy with which it is possible to set gain and impedance values.

1.4 More examples of the ideal op-amp at work

The ideal op-amp serves as a valuable starting point for a preliminary analysis of op-amp circuits. In this section we present a few more examples illustrating the usefulness of the ideal op-amp concept. Once the significance of the summing point restraints are firmly understood, ideal circuit analysis involves little more than the intelligent use of Ohm's law.

Remember that the ideal differential input op-amp, with negative feedback, will try to keep the differential input voltage close to zero. The output voltage takes on the value required to achieve this. In doing so, it causes all currents arriving at the inverting input to flow through the feedback resistor.

1.4.1 The ideal op-amp acts as a current-to-voltage converter

An ideal op-amp can act as a current-to-voltage converter. In the circuit of Figure 1.3, the ideal amplifier maintains its inverting input terminal at earth potential and forces any input current to flow through the feedback resistance. Thus $I_{in} = I_f$ and $e_o = -I_{in}R_f$.

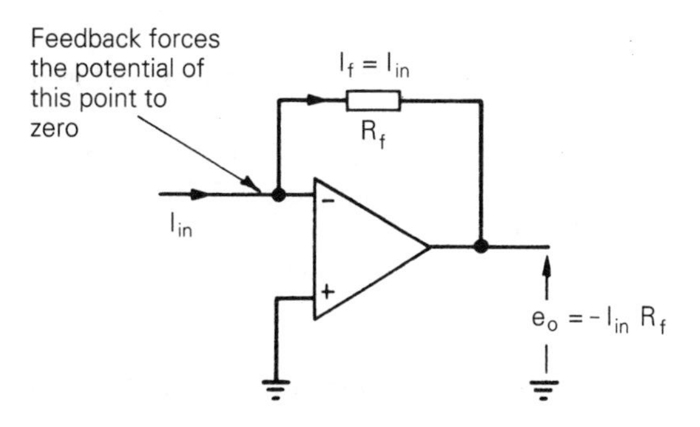

Figure 1.3 *An ideal op-amp acts as a current-to-voltage converter*

Notice that the circuit provides the basis for an ideal current measurement. It introduces zero voltage drop into the measurement circuit. The effective input impedance of the circuit, measured directly at the inverting input terminal, is zero.

1.4.2 The ideal op-amp adds voltages or currents independently

The principle involved in the current-to-voltage converter circuit of Figure 1.3 may be extended. In the ideal op-amp circuit of Figure 1.4(a) the op-amp forces the sum of the several currents arriving at the inverting input to flow through the feedback path (there is no where else for them to go). The inverting input terminal is forced to be at earth potential (a 'virtual earth') and the output voltage is thus:

$$e_o = -[I_1 + I_2 + I_3] \, . \, R_f$$

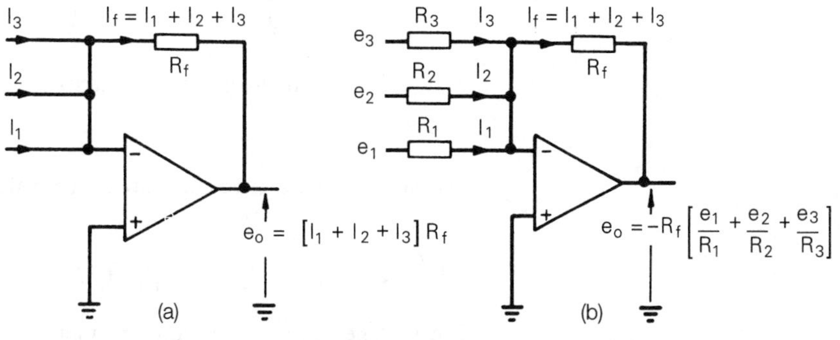

Figure 1.4 *An ideal op-amp adds currents and voltages independently*

In Figure 1.4(b) a number of input voltages are connected to resistors which meet at the inverting input terminal. The ideal op-amp maintains the inverting input at earth potential, thus input current is independently determined by each applied input voltage and series input resistor. The sum of the input currents is forced to flow through R_2 and the output voltage must take on a value that is equal to the sum of the input currents multiplied by R_2.

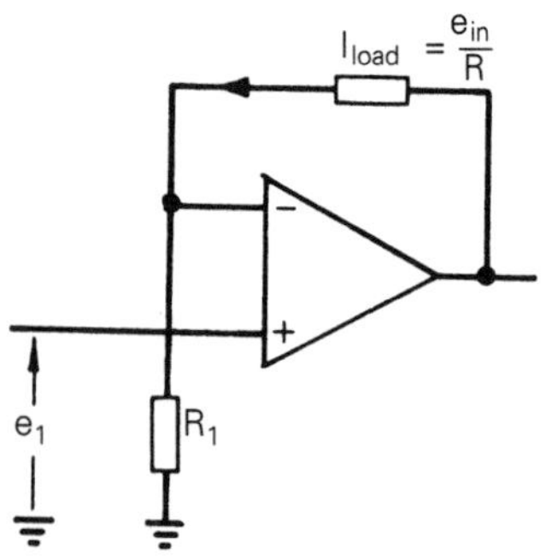

1.4.3 The ideal op-amp can act as a voltage-to-current converter

In maintaining its differential input voltage at zero, the amplifier shown in the circuit of Figure 1.5 forces a current $I = e_{in}/R$ to flow through the load in the feedback path. The value of this current is independent of the nature or size of the load.

Figure 1.5 *An ideal op-amp can act as a voltage-to-current converter*

1.4.4 The ideal op-amp can act as a perfect buffer

In the circuit of Figure 1.6 the amplifier output voltage must take on a value equal to the input voltage in order to force the differential input signal to zero. The ideal circuit has infinite input impedance, zero output impedance and unity gain, and acts as an ideal buffer stage.

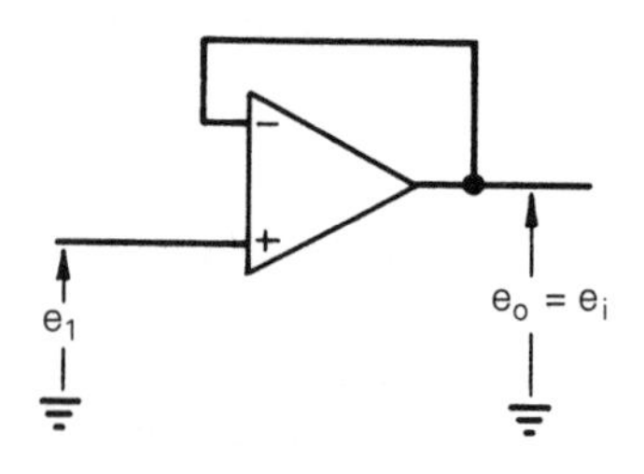

Figure 1.6 *An ideal op-amp can act as an ideal unity gain buffer*

1.4.5 The ideal op-amp can act differentially as a subtractor

The circuit shown in Figure 1.7 illustrates the way in which an op-amp can act differentially as a subtractor.

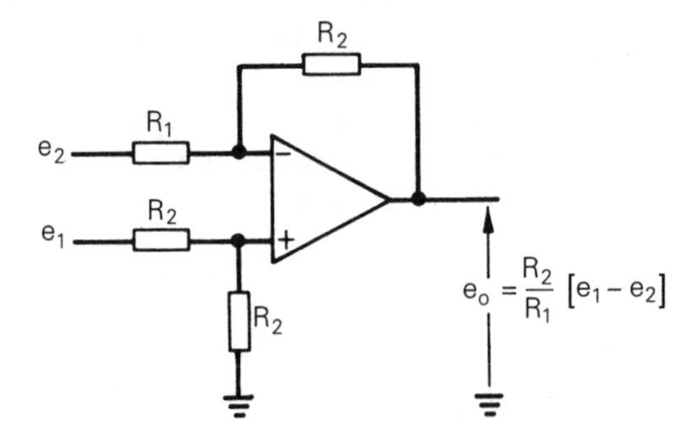

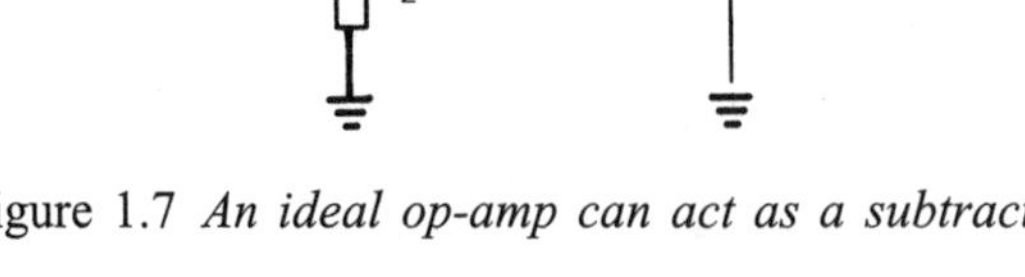

Figure 1.7 *An ideal op-amp can act as a subtractor*

The voltage at the inverting input terminal is (by superposition):

$$e^- = e_2 \frac{R_2}{R_1 + R_2} + e_o \frac{R_1}{R_1 + R_2}$$

The voltage at the non-inverting input is:

$$e^+ = e_1 \frac{R_2}{R_1 + R_2}$$

The op-amp forces $e^- = e^+$

$$\text{Thus } e_2 \frac{R_2}{R_1 + R_2} + e_o \frac{R_1}{R_1 + R_2} = e_1 \frac{R_2}{R_1 + R_2}$$

$$\text{or, } e_o = \frac{R_2}{R_1}[e_1 - e_2]$$

1.4.6 The ideal op-amp can act as an integrator

In the circuit of Figure 1.8, negative feedback is applied by the capacitor C connected between the output and the inverting input terminal. The amplifier output voltage acting via this capacitor maintains the inverting input terminal at earth potential and forces any current arriving at the inverting input terminal to flow as capacitor charging current.

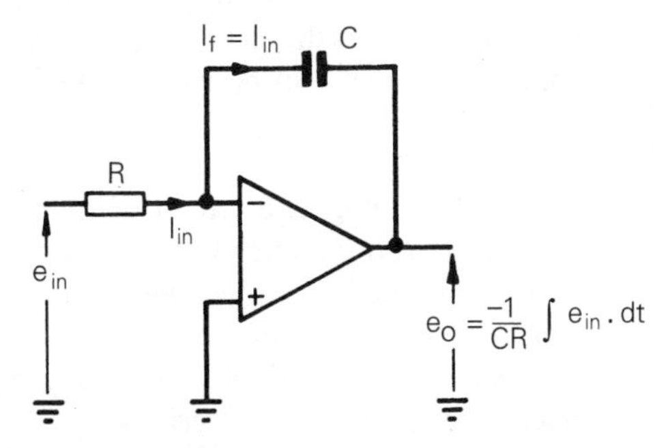

Figure 1.8 *An ideal op-amp acts as an integrator*

Thus:

$$I_{in} = \frac{e_m}{R} = C\frac{dV_C}{dt}$$

The output voltage is equal in magnitude but opposite in sign to the capacitor voltage. Therefore:

$$\frac{e_m}{R} = -C\frac{de_0}{dt}$$

$$e_0 = -\frac{1}{CR}\int e_{in}\,dt$$

The output is proportional to the integral with respect to time of the input voltage.

1.4.7 Limitations of the ideal op-amp concept

Real op-amps have characteristics that approach those of an ideal op-amp, but do not quite attain them. They have an open-loop gain, which is very large (in the region of 10^6) but not infinite. They have a large, but finite, input impedance. They draw small currents at their input terminals (bias currents). They require a small differential input voltage to give zero output voltage (the input offset voltage). And they do not completely reject common mode signals (finite common mode rejection ratio, or CMRR).

In our discussion of ideal op-amp circuits no mention has been made of frequency response characteristics. Real amplifiers have a frequency dependent gain, which can have a marked effect on the performance of op-amp circuits.

The above features of real op-amps cause the performance of circuits to differ from that predicted by an analysis based upon the assumption of ideal amplifier performance. In many respects the differences between real and ideal behaviour are quite small. In some aspects of performance, particularly those involving frequency dependent performance parameters, the differences are significant.

Chapter 2 presents detailed discussions about the parameters that are usually given on the data sheets of practical op-amps. Knowledge of these parameter values can be used to predict the behaviour of practical circuits.

1.5 Op-amp packages

Inexpensive integrated circuit op-amps are available, which are easy to use and allow working circuits to be built rapidly. The newcomer to op-amps is strongly advised to build a few of the basic op-amp circuits and practically evaluate their performance. This forms a useful learning and familiarization exercise, which is worth performing before delving more deeply into the finer aspects of op-amp performance.

A preliminary practical evaluation of op-amp applications is most conveniently carried out using a general-purpose op-amp type. There are several general-purpose amplifier types to choose from, such as the Texas Instruments TL071 or TLE2027.

As a user of op-amps, it is not necessary to have a detailed knowledge of their internal circuitry. Fortunately most general-purpose op-amps are pin compatible. It is the function of the external pin connections that the op-amp user is primarily concerned. The most common packages for op-amps are an 8-pin dual-in-line plastic package (known as DIL-8) and its surface mount equivalent, SO-8. Smaller surface-mount packages are available, these include the SOT23-5 shown in Figure 1.9.

Dual op-amps, where two op-amps are housed in the same package are available in 8-pin and 14-pin DIL or surface-mount packages. Quad op-amps, where four op-amps are housed together, are available in 14-pin DIL and surface-mount packages.

Op-amps are commonly used with dual power supplies. Input and output voltages are measured with respect to the potential of the power supply common terminal, which acts as the zero signal reference point or 'earth'. The use of dual supplies allows input and output voltages to swing both positive and negative with respect to the zero reference point.

Figure 1.9 *Op-amp packages*

Figure 1.10 shows the circuit connections which are required to make a practical form of the inverting amplifier circuit previously described in Section 1.3. Amplifier pins not shown in Figure 1.10 should be left with no connections made to them – their function will be described later.

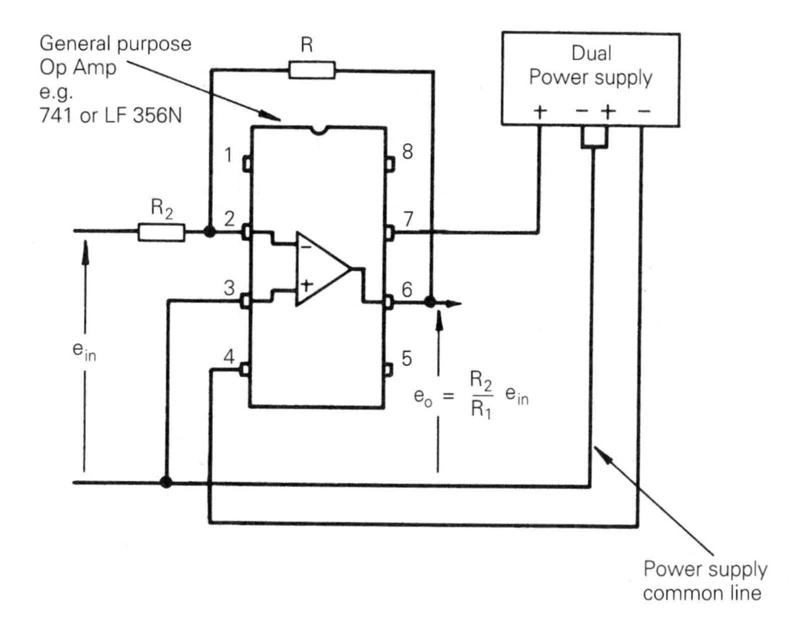

Figure 1.10 *Op-amp connections*

Particular care, however, should be taken to ensure that the power supplies are connected to the correct pins, as incorrect power supply connections can permanently damage an amplifier. Input signals should not be applied to an amplifier before power supplies are switched on, as application of input signals with no power supplies connected can damage an amplifier.

Exercises

1.1 Give component values and sketch diagrams of operational amplifier circuits for the following applications. Assume ideal op-amp performance.
  (a) An amplifier voltage gain -5 and input resistance $100\ \text{k}\Omega$.
  (b) An amplifier voltage gain -20 and input resistance $2\ \text{k}\Omega$.

(c) An amplifier voltage gain +100 with ideally infinite input resistance.

(d) An integrator with input resistance 100 kΩ and circuit performance equation

$$e_o = -100 \int e_{in} \, dt$$

(e) A circuit which when supplied by an input signal of 2 V will drive a constant current of 5 mA through a variable load resistor.

1.2 Find the value of the amplifier output voltage for each of the circuits given in Figure 1.10. In all cases assume that the operational amplifier behaves ideally.

2 Real op-amp performance parameters

There are many different op-amps to choose from. There are also different technologies such as CMOS, BiFET and bipolar. And bipolar can be separated into voltage feedback or current feedback types. Hence, some consideration of each device's performance parameters is needed. Selection of the best op-amp for a particular application is a problem, especially for the new user of op-amps. When first studying manufacturers' catalogues, the designer is faced with a huge variety of specifications and different op-amp types.

The choice of op-amp is likely to be governed by economic considerations. A general-purpose op-amp will usually cost less than a device that meets a demanding specification. This is because the manufacturer has to recover his development costs and, since general-purpose devices sell in greater quantities, these costs are spread over a larger number. The least expensive op-amp that will meet the design specifications is usually the one to choose. In order to make this choice the design objectives must be completely defined. The designer must also understand the relationship between published op-amp parameters and their effects on overall circuit performance for the intended application.

This chapter will describe the various op-amp specifications normally included in a manufacturer's data sheet. The significance of these parameters will be discussed. It is important to understand under exactly what conditions a particular parameter is defined. The important question of op-amp selection will be returned to in later chapters when op-amp applications have been described. The user should then more clearly appreciate design objectives and the way in which op-amp parameters limit their achievement.

2.1 Op-amp input and output limitations

The input circuit of an op-amp is very often a long-tailed pair. The long-tailed pair is a pair of transistors coupled together at their emitters (in the case of a bipolar input op-amp). The connection between the emitters and the supply rail is through a constant current circuit. If the base of one transistor is biased at a slightly higher potential relative to the other, it will conduct more through its collector; and the other transistor of the pair will conduct correspondingly less. The collector of each transistor is taken to the other supply rail through a resistor or, more commonly, through a constant current generator. Figure 2.1 shows the use of resistors, for simplicity.

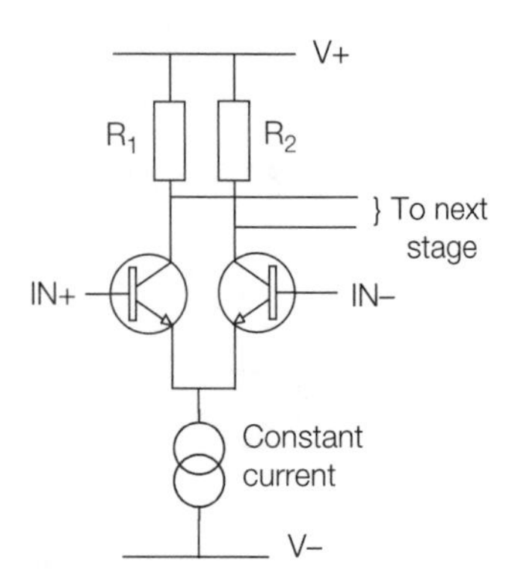

Figure 2.1 *Op-amp input circuit*

2.1.1 Maximum voltage between inputs

The voltage between the input terminals of an op-amp is maintained at a very small value, under most operating conditions, by negative feedback. If negative feedback is not used, the differential input voltage may exceed this small value and the output of the op-amp will saturate, see Figure 2.2.

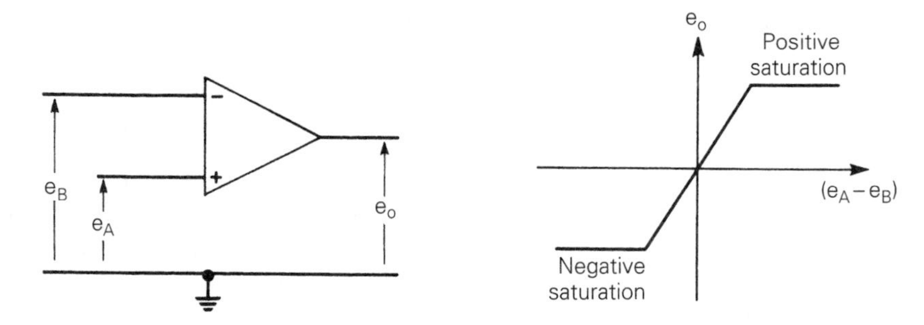

Figure 2.2 *Idealized transfer curve for an op-amp*

If the circuit design allows the application of several volts between the input terminals, care must be taken to ensure that it does not exceed the maximum allowable value, otherwise permanent damage to the op-amp may be caused. Many op-amps allow the differential input voltage to be equal to the supply voltage, and others are internally protected against input overload conditions. Where such internal protection is not provided, diodes may be connected externally to the op-amp's input terminals to provide the necessary protection.

2.1.2 Maximum output voltage swing

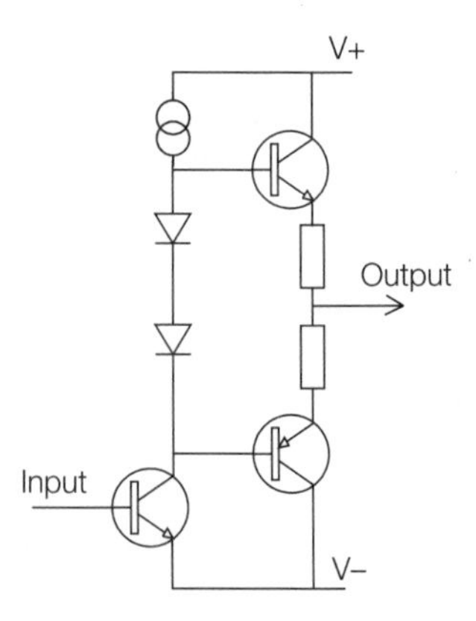

Figure 2.3 *Op-amp output circuit*

The output of an op-amp usually has two transistors, one connected to the positive rail, and the other connected to the negative rail, see Figure 2.3. This circuit controls the output voltage by increasing the drive on one transistor whilst reducing it on the other. Constant current circuits are used in the base drive of these transistors, so that quiescent supply current (the current with no signal) is minimized. Both transistors in this circuit require a certain voltage between collector and emitter, which limits the maximum output voltage swing.

The maximum output voltage swing $e_{o\ max}$ is the maximum change in output voltage (positive and negative), measured with respect to the mid-rail supply, that can be achieved without clipping the signal waveform. Values of $e_{o\ max}$ are quoted for the op-amp working into a specified load (sometimes at full rated output current) and with specified values for op-amp power supplies. Maximum values for supply voltages are normally specified and should not be exceeded. Values for $e_{o\ max}$ will be found to be dependent on the supply voltage used. BiFET and CMOS op-amps have FET outputs that allow the output voltage to be within 200 mV of the supply voltage, except when operating from high supply voltages.

2.1.3 Maximum common mode voltage

The common mode voltage is the average voltage on the two inputs relative to earth.

$$E_{cm} = \frac{e_A + e_B}{2}$$

The maximum common mode voltage, E_{cm}, is the maximum voltage that can be applied without producing saturation or non-linearity at the output. Many devices have a common mode voltage range that approaches to within 2 V of the supply rails. Single supply op-amps often use input circuits that allow the common mode voltage range to extend to the negative supply rail. In the ideal case, common mode input voltage has no effect on the output.

$$e_o = A_{OL}(e_A - e_B)$$

In practice, the common mode input voltage does affect the output.

If an op-amp is to be used under conditions in which excessive common mode voltage may cause damage, protection can be obtained by the use of a suitable pair of zener diodes. The diodes should be connected 'back to back' with their anodes joined. The two cathodes should be connected to the two op-amp inputs.

2.2 Limitations in gain, and input and output impedance

2.2.1 Non-infinite open-loop voltage gain

The open-loop voltage gain, A_{OL}, of an op-amp may be defined as the ratio

$$\frac{\text{change of output voltage}}{\text{change of input voltage}}$$

The input voltage being that measured directly between the inverting and non-inverting input terminals. A_{OL} is normally specified for very slowly varying signals and can in principle be determined from the slope of the non-saturated portion of the input/output transfer curve (Figure 2.2). The magnitude of A_{OL} for a particular op-amp depends on the op-amp load and on the value of the power supplies. Values of A_{OL} are normally quoted for specified supply voltages and load.

Op-amps are never used in an open-loop arrangement. They are occasionally used in positive feedback circuits, but much more often in negative feedback circuits that define precise operation. The significance of open-loop gain is that it determines the accuracy limits in such applications. An assessment of the quantitative effects of the open-loop gain magnitude requires a study of the principles underlying feedback op-amp operation.

In a negative feedback op-amp circuit, a signal is fed back from the output to the input. This feedback opposes the externally applied input signal. The signal that actually drives the input of the op-amp results from a subtraction process. The larger the gain of the op-amp without feedback (the open-loop gain) the smaller is the signal voltage applied between the op-amp input terminals. If the open-loop gain of the op-amp is infinite (as assumed for an ideal op-amp) negative feedback forces the op-amp's differential input signal to zero. However, with a large but finite open-loop gain, a small input signal must exist between the op-amp's input terminals. It is convenient to think of this as an input error voltage, which arises because the real op-amp has a finite open-loop gain.

2.2.2 Non-infinite input impedance

The circuit analysis based on the ideal op-amp assumed that no current flowed into the op-amp's input terminals. In practice there is a large, but finite, differential input impedance. Part of this impedance is due to input capacitance and this affects high frequency operation. For most applications it is the input resistance that can affect performance. Op-amps with FET inputs have an input resistance in the order of $10^{12}\ \Omega$. Bipolar input devices have a lower resistance, but this is usually greater than $10^{6}\ \Omega$.

The common mode input resistance to earth is much higher than this, typically 100 times greater (i.e. $10^{8}\ \Omega$ for a bipolar input device) and can be largely ignored.

2.2.3 Non-zero output resistance

Op-amps do not have zero ohm output resistance. The output resistance of a typical device is 50 Ω. This resistance restricts the maximum output voltage swing into a low resistance load, where a significant voltage drop takes place across the internal resistance. Feedback can reduce the effects of output resistance, making the op-amp generate a larger internal voltage to compensate for any reduction due to the output resistance. With feedback, the effective output impedance is typically less than 1 mΩ.

2.2.4 Effect on a non-inverting amplifier

The effects of finite open-loop gain, finite input resistance and non-zero output resistance will be considered for a non-inverting amplifier. To analyse the effects, each parameter will have to be considered separately. First we must find a few general relationships for a non-inverting amplifier in terms of the non-infinite open-loop gain.

A differential input op-amp with series negative voltage feedback applied to it is shown in Figure 2.4. The op-amp has a differential input voltage, e_ε. The op-amp's output is represented in terms of its Thévenin equivalent circuit. The output behaves like a source of EMF ($-A_{OL}e_\varepsilon$) in series with the op-amp output impedance. (Note the minus sign simply comes from the assumed positive direction of the differential input signal e_ε.)

A voltage, e_f, which is directly proportional to the output voltage, e_o, is fed back to the inverting input terminal of the op-amp (negative feedback):

$$e_f = \beta e_o$$

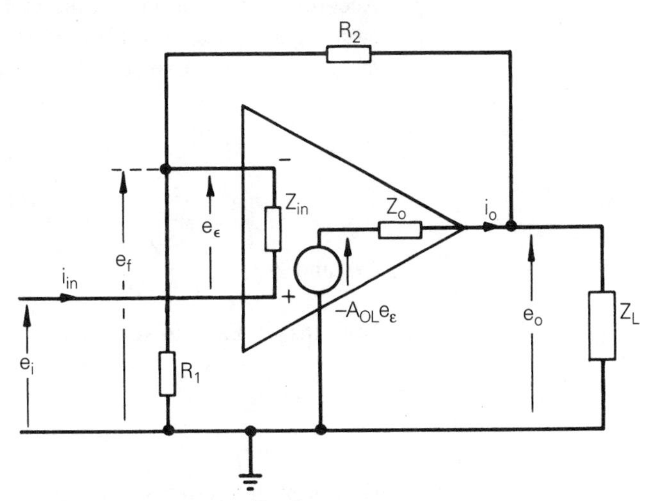

Figure 2.4 *Series voltage feedback*

The constant of proportionality β is called the voltage feedback fraction; it is an important quantity when analysing the effects of feedback.

If, in Figure 2.4, we assume $Z_{in} >> R_1$ and neglect the shunting effect of Z_{in} on R_1 we may write:

$$\beta = R_1/(R_1 + R_2) \tag{2.1}$$

Now we can examine the effects of non-zero output impedance. The output voltage of the op-amp may be written as:

$$e_o = -A_{OL}e_\varepsilon - I_oZ_o \tag{2.2}$$

It is simply a use of the general equation for the output voltage produced by a loaded source of EMF: Output voltage = Open circuit voltage − Internal volts drop. e_ε is the difference between the externally applied input signal e_i and the feedback signal e_f. Note that e_f and e_i are effectively applied in series to the differential input terminals of the op-amp.

$$e_\varepsilon = e_i - e_f \tag{2.3}$$

Substitution for e_ε in equation 2.2 gives:

$$e_o = A_{OL}(e_i - e_f) - I_oZ_o$$

Substituting $e_f = \beta e_o$ and rearrangement gives:

$$e_o = \frac{A_{OL}}{1 + \beta A_{OL}} e_i - i_o \frac{Z_o}{1 + \beta A_{OL}} \tag{2.4}$$

According to equation 2.4, the circuit behaves like an amplifier with open-circuit gain $A_{OL}/(1 + \beta A_{OL})$ and output impedance $Z_o/(1 + \beta A_{OL})$. These are the closed-loop parameters for the circuit. Thus:

$$A_{CL} = \frac{A_{OL}}{1 + \beta A_{OL}} = \frac{1}{\beta}\left[\frac{1}{1 + \dfrac{1}{\beta A_{OL}}}\right] \tag{2.5}$$

$$\text{and } Z_{oCL} = \frac{Z_o}{1 + \beta A_{OL}} \tag{2.6}$$

Note that if βA_{OL} is very large, the quantity

$$\left[\frac{1}{1 + \dfrac{1}{\beta A_{OL}}}\right]$$

is as near unity as makes no difference and the closed-loop gain is determined almost entirely by the value of the feedback fraction. The closed-loop output impedance is made very small (i.e. output voltage little affected by loading). The product of the feedback fraction and the open-loop gain is the gain around the feedback loop and it is called the loop gain. Loop gain, βA_{OL} is a most important parameter in determining the quantitative effects of feedback.

To see the effect of feedback on the output impedance, suppose that $\beta = 0.1$, so that $A_{CL} = 10$. An op-amp with $Z_o = 50\ \Omega$ and open-loop gain $A_{OL} = 10^6$ will have a closed-loop output impedance of

$$Z_{oCL} = \frac{50}{1 + 10^5} = 0.5\ \text{m}\Omega$$

At high frequencies the open-loop gain reduces, which causes the output impedance to rise.

Let us now derive an expression for the input impedance of the circuit. Note that e_f is applied in opposition to e_i (effectively it is series feedback) and tends to oppose any current into the circuit. Series negative feedback may thus be expected to increase effective input impedance. We write:

$$e_i = e_f - e_\varepsilon = \beta e_o - e_\varepsilon$$

$$\text{But } e_o = -A_{OL} e_\varepsilon \frac{Z_L}{Z_o + Z_L} \quad (\text{assuming } R_2 >> Z_L)$$

Substitution gives:

$$e_i = -e_\varepsilon\left[1 + \beta A_{OL}\frac{Z_L}{Z_o + Z_L}\right]$$

$$\text{Now } Z_{inCL} = \frac{e_i}{I_{in}} = -\frac{e_\varepsilon}{I_{in}}\left[1 + \beta A_{OL}\frac{Z_L}{Z_o + Z_L}\right]$$

$$\text{But } -\frac{e_\varepsilon}{I_{in}} = Z_{in}$$

$$\text{Thus } Z_{\text{inCL}} = -Z_{\text{in}}\left[1 + \beta A_{\text{OL}}\frac{Z_{\text{L}}}{Z_{\text{o}} + Z_{\text{L}}}\right] \tag{2.7}$$

Series voltage feedback increases input impedance to an extent determined by the loop gain βA_{OL}.

2.2.5 Effect on inverting amplifier

The effects of finite open-loop gain, finite input impedance and non-zero output impedance will be considered for the inverting amplifier. To analyse the effects, each parameter will have to be considered separately. First we must find a few general relationships for a non-inverting amplifier in terms of the non-infinite open-loop gain, A_{OL}.

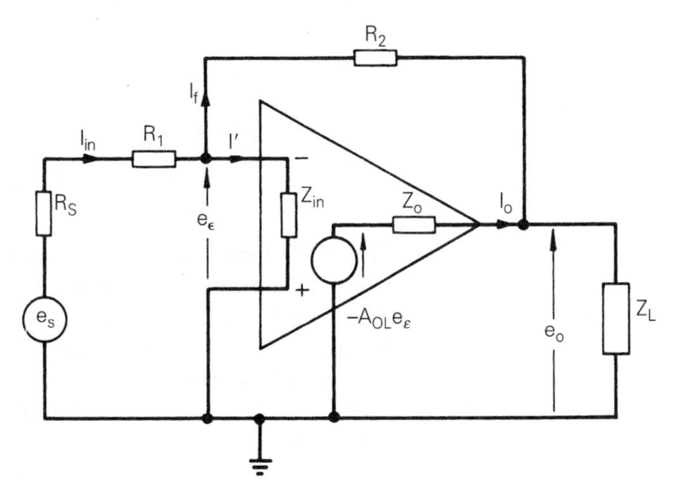

Figure 2.5 *Shunt voltage feedback*

In Figure 2.5, the externally applied input signal voltage e_{s} and the output voltage e_{o} are effectively applied in parallel to the op-amp's differential input. The signal e_ε, which drives the differential input, is a superposition of the effects of e_{s} and e_{o}.

$$e_\varepsilon = e_{\text{s}}\frac{R_2}{R_1 + R_2 + R_{\text{s}}} + e_{\text{o}}\frac{R_1 + R_{\text{s}}}{R_1 + R_2 + R_{\text{s}}} \tag{2.8}$$

It is assumed that $Z_{\text{in}} >> R_1 + R_{\text{s}}$ and that $Z_{\text{o}} << R_2$.

$$\text{The feedback fraction } \beta = \frac{R_1 + R_{\text{s}}}{R_1 + R_2 + R_{\text{s}}}$$

Let us now examine the effect of non-zero output impedance. The output voltage may be written as

$$e_{\text{o}} = -A_{\text{OL}}e_\varepsilon - I_{\text{o}}Z_{\text{o}}$$

Substitution for e_ε and rearrangement gives

$$e_o = -\frac{R_2}{R_1 + R_2 + R_s}\frac{A_{OL}}{1 + \beta A_{OL}}e_s - i_o\frac{Z_o}{1 + \beta A_{OL}}$$

The closed-loop signal gain of the circuit is thus

$$e_o = -\frac{R_2}{R_1 + R_2 + R_s}\frac{A_{OL}}{1 + \beta A_{OL}} = -\frac{R_2}{R_1 + R_s}\left[\frac{1}{1 + \dfrac{1}{\beta A_{OL}}}\right] \quad (2.9)$$

For large values of βA_{OL}, the term

$$\left[\frac{1}{1 + \dfrac{1}{\beta A_{OL}}}\right]$$

is very close to unity and the closed-loop gain is

$$\frac{R_2}{R_1 + R_s}$$

The closed-loop output impedance is

$$Z_{oCL} = \frac{Z_o}{1 + \beta A_{OL}} \quad (2.10)$$

The closed-loop output impedance of an op-amp in many circuits is a tiny fraction of the open-loop impedance, typically less than 1 mΩ at low frequencies.

Compare equations 2.9 and 2.10 with equations 2.5 and 2.6. Again, notice the importance of the loop gain βA_{OL}. If the loop gain is sufficiently large the closed-loop performance is determined by the value of the components used to fix the feedback fraction β. If $R_1 << R_s$ and the loop gain is large, the closed-loop signal gain approximates to $A_{CL} = -R_2/R_1$.

Now let us consider the input impedance. In Figure 2.5

$$I_{in} = I' + I_f$$

Now $I' = e_\varepsilon/Z_{in}$ and $I_f = (e_\varepsilon - e_o)/R_2$.

$$\text{So } I_{in} = \frac{e_\varepsilon}{Z_{in}} + \frac{e_\varepsilon - e_o}{R_2}$$

If $Z_L > Z_o$, $e_o \cong -A_{OL}e_\varepsilon$, i.e. assume that there is no voltage drop across the internal output impedance.

$$\text{By substitution, } I_{in} = e_\varepsilon\left[\frac{1}{Z_{in}} + \frac{1 + A_{OL}}{R_2}\right]$$

In terms of input impedance, we have Z_{in} and additional shunt impedance $R_2/(1 + A_{OL})$. Thus the effect of the shunt feedback is to reduce the effective differential input impedance of the op-amp. And if A_{OL} is very large, the input impedance is very small (typically < 1 Ω). The overall input impedance of the inverting op-amp circuit then effectively equals the value of the resistor R_1.

2.2.6 Summary of some of the effects of negative feedback

It is useful to summarize the effects of negative feedback as shown by the above analysis.

1. Series negative feedback increases input impedance.
2. Shunt negative feedback decreases input impedance.
3. Negative voltage feedback makes for a stable distortion-free output voltage.
4. Negative current feedback makes for a stable distortion-free output current.

2.3 Real op-amp frequency response characteristics

We have not yet considered the dynamic response of op-amps. Gain has been defined as the ratio change of output voltage to slow change of input voltage. The effect of rapid changes has not yet been considered. It is usual to distinguish between sinusoidal and transient response characteristics.

Sinusoidal response parameters describe the way in which an op-amp responds to sinusoidal signals. In particular they show how the op-amp's response depends upon signal frequency. Transient response parameters characterize the way in which an op-amp reacts to a step or square-wave input signal. An added complication is that it is necessary to distinguish between small signal and large signal response parameters; differences arise because of dynamic saturation effects that occur with large signals.

This section is concerned with small signal sinusoidal response characteristics. An ideal op-amp is assumed to have an open-loop gain that is independent of signal frequency, but the gain of a real op-amp does have frequency dependence. Both the magnitude and the phase of the open-loop gain are frequency dependent. This frequency dependence has a marked effect on closed-loop performance.

2.3.1 Bode plots

Gain/frequency characteristics are often presented graphically. It is usual to plot gain magnitude in decibels (dB) against frequency on a logarithmic (base 10) scale. Gain in dB is determined from the relationship:

$$\text{Voltage gain in dB} = 20 \log |e_o/e_{in}| \qquad (2.11)$$

The reader who is unfamiliar with the use of decibels (or dB) should get practice in working out the dB equivalents of some voltage ratios (try Exercises 2.5 and 2.6).

Examples

$e_o/e_{in} = 10$ represents a voltage gain of $20 \log (10) = 20$ dB; $e_o/e_{in} = 100$ represents 40 dB; $e_o/e_{in} = 1000$ represents 60 dB; $e_o/e_{in} = 1/10$ represents -20 dB; $e_o/e_{in} = \sqrt{2}$ represents 3 dB; $e_o/e_{in} = 1/\sqrt{2}$ represents -3 dB. Since power is proportional to $(\text{voltage})^2$, a fall in gain of 3 dB represents a halving of the output power.

Gain/frequency plots are often given as a series of straight-line approximations rather than as continuous curves. The straight lines are called Bode approximations and the graphs are called Bode diagrams.

The open-loop frequency response of many op-amps is designed to follow an equation of the form

$$A_{OL(jf)} = \frac{A_{OL}}{1 + j\dfrac{f}{f_c}} \tag{2.12}$$

where:

$A_{OL(jf)}$ is a complex quantity representing the magnitude and phase characteristics of the gain at frequency f,
A_{OL} represents the DC value of the open-loop gain and
f_c is a constant, sometimes called the break frequency.

Equation 2.12 describes what is sometimes called a first order lag response; its magnitude and phase characteristics are shown plotted in Figure 2.6. The magnitude of the response is

$$A_{OL(jf)} = \frac{A_{OL}}{\sqrt{1 + \left(\dfrac{f}{f_c}\right)^2}} \tag{2.13}$$

At low frequencies for which $f < f_c$, $A_{OL(jf)} \rightarrow A_{OL}$ and the straight line $|A_{OL(jf)}| = A_{OL}$ is the low frequency asymptote.

At high frequencies for which $f > f_c$, the response is asymptotic to the line $|A_{OL(jf)}| = A_{OL}(f_c/f)$ which has a slope of -20 dB/decade change in frequency.

For each ten times increase in frequency the magnitude decreases by 1/10, or a change of -20 dB. (Note that a slope of -20 dB/decade is sometimes expressed as -6 dB per octave; it goes down by 6 dB for each doubling of the frequency.) Gain attenuation with increase in frequency is referred to as the roll-off in the frequency response.

The two straight lines intersect at the frequency $f = f_c$ and at this frequency $|A_{OL(jf)}| = A_{OL}/\sqrt{2}$: the response is thus 3 dB down when $f = f_c$. The frequency f_c is sometimes referred to as the 3 dB-bandwidth limit.

The phase/frequency characteristic associated with equation 2.12 is determined by

$$\theta = -\tan^{-1}\frac{f}{f_c} \tag{2.14}$$

At $f << f_c$, $\theta \rightarrow 0°$; at $f = f_c$, $\theta = -45°$; and at $f >> f_c$, $\theta \rightarrow -90°$.

The Bode phase approximation approximates the phase shift by the asymptotic limits of 0° at 1/10 of f_c and $-90°$ at 10 times f_c. The asymptotes are connected by a line whose slope is $-45°$ per decade of frequency as shown in Figure 2.6. The errors involved in using the straight-line approximation for the magnitude and phase behaviour of equation 2.12 are tabulated in Figure 2.6.

Op-amp data sheets normally give values of A_{OL} and the unity gain frequency f_1, which is the frequency at which the open-loop gain has fallen

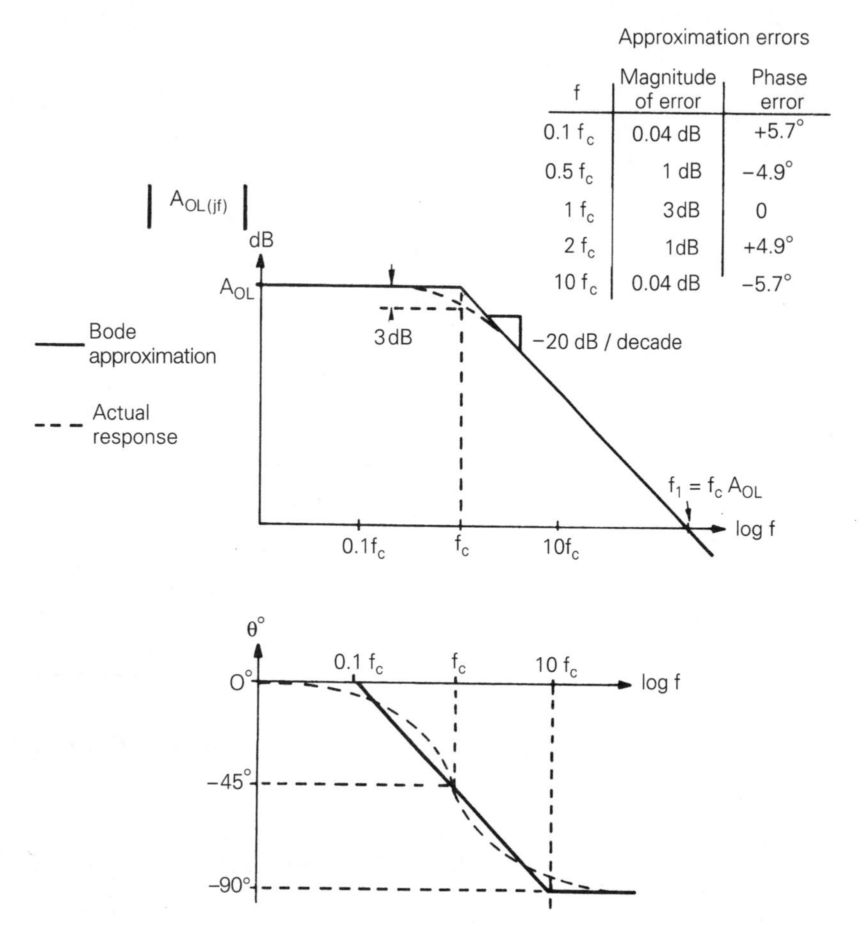

Figure 2.6 *First order low-pass magnitude and phase response and Bode approximations*

to 0 dB because of open-loop roll-off. In the case of op-amps which exhibit a first order frequency response, with a 20 dB per decade roll-off down to unity gain, the frequency f_1 is related to the 3 dB bandwidth frequency f_c by the expression $f_c = f_1/A_{OL}$.

Frequency response characteristics are readily plotted from knowledge of A_{OL} and f_1. The Bode magnitude approximations are obtained by simply drawing two straight lines, one horizontal line at the value of A_{OL} and the second through f_1 with a slope of -20 dB/decade. The two intersect at the frequency f_c.

Bode diagrams are useful in evaluating the frequency response characteristics of cascaded gain stages. The gain of a multistage op-amp is obtained as the product of the gains of the individual stages, but since gain is represented logarithmically in Bode plots, the overall response may be determined by linearly adding the Bode plots for the separate stages as shown in Figure 2.7.

Note that the final roll-off and limiting phase shift depend upon the number of gain attenuating stages. Two stages give a final gain roll-off of

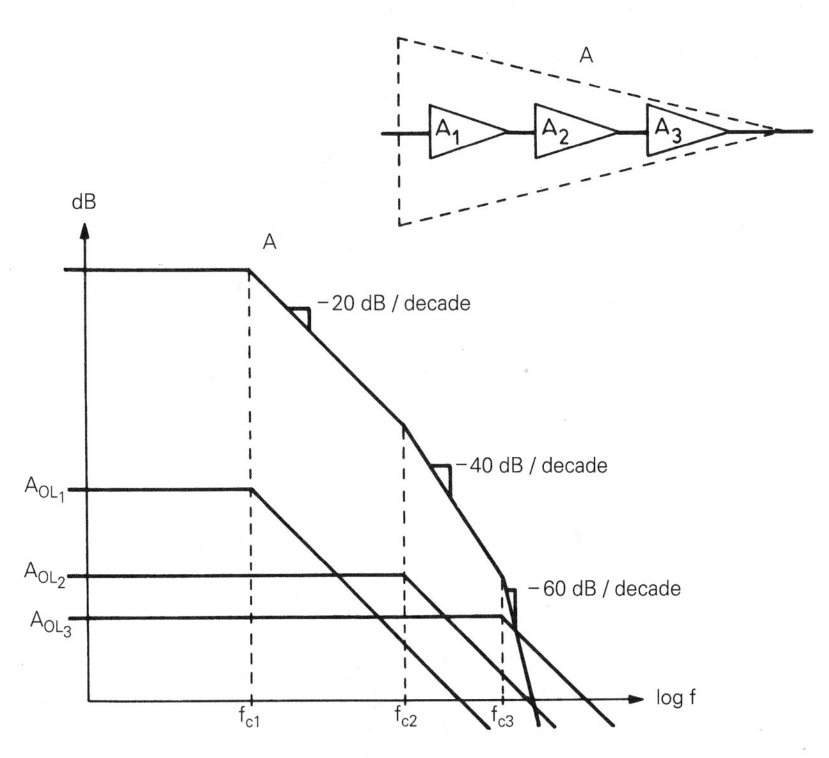

Figure 2.7 *Frequency response of cascaded gain stages*

−40 dB/decade and a limiting phase shift of 180°; three stages give a gain roll-off of −60 dB/decade and a 270° phase shift.

2.4 Small-signal closed-loop frequency response

The desirable characteristics of op-amp circuits stem from the use of negative feedback. The quantitative effects of negative feedback are related to the loop gain βA_{OL}. Real op-amps exhibit a frequency dependent A_{OL}, and in some applications the feedback fraction β is also frequency dependent. Therefore, practical op-amp circuits have a frequency dependent loop gain and this has a marked effect on closed-loop performance.

Frequency dependence implies both a magnitude change and a phase change with frequency. In a circuit using negative feedback, it only needs a phase shift of 180° in the feedback loop to make the circuit apply positive feedback; this can cause serious problems. An op-amp feedback circuit will produce self-sustained oscillations if the phase shift in the feedback loop reaches 180° while the magnitude of the loop gain is greater than unity. This should not be allowed to happen.

Phase shifts in the feedback loop of greater than 90° but less than 180° will not result in sustained oscillations. However, they can cause a frequency response that peaks up at the bandwidth limit, before it rolls off. Associated with this closed-loop gain peaking, the circuit will have a transient response that exhibits overshoot and ringing. Transient response refers to the output changes produced in response to a step or square-wave input signal.

Phase margin is a term used to express the relative stability of a closed-loop op-amp circuit. The phase margin is the amount by which the phase shift is less than 180° at the frequency where the magnitude of the loop gain is unity. A closed-loop circuit with 90° phase margin shows no gain peaking. As the phase margin is reduced, gain peaking becomes noticeable for phase margins of approximately 60° (about 1 dB peaking) and becomes more marked with further reduction in phase margin (20° phase margin gives approximately 9 dB of gain peaking).

Most general-purpose op-amps have an open-loop frequency response that follows a first order decay characteristic. The open-loop gain reduces in proportion to the signal frequency. This ensures that they are unconditionally stable under any value of resistive feedback. This type of response was discussed in the previous section; it has a 20 dB/decade roll-off down to unity gain and the phase shift associated with this never exceeds 90°. The phase margin for any value of resistive feedback is therefore never less than 90°.

The gain frequency dependence of the open-loop response also affects the closed-loop response. The effect on closed-loop gain is most conveniently demonstrated in graphical form by sketching the appropriate Bode plots. We look for the effect of A_{OL} on loop gain and then to the effect of loop gain on the gain error factor. We may write:

$$|\beta A_{OL(jf)}| = \left| \frac{A_{OL(jf)}}{\dfrac{1}{\beta_{(jf)}}} \right|$$

Which when expressed in decibel form gives:

$$\text{loop gain (in dB)} = \text{open-loop gain (in dB)} - \frac{1}{\beta}\text{ (in dB)} \qquad (2.15)$$

That is, the magnitude of the loop gain in decibels at any frequency is equal to the difference between the open-loop gain magnitude in decibels and $1/\beta$ in decibels.

As an example of the graphical approach, consider an op-amp with a first order frequency response used with resistive feedback in the follower configuration. The circuit and its Bode plots are illustrated in Figure 2.8. In order to display the frequency dependence of the loop gain we merely superimpose the plot of $1/\beta$ (in dB) on the open-loop frequency response plot of the op-amp. If feedback is purely resistive, as it is here, β is independent of frequency and $1/\beta$ is a straight line parallel to the frequency axis. In this case, the frequency dependence of the loop gain is entirely due to the frequency dependence of the open-loop gain.

As the frequency increases there is a reduction in open-loop gain, A_{OL}. There is a corresponding decrease in the loop gain, βA_{OL} and an increase in the gain error. Remember that gain error is related to the amount by which the gain error factor $[1/(1 + 1/\beta A_{OL})]$ differs from unity.

If it is required to compute the gain error at frequencies approaching or exceeding the open-loop bandwidth f_c, the phasor nature of the loop gain

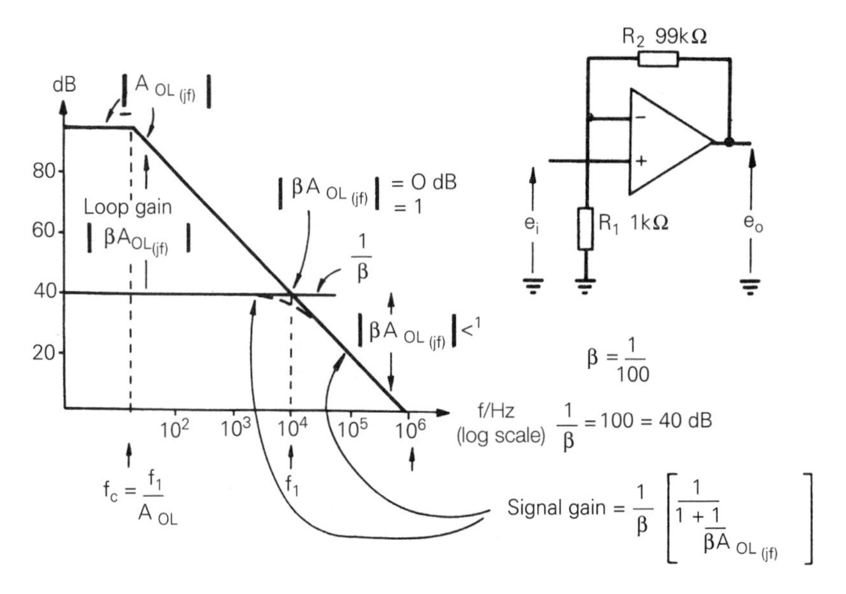

Figure 2.8 *Bode plots show frequency dependence of loop gain*

must not be forgotten. Let us evaluate the gain error for the circuit of Figure 2.8 at a frequency $f = 10^3$ Hz. At this frequency, $|\beta A_{OL(jf)}| = 20$ dB $= 10$, and the phase shift in the loop gain is close to $-90°$. Thus:

$$\left|\frac{1}{1+\dfrac{1}{\beta A_{OL(jf)}}}\right| = \left|\frac{1}{1+\dfrac{1}{-j10}}\right| = \frac{1}{\sqrt{1+0.01}} = 0.995$$

Compare this with the value obtained by neglecting the phasor nature of the loop gain, which is:

$1/(1 + 1/10) = 0.909$, a 9% gain error!

At the frequency f_1' at which the open-loop and $1/\beta$ magnitude plots intersect, the magnitude of the loop gain is unity (0 dB). The two plots close at a rate of 20 dB per decade, which is indicative of a 90° phase shift in the loop gain and a remaining 90° phase margin. The magnitude of the gain error at the frequency f_1 is:

$|1/(1 + 1/-j1)| = 1/\sqrt{2}$

The closed-loop gain magnitude is thus 3 dB down on its ideal value $1/\beta$ at the frequency f_1'. f_1' represents the closed-loop bandwidth; at frequencies greater than f_1' the magnitude of the closed-loop gain approaches the magnitude of the open-loop gain. If $\beta A_{OL(o)} >> 1$, the product of closed-loop gain and closed-loop bandwidth $= 1/(f_1') = f_1$ remains constant for different values of β. Negative feedback makes the closed-loop bandwidth greater than the

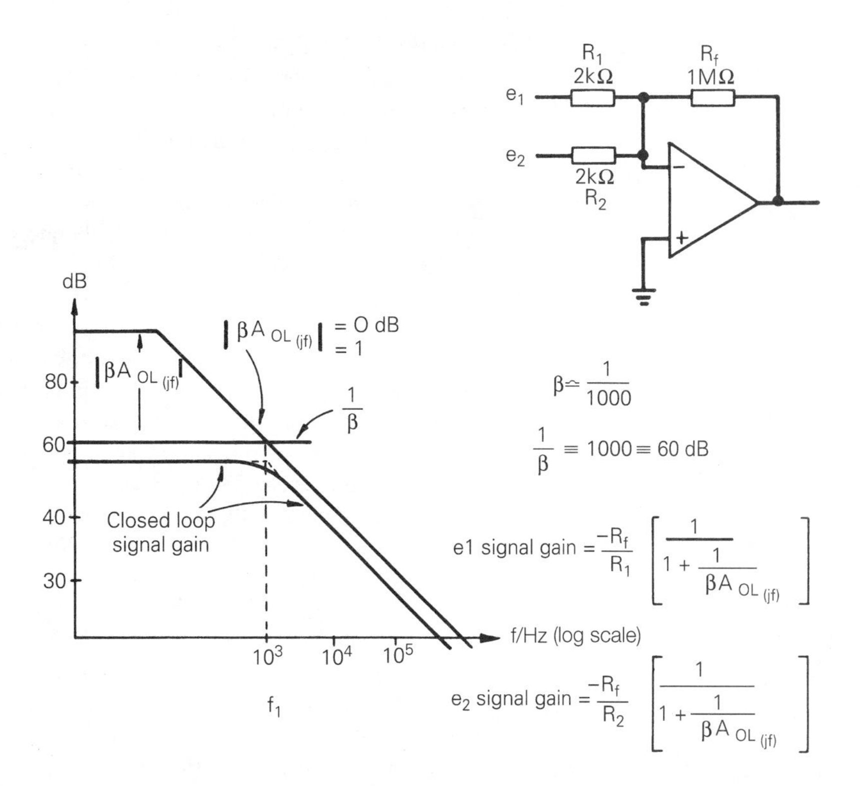

Figure 2.9 *Bode plots for inverting adder*

open-loop bandwidth. The greater β, the smaller the closed-loop gain but the wider the closed-loop bandwidth.

A second example of the graphical approach used to find closed-loop signal bandwidth is illustrated in Figure 2.9. The circuit considered here is an inverting adder application. In this type of circuit the feedback fraction is influenced by the presence of the two input resistors R_1 and R_2.

$$\beta = \frac{R_p}{R_p + R_f} \quad \text{where} \quad R_p = R_1 \,//\, R_2 = \frac{R_1 R_2}{R_1 + R_2}$$

Substituting component values gives $\beta = 1/1000$ and $1/\beta = 1000$ or 60 dB. $1/\beta$ intersects the open-loop frequency response at the frequency $f_1' = 1$ kHz. This fixes the closed-loop bandwidth at 1 kHz but note that, in this circuit, the closed-loop signal gain is not the same as the closed-loop gain ($1/\beta$) since there are two possible input signal paths. The ideal signal gain for the e_1 signal is $-R_f/R_1$ and is $-R_f/R_2$ for the e_2 signal. In this particular example $R_1 = R_2$, so the two gains are equal and the closed-loop signal bandwidth is 1 kHz.

2.5 Closed-loop stability considerations

Most op-amps are internally frequency compensated and have an open-loop frequency response with a 20 dB/decade roll-off. A response of this kind, in principle, ensures that the op-amp will be closed-loop stable under all conditions of resistive feedback. However, it is important to be aware that the use of an internally frequency compensated op-amp does not always ensure closed-loop stability.

Capacitive loading at the output of an op-amp, or stray capacitance between the inverting input terminal and earth, can cause phase shifts leading to instability – even in resistive feedback circuits. In differentiator applications, in which the feedback fraction β is deliberately made frequency dependent, an internally compensated op-amp exhibits instability.

Some op-amps exhibit a final roll-off in their open-loop frequency response of greater than 20 dB/decade; they are called externally frequency compensated op-amps. These fast roll-off op-amps are often used in circuits where both wide closed-loop bandwidth and greater than unity gain are required. They require the external connection of a capacitor to make them closed-loop stable.

The closed-loop frequency response obtained with fast roll-off (externally frequency compensated) op-amps can be explained using Bode plots. The response is related to the gain error caused by the decaying open-loop gain and the associated phase shift. For example, consider an op-amp with a response that has three gain stages, each stage having a frequency response with a different cut-off point. The magnitude and phase characteristics of the open-loop gain are illustrated in Figure 2.10.

The magnitude and phase characteristics of the loop gain for a particular feedback fraction are obtained by superimposing a plot of $1/\beta$ on the open-loop frequency response plot. With resistive feedback, β is frequency independent. The phase shift in the closed-loop gain is determined by the phase shift in the open-loop gain; this can be found by referring back to the graphs in Figure 2.5.

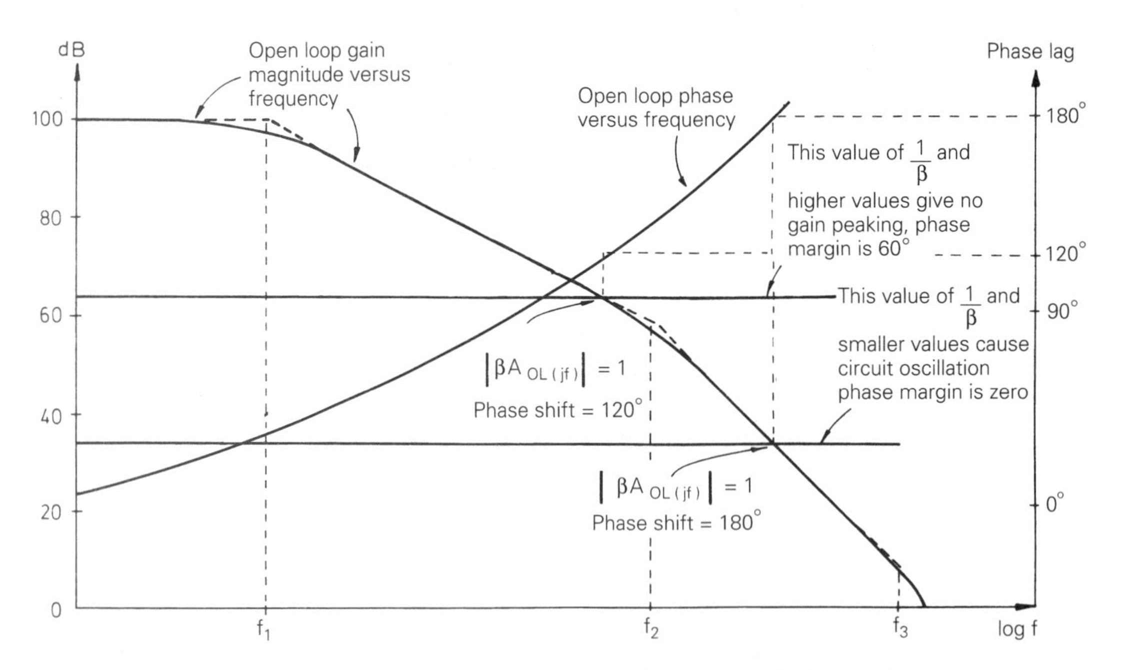

Figure 2.10 *Gain magnitude and phase characteristics of op-amp with three-pole response*

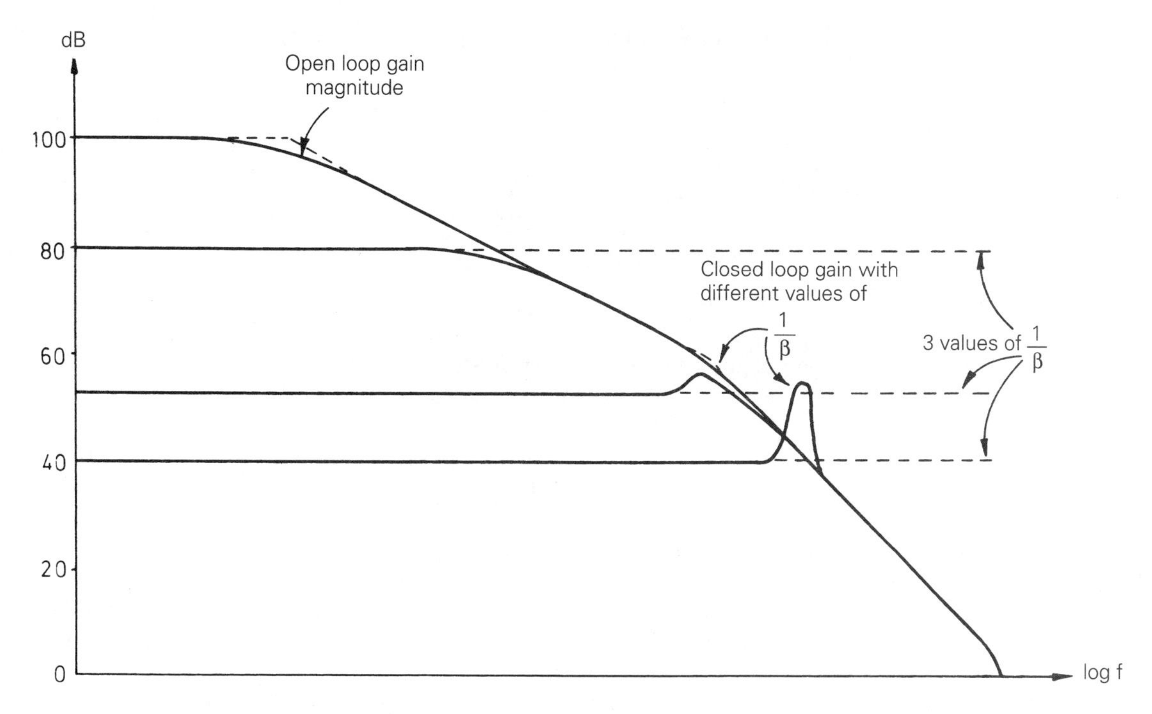

Figure 2.11 *Too much feedback gives gain peaking with uncompensated fast roll-off op-amps*

Phase margin is the amount by which this phase shift is less than 180° at the frequency at which the magnitude of the loop gain is unity (0 dB). Note that increasing β results in successively smaller phase margins. Phase margins less than 60° cause the closed-loop gain to peak up (see Figure 2.11). The gain peaking increases as the phase margin is reduced further until, at zero phase margin, the circuit breaks out into sustained oscillations.

2.5.1 Phase margin determines closed-loop gain peaking

The gain peaking occurs as a result of inadequate phase margin and is caused by positive feedback. Positive feedback occurs when the feedback signal has a component that is in phase with the externally applied input signal. If the gain is greater than unity when phase shift in the loop gain reaches 180°, the circuit oscillates.

When considering the extent of the gain peaking (obtained as a result of inadequate phase margin) we must look to the effect of the loop gain magnitude/phase behaviour on the gain error factor.

$$\beta A_{\mathrm{OL(jf)}} = |\beta A_{\mathrm{OL(jf)}}| e^{-j\theta}$$

The value of the gain error factor may then be expressed as

$$\frac{1}{1+\dfrac{1}{\beta A_{OL(jf)}}} = \frac{1}{1+\dfrac{e^{j\theta}}{|\beta A_{OL(jf)}|}}$$

$$\frac{1}{1+\dfrac{\cos\theta + j\sin\theta}{|\beta A_{OL(jf)}|}}$$

The magnitude of the gain error factor can then be written as

$$\left|\frac{1}{1+\dfrac{1}{\beta A_{OL(jf)}}}\right| = \sqrt{\left(\frac{1}{1+\dfrac{1}{|\beta A_{OL(jf)^2}|}+\dfrac{2\cos\theta}{|\beta A_{OL(jf)}|}}\right)} \qquad (2.16)$$

Since the cosine of angles lying between 90° and 180° is negative we have the possibility of a gain error factor magnitude greater than unity for values of β greater than 90°. It is this variation in the gain error factor that is responsible for closed-loop gain peaking.

Gain peaking usually arises as a result of phase shift with frequency controlled by a single first order function. Here the break frequency is greater than a decade away from other break frequencies. Two situations are illustrated in Figure 2.11. The relationship between closed-loop gain peaking and phase margin that is to be expected in a situation of this kind is also shown graphically in Figure 2.12 (see also Appendix A2).

In order to assess the phase margin in a particular circuit, the $1/\beta$ graph (in dB) is superimposed on the open-loop response. The intersection of the two curves gives the frequency f_1' at which the magnitude of the loop gain is unity. The phase shift β at this frequency is then determined from the phase/frequency variation in A_{OL}; the phase margin is $\theta_m = 180° - \theta$. The amount of gain peaking can be found from the graph. Note that the gain peaking in fact occurs at frequencies slightly less than the frequency f_1'. However, as the phase margin is reduced, the gain peak increases in amplitude; the frequency at which it occurs moves closer to the frequency f_1'.

2.6 Frequency compensation (phase compensation)

Frequency compensation or phase compensation is the name given to the process of tailoring the loop gain magnitude/phase characteristics of a feedback op-amp circuit to give an adequate phase margin. Adequate phase margin ensures closed-loop stability and freedom from closed-loop gain peaking. Bode diagrams are particularly useful in assessing the stability and frequency response of feedback circuits, and examples will be given in terms of their Bode diagrams.

General-purpose op-amps are normally internally frequency compensated and give unconditional stability with all values of resistive feedback. The phase shift in their open-loop gain is typically controlled to be 135° or less

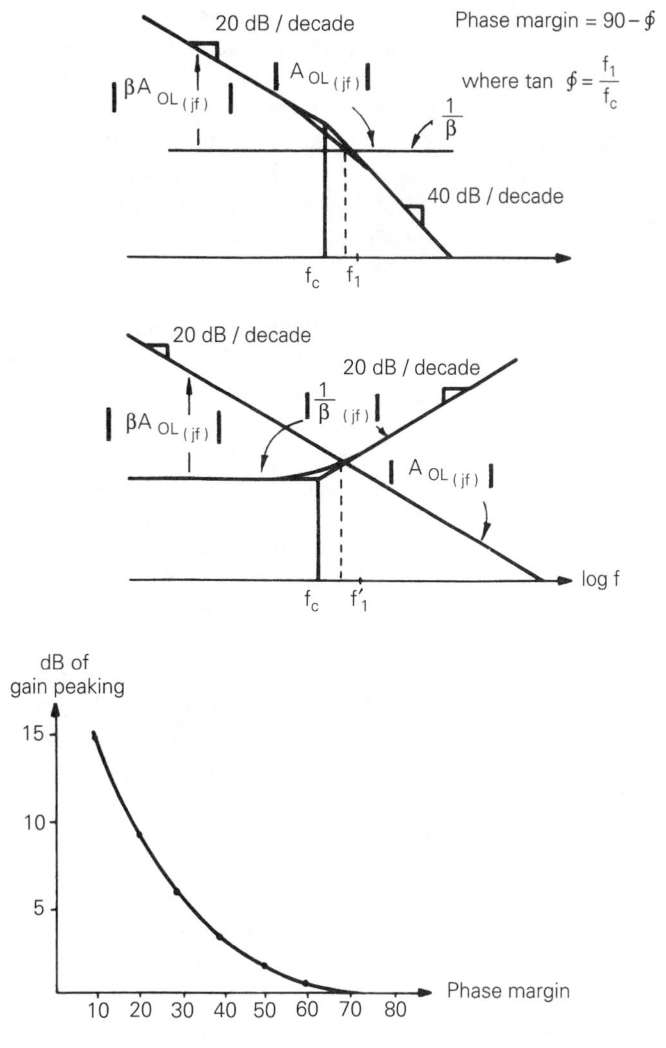

Figure 2.12 *Gain peaking versus phase margin for two commonly encountered situations*

for all frequencies where the open-loop gain magnitude is greater than unity, assuring a minimum phase margin of 45° for all values of resistive feedback. Internal frequency compensation gives user convenience at the expense of closed-loop bandwidth and speed (slew rate – see later) which would otherwise be available when the op-amp is used at higher closed-loop gains than unity.

It is important to note that even frequency compensated op-amps can become unstable if the load is sufficiently capacitive. The internal resistance and the external capacitance cause a phase shift at the op-amp's output terminal. Even though the op-amp may have a phase margin of 45° or more, the phase shift at the output can be greater than this and lead to oscillation. This is because feedback is taken from the op-amp's output. An external resistor between the op-amp's output at the load reduces the phase shift at the output, and hence in the feedback path. The subject of compensating for capacitive loads and capacitive inputs is discussed further in Section 10.5.

Op-amps without internal frequency compensation require external frequency compensating components. They allow the user to select a frequency compensating scheme appropriate to the particular closed-loop circuit. Closed-loop bandwidth, slew rate, full power response and noise performance (see later) are all affected by the frequency compensating method adopted. Compensation methods advocated for different op-amp types differ in detail because of internal circuit differences. The general principles involved in frequency compensation are the same for all op-amps.

Amplifying stages within an op-amp can achieve very high gains by using active loads. In many cases the overall gain can be sufficiently large using only two internal voltage gain stages. Op-amps of this type are normally frequency compensated by means of a single feedback capacitor connected around the second inverting gain stage in the op-amp. The technique requires only small values of frequency compensating capacitor (10 pF–30 pF). Capacitors of this size are small enough to be fabricated on the same integrated circuit chip as the rest of the op-amp circuitry. This is the method of internal frequency compensation in general-purpose op-amps: a simplified model of the internal circuit is given in Figure 2.13.

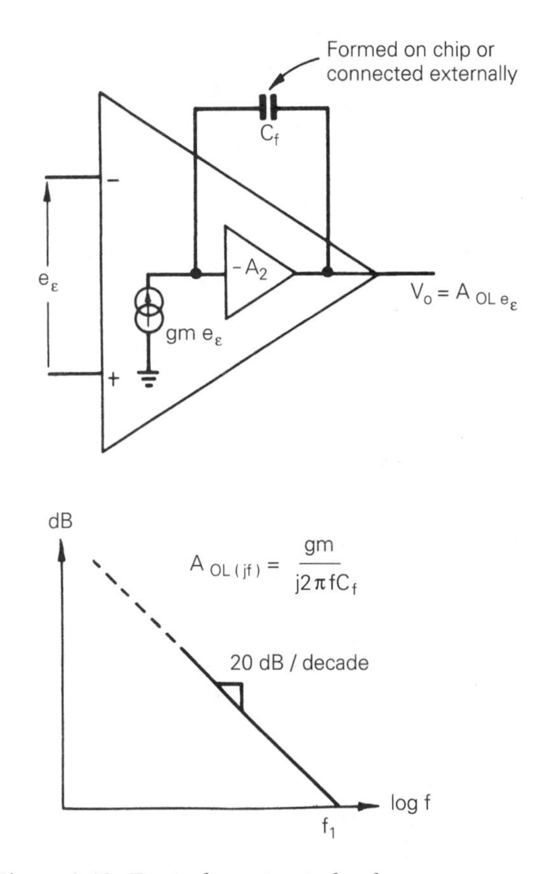

Figure 2.13 *Equivalent circuit for frequency compensation*

The differential input stage used in many op-amps has a very high gain and a very high output impedance; it provides what is essentially a current drive to the second gain stage. In Figure 2.12, its action is represented by the current generator $g_m e_\varepsilon$.

The second gain stage is inverting; a capacitor connected between input and output of the high gain inverting stage gives that stage the frequency response characteristics of an integrator. Its output voltage is proportional to the integral of the input current. Assuming ideal integrator action for the second stage, its output and the output of the complete op-amp has a frequency roll-off that can be approximated as

$$e_o = \frac{g_m e_\varepsilon}{j\omega C_f}$$

and the open-loop gain roll-off is approximated by

$$A_{OL(jf)} = -\frac{g_m}{j2\pi f C_f} \tag{2.17}$$

Equation 2.17 must be made to dominate the overall frequency response. Unity gain frequency compensation requires that the value of C_f be chosen so that the equation brings the open-loop gain down to unity at a frequency lower than the break frequency of other gain attenuating stages.

Setting $|A_{OL(jf)}| = 1$ in equation 2.17 and transposing gives equation 2.18. This gives the relationship between the unity-gain frequency (f_1) and the required unity-gain frequency compensating capacitor (C_1) as

$$f_1 = g_m/2\pi C_1 \tag{2.18}$$

In many general-purpose op-amps, the current drive supplied by the first gain stage has a frequency dependence. This is determined by the frequency response of transistors in the first stage, where the break frequency may be a few MHz. Dependent upon the unity-gain phase margin required, f_1 must be made to have the same order of magnitude. Most general-purpose monolithic op-amp designs have C_f chosen to make f_1 typically slightly less than the break frequency of the first stage transistors.

Unity-gain frequency compensation, although satisfactory, is not strictly necessary when an op-amp is used in a circuit where the closed-loop gain ($1/\beta$) is greater than unity. Externally compensated op-amps in which the frequency compensating capacitor is user connected permit the designer to apply just sufficient compensation to achieve a desired phase margin. Some internally compensated op-amps have a minimum stable gain specified; these devices have a greater gain-bandwidth product than would otherwise be achievable with a unity-gain stable device.

Use of the minimum frequency compensating capacitor, consistent with achieving adequate phase margin, gives a wider closed-loop bandwidth than would be obtained if the op-amp were unity-gain frequency compensated. In applications that are concerned only with slowly varying input signals, a wide closed-loop bandwidth is of course not required. In such cases it is often advantageous to restrict closed-loop bandwidth (in order to reduce noise) by using much greater frequency compensation than is required for closed-loop stability. These points are illustrated in Figure 2.14 which shows Bode approximations for the open-loop frequency response of a general-purpose op-amp, using values of the frequency compensating capacitor C_f larger than and smaller than the value C_1 required for unity-gain frequency compensation of the op-amp.

Adequate phase margin (60° in the case considered in Figure 2.14) requires that the minimum value of C_f be chosen so, for a particular value of B used in the circuit configuration, the magnitude of the loop gain $B*A_{OL}$ is reduced to unity at the frequency f_1. Use of equation 2.17 gives

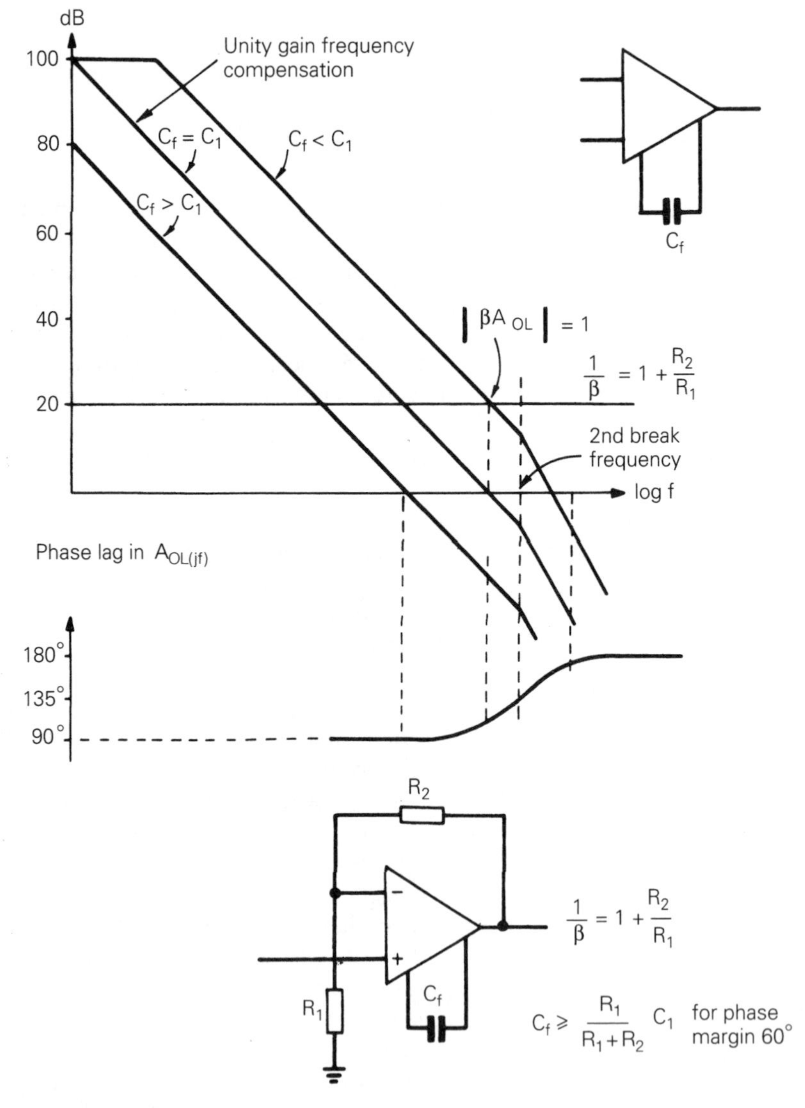

Figure 2.14 *Open-loop frequency response of op-amp with different values of compensation capacitor*

$$|\beta A_{\mathrm{OL(j}f)}| = \beta \frac{g_{\mathrm{m}}}{2\pi f C_{\mathrm{f}}}$$

We require

$$|\beta A_{\mathrm{OL(j}f)}| = 1, \text{ at } f = f_1$$

Substituting for f_1 from equation 2.23 gives the required minimum value of C_{f} as

$$C_{\mathrm{f}} \geqslant \beta C_1 \tag{2.19}$$

Remember that C_1 is the value of the frequency compensating capacitor required for unity-gain frequency compensation.

A typical value of frequency compensating capacitor is 30 pF for unity-gain frequency compensation. Lower values can be used for higher closed-loop gains. Typically, a 3 pF frequency compensating capacitor can be used if the closed-loop gain is 20 dB.

There are certain conditions in which the use of minimum values of C_{f} can lead to instability problems. These are in circuits with large resistor values, or where there is appreciable stray capacitance to earth at the inverting input terminal, or those in which the op-amp is expected to drive a capacitive load.

2.6.1 Frequency compensation and slew rate considerations

There is a limit to the rate at which the output voltage of an op-amp can change; this is called the slew rate. Slew rate is usually expressed in volts per microsecond and is defined as the maximum rate of change of output voltage produced in response to a large input step.

The basic mechanism governing slew rate is capacitor charging. The rate of change of voltage, at any point in a circuit, is limited by the maximum current available to charge the capacitance at that point. In many op-amp applications it is the charging of the frequency compensating capacitor (internal or external) that sets the output slew rate. For this reason, op-amps designed to have low supply current requirements are generally slower and bandwidth limited.

The frequency compensating capacitor of an op-amp is charged by the output current supplied by the first gain stage in the op-amp. The limitation on the charging rate is therefore determined by the first stage output current capabilities, thus:

$$\text{Slew rate} = \left|\frac{\mathrm{d}e_{\mathrm{o}}}{\mathrm{d}t_{\mathrm{max}}}\right| \cong \frac{I_{\mathrm{o}}}{C} \tag{2.20}$$

where I_{o} is the first stage operating current.

Equation 2.20 suggests that increased slew rate may be achieved by simply increasing the first stage operating current, but this is not the case for op-amps using bipolar transistor input stages. In normal bipolar transistor op-amp stages increase in operating current causes a corresponding increase in the transconductance of the stage:

$$\text{Transconductance } g_m = \frac{I_o}{\frac{2kT}{q}} \qquad (2.21)$$

where:
k is Boltzmann's constant,
T is temperature in K (Kelvin) and
q is electronic charge.

Equation 2.21 is a modification of the transconductance equation for bipolar transistors:

$$\text{Transconductance } g_m = \frac{I_C q}{kT}$$

which is ~40 I_C V^{-1} at room temperature.Thus for a transistor with a collector current of 1 mA, g_m = 40 mA V^{-1} and a V_{BE} change of 1 mV causes a collector current change of 40 μA. The value of transconductance is halved in the input stages of an op-amp because of the differential input, so a differential input of 1 mV gives rise to an output current of 20 μA.

Increase in transconductance that accompanies any increase in operating current requires a corresponding increase in C_f in order to set a particular value for f_1. Combining equations 2.20 and 2.21 with equation 2.18 gives

$$\text{Slew rate} = \left|\frac{de_o}{dt_{max}}\right| \cong \frac{2kT}{q} 2\pi f_1 \qquad (2.22)$$

Slew rate is seen to be independent of input stage current level. Our approximate treatment explains why most internally compensated general-purpose bipolar input op-amps have slew rates of the order of 1 V/μs.

Bipolar input op-amps that are externally frequency compensated have the same slew rate limitation when compensated down to unity gain. When they are frequency compensated for closed-loop gains greater than unity the smaller value of the frequency compensating capacitor which is required gives an increased slew rate. High slew rate bipolar input op-amps are available; they feature specialized input stage circuitry which provides increased current output without at the same time giving an increase in the transconductance of the stage.

FET input op-amps do not have the above limitation on slew rate because unlike bipolar transistors, FETs do not have their transconductance directly dependent upon operating current. FET input op-amps normally feature a higher slew rate than bipolar input op-amps.

2.6.2 Feed-forward frequency compensation

A few op-amps are suitable for use with feed-forward frequency compensation. This technique can provide a significant increase in bandwidth and slew rate over standard lag compensation techniques.

In most op-amps the first stage provides the greatest single contribution to the overall gain of the op-amp, but its frequency response is normally rather limited. In feed-forward frequency compensation the high-gain low-bandwidth first stage is bypassed at the higher signal frequencies and these are fed directly to the wider bandwidth second stage of the op-amp. Using this technique, the phase shift at the higher frequencies is primarily due to the wide band stage, and the phase shift due to the high-gain low-bandwidth stage is eliminated. The principle underlying the scheme is illustrated in Figure 2.15.

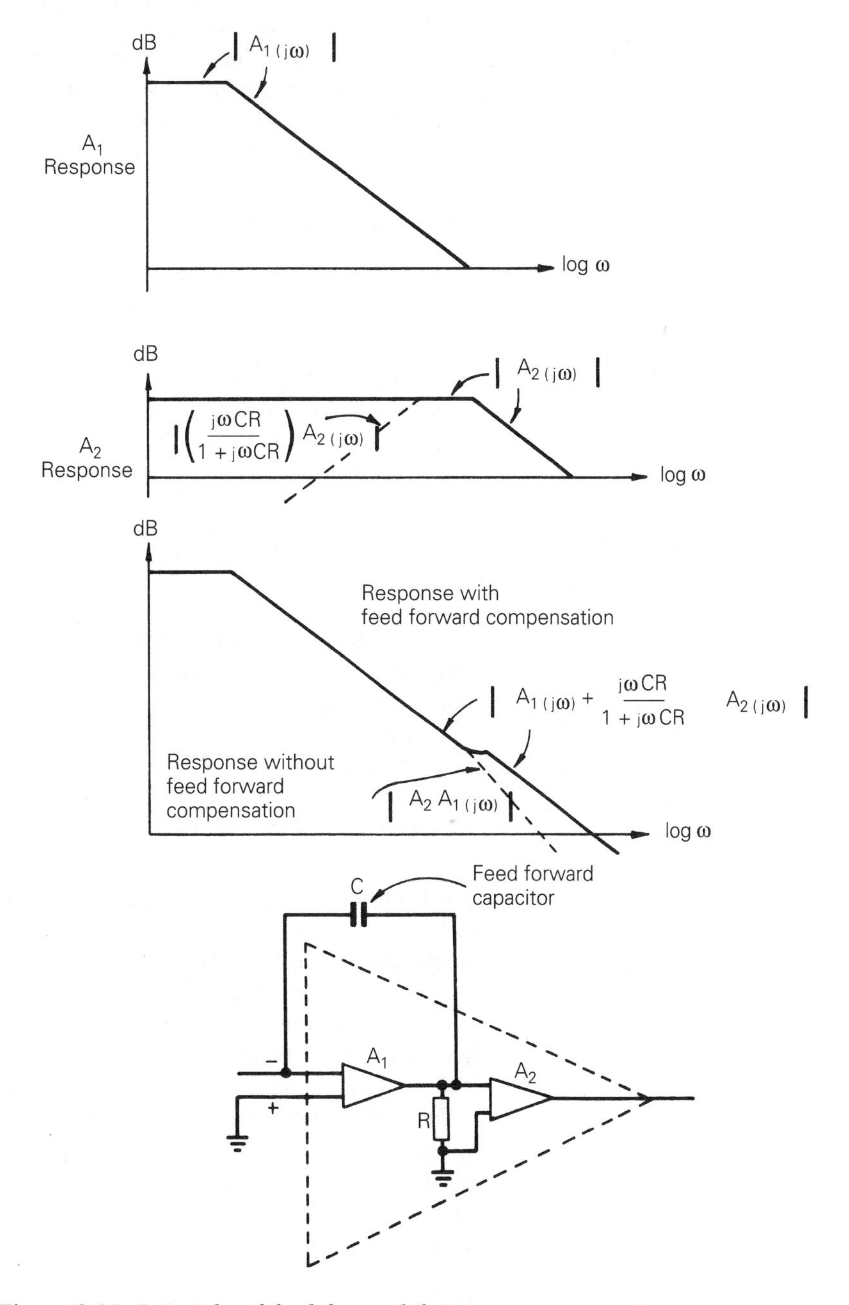

Figure 2.15 *Principle of feed-forward frequency compensation*

The overall gain due to both stages may be expressed as

$$A_{OL(j\omega)} = \left[A_{1(j\omega)} + \frac{R}{R + \dfrac{1}{j\omega C}}\right] A_{2(j\omega)} = \left[A_{i(j\omega)} + \frac{1}{1 + \dfrac{1}{j\omega CR}}\right] A_{2(j\omega)}$$

C is chosen so that when $f > 1/2\pi CR$ the gain of the first stage has fallen to below unity, making the overall gain approximately that of the second gain stage.

The second stage 20 dB/decade roll-off takes the overall gain down to unity. Bode plots of the uncompensated response and the response with feed-forward compensation are illustrated in Figure 2.15. Feed-forward frequency compensation is only applicable to inverting feedback configurations using externally compensated op-amps.

2.6.3 Lead compensation

Lead frequency compensation is a technique used to increase the phase margin. A capacitor is included in a feedback loop to introduce a phase lead, compensating for the op-amp phase lag, which would otherwise result in insufficient phase margin. A simple way of achieving this is to connect a capacitor C_f in parallel with the feedback resistance. A circuit using this method of lead compensation is shown, together with its associated Bode plots, in Figure 2.16.

We write

$$\frac{1}{\beta} = 1 + \frac{R_2}{R_1}\left(\frac{1}{1 + j\omega C_f R_2}\right) = \left[1 + \frac{R_2}{R_1}\right] 1 + \frac{1 + j\omega C_f (R_1//R_2)}{1 + j\omega C_f R_2}$$

At frequencies greater than $1/2\pi C_f R_2$ the capacitor introduces a phase lead in the feedback fraction, which approaches 90°. If C_f is chosen so that the frequency $1/2\pi C_f R_2$ is a decade below the frequency at which the $1/\beta$ and open-loop response plots intersect, a phase margin of approximately 90° is obtained. Use of a lead capacitor in parallel with a feedback resistor is a convenient way of getting extra phase margin. It is also a technique that can be used to overcome the effect of stray capacitance between the op-amp's inverting input and earth (see Section 9.5).

2.6.4 Other frequency compensating techniques

Techniques other than those described in the above sections are sometimes used for frequency compensation. Whatever technique is used, the same basic principle is involved. Frequency compensation involves attenuating the loop gain magnitude down to unity without, at the same time, introducing an excessive phase shift leading to closed-loop instability.

Frequency compensation is achieved by simply shunting a signal point in the feedback loop with a capacitor (Figure 2.17). Assuming the output resistance at the signal point is R_o the added capacitor introduces a 20 dB/decade rate of attenuation, which starts at the break frequency $1/2\pi C_1 R_o$. The maximum phase shift associated with a CR lag network is 90°. The capacitor value must be chosen so that the loop gain magnitude is attenuated down to unity at a frequency lower than other break frequencies of attenuating stages.

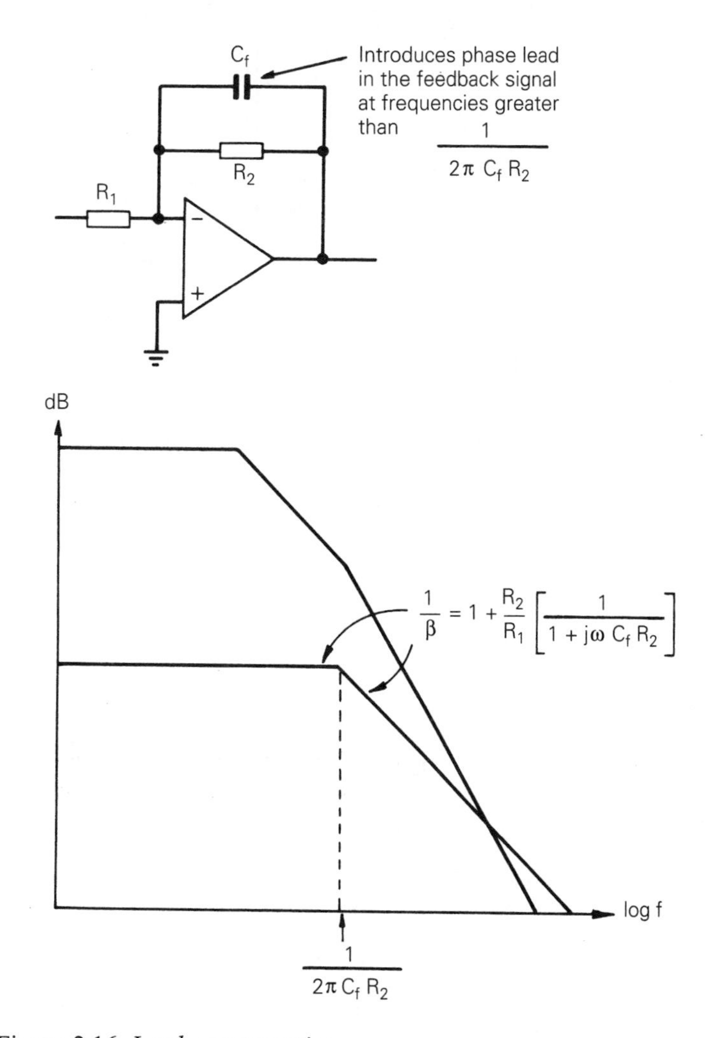

Figure 2.16 *Lead compensation*

Shunting a signal point with a capacitor resistor combination (a lag-network) is an alternative technique that allows wider closed-loop bandwidths (Figure 2.18). At frequencies above $1/2\pi C_1 R_1$ (the break-back frequency) a network of this kind produces an attenuation $R_1/(R_1 + R_o)$ but the phase shift returns to zero.

2.7 Transient response characteristics

Previous sections have been concerned with factors influencing the small-signal frequency response characteristics of op-amp feedback circuits. Attention is now directed to the factors influencing their behaviour in time, namely their transient behaviour in response to large and small input step or square-wave signals.

Students may gain a greater understanding of op-amp transient behaviour, and the terminology used to describe it, by performing transient tests. Frequency compensating component magnitude, load capacitance, input

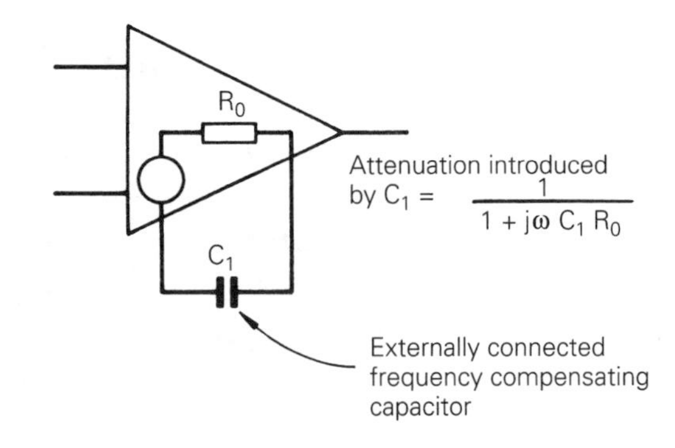

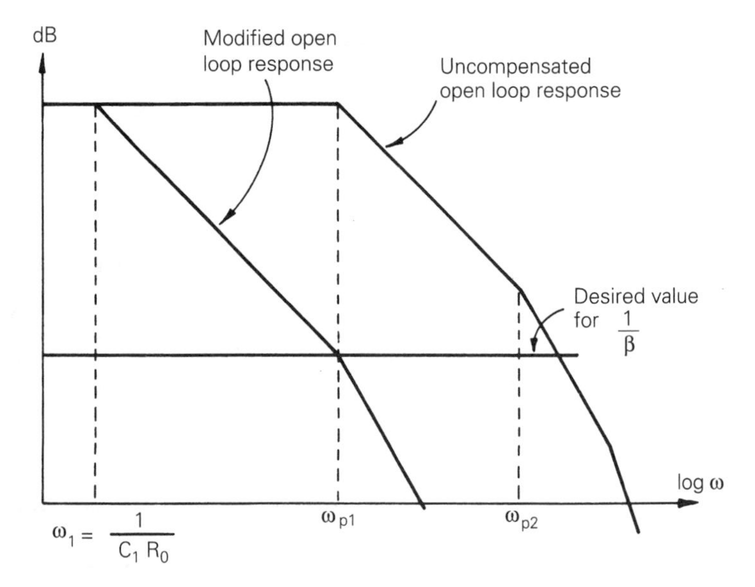

Figure 2.17 *Simple lag compensation with single capacitor*

capacitance and any stray feedback capacitance all influence closed-loop transient behaviour.

2.7.1 Small-signal transient response

Small-signal characteristics are those obtained when there are no saturation effects (no slew rate limited output) and the op-amp circuit is operating in its linear range. In small-signal operation circuit relationships are independent of the level of the output voltage and current, and of their previous history.

The small-signal transient behaviour of an op-amp feedback circuit is closely related to its small-signal sinusoidal frequency response. In our previous discussion of small-signal closed-loop frequency response we distinguished between two different closed-loop situations.

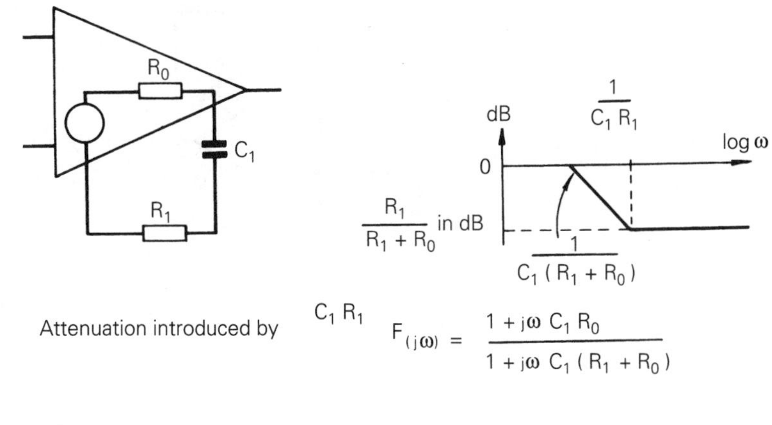

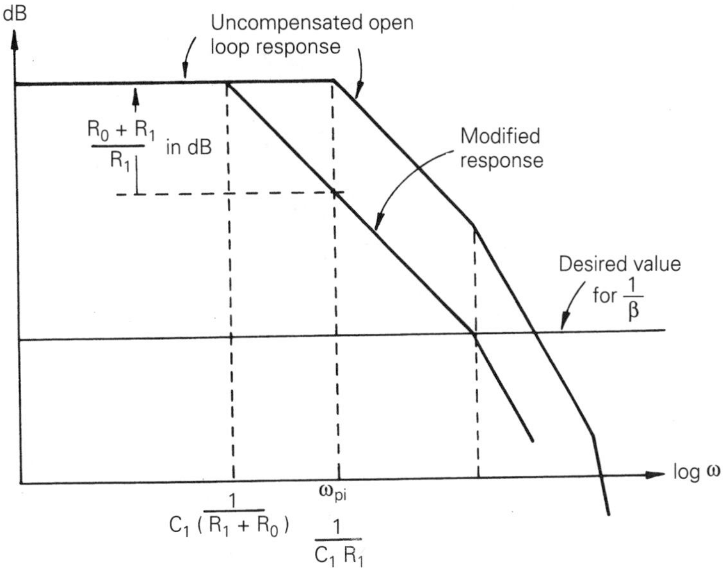

Figure 2.18 *Frequency compensation with RC shunt*

In the first situation, consider a unity gain frequency compensated op-amp with resistive feedback. Figure 2.19 illustrates the considerations governing the behaviour of such a circuit. In response to an input step signal the output follows an exponential governed by the relationship

$$V_o = V_f [1 - e^{-t/T_c}] \tag{2.23}$$

where the time constant $T_C = T_1/\beta$ and $T_1 = -1/2\pi f_1$. Notice that T_C increases for increasing values of closed-loop gain (decrease in β) and decreases for increasing values of the unity gain frequency f_1.

Rise time is a parameter that is frequently used to characterize the response of an op-amp to an input step. Rise time is defined as the time taken for the output to rise between 10 per cent and 90 per cent of its final value. Neglecting

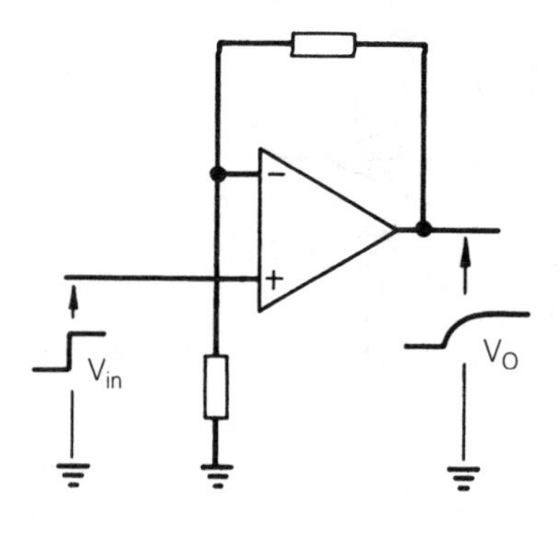

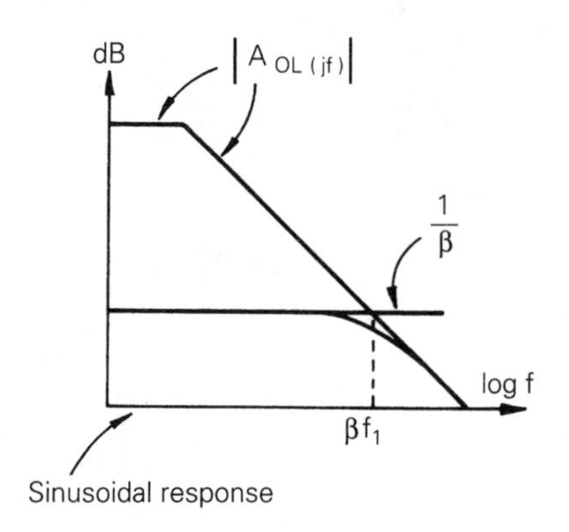

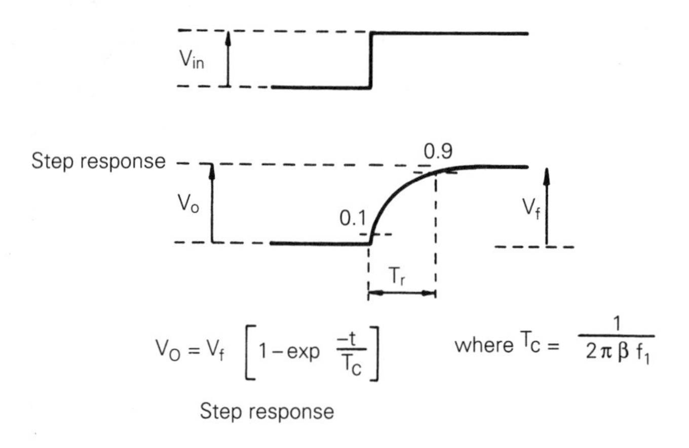

Figure 2.19 *Small signal sinusoidal and transient response for unity gain frequency compensated op-amp with resistor feedback*

the time for the initial 10 per cent rise an approximate expression for rise time can be obtained by substituting $V_o = 0.9V_f$ in equation 2.23. Thus,

$$0.9V_f = V_f[1 - e^{-T_r/T_c}]$$

giving $T_r/T_c = \ln(10)$, where $\ln(x)$ is the natural logarithm of x.

$$\text{or } T_r = \ln(10)/2\pi f(3\text{ dB})$$

$$T_r \sim 1/[3f(3\text{ dB})]$$

$f(3\text{ dB}) = \beta f_1$ is the closed-loop small-signal 3 dB bandwidth limit.

The second situation is when using a closed-loop configuration with a lightly damped transient response.

The most commonly encountered closed-loop configurations which exhibit a lightly damped transient response are those in which the frequency response is governed by two breaks, and in which the break frequencies are remote

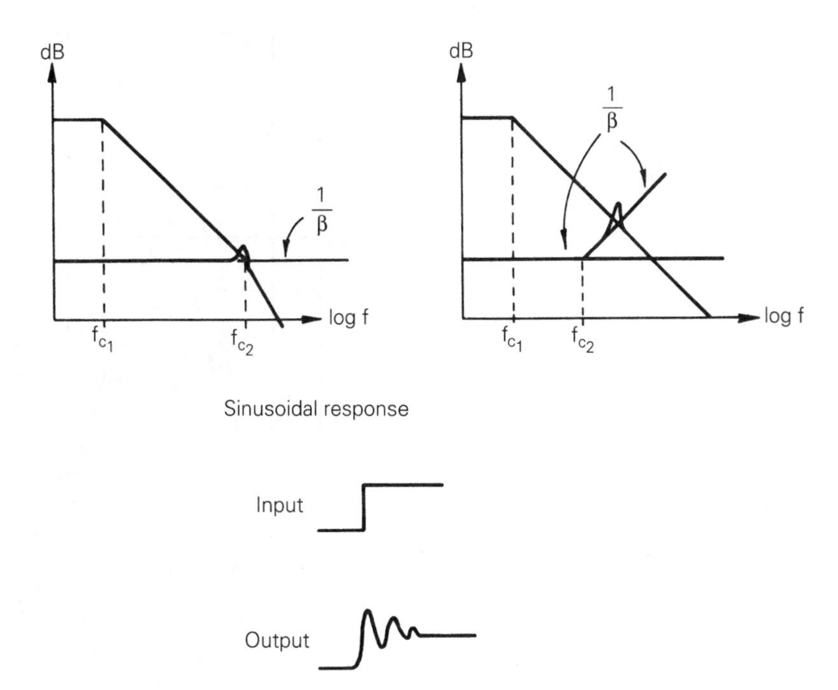

Step response

Figure 2.20 *Lightly damped closed-loop response*

from each other by at least a decade (see Figure 2.20). Systems of this kind have a response that is typical of a second order system. The response equation is obtained by substituting the frequency dependent loop gain expression into the gain error factor (see Appendix A2).

A second order system is characterized by parameters called the damping factor Ω and natural frequency f_o. A step input, V_{st}, causes overshoot and ringing. Treating $A_{CL}V_{st}$ as a scaling factor and plotting time in units of $\omega_o t$ the normalized step response for different values of the damping factor is plotted in Figure 2.21.

The step response shows an increasing overshoot and ringing as the value of the damping factor is successively reduced below unity. The damping factor is related to the break frequencies, governing the frequency response of the op-amp by the expression

$$\zeta = \frac{\sqrt{f_{c_2}}}{2\sqrt{|\beta A_{(o)}| f_{c_1}}}$$

The amount of gain peaking to be expected in the small-signal closed-loop frequency response is related to the damping factor by (Appendix A2)

$$P_{(\text{dB of peaking})} = 20 \log \left(\frac{1}{2\zeta \sqrt{1 - \zeta^2}} \right) \qquad (2.24)$$

where $\zeta = 1/\sqrt{2}$.

Note: there is no peaking in the sinusoidal response for $\zeta > 1/\sqrt{2}$.

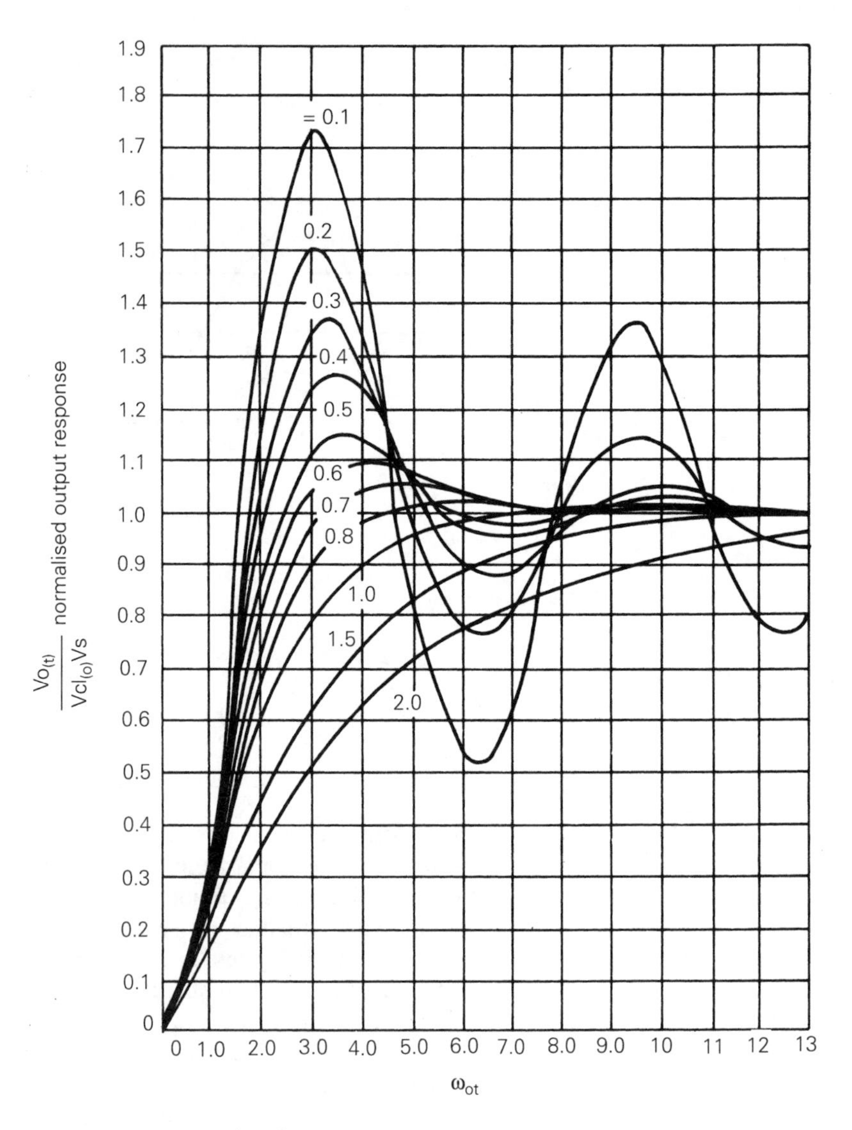

Figure 2.21 *Second order step response*

2.7.2 Overshoot

In the case of a lightly damped response ($\zeta < 1$), the amount by which the first ringing peak exceeds the final value is referred to as overshoot. Expressed as a percentage of the final value

$$\text{Overshoot \%} = 100e\left(\frac{-\zeta\pi}{\sqrt{1-\zeta^2}}\right) \tag{2.25}$$

2.7.3 Small-signal settling time

Overshoot represents the maximum output transient error following initial rise in response to a stepped input. The time taken by the output to settle within a certain accuracy (settling time) following a transient is often of greater interest. A conservative estimate of the small-signal settling time for a second order system can be made by finding the smallest value of N which satisfies:

$$100e\left(\frac{-\zeta N\pi}{\sqrt{1-\zeta^2}}\right) <= x\,\%$$

where $x\%$ represents a specified accuracy.

The settling time is found by substituting N into the following equation:

$$t = \frac{N\pi}{\omega_o \sqrt{1-\zeta^2}}$$

Small-signal settling time is clearly directly related to the value of the damping factor. For fast settling to a high accuracy, nothing is to be gained by using damping factors less than unity. Although light damping does give a faster initial rise, any ringing prolongs settling time. It is for this reason that designers of fast settling op-amps strive to have the open-loop frequency characteristic strongly dominated by a single 20 dB per decade roll-off down to unity gain in the open-loop frequency response.

2.7.4 Large-signal time response characteristics

If the op-amp is operating in its non-linear regions, the small-signal transient response characteristics discussed in the previous section no longer apply. In this section some of the effects accounting for the difference between small- and large-signal characteristics are considered.

Slew rate

Within an op-amp there is inherent semiconductor and circuit capacitance, as well as those added for frequency compensation. There is also load capacitance at the output. The rate of change of voltage at a point in the circuit depends on the available current to charge the capacitance at that point:

$$\frac{\mathrm{d}V}{\mathrm{d}t_{\max}} = \frac{I_{\max}}{C}$$

This mechanism sets an upper limit to the rate at which the output voltage of an op-amp can change. Slew rate, usually expressed in V/μs is the parameter that is used to characterize the effect. As discussed in Section 2.7.1 it is often the charging of the frequency compensating capacitor which determines the output slew rate, but there are applications in which the charging of some other circuit capacitance sets the limit, for example large capacitive loads.

Slew rate is the performance parameter which determines the maximum frequency at which an op-amp can give a full-scale sinusoidal output signal, and is one of the important factors in determining large signal settling time. Slew rate determines the maximum operating frequency in such applications as precise rectifiers.

Overload recovery

An op-amp in a saturated overload condition takes a finite time to recover to linear operation. Overload recovery defines the time required for the output voltage to recover to within its rated value from a saturated condition. Saturation takes place when an op-amp's output voltage exceeds its rated value. It also occurs during non-linear slew with the output within rated limits. Saturation causes charges within the circuit to become unbalanced. These charges must be brought back to equilibrium before the op-amp can operate normally.

In an op-amp circuit required to give a full-scale output step there is a period of recovery which is comparable to the period of slew. The recovery period may be substantially greater if many internal stages are involved. Fast slew rate, therefore, is not by itself a good indicator of a fast settling op-amp. Some op-amps with extremely large slew rates have excessive recovery time.

Large signal settling time

Settling time is defined as the time elapsed from the application of a perfect step input to the time when the op-amp's output has reached its final value (within specified tolerances). Large-signal settling time is usually specified for the condition of unity gain and a full-scale output step. The main contributions to settling time are slew rate and overload recovery.

2.8 Full power response

The inability of an op-amp's output voltage to slew faster than a limiting rate can lead to distortion of sinusoidal signals. This is true, even though their amplitude is below the maximum rated output voltage for the op-amp. Some manufacturers specify the effect by op-amp full power response, f_{p1}, defined as the maximum frequency, measured at unity closed-loop gain, for which full output can be obtained at rating load without distortion. An approximate relationship between slew rate and full power response is readily derived if it is remembered that in the case of a sinusoidal signal the maximum rate of change occurs as the signal passes through zero. Consider a sinusoidal output signal with amplitude equal to the rated output voltage E_o and frequency f_p:

$$e_o = E_o \sin(2\pi f_p t)$$

$$\frac{de_o}{dt} = 2\pi f_p E_o \cos(2\pi f_p t)$$

$$\text{Slew rate} = \left|\frac{de_o}{dt}\right|_{max} = 2\pi f_p E_o \qquad (2.26)$$

If the output amplitude is reduced, distortion due to slew rate does not occur until the frequency is increased above f_p. Op-amp data sheets sometimes give graphs which relate maximum sinusoidal output voltage obtainable without distortion to frequency; they show what is called the power bandwidth of the op-amp.

2.9 Offsets, bias current and drift

In circuits where the DC response is important, offset voltages, bias currents and drift have to be taken into account. An op-amp is normally required to give zero output voltage (referred to earth – or mid-rail) when the voltage between its input terminals is zero. When a constant DC input signal is applied, the op-amp's output should remain at a constant voltage. Parameters are defined which indicate how far real op-amps depart from this ideal behaviour.

In a circuit designed to handle AC signals, a DC path from both inputs to either earth or another DC voltage source is required. A non-inverting op-amp with a capacitively coupled input signal needs a resistor, between the input and the mid-rail supply, in order to supply bias current to the non-inverting (+) input. The feedback resistor, between the output and the inverting input, supplies bias current to the inverting (−) input.

An op-amp with its input terminals shorted together is found to give a non-zero output voltage or 'offset'. In some cases, the high gain of the op-amp will cause the output voltage to be at one of its saturated levels. It is therefore usual to specify op-amp offsets by referring them to the input of the op-amp.

The input offset voltage, V_{io}, is defined as that input voltage which would have to be applied in order to cause the op-amp output voltage to be zero. It is specified at a particular temperature.

All op-amps require some small relatively constant current at their input terminals, called an input bias current. In the case of a differential op-amp the input bias current, I_b, is defined as the average value (half the sum) of the currents at the two input terminals with the op-amp output voltage at zero. It too is specified at a particular temperature. Ideally the currents taken by the two input terminals should be the same under these conditions but in practice some degree of mismatch always exists.

The input offset current, I_{io}, is defined as the difference in the input bias currents to the two input terminals, at a particular temperature. With equal source impedances connected to the two input terminals, it is only this mismatch, or difference current, which causes an offset error. The effects of bias and offset currents tend to overshadow the effects of input offset voltage when the input source impedances are high.

Provision is normally made for balancing out the effects of initial op-amp offsets by means of a suitable potentiometer. After this adjustment has been made the output voltage of an op-amp is still found to change, even though the applied input signal is zero or a constant DC value. This slow change in the output voltage of an op-amp is referred to as drift. Drift problems do not arise in AC op-amps because any DC change in voltage level is effectively blocked off from the output by a coupling capacitor.

A specification for the drift in an op-amp's output voltage would, in itself, give little criterion for the selection of an op-amp for drift performance. The observed output drift is dominated by drift in the early stages of the op-amp, for this is magnified many times by subsequent stages before appearing at the op-amp output. It is usual to characterize drift performance by referring the drift to the input; the various contributions to drift are specified by their effects on op-amp input offsets.

2.9.1 Temperature drift

In op-amps, drift with temperature normally represents the largest single source of drift. This causes the biggest errors in many applications. It arises because of the temperature dependence of the characteristics of both active and passive components. Temperature drift may be specified by the temperature coefficients of bias current and input offsets. The coefficients, $\Delta I_b/\Delta T$, $\Delta I_{io}/\Delta T$ and $\Delta V_{io}/\Delta T$ are usually defined as the average slope over a specified temperature range.

The drift to be expected for a defined temperature change from ambient is found by multiplying the specified drift rate by the temperature excursion. The drift of bias current, the input offset current and the input offset voltage are generally a non-linear function of temperature. The drift rates are normally greater at the extremes of temperature.

2.9.2 Supply voltage sensitivity

Changes in the op-amp's power supply voltage causes changes in op-amp output voltage. The effect is usually specified by the effect of supply voltages on input bias current and input offsets. Supply voltage coefficients, $\Delta V_{io}/\Delta V$, $\Delta I_b/\Delta V$ and $\Delta I_{io}/\Delta V$ are included in most data sheets. In the case of op-amps using twin power supplies, the positive and negative supply voltage coefficients will not normally be the same. However, with regulated power supplies, drift due to power supply changes will normally be negligible compared with temperature drift.

2.9.3 Evaluating errors due to input offset voltage and bias current

In applications requiring a response down to DC, the op-amp input offset voltage and bias current, and their drift coefficients, are usually the limiting performance parameters. A general method for evaluating offset errors will now be described.

The use of error signal generators at the input of an otherwise ideal op-amp (see Figure 2.22) conveniently represents offset voltage and bias current.

Combining the effects of the separate error generators into one single generator provides further simplification. The effects of bias current are expressed

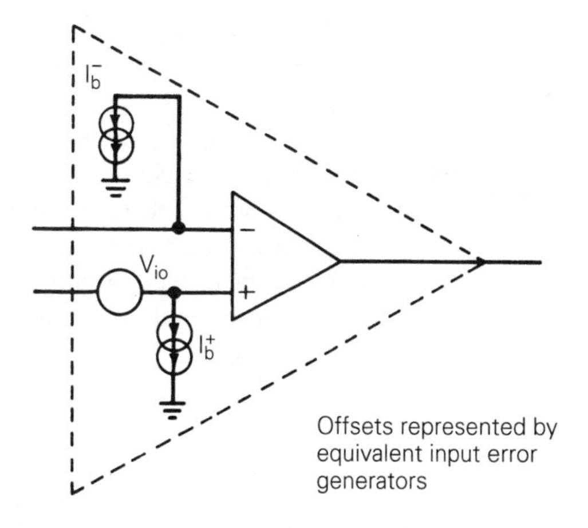

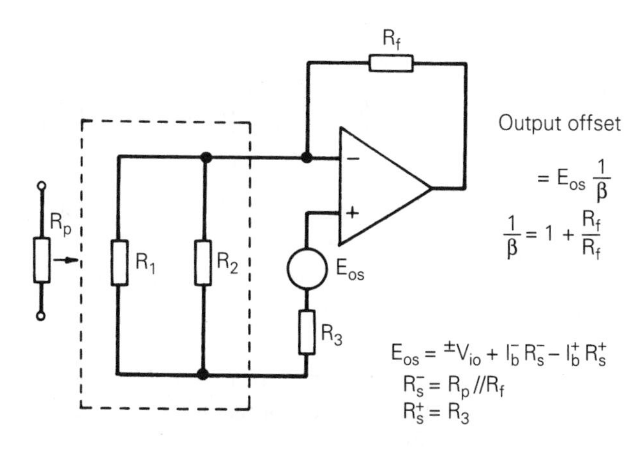

Figure 2.22 *Evaluating offset errors*

in terms of the equivalent voltages connected directly to the input terminals of the op-amp. Thus I_b^- applies a voltage $-I_b^- R_{source}^-$ to the inverting input terminal and I_b^+ applies a voltage $-I_b^+ R_{source}^+$ to the non-inverting terminal.

R_{source}^- and R_{source}^+ represent the effective source resistance connected at the inverting and non-inverting input terminals respectively. They represent the parallel combinations of all resistive paths to ground, including in the case of R_{source}^- the path through any feedback resistor and the op-amp output resistance to ground.

Since V_{io} is directly applied to the input terminal, we may represent the total equivalent input offset voltage as

$$e_{os} = \pm V_{io} + I_b^- R_{source}^- - I_b^+ R_{source}^+ \tag{2.27}$$

Drift in the total equivalent input offset voltage is obtained by substituting values of the drift coefficients of I_b and V_{io}.

Graphs showing the dependence of e_{os} drift on source resistance are given in some op-amp data sheets. e_{os} appears at the output multiplied by the 'noise gain' $1/\beta$; the resultant error may be referred to any signal input by simply dividing by the signal gain associated with that input. A numerical example should serve to clarify the evaluation of offset error.

An op-amp with $I_b = 100$ nA, $I_{io} = 10$ nA and $V_{io} = 1$ mV is to be used in the inverting summing circuit shown in Figure 2.23. Find the minimum signals that can be amplified at the two input signal points with less than 1 per cent error due to offset.

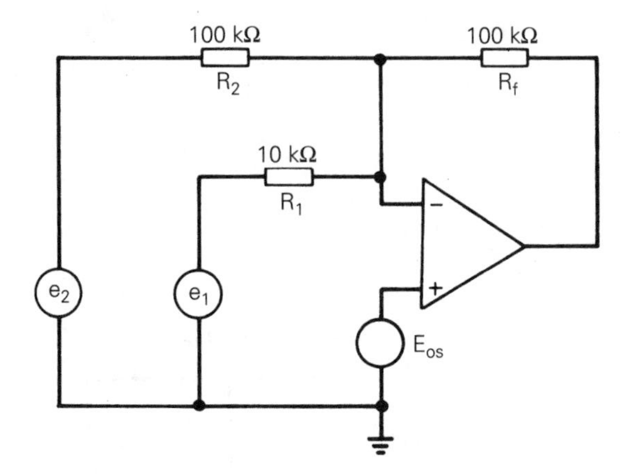

Figure 2.23 *Circuit for example of offset error evaluation*

In the circuit of Figure 2.23, the non-inverting input terminal is connected directly to earth, making R^{+}_{source} zero. The effective source resistance through which bias current must flow to the inverting input terminal is

$$R^{-}_{\text{source}} = R_1 \,//\, R_2 \,//\, R_f = 8.3 \text{ k}\Omega$$

According to equation 2.27

$$e_{os} = 10^{-3} + 10^{-7} \times 8.3 \times 10^{3} \text{ (worst case)} = 1.83 \text{ mV}$$

In this circuit

$$\frac{1}{\beta} = 1 + \frac{R_f}{R_1 \,//\, R_2} \cong 12$$

The output offset error is thus $e_{os}/\beta = 22$ mV. Referring this error to the e_1 input the equivalent input error is $22/10 = 2.2$ mV. Referring the output error to the e_2 input the equivalent input error is $22/1 = 22$ mV. The smallest input voltage for less than 1 per cent error is thus 220 mV at the e_1 input, or 2.2 V at the e_2 input.

The input offset error due to bias current can be reduced, by connecting a resistor equal in magnitude to R^{-}_{source} between the non-inverting input and earth. This makes

$$e_{os} = \pm V_{io} + I_{io}R_{source}$$

Accuracy is still relatively low, but can be improved if the initial offset is balanced out by using one of the offset balancing methods discussed in Section 9.6. An evaluation of subsequent offset error would then require knowledge of the temperature coefficient of I_b and V_{io}, and an estimate of the possible ambient temperature variations (δT).

Values $V_o = \dfrac{\Delta V_{io}}{\Delta T}\delta T$ and $I_b = \dfrac{\Delta I_b}{\Delta T}\delta T$

should then be substituted in equation 2.27 in order to find the equivalent input error due to temperature drift.

2.10 Common mode rejection ratio (CMRR)

An ideal differential op-amp responds only to the difference in the voltages applied to its input terminals and produces no output for a common mode input voltage. In practical op-amps, common mode input voltages are not entirely subtracted at the output due to slightly different gains between the inverting and non-inverting inputs. The gain of an op-amp for common mode input voltage is known as the common mode response. The ratio of the gain with the signal applied differentially to the common mode response is called the common mode rejection ratio, CMRR. It is often expressed in decibels (dB) by taking 20 times logarithm (base 10) of the ratio.

Common mode rejection presents no problem for op-amps used in the inverting configuration. This is because, with one input earthed, the input common mode voltage e_{cm} must be zero.

In non-inverting circuits, feedback causes the voltage at the inverting input to follow that at the non-inverting input. The input common mode voltage thus varies directly with the input signal. With finite CMRR an output signal is produced in response to this common mode input signal. Thus an error is introduced which affects the overall circuit accuracy.

The common mode error is conveniently represented in terms of equivalent input common mode error voltage, $e_{\varepsilon cm}$, where this is the common mode output divided by the differential gain. If the op-amp is considered to have $e_{\varepsilon cm}$ applied to its non-inverting input terminal, along with the input signal, it may then be treated as though it completely rejected the actual input common mode signal e_{cm}. The relationship between input common mode error voltage and input common mode voltage is readily obtained as

$$\frac{e_{\varepsilon cm}}{e_{cm}} = \frac{1}{\text{CMRR}} \qquad (2.28)$$

For example, consider an op-amp with CMRR = 1000 (60 dB) used in the non-inverting configuration with an input signal of 1 V. The input common mode voltage e_{cm} would also be 1 V. The input common mode error voltage is seen to be 1 mV and this represents a 0.1 per cent measuring error. The op-amp is illustrated in Figure 2.24.

It is not always possible to compensate for common mode errors. This is because the CMRR for some op-amps shows a dependence on the magnitude

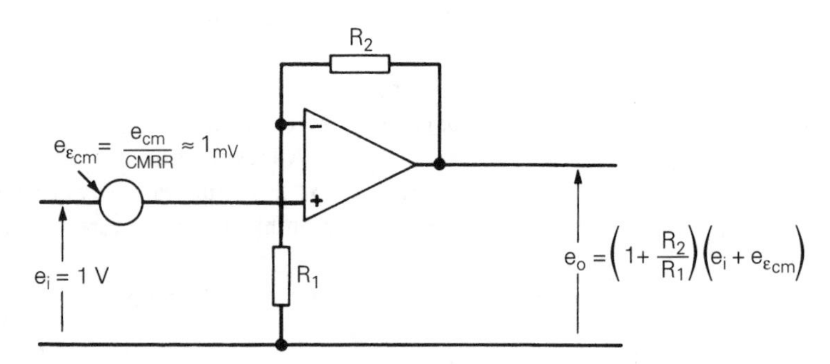

Figure 2.24 *Representation of common mode error*

of the input common mode signal. Also, the common mode error voltage is a non-linear function of common mode voltage and there is also an added complication of temperature dependence. Since linearity of common mode error voltage with common mode voltage is really more important than the actual value of the CMRR, a graph illustrating this relationship is valuable if an op-amp is to be used in an application which is critically dependent on common-mode performance. Figure 2.25 illustrates an example.

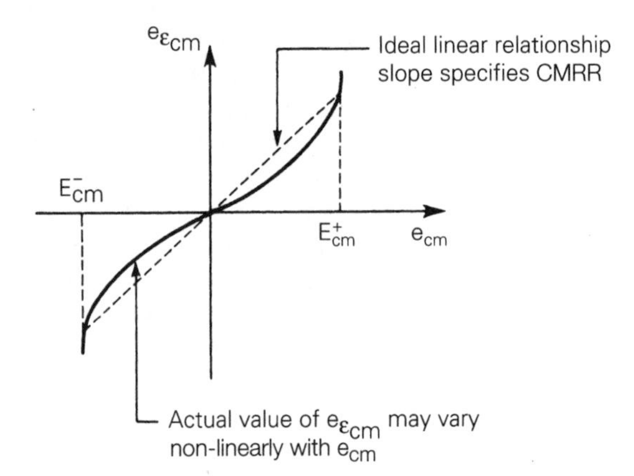

Figure 2.25 *Common mode error voltage as a function of common mode input voltage*

Specified values of CMRR where non-linearities exist are usually average values, assuming a measurement of $e_{\varepsilon cm}$ at the end points corresponding to the maximum common mode voltage E_{cm}. It is important to note that published common mode specifications generally apply to DC input signals; CMRR is usually found to decrease at the higher frequencies.

2.11 Noise in op-amp circuits

The output of an op-amp is always found to contain random signals that are unrelated to the input signals. These unwanted signals are called noise. Errors such as drift error can, theoretically at least, be reduced to negligible proportions (by, say, using a temperature-controlled environment), but there always remains a noise error that limits the attainable accuracy and resolution. Noise should be taken into account if the circuit being designed must process low-level signals with high accuracy.

There are two basically different types of noise in a circuit. Interference noise is picked up from outside the circuit. Inherent noise arises within the circuit itself.

Sources of interference noise are many and varied. They include electromagnetic or electrostatic pickup from power sources at mains frequency, broadcast radio, electrical arcing at switch contacts and signals radiated from digital electronic circuits. Fortunately the circuit designer can usually minimize interference noise by suitable shielding and guarding and the elimination of earth loops (see Section 9.4) and by proper attention to mechanical design.

Inherent noise is a function of a particular op-amp and the circuit in which it is used. The only way in which the designer can influence inherent noise is through his choice of op-amp and circuit components. The noise in an op-amp can vary by several orders of magnitude.

2.11.1 Characterization of random noise sources

The total noise that is inherent in an op-amp circuit (or any circuit for that matter) can be thought of as a combination of the effects of several, separate, noise sources. These inherent noise sources are essentially random signals. They give an electrical signal whose waveform has no defined shape, amplitude or frequency. They may be thought of as a superposition of signals at all possible frequencies, with amplitude and phase varying in a completely random fashion.

Root mean square (RMS) value of a noise source

It is a characteristic of most forms of random noise source that averaged over a sufficiently long time interval their RMS value in a specified bandwidth remains constant. The RMS value in a specified bandwidth is thus a useful and meaningful way of characterizing a random noise source. The general defining equation is

$$N_{\mathrm{RMS}} = \sqrt{\frac{1}{T}\int_0^{\mathrm{T}} n_{\mathrm{i}}^2 \, \mathrm{d}t} \qquad (2.29)$$

where n_{i} is the instantaneous noise amplitude (current or voltage), and N_{RMS} is the RMS value of the noise source. In order to be meaningful, the RMS value of a noise source must have the bandwidth clearly defined. The wider the bandwidth: the greater is the RMS value of the noise.

Combining noise sources

The combined effect of several random noise sources is found by root sum of the square addition of the RMS values of the separate noise sources.

Thus, if E_1, E_2, E_3 are the RMS voltage values of three separate voltage noise generators their combined effect when connected in series is equivalent to a single noise voltage generator of RMS value

$$E = \sqrt{E_1^2 + E_2^2 + E_3^2} \tag{2.30}$$

2.11.2 Peak-to-peak noise

In some applications it is peak-to-peak noise which really sets the limit to a system performance. Peak-to-peak noise is the difference between the largest positive and negative peak excursions to be expected during some arbitrary time interval. Random noise is, for all practical purposes, Gaussian in amplitude distribution; the highest noise amplitudes having the smallest (yet not zero) probabilities of occurring.

Peak-to-peak noise is thus difficult to measure repeatedly, but a useful rule of thumb for converting from an RMS noise value to a peak-to-peak value is to multiply the RMS value by a factor of 6. The amplitude obtained is exceeded less than 0.25 per cent of the time by a random noise signal of the given RMS amplitude.

2.11.3 Noise density spectrum

The noise generated by any random noise source exists in all parts of the frequency spectrum. The amount of noise contributed by a source varies with the range of frequencies over which the observation is made:

$$N_{\mathrm{RMS}(f_1 - f_2)} = \sqrt{\int_{f_1}^{f_2} n^2 \, \mathrm{d}f} \tag{2.31}$$

A noise density spectrum shows the way in which the noise produced by a given source is distributed over the frequency spectrum. Noise density n is shown as a function of frequency, usually on log–log axes. Examples of noise spectra are given in Figures 2.26, 2.28 and 2.29.

In the spectral regions of interest in op-amp applications, the noise sources encountered often have spectral distribution belonging to one of two types: in one, n is constant as a function of frequency; and in the other, n varies inversely with the square root of frequency.

2.11.4 White noise

Noise for which n is constant with change in frequency is called white noise. The noise from a white noise source is distributed uniformly throughout the frequency spectrum.

Thermal agitation of electrons in a resistor causes random voltages to appear across it. The spectrum of this noise voltage is characterized by a noise density that is constant as a function of frequency. Resistance noise

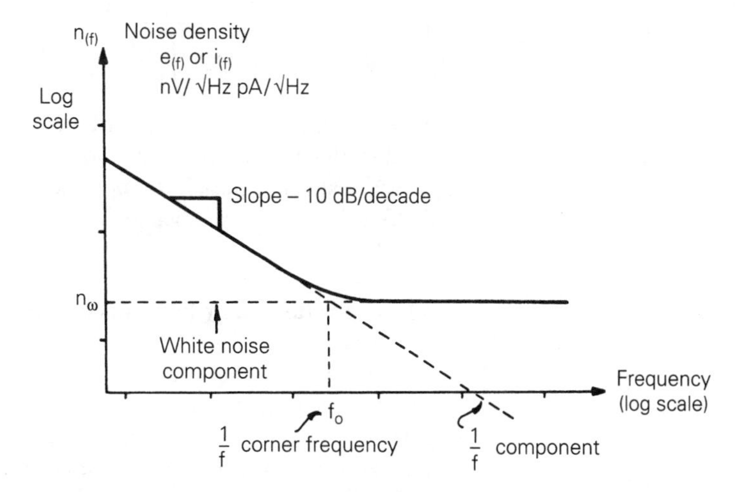

Figure 2.26 *Noise density spectrum for op-amp generated noise*

(Johnson noise) is thus an example of white noise. The noise voltage associated with a resistor has a noise density:

$$\text{Resistance noise } e = \sqrt{(4kTR)} \text{ V per } \sqrt{\text{Hz}} \tag{2.32}$$

where:
k = Boltzmann's constant = 1.37×10^{-23} J/K,
T = the temperature in K and
R = the resistor value in Ω.

The RMS noise voltage generated by a resistor R, in the range of frequencies f_1 to f_2, is:

$$\text{Resistance noise } e_{\text{RMS}(f_1 - f_2)} = \sqrt{4kTR\,(f_1 - f_2)} \text{ volt RMS} \tag{2.33}$$

2.11.5 1/*f* or 'pink' noise

Noise, which has a density that varies inversely with the square root of frequency, is referred to as 1/*f* noise. This is sometimes called 'pink' noise. The noise density for a pink noise source is determined by an equation of the form

$$\text{Pink noise } n = K\sqrt{\frac{1}{f}} \tag{2.34}$$

where K is the value of n at f = 1 Hz.

A graph of n against frequency for a pink noise source when shown as a log–log plot is a straight line of slope −10 dB/decade. A graph of n^2 against frequency gives a straight line of slope −20 dB/decade.

The contribution which a pink noise source makes to the RMS value of the noise in a frequency range f_1 to f_2 may be found by substituting equation 2.34 into the general equation 2.31. Thus:

$$N_{\mathrm{RMS}(f_1 - f_2)} = K\sqrt{\int_{f_1}^{f_2}\frac{df}{f}} = k\sqrt{\ln\left(\frac{f_2}{f_1}\right)} \quad (2.35)$$

Note that the RMS noise contributed by a pink noise source in a particular bandwidth depends upon the ratio of the frequencies defining that bandwidth. Every frequency decade of noise from a pink noise source has the same RMS value as every other decade.

2.11.6 Evaluation of RMS noise from a noise density spectrum

The contribution that a particular noise source makes to the RMS noise in any specified bandwidth can, in principle, be found by evaluating the integral in equation 2.31. Equations 2.33 and 2.35 are the results of such evaluations for the particular cases of a white noise source and a pink noise source. Note that in order to evaluate the integral the equation defining the noise density as a function of frequency must of course be known.

In the spectral regions of interest, the noise generators used to represent the effect of internally generated op-amp noise often exhibit a noise density spectrum of the form shown in Figure 2.26. A spectrum of this kind can be thought of as consisting of two components; a white noise component, which is the predominant noise component at high frequencies, and a $1/f$ component which predominates at low frequencies. The RMS value of the noise contributed by the source in any bandwidth can be found by a root sum of the squares addition of the RMS contributions of the two separate components in that bandwidth.

2.11.7 Op-amp noise specifications

The noise present at the output of an op-amp is a combination of the amplified noise present at its input and noise generated internally inside the op-amp. Noise produced internally by an op-amp is conveniently modelled as shown in Figure 2.27, by a noiseless op-amp with a noise voltage and a noise current generator at its input terminal.

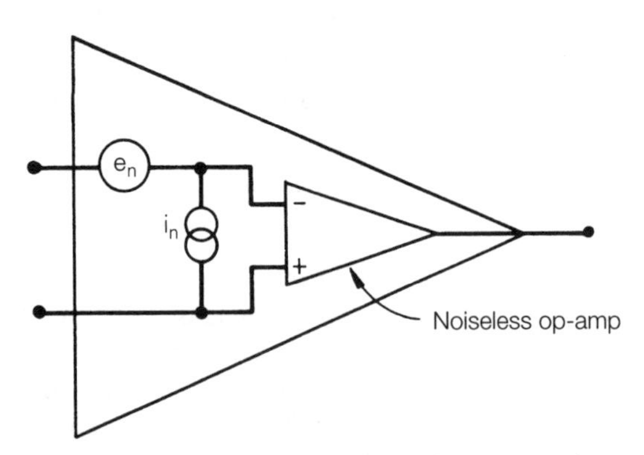

Figure 2.27 *Op-amp model for internally generated noise*

Equivalent noise generators connected to the op-amp's input terminals could be used to represent noise generated by resistors at the input. The several input contributions to the noise can be combined as a single resultant input-referred noise source and in closed-loop op-amp applications this total input-referred noise appears at the output multiplied by the closed-loop noise gain $1/\beta$.

The technique for noise evaluation, just described, is similar to the technique for evaluating offset and drift errors, previously described in Section 2.9.3. The main difference between a drift error evaluation and the evaluation of noise errors is the dependence of noise on bandwidth. In making a noise assessment of an op-amp circuit, the designer must use noise data from the op-amp data sheet.

Op-amp noise data will be found presented in both graphical and numerical form. An example of graphical data is given in Figure 2.28, where the frequency dependent nature of the noise can be seen. Numerical noise data is usually given in terms of voltage noise and current noise, as modelled in Figure 2.27.

Some manufacturers specify typical peak-to-peak input voltage and current noise in a low frequency band (say 0.01 to 1 Hz). A peak-to-peak specification of this kind is particularly useful in assessing accuracy limits (as limited by noise) in applications in which the signals of interest are essentially DC, or very slowly varying quantities. Wide bandwidth noise will of course be present in the op-amp output but it can be removed by following the op-amp with a suitable low-pass filter.

RMS values of noise can be used as a means of estimating peak-to-peak values. The rule of thumb multiplication factor of 6, mentioned previously, is used.

2.11.8 Evaluating noise errors using noise specifications

The problem facing the circuit designer is to assess accuracy and resolution limits as determined by noise. Clearly if the noise level at the output is comparable to the signal level, the signal is obscured by the noise. In wide band applications signal-to-noise ratio (SNR) is a useful figure of merit in describing how well the signal 'stands out' from the noise. The signal-to-noise ratio at the output is defined as

$$\text{SNR} = \text{signal power out/noise power out}$$

The ratio is sometimes expressed in decibels (dB) by the relationship

$$\text{SNR(dB)} = 10\log(P_s/P_n) \tag{2.36}$$

In DC and low frequency applications, accuracy limits determined by noise can be related to peak-to-peak noise. Noise error is expressed as a percentage, from the relationship:

$$\text{Noise error} = \frac{\text{peak-to-peak value of output noise}}{\text{peak-to-peak value of output signal}} \times 100\% \tag{2.37}$$

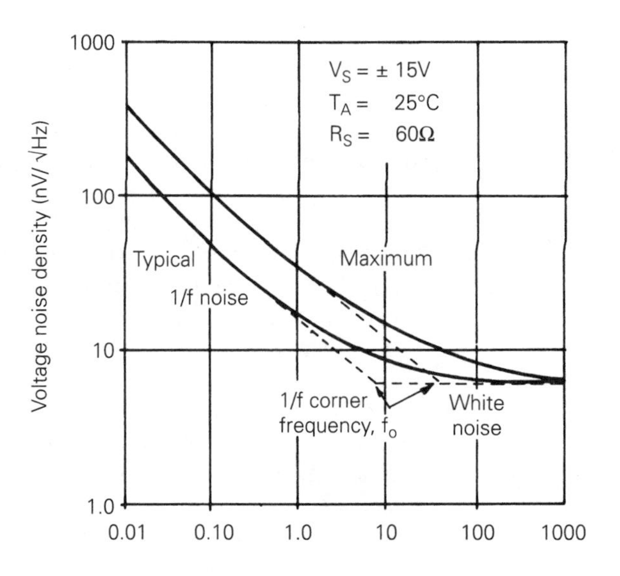

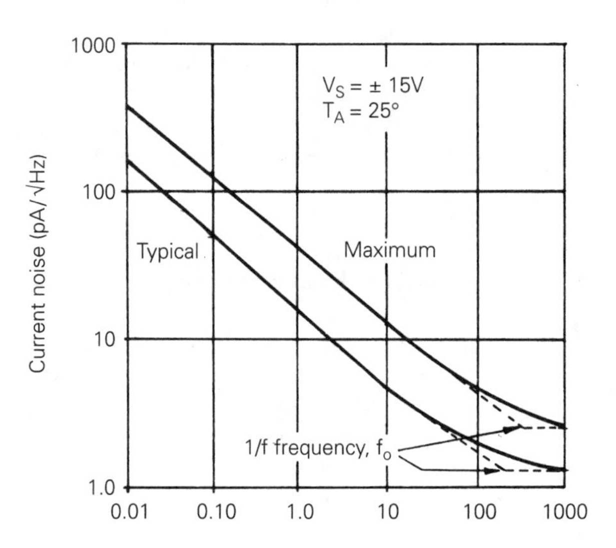

Figure 2.28 *Noise spectral density for a typical op-amp*

An estimate of the amount of noise to be expected in a given circuit is made by using the noise data for the op-amp used. Rigorous noise evaluations are time consuming and can be of dubious practical value if they are based upon 'typical' noise data. In many applications the effect of a single noise source can be dominant, and the ability to identify the most significant noise contributions allows a rapid order-of-magnitude noise assessment.

As a starting point in any noise evaluation, the signal gain and the noise gain in the circuit should be found. Bode plots giving their frequency dependence help to show up the spectral regions where significant noise contributions are to be expected. An evaluation of the noise performance of

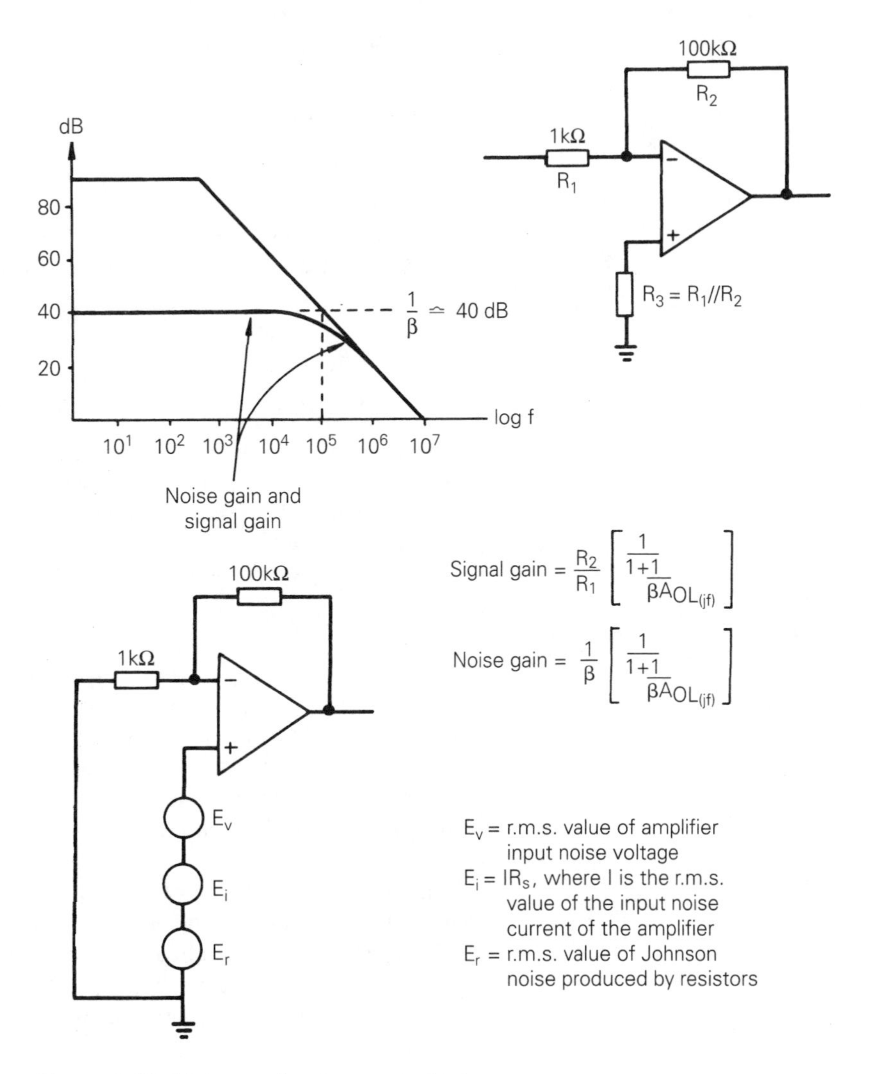

Figure 2.29 *Noise evaluation example 1*

an op-amp used in a basic resistive feedback configuration is taken as a first example, Figure 2.29. Input noise sources appear at the op-amp output multiplied by the noise gain; in this example the noise gain is 100, its 3 dB bandwidth limit is 10^4 Hz and it rolls off at 20 dB/decade beyond this frequency.

Using equivalent input noise generators

Noise voltage and noise current can be used to find the total noise. In order to calculate the total noise, both noise sources need to be described in the same form. Noise voltage is normally used. To convert the noise current (I_n) into voltage form it is multiplied by the resistance presented to the op-amp's inputs.

Thus we have

$$e_{n1}^{+} = I_{in}R_{source}^{+}$$

$$e_{n2}^{-} = I_{in}R_{source}^{-}$$

Simplifying into one noise generator, where R_s is the effective resistance between the op-amp's two inputs (given by $S_S = \sqrt{(R_1//R_2)^2 + R_3^2}$ in Figure 2.29), we have

$$e_{in} = I_{in}R_s$$

The total intrinsic noise is $\sqrt{[e_n^2 + (I_nR_s)^2]}$. The voltage and current generators have equal contribution to the intrinsic noise when $e_n = I_nR_s$. This occurs when $R_s = R_n = e_n/I_n$, and R_n is known as the noise resistance. If the resistance presented to the op-amp's inputs is much lower than e_n/I_n, the intrinsic noise can be considered due to the voltage generator alone. Conversely, if the resistance is much higher than e_n/I_n, the intrinsic noise is due to the current generator alone.

Lowering an AC op-amp's noise figure

The noise figure of the op-amp is described as the signal-to-noise ratio at the output divided by the signal-to-noise ratio at the input. This is often measured as a voltage, but expressed in decibels (dB) by taking 20 log(voltage ratio). In other words, it is the amount of noise added to the signal by the op-amp.

If the resistance presented to the op-amp's input is lower than the noise resistance, the op-amp's voltage noise will dominate. The noise power is this voltage squared, multiplied by the resistance presented to the input. One way to minimize the noise contribution of an op-amp is to transformer couple the input signal with a step-up transformer (1:*N*).

When a step-up transformer is used, the signal voltage is multiplied by the turns-ratio (*N*). The effective impedance seen by the op-amp input is the source resistance multiplied by *N* squared. Assume that the op-amp's input impedance must match the source impedance. First, we choose a value of *N* such that $N^2 = 2R_n$, for the signal impedance given. Second we terminate the secondary of the transformer with a load resistor equal to $2R_n$. The op-amp's inputs now see impedance R_n, and both V_n and I_n contribute to the total noise. The equivalent voltage noise is now $(\sqrt{2})e_n$, or 3 dB above e_n.

Suppose $R_n = 3\ \text{k}\Omega$ and the signal impedance is 600 Ω. We choose a transformer ratio of 1:3.16. The 600 Ω primary impedance is transformed up to 6 kΩ at the secondary. A load of 6 kΩ is then applied across the secondary winding. The input voltage is increased by 10 dB (20 times log(3.16)). Since the impedance at the op-amp's input equals R_n, the total input noise is only increased by 3 dB. Therefore this circuit has raised the signal-to-noise ratio by 7 dB, compared to a circuit with a 600 Ω resistor across its input. This is shown in Figure 2.30.

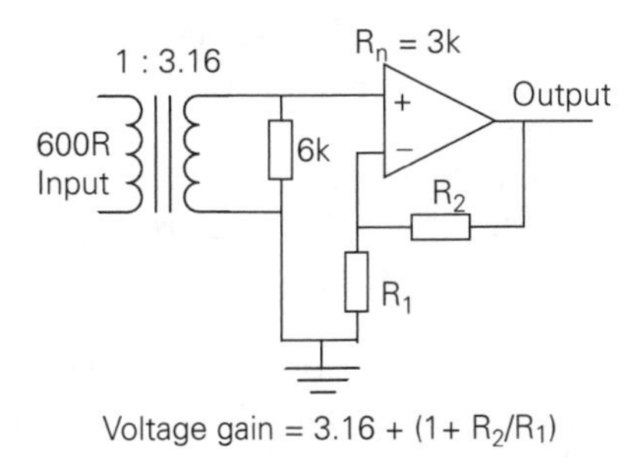

Figure 2.30 *Transformer coupled op-amp input*

In some circuits, impedance matching is not required. This is usually where the source and load are physically close. If there is a transmission line between source and load, matching is often necessary. Unmatched transmission lines suffer from reflections and crosstalk from other signal carrying circuits.

The important notion of noise resistance can lead to a misunderstanding. The resistance of a source should not be artificially raised to be equal to the noise resistance. This will increase thermal noise due to the extra resistance in series with the signal path. Transformer coupling, to increase the effective source resistance, works because the signal voltage is also raised in the process.

The use of a transformer gives additional benefits. Transformer coupling, using an earthed screen between primary and secondary windings, increases the rejection of common mode signals. Transformers are designed to cover a specified range of frequencies and tend to reduce the amplitude of signals outside this range; hence they act as first order bandpass filters.

Exercises

2.1 An op-amp is to be used in the inverting feedback configuration with a closed loop signal gain of 100 and an input resistance of 10 kΩ.
   (a) Assuming ideal amplifier performance what values of input and feedback resistor should be used?
   (b) If the op-amp is assumed ideal except for a finite loop gain of 10^4, by how much will the signal gain differ from 100?
   (c) If the open-loop gain of the amplifier changes by 5 per cent what effect will this have on the closed-loop signal gain?

2.2 The amplifier used in the circuit of Figure 2.5 has an open-loop gain 5×10^4 and differential input resistance 100 kΩ, resistor R_1 = 1 kΩ, R_2 =3.9 kΩ. Find the closed-loop gain and the effective input resistance of the circuit. Assume that the common mode input impedance of the amplifier is infinite and that its output resistance is negligible.

2.3 Write expressions for the feedback fraction β and the closed-loop gain $1/\beta$ for the circuits given in Figures 1.3, 1.4(b), 1.6, 1.7 and 1.8.

2.4 Express the following voltage ratios in decibels to the nearest whole dB. (a) 1, (b) 2, (c) 3, (d) 10, (e) 100, (f) 1000, (g) 10^6.

2.5 Without using log tables, using only the results of Exercise 2.4, calculate the dB equivalents of the following voltage ratios to the nearest whole dB.
(a) 6, (b) 15, (c) 3.33, (d) 333, (e) 9, (f) 0.01, (g) 0.05, (h) $\sqrt{2}$, (i) $1/\sqrt{2}$.
(Hint: 3.33 = 10/3. Thus (3.33 expressed in dB) = (10 expressed in dB) − (3 expressed in dB).)

2.6 An op-amp has an open-loop frequency response that exhibits a 20 dB/decade roll-off down to unity gain. Its open loop gain at zero frequency is 100 dB and its unity-gain frequency is 1 MHz. Sketch the open-loop frequency response on a dB/log f plot:
(a) The amplifier is connected as a non-inverting feedback amplifier (a follower) with closed-loop gain (i) 2, (ii) 10, (iii) 50. Find the small-signal closed-loop bandwidth in each case and sketch the appropriate Bode plots.
(b) The amplifier is connected as an inverting adder, as in Figure 1.4(b), so as to form the weighted sum of three separate signals. Input resistors R_1 = 27 kΩ, R_2 = 39 kΩ, R_3 = 56 kΩ, and a feedback resistor R_f = 120 kΩ, are used. Find (i) the ideal performance equation, (ii) the value of $1/\beta$ for the circuit, (iii) the small-signal closed-loop bandwidth, (iv) by how much the ideal performance equation is in error at a frequency 20 kHz (see Section 2.4).

2.7 An op-amp has a slew rate of 0.5 V/μs. What is the maximum frequency for which the amplifier will give an undistorted sinusoidal output signal of (a) 20 V peak-to-peak; (b) 10 V peak-to-peak? (see Section 2.8).

2.8 An op-amp employing the frequency compensating technique discussed in Section 2.6 has a frequency compensating capacitor of value 10 pF connected to it. When it is used as a unity-gain follower, it has a closed-loop frequency response that exhibits 5 dB of gain peaking. What damping factor and overshoot do you expect in the small-signal step response of the follower?

What minimum value of frequency compensating capacitor would be required for no gain peaking, and what overshoot would result in the small-signal step response if this value of capacitor were connected?

2.9 The open-loop gain of an op-amp is 100 dB at DC and its open-loop frequency response exhibits two breaks at frequencies f_{c1} = 100 Hz, f_{c2} = 4 MHz. The amplifier is connected as a unity gain follower. At what frequency is the magnitude of the loop gain unity? What is the phase margin in the circuit? By how much does the closed-loop gain peak and at what frequency does the gain peak occur? Estimate the settling time to 0.1 per cent if a small input step signal is applied. Find the minimum closed-loop gain for which the amplifier will exhibit (a) no closed-loop gain peaking; (b) a critically damped response with no overshoot in the transient response.

Find the closed-loop 3 dB bandwidth for each of these values of closed-loop gain (see Section 2.7 and Appendix A2).

2.10 An op-amp has the following offset and temperature drift specifications:

$V_{io} = 2$ mV; $\Delta V_{io}/\Delta T = 10$ μV/°C; $I_B = 500$ nA; $\Delta I_B/\Delta T =$ 1 nA/°C; $I_{io} = 50$ nA; $\Delta I_{io}/\Delta T = 0.1$ nA/°C.

The amplifier is connected as a simple inverter with $R_1 = 10$ kΩ, $R_f = 1$ MΩ, and is supplied by a signal source of negligible resistance. Find:

(a) the output offset voltage;
(b) the change in output offset voltage to be expected from a temperature change of 10°C;
(c) assuming initial offset balanced, the smallest input signal that can be amplified with less than 1 per cent error, due to a 10°C temperature change;
(d) the value of a resistor R_c that should be connected between the non-inverting input terminal and earth to reduce the offset error due to amplifier bias current.

Repeat parts (a), (b) and (c), assuming that the resistor R_c is connected in the circuit. In all cases assume worst case errors (see Section 2.9.3).

2.11 An op-amp with the offset and temperature drift specifications given in Exercise 2.10 is to be used as a follower with a feedback resistor of 10 kΩ and a resistor of 1 kΩ connected between the inverting input terminal and earth. The circuit is supplied by a signal source of internal resistance 100 kΩ. Find:

(a) the output offset error with no offset balance;
(b) the smallest input signal that can be amplified with no more than 1 per cent error if initial offsets are balanced and the temperature changes by 10°C (see Section 2.9.1).

2.12 A differential input op-amp, assumed ideal except for finite open-loop gain and finite CMRR, has an open loop gain of 5×10^4. When the op-amp inputs are connected together, and a signal of 1 V with respect to earth is applied to them, the output voltage of the amplifier is found to be 5 V. Find the CMRR of the amplifier and the measurement error due to common mode signals (expressed as a percentage), when the amplifier is used as a non-inverting feedback amplifier.

2.13 A random noise voltage source has a noise density function which varies inversely with frequency; the RMS value of the noise voltage produced by the source is 2 μV in the frequency range 20 Hz to 100 Hz. Find the RMS noise voltage produced by the source in the frequency range: (a) 1 Hz to 10 Hz, (b) 10 Hz to 100 Hz, (c) 1 Hz to 1 kHz (use equation 2.35).

2.14 The input connected noise voltage and noise current generators that are used to represent the noise generated by an op-amp have noise density spectra consisting of white noise and $1/f$ components. The noise voltage generator has a white noise component with density 20 nV/$\sqrt{\text{Hz}}$ and a

$1/f$ corner frequency of 50 Hz. The current generator has a white noise component with density 0.3 pA/√Hz and a $1/f$ corner frequency of 1 kHz. Sketch the noise density spectra. Find:

(a) the RMS value of the noise voltage generator;

(b) the RMS value of the noise current generator, in the frequency ranges (i) 0.1 Hz–10 Hz, (ii) 1 Hz–100 Hz, (iii) 1 Hz–1 kHz, (iv) 1 Hz–10 kHz;

(c) the RMS value of the total input referred noise voltage in the above frequency ranges for source resistance 1 kΩ, 10 kΩ, and 100 kΩ.

3 Analogue integrated circuit technology

This chapter describes the technology used within analogue integrated circuits, concentrating on op-amps. It will also describe the differences between voltage feedback and current feedback.

The technology is determined by the type of transistor used in the integrated circuit (IC). The types include bipolar, bipolar with JFET inputs, LinCMOS (linear CMOS) and BiCMOS (incorporating bipolar and CMOS transistors).

The majority of op-amps are designed to use voltage feedback. However, current feedback devices are now employed in radio frequency and video signal processing because they have a very wide bandwidth capability. It may be useful to define (i) voltage feedback, and (ii) current feedback.

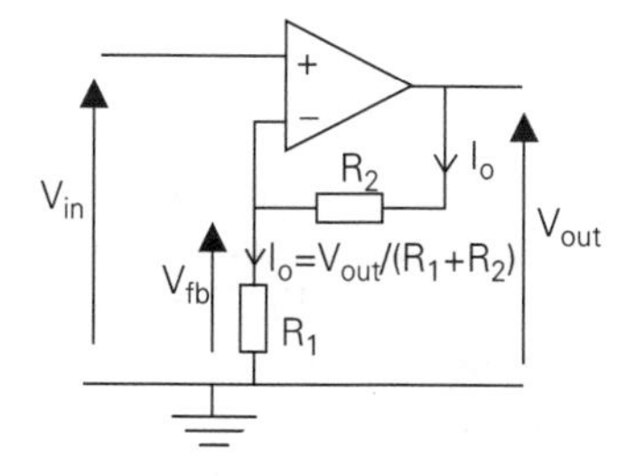

Figure 3.1 *Voltage feedback amplifier*

Voltage feedback refers to a closed-loop configuration in which the error signal is in the form of a voltage (see Figure 3.1). Traditional op-amps use voltage feedback and produce an output voltage in response to a difference in voltage at their inputs. In other words, their inputs respond to voltage changes. The ideal voltage feedback op-amp has high impedance inputs and zero input current. Voltage feedback is used to maintain zero differential input voltage.

The transfer function of a non-inverting voltage feedback amplifier is given by:

$$\frac{V_o}{V_{IN}} = \left(1 + \frac{R_2}{R_1}\right)\frac{1}{1 + \dfrac{1}{LG}}$$

Here loop gain (LG) is given by $\dfrac{A_{OL}}{1 + \dfrac{R_2}{R_1}} = \beta A_{OL}$

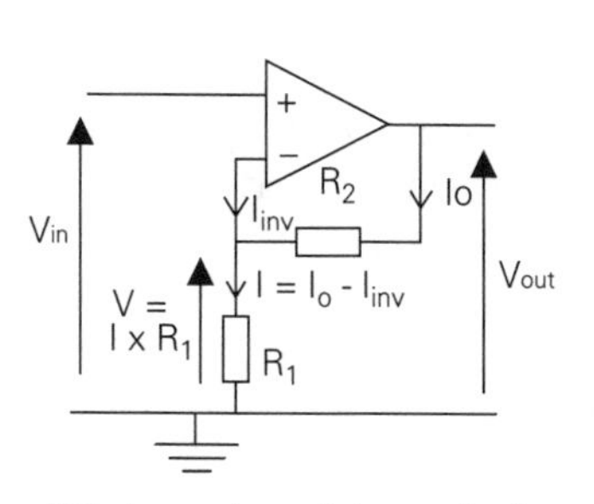

Figure 3.2 *Current feedback amplifier*

Current feedback refers to any closed-loop circuit that uses an error signal in the form of a current (see Figure 3.2). Unlike voltage feedback op-amps, current feedback devices have a low impedance inverting input. The low impedance allows current to flow into and out of the inverting input. Any current flow at this input is an error current, and the op-amp produces an output voltage in proportion to its magnitude. Current feedback is used to maintain zero error current at the inverting input. The non-inverting input is high impedance, like that of a voltage feedback op-amp.

The heart of a current feedback op-amp is a transimpedance amplifier. The transimpedance amplifier produces a voltage output from a current input.

As the function implies, the open-loop 'gain', V_o/I_{in}, is expressed in ohms. We will consider this impedance in its separate resistive (R_m) and capacitive (C_C) forms, or as a complex impedance $Z(s)$. Hence a current feedback op-amp is sometimes called a transimpedance amplifier.

The transfer function of a non-inverting current feedback amplifier is given by:

$$\frac{V_o}{V_{IN}} = \left(1 + \frac{R_2}{R_1}\right)\frac{1}{1 + \dfrac{1}{\text{LG}}}$$

In this case loop gain (LG) is given by $\text{LG} = \dfrac{Z(s)}{R_2}$

The closed-loop gain now depends on just R_2 and the op-amp's trans-impedance $Z(s)$.

3.1 Voltage feedback op-amps

Op-amp integrated circuits employing bipolar transistors have been used since 1965. These op-amps were a great improvement on discrete 'operational amplifiers' that were built using individual transistors, resistors and capacitors. They were smaller, low cost and simple to use. Field effect transistors were later used in some op-amps to improve certain aspects of bipolar op-amp performance.

Devices with JFET input transistors, but otherwise using bipolar transistors throughout, were developed in the late 1960s. The advantage of the JFET input op-amp was reduced input bias current requirement. JFET input op-amps were called 'BiFET' op-amps by Texas Instruments; this name is very descriptive since it employs bipolar and FET transistors.

Op-amps that use CMOS transistors throughout have been used more recently to allow very low power operation. Further development has produced BiCMOS op-amps, employing both bipolar transistors and complementary MOSFET transistors.

3.1.1 Bipolar op-amps

The problem with the original (1965) op-amp designs was that the input impedance was much lower than the ideal (about 200 kΩ). Also the input bias current and offset voltages were significant. Continual development work by several semiconductor manufacturers has enabled the performance of bipolar op-amps to improve steadily.

Bipolar op-amps have a typical input impedance of about 10 MΩ, but negative feedback can increase this to 1 GΩ or more. The greater problem is that input bias currents are in the order of 10 nA, and the resulting noise current is in the order of 0.3 pA/√Hz. These levels of bias and noise current make bipolar op-amps unsuitable for use with high impedance sensors.

Bipolar op-amps use fast npn and pnp transistors. Typical input and output circuits were described in Chapter 2 (Figures 2.1 and 2.3). These allow the device to have a gain–bandwidth product of 10 MHz or more. Good matching between the input transistors not only reduces the input offset voltage, but also reduces drift with temperature and with time. Op-amps with a bipolar input stage have the greatest long-term stability of all existing technologies. One reason is that collector currents flow vertically through the semiconductor, and hence are not subject to lateral stress and strain due to temperature.

The noise voltage in a bipolar transistor is due to the emitter resistance, and this is far lower than the equivalent resistance in JFETs or MOSFETs. Hence the noise voltage is lower in a bipolar input stage, compared with a FET input stage. Typical noise voltage is 15 nV/√Hz, although low noise types have a noise voltage below 5 nV/√Hz.

3.1.2 Complementary bipolar (Excalibur) technology

There are thousands of different types of integrated circuit op-amps available commercially, and the number is increasing almost daily. The many manufacturers of these different types are trying to produce the best devices by giving them improved DC precision, faster AC performance and lower power consumption.

One difficulty in improving the performance of op-amps is the relative slow response of the pnp transistor compared with that of the npn. The reason for this is the low mobility of the majority carrier 'holes' in the pnp transistor, compared with that of the electrons in the npn transistor. Many op-amp designs use pnp transistors, particularly in input stages that use differential pair transistors; they are also used in emitter follower outputs. However, the pnp transistor has a typical f_T of only 5 MHz compared with 150 MHz for the npn transistor. One method of increasing the speed of slow complementary bipolar circuits is to increase the speed of the npn element, so the overall speed increases. This is the approach of several semiconductor manufacturers.

Conventionally, a vertical structure in the silicon die is used to create transistors. An npn transistor will have a p-type substrate and an n-type epitaxial layer. In theory, faster pnp transistors can be achieved by simply reversing this design, with a substrate of n-type silicon. Although this method successfully increases the pnp transistor speed, it reduces the speed of the complementary npn fabricated in the same IC.

In contrast to this, Texas Instruments has developed a manufacturing process for a fast vertical pnp device structure that retains the speed of the npn devices. Texas Instruments call this their Excalibur process and it uses a deeply submerged n-region as an 'artificial substrate' in which a buried p-region becomes the collector. The fast pnp Excalibur transistor can be integrated directly into the signal path without fear of limiting the bandwidth or slew rate. This often has the added benefit of requiring less supply current than its predecessors.

An example of the Excalibur range of op-amp is the TLE2021, which is a low power, precision op-amp. The TLE2021 achieves a unity-gain bandwidth in excess of 2 MHz and a slew rate of 0.9 V/μs. The supply current is less than 200 μA with an input voltage offset of less than 100 μV.

A common problem with many op-amps is that they suffer from 'phase inversion'. This happens if the input swings close to the power rail potential, causing the output to change state and swing to the opposite rail. Many of the older bipolar and BiFET op-amps are known to have this problem. Many newer products, such as the TLE2021 family, have been designed to avoid this.

3.1.3 'Chopper' stabilization

The TLC265X family of CMOS technology op-amps offers enhanced DC performance using a technique known as chopper stabilization. The chopper op-amp is designed continuously to undertake self-calibration to provide an ultra low offset voltage, which is extremely time and temperature stable. At the same time, the CMRR is increased and the 1/*f* noise content is reduced. Figure 3.3 shows a typical chopper stabilized op-amp.

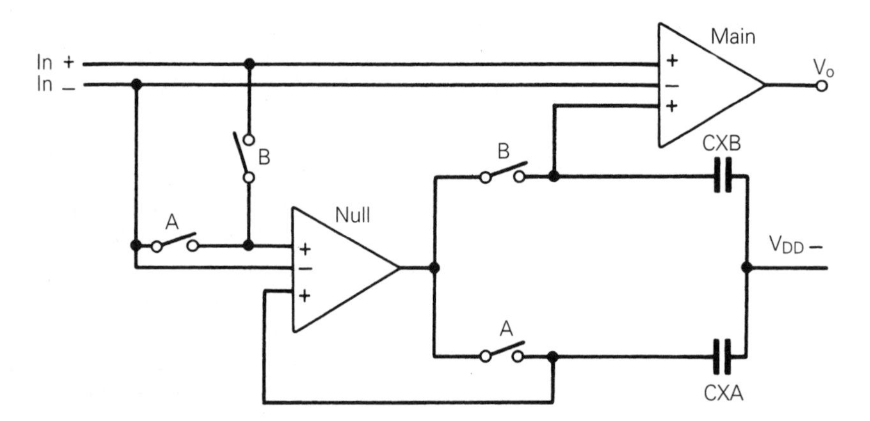

Figure 3.3 *Chopper stabilized op-amp*

Basically, the enhanced performance of a chopper stabilized op-amp is achieved by using two op-amps. A nulling op-amp and a main op-amp are used together with an oscillator, switches and two external (or internal) capacitors to create a system that behaves as a single op-amp. With this approach, the TLC2652 op-amp achieves a sub-microvolt input offset voltage and a sub-microvolt input noise voltage. Offset variations with temperature are in the nV/°C range.

The on-chip control logic produces two dominant clock phases: a nulling phase and an amplifying phase. During the nulling phase, switch 'A' is closed, shorting the nulling op-amp inputs together. This allows the nulling op-amp to reduce its own input offset voltage, by feeding its output signal back to an inverting input node. Simultaneously, the external capacitor, C_{XA}, stores the nulling potential, to allow the offset voltage of the op-amp to remain nulled during the amplifying phase.

During the amplifying phase, switch 'B' is closed. This connects the output of the nulling op-amp to the non-inverting input of the main op-amp. In this configuration, the input offset voltage of the main op-amp is nulled. Also, the external capacitor, C_{XB}, stores the nulling potential, to allow the offset of the main op-amp to remain nulled during the next phase.

This continuous chopping process allows offset voltage nulling during variations in time and temperature. The nulling process works over both the common mode input voltage range and the power supply voltage range. Additionally, because the low frequency signal path is through both the nulling and main op-amps, an extremely high gain is obtained.

The level of low frequency noise output from the chopper op-amp depends upon the magnitude of component noise prior to chopping. It also depends upon the capability of the circuit to reduce this noise while chopping. Increasing the chopping frequency reduces the low frequency noise. Limiting the input signal frequencies to less than half the chopping frequency reduces the effects of intermodulation and aliasing.

3.1.4 JFET input op-amps

Junction field effect transistors (JFETs) were introduced into op-amp input stages in an attempt to increase the input impedance and reduce the bias current. The intermediate and output stages of the op-amp continued to use bipolar transistors, as shown in Figure 3.4.

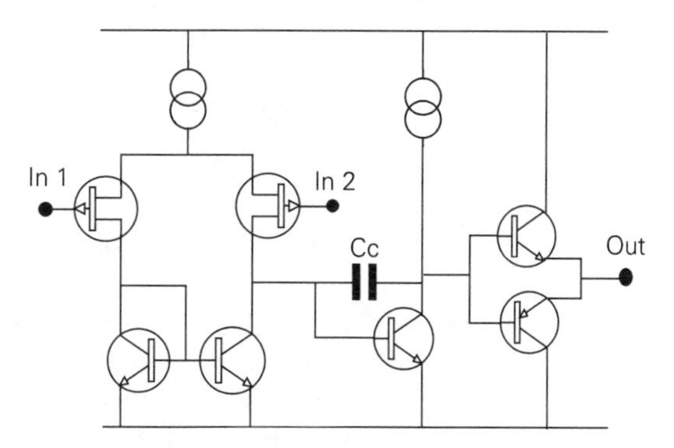

Figure 3.4 *JFET input op-amp circuit*

Op-amps with JFET inputs have very high input impedance, typically 1 TΩ. Their input bias current is typically 50 pA and their noise current is about 10 fA/√Hz. Noise voltage is higher in JFET input op-amps than in bipolar devices, due to the high channel resistance. The noise voltage is typically 20 nV/√Hz.

Input offset voltages in JFET input op-amps (typically 500 μV) are about ten times that for bipolar input op-amps. The stability of a JFET input op-amp is also far worse than for a bipolar input device. Current flow through a lateral JFET channel is subject to stress and strain due to temperature, which results in changes in the channel current. Special circuits that use bipolar transistors to reduce input offset voltage drift are used in the 'Texas Instruments' Excalibur process.

The JFET at the input does not allow as much gain as in a bipolar stage. Because of this, greater slew rates can be achieved than for a bipolar input op-amp having the same gain-bandwidth product. A fast slew rate makes

JFET input op-amps suitable for use in rectifier circuits, peak detector circuits, pulse amplifying circuits and sample and hold circuits.

3.1.5 CMOS op-amps

Digital electronics has employed CMOS transistors for many years, in order to reduce the size and power consumption of circuits. Power consumption has been reduced by the combination of low voltage and low quiescent (steady state) current requirements. Now analogue op-amps use CMOS for similar reasons, as shown in Figure 3.5.

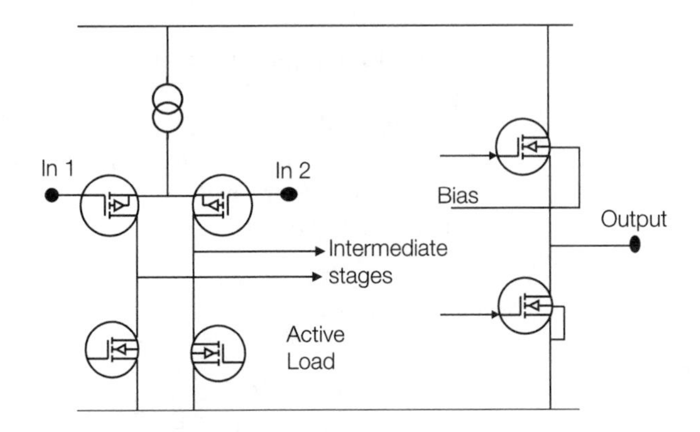

Figure 3.5 *CMOS op-amp circuit*

Op-amps are available that draw just 1 μA quiescent current from their supply rails. Devices that operate from supply voltages as low as 1.4 V are also available. Most CMOS op-amps are unable to operate with supply voltages greater than 16 V and many are limited to about 6 V operation.

As a result of using p-channel MOSFETs on their input, CMOS op-amps can operate correctly with input voltages down to the negative supply rail. This makes them suitable for use in single supply circuits where the input voltage is referenced to the negative rail. The MOSFET input provides high input impedance, with low offset and bias currents.

The input bias current is typically about 100 fA. However, like the JFET input, this current doubles for every 10°C rise in temperature. The CMOS op-amp is therefore susceptible to drift with temperature.

Offset voltages are typically 1 mV, although some op-amps are designed for offset voltages as low as 200 μV. This is better than many JFET input op-amps, but not as good as can be achieved with bipolar devices. Chopper stabilized CMOS op-amps achieve high DC precision, with maximum offset voltage in the order of 1 μV. The offset voltage stability is generally better in CMOS op-amps than in JFET input devices.

Unfortunately, CMOS op-amps suffer from high noise voltage. Noise voltage is typically 30 nV/√Hz, although some devices have been designed for low noise and produce a noise voltage of about 10 nV/√Hz. This level of noise is lower than that of JFET input op-amps, and lower than some bipolar types.

The single supply, low voltage and low quiescent current requirements make CMOS op-amps ideal for portable equipment. The high input impedance and low bias current make them suitable for interfacing high impedance transducers.

To prevent damage due to electrostatic discharge (ESD), care has to be taken with all electronic devices. Bipolar and JFET inputs will conduct when a high voltage is applied. If reverse biased, a pn junction will break down temporarily, like a zener diode. Provided that current flow is limited, no permanent damage occurs. The very high impedance of MOSFET input op-amps makes them more susceptible to ESD damage. Overvoltage applied to an input will permanently damage a MOSFET gate by burning a hole in its surface.

3.1.6 BiCMOS op-amps

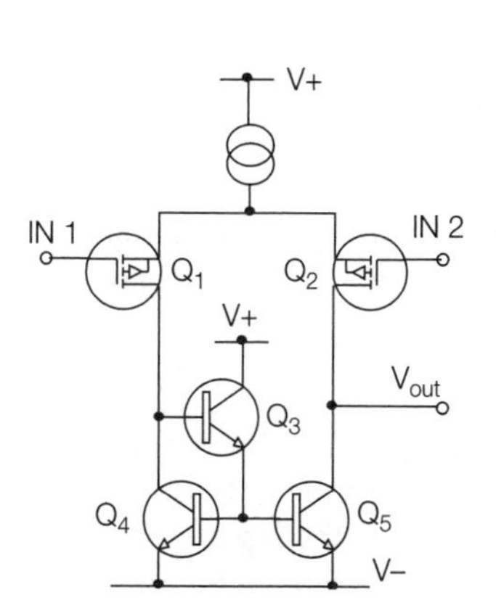

Figure 3.6 *BiCMOS op-amp input stage*

BiCMOS technology has been used in logic integrated circuits for a few years, but until the late 1990s there were few BiCMOS analogue devices. However, the move to BiCMOS has been encouraged by the need to use single-rail low-voltage power supplies. Op-amps are available that operate from a single-rail supply, typically between 2.7 V and 12 V, and draw very little current. They are therefore suitable for battery-operated equipment.

The input stage of a BiCMOS op-amp is illustrated in Figure 3.6. The current source connected to the V+ supply limits the current into the circuit, and the bipolar transistors form an amplified current mirror. If the voltage at input IN1 is made negative to increase the current flow through the MOSFET Q_1, then more current flows through the base of Q_3. This current is amplified by Q_3 and then used to drive the bases of Q_4 and Q_5. An increase in current through Q_5 lowers the output voltage. Thus the gain of this stage is very high due to the amplified current mirror.

Like CMOS op-amps, the use of p-channel MOSFETs at the input allows the BiCMOS op-amp to operate down to the negative supply rail. They are ideal for single supply operation where the input voltage is referenced to the negative supply rail. The MOSFET input gives very high input impedance and bias currents of 1 pA are typical.

One advantage of BiCMOS is the ability to produce a good output bandwidth and slew rate, whilst drawing little current from the supply. One device, the TS951, is intended for use in mobile phones. This op-amp draws 0.9 mA from the supply and delivers a gain-bandwidth product of 3 MHz. Another op-amp, the LMV321, requires just 0.1 mA to deliver a gain-bandwidth product of 1 MHz.

3.2 Comparison of voltage feedback op-amps

3.2.1 DC considerations

Table 3.1 shows typical DC performance figures and highlight some of the features and benefits for the three basic types of op-amp technology: bipolar, BiFET and CMOS.

Table 3.1 *DC comparison of voltage feedback op-amps*

DC parameter	Bipolar	BiFET	CMOS/BiCMOS
Input offset voltage	10 μV–7 mV	500 μV–15 mV	200 μV–10 mV
Input offset voltage drift	0.1–10 μV/°C	5–40 μV/°C	1–10 μV/°C
Input bias current	100–50 000 pA	1–100 pA	0.1–10 pA
Input bias current drift	Fairly stable with temperature change	Doubles for every 10°C increase	Doubles for every 10°C increase

Bipolar features

Very low offset and drift; allow low level signal conditioning.
Stable bias current; remains low at high temperature.
High voltage gain; ensures accurate amplification.
Lowest voltage noise; wide dynamic range.

BiFET features

Very low input bias and noise current; matches high impedance circuits.
Good AC performance; useful for combined AC and DC applications.

CMOS features

Very low input bias and noise current; matches high impedance circuits.
Single supply operations; can be operated from battery or 5 V supply.
'Chopper' stabilizing techniques are used to overcome the large input offset drift prevalent in standard MOS devices; can be used to amplify very small signals and provide a wide dynamic range.

BiCMOS features

DC operation of BiCMOS is similar to that of CMOS. Very low input bias and noise current that matches high impedance circuits and the ability to operate from low voltage, single rail supplies.

Comparing DC errors for the different types of op-amp shows that the newer bipolar designs are better than the older LM741 and LM301 devices. The input offset and bias current have been greatly reduced, while the open-loop gain has been increased dramatically.

BiFET op-amps normally have higher input offset voltage and drift, compared with bipolar devices. However, the input bias current of FET op-amps is insignificant when compared with that of bipolar devices. The FET bias current doubles for every 10°C temperature increase. Note that some bipolar designs actually have lower bias currents at higher temperatures than FET input op-amps.

Silicon gate CMOS technologies such as LinCMOS have reduced the problem of unstable offsets in CMOS designs. The TLC2201, designed using LinCMOS, is an example of the new breed of CMOS devices. It offers extremely low and stable offsets while simultaneously featuring the high input impedance and low noise current found in the best of JFET devices. For high order DC precision, the chopper stabilized op-amps, such as the TLC2652, provide the lowest input offset and drift.

3.2.2 AC considerations

Op-amps require both AC and DC accuracy. While input offset voltages and bias currents are relevant in DC applications, other parameters must be considered for AC circuits.

The op-amp is designed to perform an amplification function. Unfortunately, amplification of the signal is usually accompanied by a phase shift, which can lead to instability of the device. To prevent instability, a Miller compensation capacitor is used, but only at the expense of the slew rate and gain of the op-amp. Ultimately this compensation capacitor defines the unity-gain bandwidth. The AC performance of an op-amp is determined by the process technology used in manufacture, or the design techniques employed. Wide bandwidth devices usually have high supply current requirements.

Bipolar op-amps offer good gain and bandwidths but their slew rate for a given bandwidth is slow. This is a limitation of the bipolar technology process (high transconductance, g_m) and is not easily designed out. As discussed in Section 3.1.1, the use of pnp transistors in bipolar op-amps provides a speed limitation.

BiFET op-amps are designed using a combination of bipolar and JFET structures. P-channel JFETs (with much lower transconductance than have bipolar transistors) are used in the input stage. The remaining circuit is designed using bipolar transistors. This combination has produced op-amps with significantly higher slew rates than purely bipolar designs.

CMOS technologies such as LinCMOS have a similar performance to designs using BiFETs, but are of particular benefit for low power or single supply applications.

BiCMOS techniques have many of the same qualities as CMOS, except that for a given supply current they have much higher dynamic response. The slew rate and gain-bandwidth product of BiCMOS op-amps is generally greater than a CMOS op-amp that draws the same current. They are widely used in the audio amplifier circuits of mobile telephones.

3.2.3 Noise considerations

An op-amp's input voltage noise is described in terms of nanovolts per root hertz (nV/√Hz). This level is higher at very low frequencies than across the majority of the op-amp's operating bandwidth. The break point where the noise level 'flattens out' is known as the $1/f$ corner frequency. Below the $1/f$ corner frequency, the noise level rises inversely proportional to frequency. In a bipolar op-amp, the $1/f$ corner frequency can be as low as 100 Hz, but in FET input op-amps this frequency can be much higher (several kHz).

Bipolar devices offer the lowest voltage noise among those commercially available, typically 15 nV/√Hz, although some devices have much lower voltage noise levels (< 5 nV/√Hz). The voltage noise from a bipolar input stage, in the flat part of the band, is dominated by thermal noise from the base-spread resistance and the emitter resistance. Unfortunately, the shot noise arising from the input bias current can be significant. Therefore, in order to reduce this current noise, bias current cancellation circuits are sometimes used.

The input noise current of FET input op-amps is caused by shot noise, due to the gate current. This is very low at temperatures around 25°C compared with the base current in bipolar inputs. Consequently, FET input op-amps have negligible input current noise and provide a superior noise performance with high impedance sources.

A FET input stage has higher voltage noise and higher $1/f$ corner frequency than a bipolar input stage. The gate current is negligible and the input current is reduced to leakage current by the input protection network. The noise sources of a MOSFET input device are similar to those of the junction FET. A common disadvantage of MOSFET input op-amps (i.e. CMOS and BiCMOS op-amps) is their relatively high voltage noise and high $1/f$ frequency.

In contrast, some devices (e.g. the Texas Instruments LinCMOS op-amp, TLC2201) offer current noise levels similar to the very best junction FET input op-amps. They also have voltage noise levels comparable to many bipolar designs. The TLC2201 features low input offset voltages coupled with a very low drift with time and temperature change. Additionally, the device features operation from a single 5 V supply and rail-to-rail output swing.

In summary, bipolar input stages give the lowest voltage noise and lowest $1/f$ corner frequency and are well suited for interfacing with low impedance sources. JFET, CMOS and BiCMOS input stages have negligible input current noise, allowing them to be used with extremely high source impedances. Noise current is related to the input bias current and it will increase by V^2 for every 10°C rise in temperature.

3.2.4 Power supply considerations

Op-amps are required to operate with many different supplies. Circuits may be battery powered and required to run off a single 1.5 V supply or they may have a ±22 V or more supply in an instrumentation application. The available power affects the choice of op-amp.

Single supply op-amps will normally need to operate from a low voltage supply, maybe as low as 1.5 V if used with battery powered equipment. They will also require an input common mode range down to the negative rail and an output that swings near to ground. These restrictions do not usually apply to dual supply op-amps.

An op-amp with a common mode range down to the negative rail can easily be designed using a bipolar process. PNP input transistors ensure that the input can swing down to the negative rail, or below, without causing problems.

A good output swing is not so easy to achieve. Many bipolar devices are optimized for dual supply operation, that is, capable of sinking and sourcing current. This therefore means that the output will not normally swing down to the negative rail. If a device is optimized for single supply operation, its output suffers from crossover distortion when it is operated with dual supplies. Bipolar devices suitable for both single and dual power supply operation are uncommon.

BiFET op-amps have been designed for dual supplies and are generally unsuitable for single supply operation. Their common mode input range reaches (and sometimes exceeds) the positive supply rail potential. The output will normally swing to within 2.5 V of each supply voltage. They operate from a wide range of supplies (±3 V to ±22 V) and are optimized for AC performance.

CMOS devices (LinCMOS) and BiCMOS have been specifically designed for single supply operation. The supply voltage range is typically +2 V to +16 V. The input voltage range extends below the 0 V supply. The output voltage swing can approach the power rails for high impedance loads. Unlike bipolar designs, CMOS devices with push–pull outputs can also perform well with dual supplies. However, a limitation in some dual supply applications is the limited operating voltage.

3.3 Current feedback op-amps

The disadvantage of voltage feedback op-amps is that the gain-bandwidth product is constant. Extending the bandwidth is at the expense of gain. Current feedback op-amps have an entirely different gain-bandwidth product relationship. In fact, the bandwidth is almost constant, irrespective of gain.

The simplest model of a current feedback amplifier is shown in Figure 3.7. In this model, the input of a current feedback amplifier is a buffer, connected between the non-inverting and inverting inputs. The non-inverting input is connected to the buffer input and has high impedance. The inverting input is connected to the buffer output and has low impedance. Current can flow in and out of the low impedance inverting input. An internal amplifier senses the current flow and produces a voltage output proportional to the current. Current flowing out of the inverting input produces a positive output voltage. Current flowing into the inverting input produces a negative output voltage.

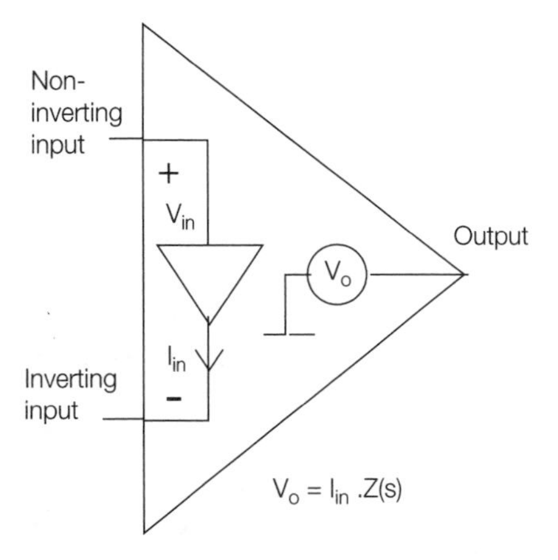

Figure 3.7 *Simple current feedback op-amp model*

3.3.1 AC performance

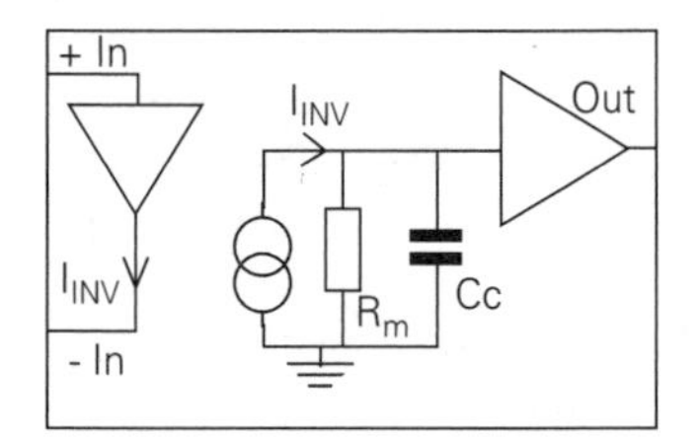

Figure 3.8 *The small signal current feedback op-amp model*

The bandwidth relationship of a current feedback op-amp can be explained by studying the amplifier in more detail. Figure 3.8 shows the small-signal model.

The input buffer has its output connected to a current mirror. When current flows from the buffer, out of the inverting input, an equal current flows out of the current mirror. A resistive load (R_m) across the current mirror converts the current flow into a proportional voltage. A small frequency compensating capacitor (C_c) is connected across R_m to ensure stability of the op-amp. The voltage across R_m is output via a second buffer.

The node equations for this model can now be worked out to find the frequency response. Since the input buffer forces the inverting input to have the same voltage as the non-inverting input, $V_1 = V_{in}$.

The voltage across R_m is $V_2 = \dfrac{IR_m}{1 + j\omega C_C R_m}$

And $I = V_1\left(\dfrac{1}{R_1} + \dfrac{1}{R_2}\right) - \dfrac{V_{out}}{R_2}$

And since V_{out} is a buffered version of V_2, $V_{out} = V_2$.

Now these can be combined to find V_{out}.

$$V_{out} = \frac{\left[V_{in}\left(\frac{1}{R_1} + \frac{1}{R_2}\right) - \frac{V_{out}}{R_2}\right]R_m}{1 + j\omega C_C R_m}$$

If $R_m >> R_2$, this can be simplified to:

$$\frac{V_{out}}{V_{in}} = \frac{1 + \frac{R_2}{R_1}}{1 + j\omega C_C R_2}$$

The closed-loop gain is $1 + R_2/R_1$, therefore:

$$\frac{V_{out}}{V_{in}} = \frac{A_{VCL}}{1 + j\omega C_C R_2}$$

When $1 = 2\pi f C_C R_2$, the closed-loop gain falls by 3 dB, thus:

$$f_{(-3dB)} = \frac{1}{2\pi C_C R_2}$$

Thus the bandwidth is dependent upon R_2 and the internal compensation capacitor. The bandwidth is not dependent upon R_1 and is therefore not dependent upon the closed-loop gain.

If we consider the gain in terms of the complex impedance $Z(s)$, we find:

$$\frac{V_O}{V_{IN}} = \left(1 + \frac{R_2}{R_1}\right)\frac{1}{1 + \frac{1}{LG}}$$

In this case loop gain (LG) is given by $LG = \dfrac{Z(s)}{R_2}$

But what about the circuit design assumptions that were made for ideal op-amps, in Chapter 1? Figure 3.9 shows a simple current feedback op-amp model using the complex transimpedance $Z(s)$, within a non-inverting amplifier circuit.

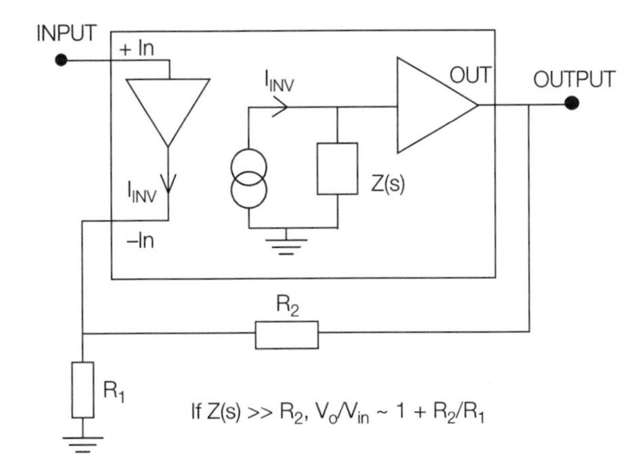

Figure 3.9 *Non-inverting amplifier with ideal current feedback op-amp*

First consider what happens when the input voltage V_{in} is raised above 0 V. The input buffer responds to the increasing input voltage by raising the voltage at the inverting input. A current then flows out the inverting input. The current flow is sensed by the transimpedance stage and the output voltage rises. The output voltage ceases to rise when a balance is reached; this is when the current fed back through R_2 is equal to the current flowing through R_1. Feedback current thus replaces the current from the inverting input. In steady state conditions, the current from the input buffer can be very small; this is dependent upon the gain of the transimpedance stage.

If the transimpedance stage has high gain (high $Z(s)$), the current from the inverting input can be assumed close to zero. The voltage gain of the input buffer is close to unity, which means that the differential voltage between inverting and non-inverting inputs can be assumed to be close to zero. Thus the ideal model can be used to determine gain; in this case $A_{VCL} = 1 + R_2/R_1$.

In practice, the input buffer's non-ideal output resistance (R_o) will be typically about 20 Ω to 40 Ω, as shown in Figure 3.10. This additional resistance will modify the response, because the two input voltages will not be exactly equal while an error current flows. Some small voltage will be dropped across R_o.

The additional resistance in the feedback path means that the loop gain will actually depend somewhat on the closed-loop gain of the circuit. At low gains, R_2 dominates, but at higher gains, the internal resistance has a greater effect and this reduces the loop gain, thus reducing the closed-loop bandwidth.

The transfer function of a non-ideal, non-inverting current feedback amplifier is given by:

$$\frac{V_O}{V_{IN}} = \left(1 + \frac{R_2}{R_1}\right)\frac{1}{1 + \frac{1}{LG}}$$

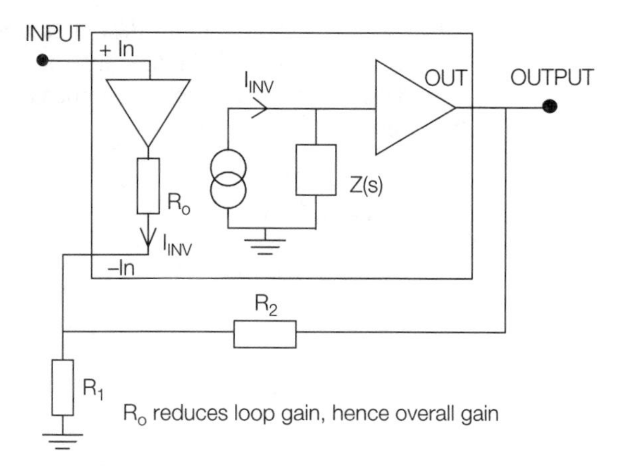

Figure 3.10 *Non-inverting amplifier with non-ideal current feedback op-amp*

Loop gain (LG) is modified by the introduction of R_o, and becomes:

$$\mathrm{LG} = \frac{Z(s)}{R_2 + R_o\left(1 + \dfrac{R_2}{R_1}\right)}$$

The gain error due to R_o is $\dfrac{R_o\left(1 + \dfrac{R_2}{R_1}\right)}{Z(s)}$

High-frequency circuits

Current feedback op-amps can be used in most applications where voltage feedback op-amps are used. They have the advantage of having very high slew rates at low supply currents. Slew rates of 1000 V/μs or more are common. The availability of high slew rates means that current feedback op-amps are often found in video amplifier and cable driver circuits.

The low impedance inverting input of a current feedback op-amp allows fast transient currents to flow into the amplifier as needed. The internal current mirrors convey this input current to the compensation node, allowing fast charging and discharging. The actual slew rate will be limited by saturation of the current mirrors, typically at 15 mA. The overall slew rate is also limited by the slew rate limit of the input and output buffers.

Using current feedback op-amps in Sallen and Key low pass filters enables much higher frequencies to be used, compared to voltage feedback designs. However, the group delay of the op-amp becomes significant if the −3 dB bandwidth of the op-amp is less than ten times that of the filter.

Sallen and Key filters usually use op-amps as unity gain buffers. These have a direct connection between output and the inverting input. Current feedback op-amps cannot be connected in this way because the inverting input of the op-amp is actually the output of a buffer. Large amounts of current would flow between the two outputs if they were connected together.

Instead a resistor can be connected between them, to limit the current flow. Alternatively, the op-amp can be given some gain using feedback and shunt resistors. Note that the filter circuit component values are dependent upon the amplifier gain.

Filter topologies that use reactive feedback, such as multiple feedback types, are not suitable for current feedback op-amps. Sallen and Key filters are feasible because the op-amp is used as a fixed gain block. In general, it is not desirable to add capacitance across the feedback resistor of a current feedback op-amp circuit.

Stability

Current feedback op-amps are like voltage feedback op-amps, because they both suffer greater phase shifts at higher frequencies. Instability can be produced with phase shifts approaching 180°. Because the optimum value of R_2 will vary with closed-loop gain, op-amp manufacturers usually supply a Bode plot and tables that give the bandwidth and phase margin for various gains. High values of closed-loop bandwidth can be obtained at the expense of a lower phase margin, which results in peaking in the frequency domain, and overshoot and ringing in the time domain.

With a voltage feedback op-amp, shunt capacitance at the inverting input (C_{IN}) generates an excessive phase shift that can lead to instability. The same effect occurs with a current feedback op-amp, but the problem may be less pronounced. This is because the phase shift occurs at higher frequencies due to the inherently low impedance of the inverting input.

Consider an amplifier circuit that employs a wide-band voltage feedback op-amp with $R_1 = 680\ \Omega$, $R_2 = 680\ \Omega$, and $C_{IN} = 10$ pF. The phase shift reaches 90° (and is thus unstable) at $1/[2\pi C_{IN}(R_1//R_2)]$, which is roughly 47 MHz. Let us now replace the voltage feedback op-amp with a current feedback device having an inverting input resistance (R_o) of 40 Ω. The frequency where 90° phase shift occurs is now given by $1/[2\pi C_{IN}(R_1//R_2//R_o)]$, this is about 445 MHz.

If the unity-gain bandwidth of both amplifiers is 500 MHz, the voltage feedback op-amp will require a feedback capacitor for compensation. Although this will reduce the effect of C_{IN}, it will also reduce the amplifier's bandwidth. Using the current feedback device will give reduced phase shift, because the break point is about a decade higher in frequency. This means that the amplifier's bandwidth will be greater because a compensating capacitor will not usually be required, unless to flatten the passband or to give optimum pulse response.

3.3.2 DC performance

The DC gain accuracy of an amplifier using a current feedback op-amp can be calculated from its transfer function.

The gain error due to R_o is $$\frac{R_o\left(1 + \frac{R_2}{R_1}\right)}{Z(s)}$$

Using a typical transimpedance of 1 MΩ, a feedback resistor of 1 kΩ, and an inverting input resistance of 40 Ω, the gain error at unity gain is 0.004 per cent (basically $R_0/Z(s)$). At higher gains, gain accuracy degrades significantly. Current feedback amplifiers are rarely used for high gains, particularly when gain accuracy is required.

For many applications, settling times are more important than gain accuracy. Although current feedback amplifiers have very fast rise times, many data sheets will only show settling times to 0.1 per cent. This is because of thermal settling tails, which are a major contributor to lack of settling precision.

Thermal tails are caused by temperature differences between input stage transistors. Power dissipation of each transistor occurs in a very small area, which is too small to achieve thermal coupling between devices. Thermal errors are significant in non-inverting circuits because these have a common mode input voltage and are thus more sensitive to differences in performance. Errors can be reduced by using the op-amp in the inverting configuration, because the common mode input voltage is eliminated.

Thermal tails do not occur instantaneously; the thermal coefficient of the transistors (which is process dependent) will determine the time it takes for the temperature change to occur and alter parameters – and then recover. Amplifiers do not usually exhibit significant thermal tails for input frequencies above a few kHz, because the input signal is changing too fast. Step waveforms, such as those found in imaging applications, can be adversely affected by thermal tails when DC levels change. For these applications, current feedback amplifiers may not offer adequate settling accuracy.

One application that is difficult for current feedback op-amps is an integrator circuit. The problem is that direct capacitive connection between the output and the inverting input can cause instability. Instability is a result of phase shifts in the feedback path. The frequency compensating capacitor produces up to 90° of phase shift. The gain and phase shift in the feedback circuit are frequency dependent due to the feedback capacitor. The circuit oscillates if the signal fed back to the inverting input approaches 180° at a frequency where the gain is greater than unity.

The integrator circuit has to be modified to prevent instability when using current feedback op-amps. A resistor (R_f) has to be inserted between the feedback capacitor (C_1) and the inverting input. This resistor ensures that a minimum value of resistance is always in the feedback path, which limits the gain. A resistor (R_2) in parallel with capacitor C_1 determines the minimum frequency (F_c) at which the integrator is effective.

$$F_c = \frac{1}{2\pi R_2 C_1}$$

3.3.3 Current feedback noise considerations

When amplifying low level currents, higher feedback resistance means higher signal-to-noise ratio. This is because signal gain increases in proportion to R, whilst resistor noise increases in proportion to $\sqrt{R}$. Doubling the feedback resistor value doubles the signal gain and increases resistor noise by a only factor of 1.4.

However, doubling the feedback resistor value causes the contribution from current noise to be doubled and the signal bandwidth to be halved. Therefore, the higher current noise of current feedback op-amps may rule out their use in photodiode amplifier circuits. In circuits where noise is less critical, select the feedback resistor based on bandwidth requirements. If more gain is required, use a second stage.

The number of applications for current feedback amplifiers will be limited because of current noise. The inverting input current noise can be about 30 pA/√Hz. However, the input voltage is somewhat lower than in voltage feedback op-amps, 2 nV/√Hz or less. The feedback resistor will usually be under 1 kΩ and, in a unity-gain circuit, the dominant noise source will be the inverting input noise current flowing through the feedback resistor.

Let the input noise current be 25 pA/√Hz in a unity gain circuit with a feedback resistor value of 680 Ω, this gives 17 nV/√Hz noise at the output. If the input noise voltage is 2 nV/√Hz, the noise current is the dominant noise source.

Let the gain of the circuit now be increased, by reducing the input resistor value. The output noise due to input current noise will not increase, because it is determined by the feedback resistor value. Now the amplifier's input voltage noise will dominate. When the closed-loop gain reaches 10, the contribution from the input noise current is only 1.7 nV/√Hz when referred to the input. The two noise sources are combined using the equation $\sqrt{(I_n^2 + V_n^2)}$ to give an input-referred noise voltage of only 2.6 nV/√Hz (neglecting the resistor's thermal noise). The current feedback op-amp is thus useful in low noise amplifiers having a moderate gain.

3.3.4 Using current feedback op-amps

The inverting amplifier circuit works because of the low impedance node created at the inverting input. The summing junction of a voltage feedback amplifier (inverting input of the op-amp) has low impedance. A current feedback op-amp will operate very well in the inverting circuit because it has inherently low inverting-input impedance. The internal buffer holds the summing node at the same potential as the non-inverting input.

In the inverting circuit, voltage feedback amplifiers suffer from voltage spikes at the summing node in high speed applications. This is because the feedback loop takes time to settle and, until the loop has settled, the summing node impedance is not low. Current feedback op-amps do not produce these voltage spikes because the summing node is low impedance irrespective of the feedback loop. Other advantages of the inverting circuit include maximizing the input slew rate and reducing thermal settling errors.

Current feedback op-amps can be used in current-to-voltage converters, by applying the input current into the op-amp's inverting input. There are limitations introduced by this arrangement: the amplifier's bandwidth varies directly with the value of feedback resistance; and the inverting input current noise tends to be high.

Although the inputs of a current feedback op-amp are not matched, the transfer function for the ideal difference amplifier is still valid. At low frequencies, the differential amplifier's CMRR is limited by the matching of the external resistor ratios, with 0.1 per cent matching yielding about 66 dB. At high frequencies, what matters is the matching of time constants formed by the input impedances. High speed voltage feedback op-amps usually have well-matched input capacitance, achieving a CMRR of about 60 dB at 1 MHz.

Because the current feedback op-amp's input stage is unbalanced, the input capacitance will not be matched. Low value external resistors (100 Ω to 200 Ω) must be used on the non-inverting input of some amplifiers to minimize the mismatch in time constants. With careful attention given to resistor selection, an amplifier using a current feedback op-amp can yield a high frequency CMRR equal to that obtained using a voltage feedback op-amp. Both voltage feedback and current feedback amplifiers can further benefit from additional trimmer capacitors, but this reduces the signal bandwidth.

If higher performance is needed, the best choice would be a monolithic high speed difference amplifier, such as the AD830. It requires no resistor matching and has a CMRR > 75 dB at 1 MHz, which reduces to about 53 dB at 10 MHz.

Load capacitance presents the same problem with a current feedback amplifier as it does with a voltage feedback amplifier. It causes increased phase shift of the error signal, which results in reduction of phase margin and possible instability. The most popular method of dealing with capacitive loads is a resistor in series with the output of the op-amp. The resistor should be outside the feedback loop, but in series with the load capacitance. A current feedback op-amp also gives the option of increasing R_2 to reduce the loop gain. All methods produce a reduction in bandwidth, slew rate and settling time.

Care has to be taken with all circuit layouts to prevent instability. Stray and parasitic capacitance can reduce the phase margin and lead to oscillation. It is not a good idea to use sockets for the op-amp; these increase the capacitance between device pins. Capacitance can be reduced by removing the printed circuit ground plane from the area around the input pins.

Low value feedback resistors are advisable to reduce capacitive effects. Many current feedback op-amps have recommended feedback resistor values quoted on their data sheets. Often graphs of frequency response versus feedback resistor values are given to show how the flatness of the response varies with resistance. As we know, the feedback resistor determines the bandwidth of the amplifier circuit.

3.3.5 Power supplies for current feedback amplifiers

Current feedback op-amps cannot be used for single supply operation. Op-amps that are designed to deliver good current drive and have a voltage swing that approaches the supply rails usually use common emitter output stages, rather than the usual emitter followers. Common emitter circuits allow the output voltage to swing almost to the supply rail (less the output transistor's collector–emitter saturation voltage). This type of output stage is slower than emitter followers, due to the increased circuit complexity and higher output impedance. Because current feedback op-amps are specifically developed for the highest speed and output current, they feature emitter follower output stages.

Higher speed processes have produced a common emitter output stage with 160 MHz bandwidth and 160 V/μs slew rate. An example of a device using this technology is the 'Analog Devices' AD8041. This voltage feedback op-amp is powered from a single 5-volt supply.

Single supply input stages use pnp differential pairs. This arrangement allows the common mode input range to extend down to the lower supply rail (usually ground). Such an input stage is impossible with current feedback op-amps. Note that even in circuits using 'rail-to-rail' voltage feedback devices, the output voltage will not be near the supply rails if driving a low impedance load; this is due to the voltage drop across the output stage's internal resistance.

Current feedback op-amps can be used in single supply circuits provided that the input and output voltages are not allowed to approach the supply rails. This may require level shifting or AC coupling. The non-inverting input must be biased to the middle of its working range, but this is already a requirement in most single supply systems.

Decoupling capacitors across the power supplies are very important. As with all high frequency circuit design, capacitors suitable for all encountered frequencies are needed. As a rule of thumb, a 10 μF tantalum capacitor in parallel with a 10 nF or 100 nF ceramic capacitor should be used. The capacitors should be connected close to the op-amp's power supply pins.

Exercises

3.1 In an inverting amplifier circuit, using a current feedback op-amp, the transimpedance is 1 MΩ and the feedback resistor R_2 is 750 Ω. Find the loop gain (LG). Resistor R_1, connecting the inverting input to ground has a value 100 Ω. What is the closed-loop gain V_O/V_{IN} at low frequencies? (Assume an ideal amplifier.)

3.2 The circuit in Exercise 3.1 is now used with a non-ideal amplifier, having a non-inverting input resistance of $R_o = 30\ \Omega$. What is the closed-loop gain V_O/V_{IN} in this case?
What is the gain error due to the introduction of R_o?

3.3 A current feedback amplifier has a voltage noise of 2 nV/√Hz and inverting input current noise of 25 pA/√Hz. With a feedback resistor of 750 Ω and a gain of 20, what is the input referred noise?

4 Applications: linear circuits

4.1 Introduction

This chapter concentrates on linear circuit applications, including inverting and non-inverting amplifiers, differential amplifiers, buffers, current-to-voltage converters and voltage-to-current converters. Modifications to these basic circuits can be found in later chapters. A further collection of op-amp circuits will be found in Appendix A1. The usefulness of the op-amp approach is the many variations of a basic circuit that are possible.

Op-amps are used extensively in analogue circuit design. The approach involves breaking down the circuit, or system function, into a series of specific operations. A separate op-amp circuit can then perform each operation. The requirements of each circuit may vary considerably, but the specific operations required in the different systems are common to many systems. The designer should be able to pick out, from the many circuits given, those appropriate to their own particular system.

The circuits presented in this chapter do not generally refer to particular op-amp devices. Most applications will function with any op-amp type. The particular op-amp used in a circuit determines the errors and performance limits of the application. In order to make a working circuit from those given in the text, all that is normally required is to add power supply connections to the op-amp. Only those applications requiring very low noise, or wide bandwidth, or very fast slew rate, will normally require the use of more specialized (and more expensive) op-amps.

Passive external components are connected to the op-amp in order to define a precise circuit operation. The circuit designs given do not generally give component values, and the designer must choose these for himself. As a general guideline to resistor value selection, choose the lowest value that does not significantly load the op-amp's output. Most op-amps are designed to supply a load at their output terminal that should be no less than 1 kΩ. Large input resistor values increase the offset errors due to bias current (see Chapter 2); and they are shunted by stray capacitance, which limits the operating bandwidth. It should be remembered that the feedback resistor also contributes to the op-amp's load. The effective load is the parallel combination of an external load and feedback resistor.

In inverting amplifier circuits, the inverting input of the op-amp is effectively at earth potential. The input resistors provide a load to the signal source, and their value must be chosen with care. In some instances, impedance matching may be required and the input resistor should have a value equal to the characteristic impedance (often equal to the source impedance). Most often, input resistors are chosen to have a higher value than the input signal source, so that they do not significantly load it.

4.2 Voltage scaling and buffer circuits

Ideal forms of the basic voltage scaling and buffer circuits have already been dealt with in Chapter 1. The circuits are for convenience shown again in Figure 4.1.

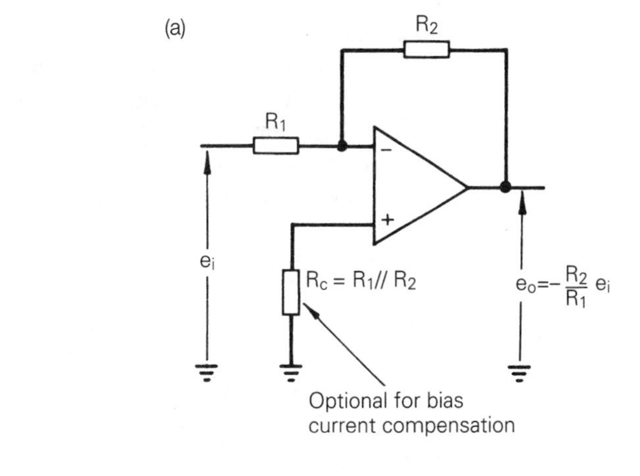

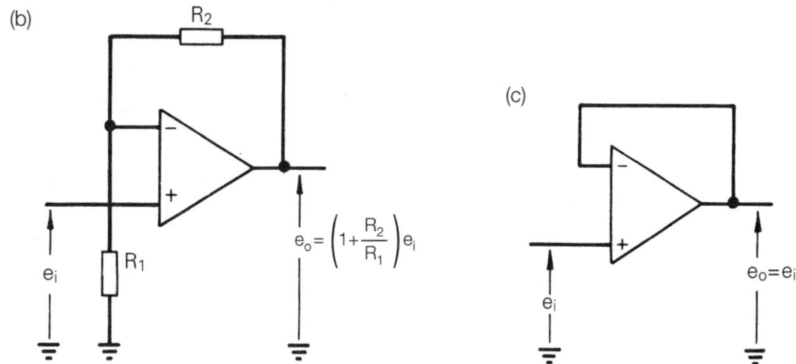

Figure 4.1 *Basic voltage scaling applications (a) Inverting amplifier. (b) Non-inverting amplifier. (c) Unity-gain follower (buffer)*

The great attraction of all op-amp circuits lies in the ability to set a precise operation with a minimum number of precise components. In Figures 4.1(a) and (b), closed-loop gain is determined by simply selecting two resistor values. The accuracy of this gain depends almost entirely upon the resistor value tolerance.

The inverting circuit in Figure 4.1(a) can be given any gain from zero upwards. The lower limit of the gain for the non-inverting circuit Figure 4.1(b) is unity. In both configurations, the practical upper limit to the gain depends on the requirement for maintaining an adequate loop gain, so as to minimize gain error (see Section 2.3.1). Also, closed-loop bandwidth decreases with increase in closed-loop gain. If high closed-loop gains are required, it is often better to connect two op-amp circuits in cascade rather than to use a single op-amp circuit.

Both inverting and non-inverting amplifier circuits feature low output impedance. This is a characteristic of negative voltage feedback (see Chapter 2).

The main performance difference between them, apart from signal inversion, lies in their input impedance. In the case of the inverter, resistor R_1 loads the signal source driving the circuit. The non-inverting amplifier presents very high input impedance, which ensures negligible loading in most applications.

The main limitation of the inverting circuit is that its input impedance is effectively equal to the value of the input resistor R_1. The application may require high input impedance, to minimize signal source loading. This demands a large value for the resistor R_1 and an even larger value for R_2, dependent upon the gain required. Large resistor values inevitably give increased offset errors due to op-amp bias current. Also, stray capacitance in parallel with a large feedback resistor limits bandwidth.

For example, assume that it is required to use the inverting circuit with closed loop gain 100 and input resistance 1 MΩ. In Figure 4.1(a) $R_1 = 1$ MΩ and $R_2 = 100$ MΩ is required. Stray capacitance C_s in parallel with R_2 would limit the closed-loop bandwidth to a frequency $f = 1/(2\pi C_s R_2)$. With C_s say 2 pF, the closed-loop bandwidth would be limited to 800 Hz – a severe restriction!

Stable very high value resistors are not freely available. If the inverting configuration must be used, the need for a very high value feedback resistor can be overcome by the use of a *T* resistance network as shown in Figure 4.2. This is at the expense of a reduction in loop gain and an increase in noise gain ($1/\beta$).

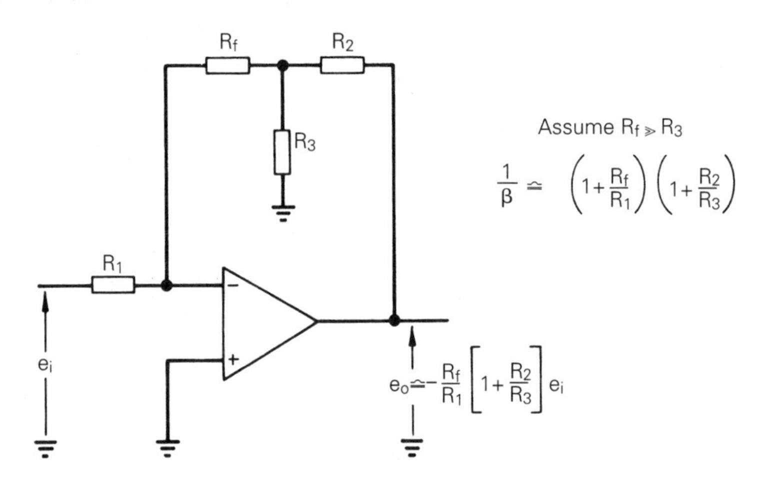

Figure 4.2 *Inverter circuit using resistive T feedback network*

The non-inverting circuit achieves high input impedance without the use of large value resistors. This is an advantage in applications requiring wide bandwidth, since large value resistors and stray circuit capacitance interact to cause bandwidth limitations. The effective input impedance of the non-inverting configuration was described in Section 2.3.3. This was shown to be equal to the differential input impedance of the op-amp, multiplied by the loop gain in the circuit ($Z_{in}\beta A_{OL}$). Practical op-amps have their common mode input impedance Z_{cm} between their non-inverting input terminal and earth. This shunts $Z_{in}\beta A_{OL}$ and so reduces its value. The effective input impedance of the follower configuration is thus Z_{cm}.

The high input impedance of the non-inverting circuit makes it a better choice than the inverting circuit for use in many applications. However, the non-inverting circuit is subject to common mode errors (see Section 2.11). Also, the voltage applied to the non-inverting input must not be allowed to exceed the maximum common mode voltage for the op-amp (since feedback will force both inputs to have the same potential). These points do not usually impose too serious a restriction.

The buffer circuit in Figure 4.1(c) has high input impedance and low output impedance. It is often used to prevent interaction between a signal source and load, e.g. for unloading potentiometers, or buffering voltage references. Buffers are used in Sallen and Key filter circuits (see Chapter 9) to prevent interaction between filter stages and to allow simple design rules.

4.2.1 Variable gain control

Instead of using fixed value resistors to set the gain, potentiometers may be used to give variable gain control. The arrangement shown in Figure 4.3(a) allows a variation of gain from zero to a very high value. However, the scaling factor does not vary linearly with respect to potentiometer rotation. A second disadvantage is that the input impedance falls as the gain is increased.

The circuit of Figure 4.3(b) gives a narrower range of scale factor variation from zero to R_2/R_1, but the gain variation is linear with respect to potentiometer setting and the input impedance remains constant (equal to R_1).

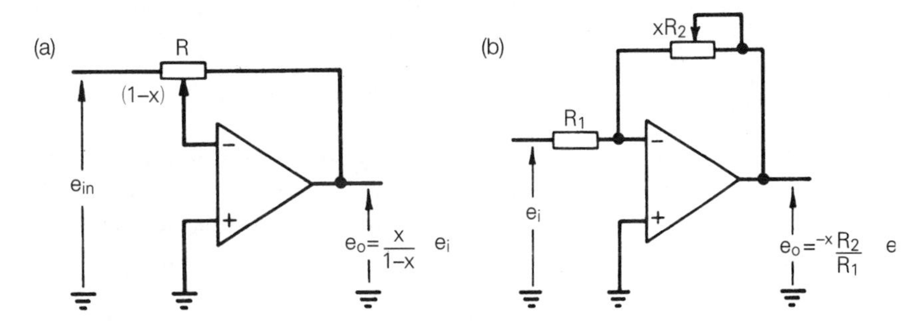

Figure 4.3 *Variable scale factor. (a) Non-linear gain control. (b) Linear gain control*

Changing the closed-loop gain inevitably changes the closed-loop bandwidth. Also, changes in the values of gain setting resistors produce a change in the offset error due to op-amp bias current. Offset errors due to bias current can be minimized by using a low bias current FET input op-amp.

4.2.2 Switched scaling factor

Gain setting resistors can be switched into circuit. Different values of scale setting resistors are switched into the signal path, thus allowing switching of the gain between preset values. Switching can be performed by a manual control of a mechanical switch, by an electromechanical switch or by means of some form of solid state switch. The circuit given in Figure 4.4 illustrates the use of an analogue switch in a programmable gain circuit.

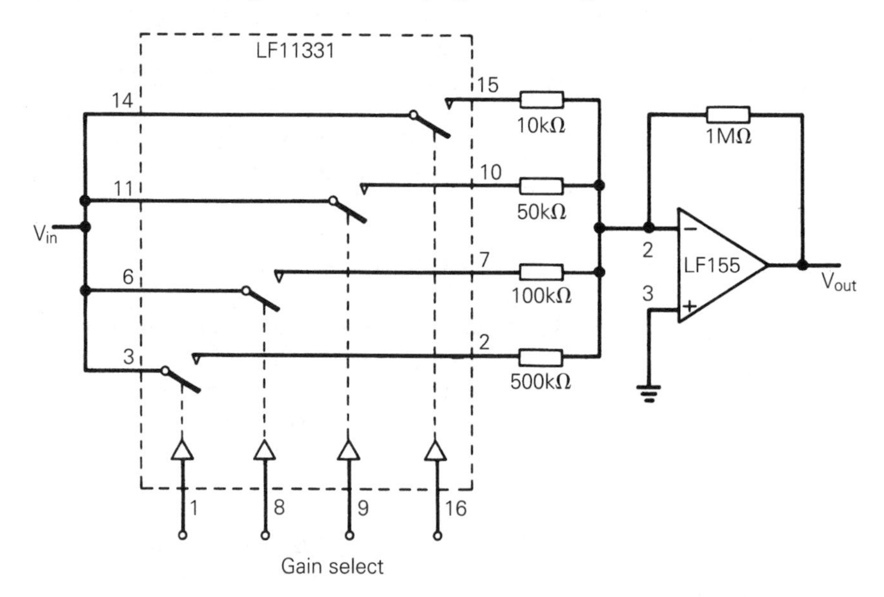

Figure 4.4 *Programmable gain operational amplifier*

4.2.3 Voltage controlled gain

Voltage control of an op-amp's gain requires a voltage controlled resistive element. Junction gate FETs when operated below pinch-off behave as linear resistors with channel resistance (r_{ds}) determined by the value of the gate source voltage. For small values of drain source voltage they exhibit a bi-lateral characteristic.

Linear voltage control of gain can be obtained by using a feedback arrangement between the drain and gate of the FET. The voltage swing across the FET can be kept small by including it in a T-network as shown in the circuit of Figure 4.5. The effective resistance of the resistive T when connected to the op-amp summing point is

$$R_e = R_2 + R_3 + \frac{R_2 R_3}{r_{ds}}$$

Where r_{ds} is the drain source resistance of the FET, which is determined by the relationship

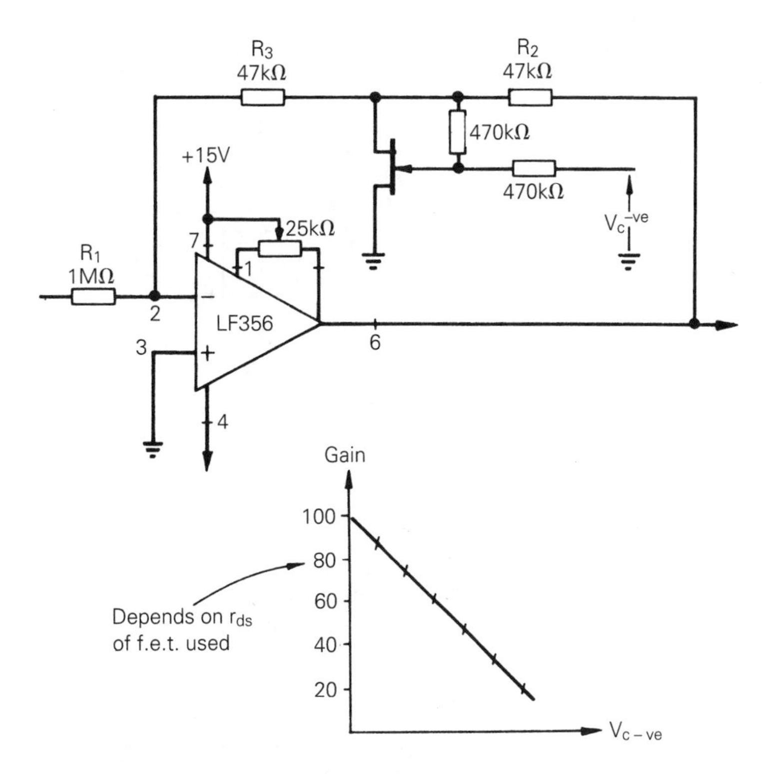

Figure 4.5 *Voltage controlled scaling factor*

$$r_{ds} = \frac{r_o}{1 - \dfrac{V_c}{2V_p}}$$

where:
r_o is the drain source resistance for $V_{ds} = 0$; $I_{ds} = 0$,
V_p is the pinch-off voltage and
V_c is the control voltage applied to the gate of the FET via a series resistor.

Substitution gives

$$R_e = R_2 + R_3 + \frac{R_2 R_3 \left(1 - \dfrac{V_c}{2V_p}\right)}{r_o}$$

Which is a linear function of V_c.

The closed-loop signal gain of the circuit, $-R_e/R_1$, also varies linearly with the value of V_c. The range of gain variation obtainable depends upon the r_0 of the FET used in the circuit. A practical circuit with the component values shown in Figure 4.5 gives the gain control shown by the graph.

4.3 Voltage summation

The voltage summing property of an ideal op-amp has been treated in Chapter 1. The behaviour of a practical summing circuit is now discussed.

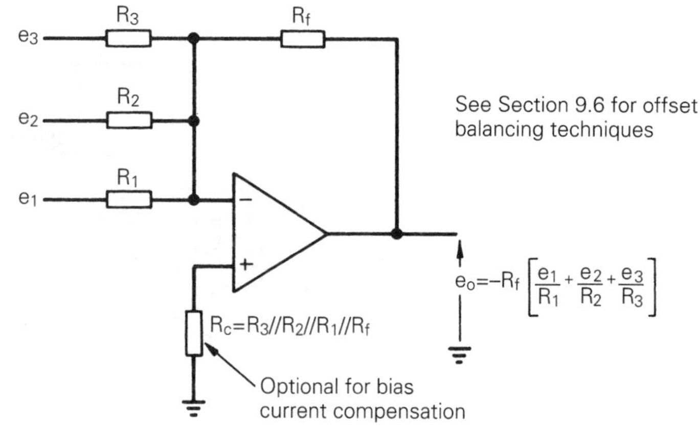

See Section 9.6 for offset balancing techniques

$e_o = -R_f \left[\frac{e_1}{R_1} + \frac{e_2}{R_2} + \frac{e_3}{R_3} \right]$

$R_c = R_3 // R_2 // R_1 // R_f$

Optional for bias current compensation

Figure 4.6 *Voltage summation*

A summing configuration is shown in Figure 4.6. In this circuit the input signals are effectively isolated from one another by the 'virtual earth' at the inverting input terminal of the op-amp.

Some consideration has to be given about the resistor values used in this circuit. The resistors should have a high enough value to prevent signal source loading, but low enough to prevent input bias current from causing offset errors. If large values of input resistor are necessary, use a low bias current FET input op-amp in order to minimize the offset error.

In the ideal circuit there is no limit to the number of input voltages that can be summed, but in the practical circuit the number of inputs is limited by the need to maintain an adequate loop gain. All paths to the inverting input terminal of the op-amp should be taken into account when assessing loop gain, closed-loop signal bandwidth and drift error. Note that the closed-loop gain $1/\beta$ for the circuit is

$$\frac{1}{\beta} = 1 + \frac{R_f}{R_1 \,//\, R_2 \,//\, R_3} \qquad (4.1)$$

4.4 Differential input amplifier configurations (voltage subtractor)

A differential amplifier circuit is commonly used to amplify or buffer differential signals whilst rejecting common mode signals. A differential signal is presented across two terminals; the voltage on one terminal rises as the voltage on the other terminal falls (relative to earth). A common mode signal is one where the voltages on both terminals rise and fall together.

An example use of a differential amplifier is terminating transmission lines where signals common to both wires are due to induction from external sources, such as mains power supplies. In many cases, the wanted differential signal is smaller in amplitude than the common mode signal, but the differential amplifier is able to extract the wanted signal because of the common mode rejection by the amplifier.

Differential amplifiers also allow one signal to be subtracted from another. Figure 4.7 shows the type of circuit configuration that is employed. An ideal

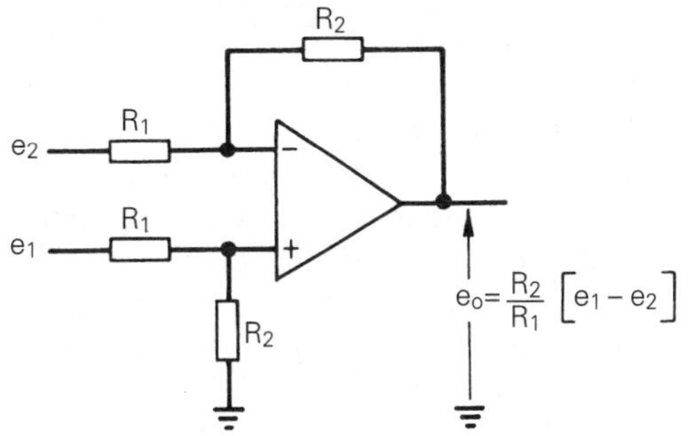

Figure 4.7 *Single op-amp differential amplifier*

analysis of this circuit was given in Chapter 1; some of its practical limitations are now discussed.

A prime requirement of a differential input amplifier circuit is that it should have a high common mode rejection ratio (CMRR). According to the ideal performance equation of the circuit in Figure 4.7, the output is zero if the two input signals e_1 and e_2 are equal. The ideal circuit has an infinite CMRR – not the case with practical circuits. In a practical circuit any mismatch in the resistor ratio values connected to the op-amp input terminals causes a common mode signal ($e_1 = e_2 = e_{cm}$) to inject a differential signal to the amplifier. This differential signal is amplified to produce a non-zero output signal. CMRR is thus degraded unless the resistor values are exactly matched.

In assessing the common mode characteristics of differential amplifiers, care must be taken in distinguishing between the characteristics of the circuit and those of the op-amp used in it. The CMRR of the circuit is defined as:

$$\text{CMRR} = \frac{\text{Differential gain of circuit}}{\text{Common mode gain of circuit}} \qquad (4.2)$$

In the circuit of Figure 4.7, CMRR depends both upon resistor matching and upon the CMRR of the op-amp. The CMRR of the circuit due to resistor mismatch using resistor values with tolerance x is in the worst case:

$$\text{CMRR (due to resistor tolerance)} = \frac{1 + \dfrac{R_2}{R_1}}{4x} \qquad (4.3)$$

(see Appendix A3)

For example, a single op-amp differential input circuit (such as Figure 4.7) with differential gain 10 ($R_2/R_1 = 10$), using resistors of 1 per cent tolerance ($x = 0.01$) in the worst case has:

$$\text{CMRR (due to resistor tolerance)} = 11/0.04 = 2.75, \text{ or } = 49 \text{ dB}$$

The overall CMRR of the circuit due to both resistor mismatch and the finite CMRR of the op-amp is:

$$\text{Total CMRR} = \frac{\text{CMRR}_{(R)}\,\text{CMRR}_{(A)}}{\text{CMRR}_{(R)} \pm \text{CMRR}_{(A)}} \quad \text{(see Appendix A3)}$$

The common mode errors due to the two effects may be of the same or opposite sign, so that the total CMRR may be greater than or less than the CMRR of the op-amp used in the circuit.

It is of course possible to trim one of the external resistors in Figure 4.7. This enables the common mode gain, due to resistor tolerance, to be equal in magnitude but opposite in sign to the common mode gain of the circuit (due to the non-infinite CMRR of the op-amp alone). In theory, an infinite CMRR can be attained in this way. In practice, resistor trimming can give a 10 to 100 times increase in CMRR for the circuit over the CMRR of the op-amp used in it. A high CMRR achieved in this way is unfortunately not maintained: resistor values change with temperature, and also the CMRR of an op-amp does not remain stable.

In many applications a requirement of differential input amplifiers is that they have high differential and common mode input impedance. The input impedance of the circuit in Figure 4.7 is determined by the resistor values. Its differential input resistance is $2*R_1$ and it has an effective common mode input resistance at each input point of $R_1 + R_2$. If large resistor values are used in the circuit, to give a high input resistance, this can have side effects. One effect is stray capacitance that causes degradation in CMRR at the higher frequencies. Another effect is to give an increased offset error because of op-amp bias current.

The single op-amp differential amplifier has limitations in its performance. Despite this, it is often used (because of its simplicity) in non-critical differential applications. Improved performance can be obtained with circuit configurations using two or more op-amps; or by using application specific integrated circuits which have the functionality of differential amplifiers.

The circuit shown in Figure 4.8 is a differential input amplifier. It uses two coupled followers to attain high input impedance without the use of high value resistors. It also provides the possibility of gain setting with a single resistor.

Treating each op-amp and its associated input and feedback resistors separately, we can derive the ideal performance equation for the circuit in Figure 4.8. Analysing the circuit we can see that op-amp A_2 has two input signals applied to it. These are signal e_2 and the output from A_1, which is signal e_1 multiplied by $[1 + R_1/R_2]$. The output of A_2, due to input e_1 alone, is the signal from A_1 multiplied by $-R_2/R_1$ (see Chapter 2).

Thus the output from A_2, in terms of input e_1, is:

$$e_o' = e_1\left[1 + \frac{R_1}{R_2}\right]\left(\frac{-R_2}{R_1}\right)$$

Now considering the output from A_2, due to input e_2 alone, we have:

$$e_o'' = e_2\left[1 + \frac{R_2}{R_1}\right]$$

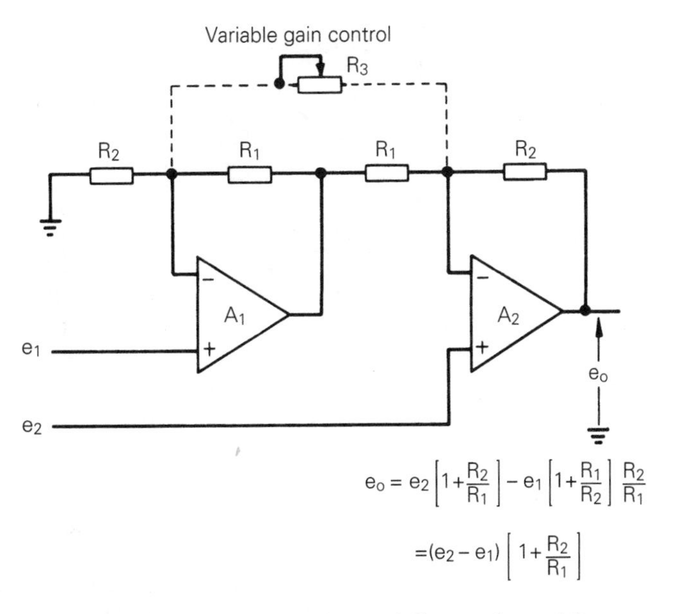

Figure 4.8 *High input impedance differential amplifier*

The total output from A_2 (e_o), in terms of inputs e_1 and e_2 applied together, is found by adding the two expressions:

$$e_o = e_o'' + e_o'$$

$$e_o = e_2\left[1 + \frac{R_2}{R_1}\right] - e_1\left[1 + \frac{R_1}{R_2}\right]\left(\frac{R_2}{R_1}\right)$$

This becomes:

$$e_o = (e_2 - e_1)\left[1 + \frac{R_2}{R_1}\right] \tag{4.4}$$

A practical circuit based upon Figure 4.8 has a CMRR that depends upon both resistor tolerances and upon the CMRR of the op-amps used in it. The input common mode range for the circuit is equal to that of the op-amps. With the gain setting resistor R_3 in circuit the output voltage in the ideal case is determined by the equation:

$$e_o = (e_2 - e_1)\left[1 + \frac{R_2}{R_1} + 2\frac{R_2}{R_3}\right] \tag{4.5}$$

In deriving this equation it should be remembered that, with R_3 in circuit, R_2 and the parallel combination of R_1 and R_3 now determine the value of $1/\beta$ for op-amp A_2.

Another differential amplifier configuration that is often used is shown in Figure 4.9. This circuit has two stages: a differential input stage and a subtractor stage. The differential input stage presents high impedance to both inputs.

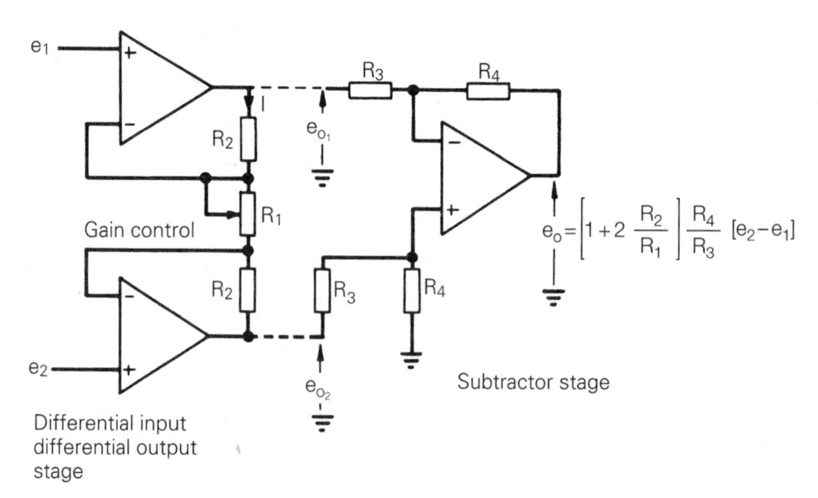

Figure 4.9 *High input impedance differential amplifier configuration*

Two coupled non-inverting amplifiers form the differential input stage. This stage produces a differential output voltage in response to a differential input signal. Assuming that the op-amps in the input stage take no current at their input terminals, the same current must flow through the three resistors (labelled R_1 and R_2). If we make the further usual assumption of negligible voltage between op-amp input terminals then this current

$$I = \frac{e_{o_1} - e_1}{R_2} = \frac{e_1 - e_2}{R_1} = \frac{e_2 - e_{o_2}}{R_2}$$

$$\text{Thus } e_{o_1} = \left(1 + \frac{R_2}{R_1}\right) e_1 - \frac{R_2}{R_1} e_2$$

$$\text{And } e_{o_2} = \left(1 + \frac{R_2}{R_1}\right) e_2 - \frac{R_2}{R_1} e_1$$

The input stage has a differential output, given by:

$$(e_{o_1} - e_{o_2}) = (e_1 - e_2)\left[1 + 2\frac{R_2}{R_1}\right] \tag{4.6}$$

Note that if $e_1 = e_2 = e_{cm}$ then $e_{o_1} = e_{o_2} = e_{cm}$. The input stage passes common mode input signals at unity gain. If the input stage used separately connected follower circuits, these would pass both common mode *and* differential signals at the same gain. The advantage of a cross-connected differential input stage, which is configured to provide some voltage gain, is that it amplifies differential input signals but not common mode signals.

An isolated load, such as a meter, can be driven directly by the differential output from the input stage. This has a theoretically infinite CMRR unaffected by resistor tolerance and the possibility of gain setting by means of a single resistor value (R_1). In practice CMRR is not infinite, because of differences in the internal common mode errors of the two op-amps. Dual op-amps can be used in this type of circuit, with the possibility of drift error cancellation (if the temperature drift coefficient on the two op-amps matches and tracks).

Monolithic dual op-amps have the advantage of maintaining both op-amps at the same temperature. However, despite the monolithic construction, the op-amp parameters are not matched. Improved performance can be obtained by using dual op-amp devices in which two separately matched op-amp chips are assembled into a single dual-in-line package.

To drive an earth-referred load, a single ended output is required. The differential output produced by the cross-coupled followers can be converted into a single ended output by using the differential amplifier circuit of Figure 4.7, which uses a single op-amp. The overall CMRR obtained with a circuit that uses three op-amps is greater than that of the single op-amp circuit, by a factor equal to the differential gain of the input stage.

The resistor values in the single op-amp circuit should be well matched, to give good common mode rejection. The input common mode range of the circuits of Figures 4.8 and 4.9 is limited to that of the op-amps used in the circuits. A differential input circuit configuration using two inverting amplifiers (see Appendix A1, Figure A1.2) can be given a larger input common mode range but with the disadvantage of lower input resistance in the inverter configuration.

In the presence of large or potentially dangerous common mode signals, consideration should be given to the use of an isolation amplifier.

4.5 Current scaling

The op-amp circuits considered up to now are suitable for scaling input signal voltages. In many systems there is a need to scale the output from current sources, such as light-sensitive diodes. Light-sensitive diodes provide a reverse leakage current proportional to the light intensity at their *pn* junction. The circuits considered in this section are designed to process an input current, rather than an input voltage.

4.5.1 Current-to-voltage conversion

In Chapter 1, it was shown that an ideal op-amp could provide an ideal, zero voltage drop, current-to-voltage conversion. There are two things that must be considered: (1) the possibility of closed-loop instability and (2) the reduction of drift errors that determine conversion accuracy.

(1) Closed-loop stability

In practice, the stability problem does not usually present too serious a difficulty. An externally connected capacitor C_f (see Chapter 2) connected in parallel with the scaling resistor R_f normally assures closed-loop stability. The importance of offset is dependent upon the size of the current to be measured and the processing accuracy required.

In Figure 4.10, an op-amp current-to-voltage converter is supplied with signals by a current source. Stability is determined by the source capacitance. Source capacitance causes a phase lag in the feedback signal at the higher

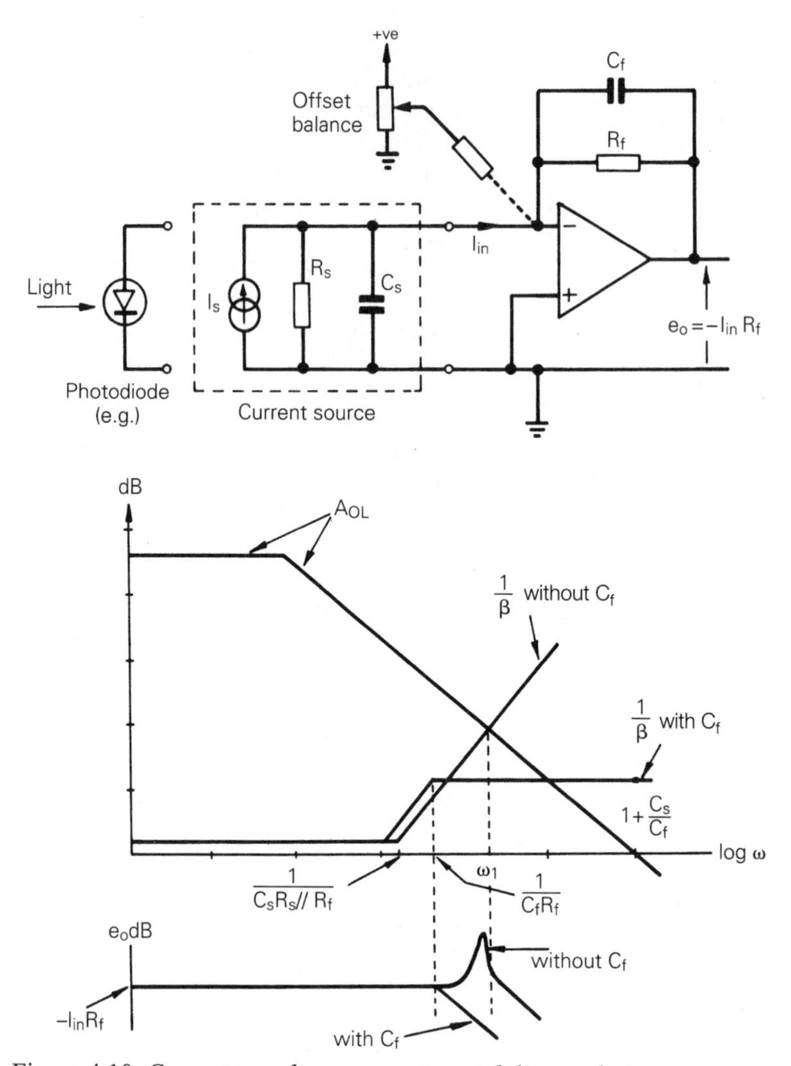

Figure 4.10 *Current-to-voltage converter–stability analysis*

frequencies, which can lead to insufficient phase margin. Closed-loop stability is most conveniently examined in terms of the appropriate Bode plots. The Bode plot for $1/\beta$ is superimposed upon the open-loop Bode plot in order to examine the frequency dependence of the magnitude and phase of the loop gain (see Chapter 2).

The value of $1/\beta$ for the circuit of Figure 4.10 without the capacitor C_f in the circuit is

$$\frac{1}{\beta} = \left[1 + \frac{R_f}{R_s}\right][1 + j\omega C_s (R_s // R_f)] \tag{4.7}$$

This breaks up at the angular frequency $\omega_c = C_s(R_s//R_f)$. If this frequency occurs before the frequency at which $1/\beta$ and A_{OL} intersect, the two plots will have a rate of closure of 40 dB/decade. This means that there may be insufficient phase margin. Note that at frequency ω_1 the phase shift in β is $\theta = \tan^{-1}(\omega_1/\omega_c)$ and the phase margin in the circuit is $90° - \theta$ (see Chapter 2). If C_f is connected in circuit it introduces a phase lead into the feedback loop which offsets the lag due to C_s. With C_f in circuit, the value of $1/\beta$ becomes

$$\frac{1}{\beta} = \left[1 + \frac{R_f}{R_s}\right]\left[\frac{1 + j\omega(C_s + C_f)(R_s//R_f)}{1 + j\omega C_f R_f}\right] \tag{4.8}$$

The $(1/\beta)\log(f)$ plot breaks back at the angular frequency $1/(C_f R_f)$ and if this frequency is suitably chosen the $1/\beta$ and A_{OL} plots close at 20 dB/decade thus ensuring an adequate phase margin. The closed-loop signal bandwidth is fixed by the value used for C_f at the frequency $f = 1/(2\pi C_f R_f)$.

(2) Conversion accuracy

In many practical applications of the current-to-voltage converter, R_s will be greater than the value of the scale setting resistor R_f, making the value of $1/\beta$ approximately unity at low frequencies. If the impedance of the source current is lower than R_f, the noise gain $1/\beta$ will be greater and the loop gain smaller. Consequently, there will be a decrease in accuracy, and an increase in drift error due to op-amp input offset voltage temperature dependence.

Offset and drift error may be estimated by applying the general method outlined in Section 2.10.4. An expression for the total equivalent input offset voltage is:

$$E_{os} = V_{io} + (R_f // R_s)I_b^-$$

This appears at the output multiplied by $1/\beta$.

$$\text{Output offset voltage} = E_{os}(1 + R_f/R_s)$$

In order to assess accuracy this may be referred to the input (by dividing by R_f) as an equivalent input error current.

$$I_{os} = \left[1 + \frac{R_f}{R_s}\right]\frac{E_{os}}{R_f} = \pm\frac{V_{io}}{R_s//R_f} + I_b^-$$

Op-amp bias current is normally the main error component. The large resistor values commonly used in current-to-voltage converters make the error due to op-amp input offset voltage negligible. Initial offset can be zeroed using a high value resistor, to feed a small adjustable current to the inverting input terminal of the op-amp.

The temperature drift of the op-amp bias current is then the limiting factor in determining accuracy. A low bias current op-amp should be chosen for accurate measurements of small currents, e.g. a FET input type. Measurement of currents in the pico-amp range requires particular attention to the avoidance of stray leakage currents otherwise the performance capabilities of low bias current op-amps cannot be realized (see Section 9.4).

High value resistors are necessary to set the scaling factor in small current measurements. Unfortunately, high value resistors tend to be less stable than commonly available devices. Sensitivity of the circuit can be increased without using very high value resistors by using a resistive T network as shown in Figure 4.11, but note that this is at the expense of a decrease in loop gain and an increase in offset and noise gain.

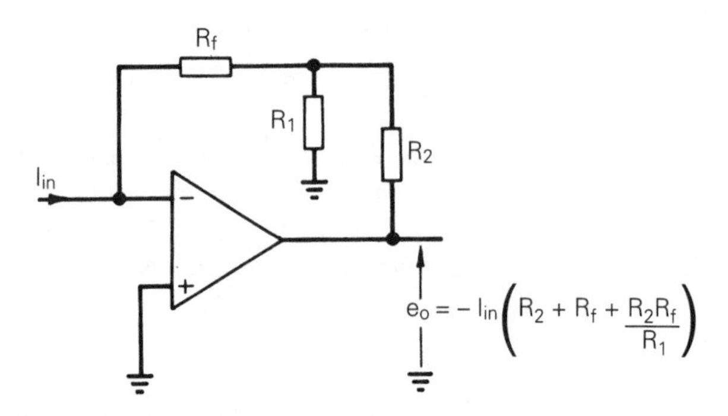

Figure 4.11 *Resistive T network gives increased sensitivity without high value feedback resistor*

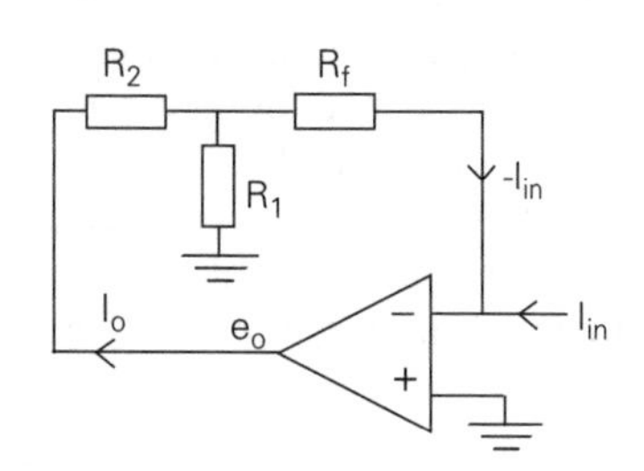

Figure 4.12 *Current flow to derive transfer function*

The transfer function for this circuit can be derived from consideration of currents in the feedback loop; see Figure 4.12.

Negative feedback forces the op-amp's inputs to be at the same (earth) potential. A voltage at the output, e_o, causes current I_o to flow through R_2 and the parallel combination of R_1 and R_f. Thus:

$$e_o = I_o \left[R_2 + \frac{R_1 R_f}{R_1 + R_f} \right] \tag{4.9}$$

Current I_o flows through the parallel combination of R_1 and R_f. The share passing through R_f is equal to I_{in}, but with opposite polarity, since no current flows into the op-amp's inverting input. The current through R_f is given by:

$$-I_{in} = I_o \left[\frac{R_1}{R_1 + R_f} \right]$$

Transposing this to find I_{in} in terms of I_o, we get:

$$I_o = -I_{in} \left[\frac{R_1 + R_f}{R_1} \right] \tag{4.10}$$

Combining equations 4.9 and 4.10, we get:

$$e_o = -I_{in} \left[\frac{R_1 + R_f}{R_1} \right] \left[R_2 + \frac{R_1 R_f}{R_1 + R_f} \right]$$

$$e_o = -I_{in} \left[\frac{R_1 + R_f}{R_1} \right] [R_2 + R_f]$$

$$e_o = -I_{in} \left[R_2 + R_f + \frac{R_2 R_f}{R_1} \right] \tag{4.11}$$

A current-to-voltage converter overcomes the problem of the finite resistance of moving-coil meters when used for current measurement. Possible circuit configurations are shown in Figure 4.13. Battery operation of the op-amp allows non-earth referred measurements to be made as in Figure 4.13(c). Note that equation 4.10 is used, except that R_f is replaced by R_2 and $-I_o$ becomes I_m. What was R_2 in Figure 4.11 is now the meter resistance, which does not affect the transfer function.

$$-I_o = I_m = I_{in}\left[\frac{R_1 + R_2}{R_1}\right] = I_{in}\left[1 + \frac{R_2}{R_1}\right]$$

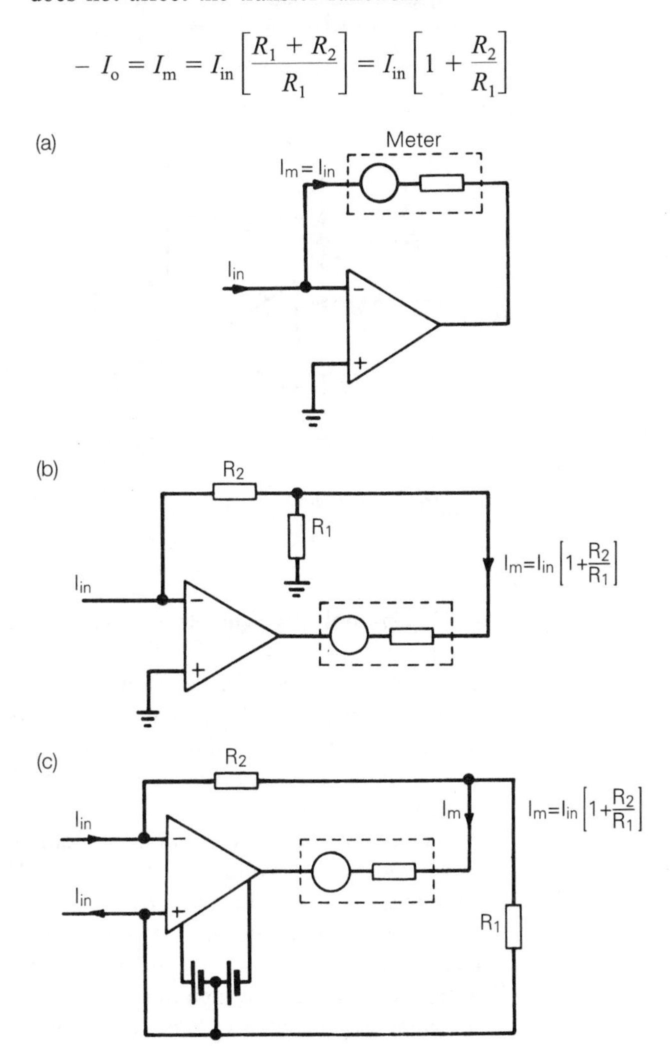

Figure 4.13 *Current measurement circuits. (a) Simple measurement. (b) Current measurement with increased sensitivity. (c) Current measurement not referred to earth (battery supplies)*

4.5.2 Current summation

The basic current-to-voltage converter circuit of Figure 4.11 can be used to sum currents to earth from separate signal sources. All that is required is to add the extra input paths to the inverting input terminal of the op-amp. The circuit shown in Figure 4.14 illustrates the principle. In order to ensure adequate phase margin, the value required for the feedback capacity C_f is now governed by the total capacitance to ground at the inverting input terminal ($C_{s_1} + C_{s_2}$ etc.).

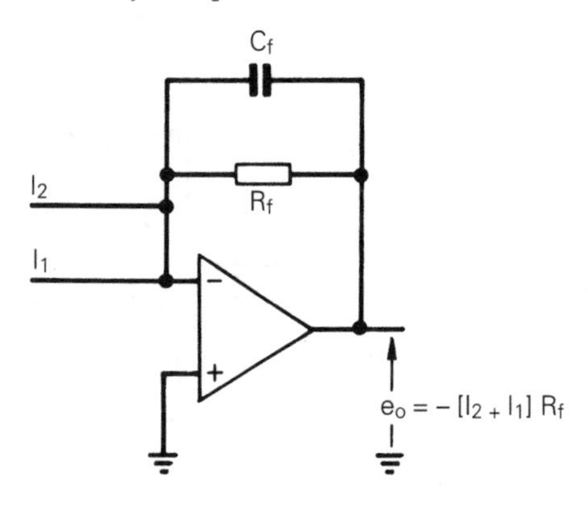

Figure 4.14 *Current summing circuit*

4.5.3 Current difference-to-voltage conversion

Op-amps allow the measurement of current with no voltage drop in the measurement circuit. Current is supplied to the inverting input terminal of an op-amp and its non-inverting input terminal is earthed. The feedback resistor provides a path for the current whilst the inverting terminal is held at earth potential.

A current difference measurement requires the use of two op-amps in order to satisfy the zero voltage drop criterion. The circuit shown in Figure 4.15 combines the summing property of one op-amp with a current inversion performed by a second op-amp. Op-amp A_2, with equal value resistors (R_1) connected between its output and its two input terminals, forces equal currents to flow towards its two input terminals in order to maintain them at the same potential. The inverted current I_2 is supplied to the summing op-amp A_1 via a very high effective output impedance obtained as a result of the positive feedback applied to amplifier A_2.

In cases where a voltage intrusion into the measurement circuit is allowable, a single op-amp can be used to perform a current difference conversion. The circuit shown in Figure 4.16 gives an output voltage that is proportional to the difference in the two input currents, I_1 and I_2. Note that, in this circuit, a voltage drop, $V = I_2R$, is introduced into the measurement path. This voltage drop represents a common mode input to the amplifier. Subtraction of equal input currents requires accurate matching of resistor values.

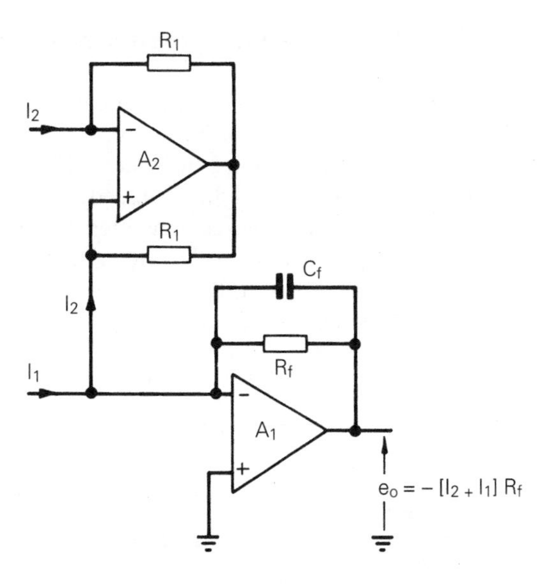

Figure 4.15 *Current difference-to-voltage conversion*

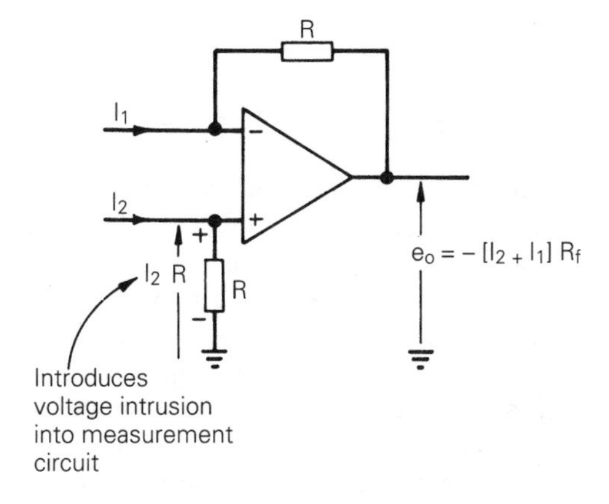

Figure 4.16 *Single op-amp for current difference-to-voltage conversion*

4.6 Voltage-to-current conversion

Some loads require a current drive rather than a voltage drive. In such cases, an op-amp circuit configuration is required that will give a linear voltage-to-current conversion. Voltage controlled current sources are very useful in a variety of measurement applications, such as resistance measurement. They can also be used to drive inductive loads for the production of controlled magnetic fields.

There are several ways in which an op-amp may be used to produce a voltage-to-current conversion. The circuit configuration adopted is determined by the operating requirements of the load. For example, is the load to be earthed or can it float, is a unidirectional or bi-directional current drive required?

4.6.1 Voltage-to-current converters – floating load

The simplest current-to-voltage converter circuits are those for which the load is allowed to float. Basic inverting and non-inverting voltage-to-current converters are illustrated in Figure 4.17. In each case the ideal performance equation, $I = e_{in}/R_1$, follows directly from the usual ideal op-amp assumptions. In the inverting configuration the input signal source must supply a current equal to the load current. In the non-inverting circuit, negligible current is drawn from the signal source but common mode limitations and errors must be considered.

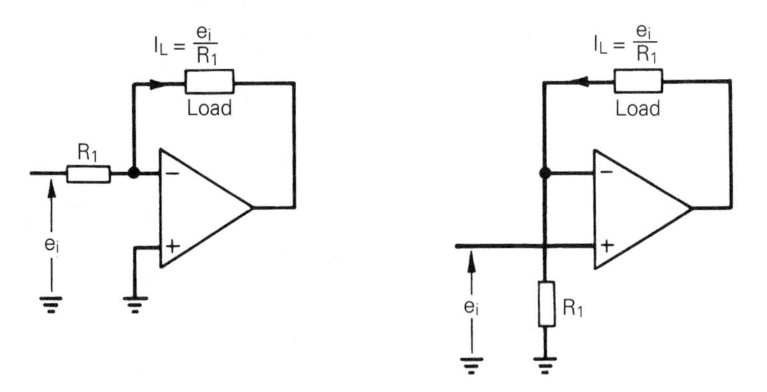

Figure 4.17 *Simple voltage-to-current converter load floating*

In all voltage-to-current conversions, the op-amp used in the circuit must be capable of providing the desired maximum load current. Also, the output voltage that is required for maximum load current must not exceed the op-amp's rating. Remember that some form of booster circuit (see Chapter 9) can always increase op-amp output limits.

Inductive loads (coil driving) require particular attention, in terms of the op-amp's maximum output limits and in achieving closed-loop stability. An inductive load introduces an extra phase lag in the feedback loop. This can lead to an inadequate phase margin, even when the op-amp used in the circuit is frequency compensated for unity-gain operation. Closed-loop stability can often be achieved by connecting a resistor in series with the inductive load, and by adding a lead capacitor directly between op-amp output and phase inverting input. Bandwidth is inevitably limited by these added components.

4.6.2 Voltage-to-current converters – earthed load

Simple circuits can be used for current drive of an earthed load provided that either the controlling input signal voltage or the power supplies to the op-amp can be floated. The input signal must float in the circuit of Figure 4.18. Negative feedback forces the differential input terminals of the op-amp to be at the same potential and in doing so produces a voltage across the resistor R that is equal to e_{in}. The current through R, except for the small op-amp bias current, passes through the load, and there is negligible loading

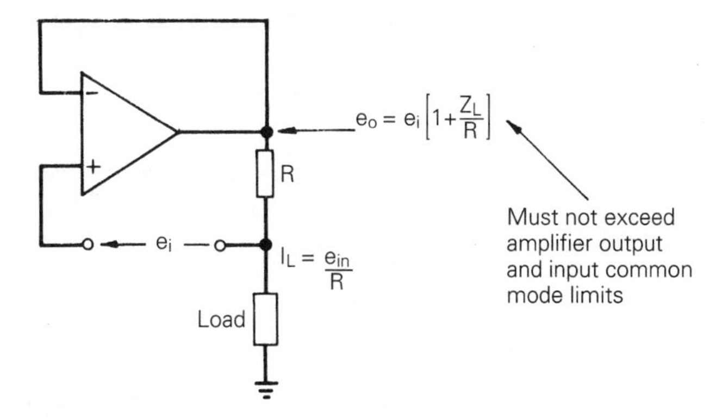

Figure 4.18 *Voltage-to-current converter floating signal source*

of the input voltage signal. Note that the voltage that appears across the load represents a common mode input voltage to the amplifier and common mode limitations and errors must therefore be considered.

4.6.3 Voltage-to-current converter – earthed load and power supplies

The circuit shown in Figure 4.19 can be used to supply a bi-directional current to an earthed load.

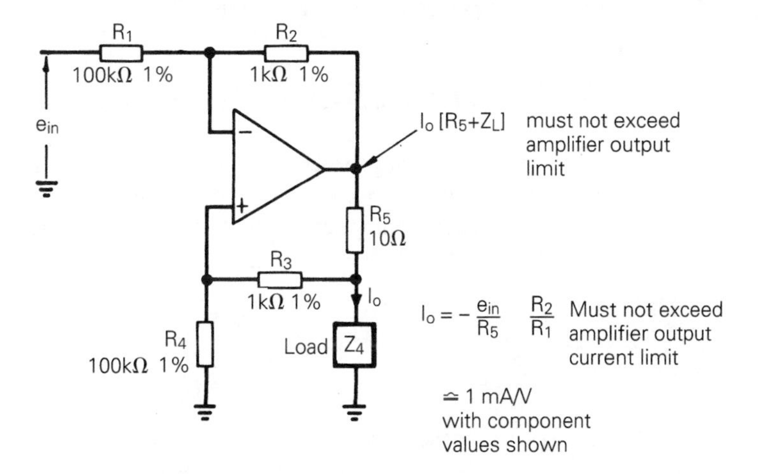

Figure 4.19 *Voltage-to-current converter (earthed load and power supplies)*

In the circuit of Figure 4.19, the current is controlled directly by a single ended input voltage. Making use of the usual ideal op-amp assumptions, we can derive the ideal performance equation for the circuit. Thus the signal at the inverting input terminal is:

$$e^{-} = e_{in}\frac{R_2}{R_1 + R_2} + e_o\frac{R_1}{R_1 + R_2}$$

and the signal at the non-inverting input is:

$$e^+ = [e_o - I_o R_5]\frac{R_4}{R_3 + R_4}$$

It is assumed that $[R_3 + R_4] >> R_5//R_L$. The op-amp forces $e^- = e^+$. Resistor values are chosen so that $R_2/R_1 = R_3/R_4$. Making these substitutions the performance equation simplifies to:

$$I_o = -\frac{e_{in}R_2}{R_5 R_1}$$

The offset error for the circuit when referred to the signal input is:

$$V_{in(offset)} = E_{os}\left[1 + \frac{R_1}{R_2}\right]$$

where $E_{os} = \pm V_{io} + I_b^- R_s^- - I_b^+ R_s^+$ (see Chapter 2).

The load current is supplied by very high effective impedance. The value of this impedance depends upon accurate matching of resistor ratios in the circuit. Accurate matching of resistor ratios provides stability of load current against fluctuations in load impedance. Trimming the value of resistor R_4 (by the use of a small potentiometer in series with it) allows the circuit to produce near constant output current with variations in load. A preferable alternative to a trimming potentiometer would be to use close tolerance (<< 1 per cent) metal film resistors.

4.6.4 Unidirectional current sources and sinks

The voltage-to-current sources considered so far have provided a bi-directional output current. Unidirectional current sources can be formed using a simple circuit comprising a transistor, a resistor and an op-amp, as shown in Figure 4.20.

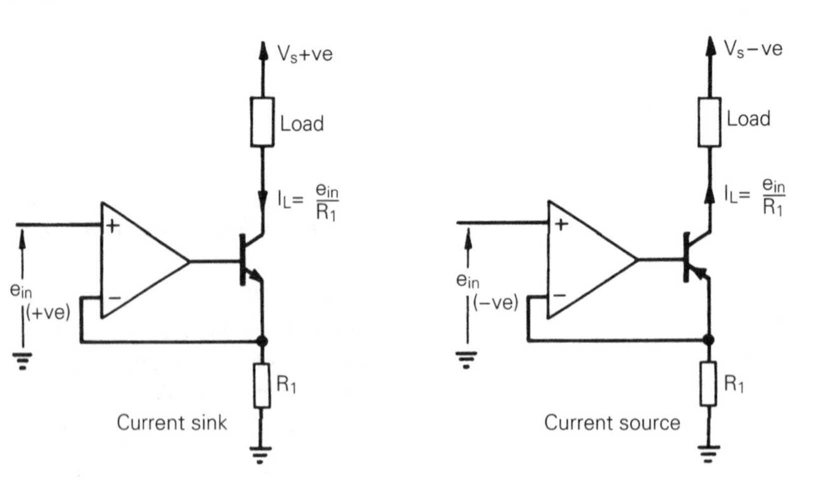

Figure 4.20 *Current source and sinks*

Resistor R_1 is a current sensing resistor. Feedback around the op-amp forces the current through resistor R_1 to take on a value such that $I_1R_1 = e_{in}$. The current I_1 is the emitter current of the transistor, less the very small bias current of the op-amp. The collector current of the transistor, which is almost equal to its emitter current, forms the stable output current to the load.

$$I_o = \alpha\left[\frac{e_{in} \pm V_{io}}{R_1} + I_b\right]$$

Output currents greater than the capability of the op-amp are possible, since the op-amp need only supply the output transistor's base current. A current limit is set by transistor saturation caused by the voltage that appears across the load.

Departure from linearity in voltage-to-current conversion is likely at low current levels. This is because the gain of a bipolar transistor falls at low values of collector current. The linearity dependence on transistor current gain, exhibited by the current sources of Figure 4.20, can be overcome by using a FET in place of the bipolar transistor. However, output current is then limited to the I_{dss} of the FET. The output current limit can be overcome by combining an n-channel FET and a bipolar npn transistor as shown in Figure 4.21.

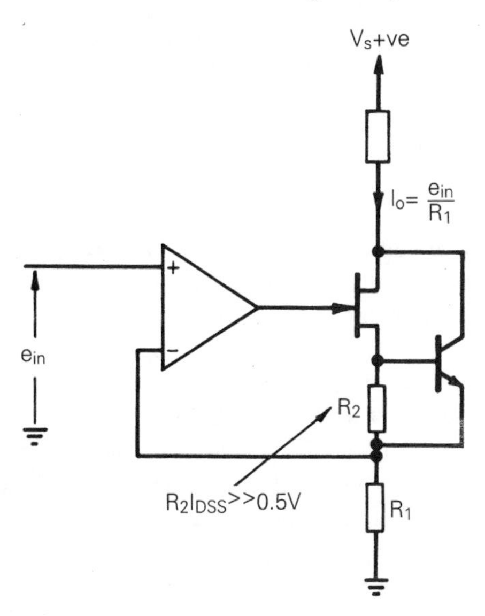

Figure 4.21 *Precise current sink*

In the circuit given in Figure 4.21, virtually all the current through the sensing resistor R_1 flows as output current. The only error contributions being the very small gate leakage current and the op-amp bias current.

4.7 Voltage regulators

A simple voltage regulator is shown in Figure 4.22. This uses an op-amp to drive the base of a power transistor, to turn on the transistor and pass current from the input (V_{in}) to the output (V_{out}). The accuracy of the output voltage depends upon both the gain of the op-amp and the transistor, as well as the reference voltage V_{ref}. This reference voltage is applied to the non-inverting input of the op-amp. A potential divider, comprising R_1 and R_2, applies a

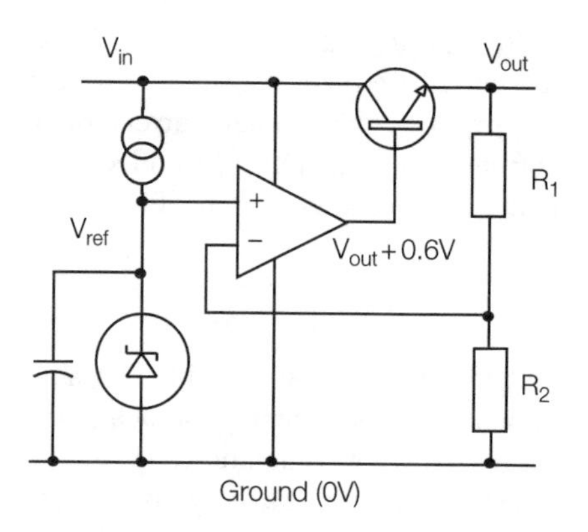

Figure 4.22 *Simple voltage regulator*

fraction of the output voltage to the inverting input of the op-amp. The output voltage is stable when the feedback voltage equals the reference voltage.

The output voltage equals the reference voltage (developed across R_1) plus the voltage drop across R_2. The current through R_2 is V_{REF}/R_1, therefore:

$$V_{OUT} = V_{REF} + R_2\left[\frac{V_{REF}}{R_1}\right]$$

This simplifies to:

$$V_{OUT} = V_{REF}\left[1 + \frac{R_2}{R_1}\right]$$

Integrated circuit voltage regulators are popular, because they have current limiting outputs and over-temperature shutdown circuits built into them. One such regulator is the LM317, which is produced by a number of manufacturers. This uses external resistors to set the output voltage and a typical circuit is shown in Figure 4.23. The reference voltage is a band-gap reference set at 1.25 V. Resistor R_1 is usually set at 240 Ω, to give a nominal 5 mA through the potential divider R_1 and R_2. This level of current overcomes any errors due to small current leakage out of the LM317 reference terminal. The output voltage is given by the equations above.

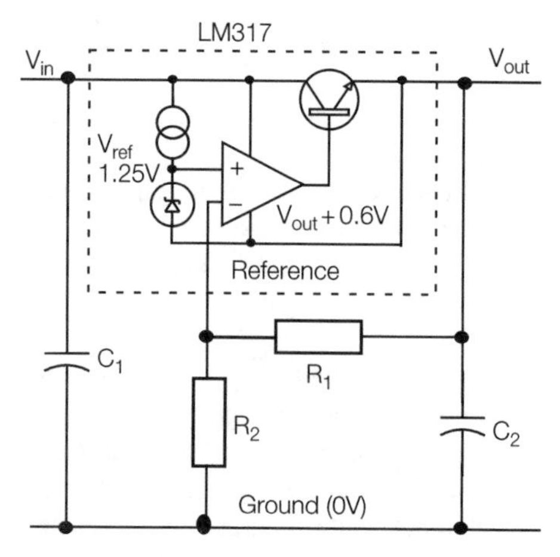

Figure 4.23 *Standard LM317 voltage regulator circuit*

Capacitors C_1 and C_2 connected from the input and output terminals to ground are necessary to prevent oscillation.

The disadvantage of the LM317 type of regulator is that the device needs at least 3 V between the input and the output in order to provide power to the internal circuits. Fixed voltage regulators are slightly better in this regard because their internal circuits are able to use the higher potential difference between input and ground. However, in a fixed voltage regulator with an integral output driving NPN transistor, the minimum potential difference between V_{in} and V_{out} must be about 1 V. This is because the base of the NPN transistor must be at least 0.6 V above the output voltage and a fraction of a volt will be dropped across the op-amp output stage.

A modification to the simple regulator circuit is given in Figure 4.24. This uses a PNP transistor Q_1 to drive the output and a second PNP transistor Q_2 to buffer the op-amp output. The op-amp output is held at about $V_{in} - 1.2$ V. Note that the input terminals of the op-amp are reversed, compared to the regulators shown in Figures 4.22 and 4.23. If the voltage at the junction of resistors R_1 and R_2 is higher than V_{ref}, the output of the op-amp goes positive, reducing the base current through transistor Q_2, which in turn reduces the base drive to transistor Q_1. This circuit arrangement allows the minimum voltage between V_{in} and V_{out} to become very low (about 0.2 V). This is because the transistor base drive from the op-amp is relative to V_{in}, instead of V_{out}. The maximum voltage across a voltage regulator is usually limited to 37 V − 60 V.

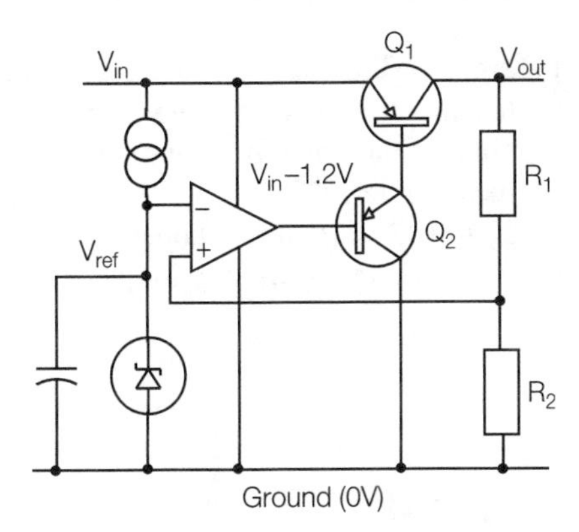

Figure 4.24 *Simple low drop-out (LDO) regulator*

However, increasing the working voltage range of a regulator is possible using a high voltage depletion mode MOSFET. The circuit shown in Figure 4.25 has the MOSFET Q_1 preceding the regulator IC_1. The drain of Q_1 is connected to the high voltage supply, the source is connected to IC_1 input and the gate is connected to IC_1 output. A depletion mode MOSFET conducts between drain and source until a voltage that is about 3 V negative, with respect to the source, is applied to the gate. This means that if the regulator has more than about 3 V dropped across it, the MOSFET conduction decreases and a greater proportion of the applied voltage is dropped across the MOSFET. However, heat dissipation limits the current that can be supplied from such a circuit, because most of the power is dissipated by the MOSFET. The LR8 high voltage regulator from Supertex uses this technique to allow supply voltages of up to 450 V.

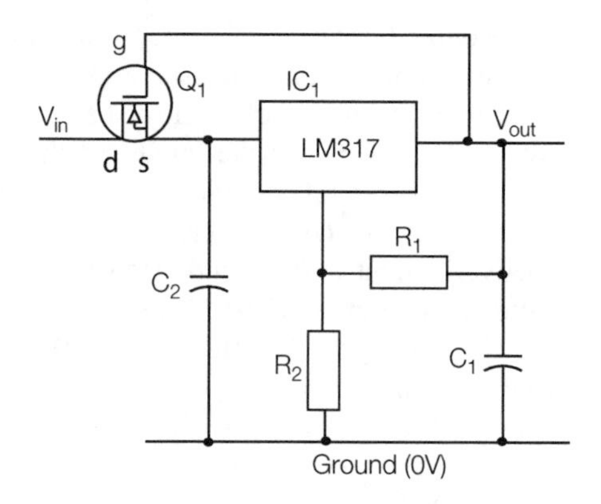

Figure 4.25 *High voltage regulator*

4.8 AC amplifiers

Op-amps are basically high gain DC amplifiers, but they are often used in applications not requiring a DC response. When used for AC amplification, DC blocking capacitors are placed in the signal path. The op-amp offset and drift specifications are not as important in AC applications, and are often ignored. Operation from a single rail supply is often used, with mid-rail biasing, to avoid the need for separate positive and negative power supplies.

4.8.1 Phase inverting AC amplifier

Figure 4.26 illustrates the basic inverting amplifier with a capacitor C_1 connected in series with the input resistor. Bias current to the inverting input terminal of the op-amp is supplied through the feedback resistor R_2. The gain of the amplifier is R_2/R_1, with the low frequency 3 dB fall in gain occurring at $f_{-3\text{ dB}} = 1/(2\pi C_1 R_1)$. The upper frequency limit of this circuit will depend on the compensated open-loop frequency response of the particular op-amp used.

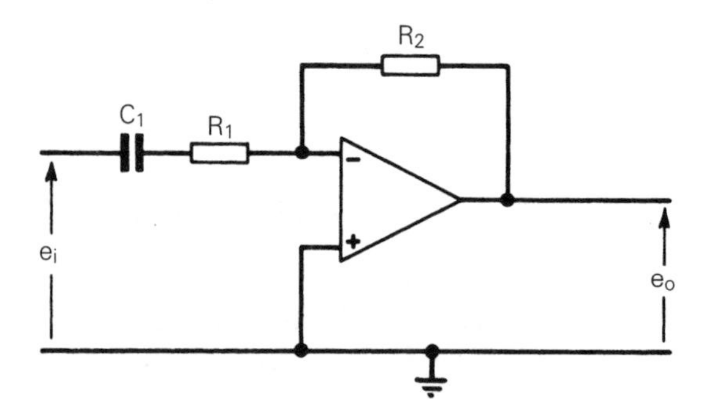

Figure 4.26 *Phase inverting AC amplifier*

4.8.2 Non-inverting AC amplifier

The circuit illustrated in Figure 4.27 is basically a non-inverting amplifier, with the addition of DC blocking capacitors and the DC bias path R_3. The

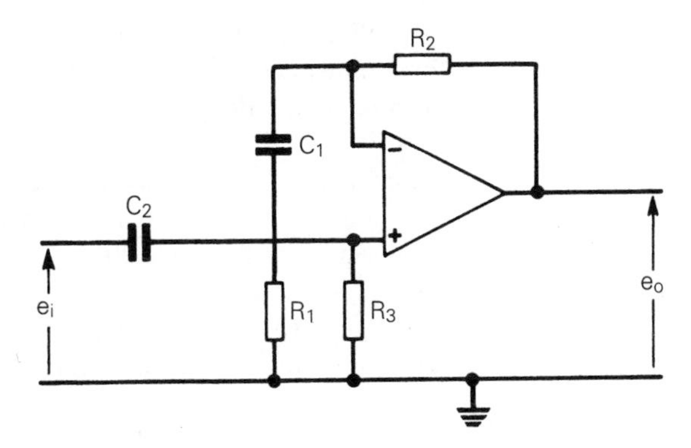

Figure 4.27 *Non-inverting AC amplifier*

closed-loop gain of the circuit is $1 + (R_2/R_1)$. The closed-loop low frequency response will show two breaks at $f_1 = 1/(2\pi C_1R_1)$ and $f_2 = 1/(2\pi C_2R_3)$. Bias resistor R_3 determines the input impedance of the circuit.

4.8.3 High input impedance AC amplifier (bootstrapped input)

The non-inverting amplifier, being a voltage follower, is intrinsically capable of high input impedance. Input impedance in the simple follower of Figure 4.27 is reduced by the need to provide a DC bias path (R_3). In the circuit illustrated in Figure 4.28, high effective input impedance is obtained because of positive feedback applied via R_2, C_1 and R_1 to the 'earthy' end of R_3. The technique of raising the apparent value of an impedance by driving its low potential end with a voltage in phase with, and almost as large as, the voltage at its high potential end is known as 'bootstrapping'. The effective value of R_3 is increased by a factor equal to the loop gain.

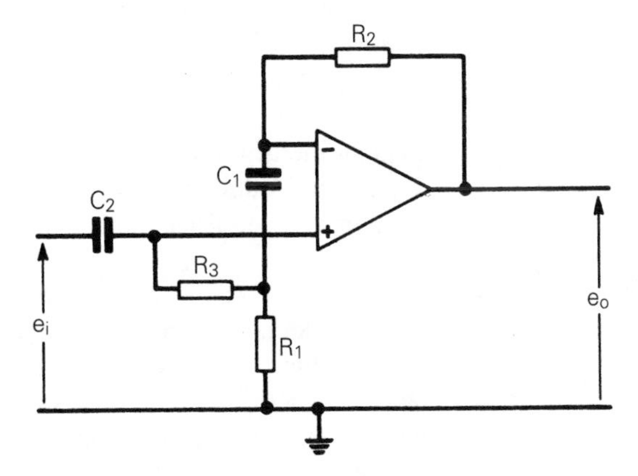

Figure 4.28 *High input impedance AC amplifier*

Exercises

4.1 In the circuit shown in Figure 4.6, input signals e_1, e_2 and e_3 are applied through input resistors 100 kΩ, 47 kΩ and 10 kΩ respectively. The feedback resistor has a value 100 kΩ. Write down the ideal expression for the output signal. If the operational amplifier is assumed ideal except for a finite open-loop gain of 80 dB what is the percentage error involved in the output sum?

4.2 An internally frequency compensated operational amplifier has an open-loop gain 100 dB, unity-gain frequency 4×10^6 Hz, input offset voltage 2 mV and bias current 100 pA. It is used in the circuit of Figure 4.2, with $R_1 = 1$ MΩ, $R_f = 1$ MΩ, $R_3 = 1$ kΩ, $R_2 = 100$ kΩ. Find: (a) the signal gain, (b) $1/\beta$, (c) the closed-loop bandwidth, (d) the output offset. (Hint, consult Sections 2.4 and 2.9.3.)

4.3 Resistors $R_1 = 10$ kΩ, $R_2 = 1$ MΩ, with tolerance 1 per cent are used in the circuit of Figure 4.7. The operational amplifier has $A_{OL} = 100$ dB, CMRR = 80 dB, unity-gain frequency $f_1 = 10^6$ Hz, input offset voltage $V_{io} = 2$ mV, input difference current $I_{io} = 50$ nA. Find:
(a) the worst case CMRR of the circuit (use equations 4.3 and 4.4);
(b) the closed-loop bandwidth (consult Section 2.5);
(c) the output offset.
(Hint, consult Section 2.9.3.)

4.4 Resistors $R_1 = 1$ kΩ, $R_2 = 100$ kΩ, of 2 per cent tolerance, are used in the circuit of Figure 4.8. What is the worst case CMRR of the circuit due to resistor mismatch? If the operational amplifiers have an open-loop gain bandwidth product of 4×10^6 Hz, what is the closed-loop signal bandwidth? (See Section 2.4.)

4.5 An internally frequency compensated operational amplifier with unity-gain frequency 10^6 Hz is used as a current-to-voltage converter and is supplied by a current source of very high internal resistance and capacitance C_s = 5 pF. A feedback resistor of value 1 MΩ is used. Initially no feedback capacitor is connected, but the circuit is found to be very lightly damped. Explain this fact and estimate the phase margin in the circuit.

The problem of the lightly damped response is overcome by connecting a capacitor of value 10 pF in parallel with the feedback resistor. Explain the action of this capacitor and estimate the phase margin and signal bandwidth with the capacitor connected. Illustrate your answer with appropriate Bode plots. (Consult Sections 4.5.1 and 2.5.)

4.6 A current-to-voltage converter has a feedback resistor of value 1 MΩ. Initial offset in the circuit is balanced by means of an adjustable current bias supplied through a resistor of value 10 MΩ. Assuming a temperature change of 10°C, estimate the smallest current which can be converted with an error no greater than 1 per cent:
(i) (a) using a bipolar transistor input operational amplifier with $\Delta I_B/\Delta T = 1$ nA/°C and $\Delta V_{io}/\Delta T = 10$ μV/°C; (b) using a FET input operational amplifier with $I_B = 50$ pA, doubling for a 10°C rise in temperature and $\Delta V_{io}/\Delta T = 40$ μV/°C. Assume the input signal source has a resistance $R_s = 10$ MΩ.
(ii) Repeat the question assuming a source resistance $R_s = 100$ kΩ.
(Hint, consult Section 4.5.1.)

5 Logarithmic amplifiers and related circuits

In Chapter 4, amplifier circuits were described that had a linear and frequency independent relationship between input and output. This relationship arose because of the use of linear resistors for input and feedback components. Later, Chapter 6 will describe circuits that use reactive components to give frequency dependent relationships. This chapter discusses the use of non-linear components to give non-linear relationships.

Non-linear circuits find many applications in signal processing. In particular, logarithmic (log) amplification has many uses; principally to increase the voltage range of signals that can be handled. Conversely, a power law (anti-log) relationship allows an expansion of a narrow range of voltages. Operations such as multiplication, division, and the taking of powers or roots may also be performed using log and anti-log circuits.

5.1 Amplifiers with defined non-linearity

Defined non-linear amplification requires defined non-linear element characteristics to produce the desired input–output relationship (or transfer function). In the general case, a non-linear element may comprise one or many non-linear components.

The non-linear element is connected as either the input or feedback path in an op-amp circuit. In Figure 5.1 a non-linear element is shown replacing the normal input resistor used in the inverting amplifier circuit.

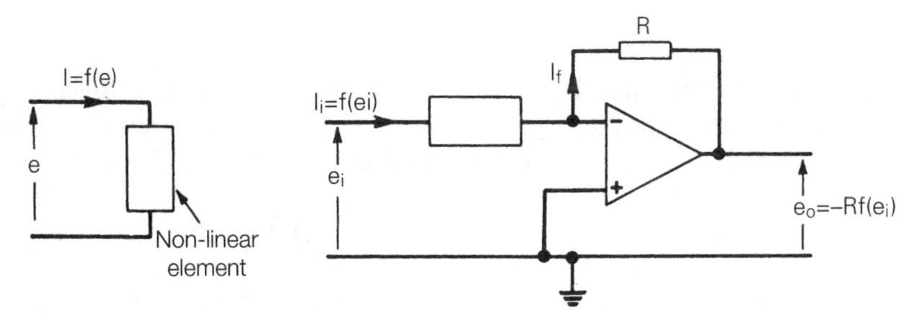

Figure 5.1 *Op-amp with non-linear input path*

The usual summing point restraints applied to the circuit give:

$$I_i = f(e_i) \text{ and } I_i = I_f = -e_o/R$$

Thus $e_o = -Rf(e_i)$

In Figure 5.2, the positions of resistor and non-linear element are interchanged. The op-amp output voltage drives the non-linear element and the feedback current is thus related to the output voltage in the defined non-linear manner.

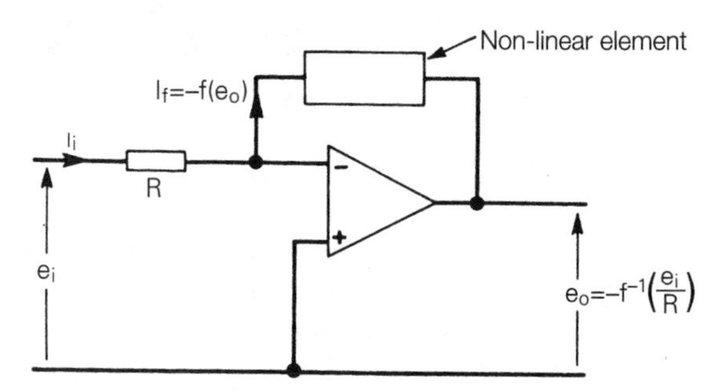

Figure 5.2 *Op-amp with non-linear feedback path*

$$I_f = -f(e_o)$$

and $I_i = I_f = e_i/R$

Thus $e_o = -f^{-1}(e_i/R)$

The circuit develops the required inverse function f^{-1}.

The main problem is finding a component that has the desired non-linear characteristic. Such components are required to show this non-linear characteristic over the widest possible range of current. They should also be insensitive to temperature changes.

The techniques used to achieve non-linear amplification generally fall into two categories. In one method, the desired non-linear response is synthesized by a network using a number of semi-linear elements (piecewise linear). The other method makes use of the inherent non-linearity of semiconductors.

5.2 Synthesized non-linear response

In graphical terms, any non-linear function can be approximated by a series of straight-line segments, each tangential to the desired function. The process is illustrated by the graph in Figure 5.3 in which the currents are:

$I_1 = k_1(e_i - e_1)$, for $e_i > e_1$;

$I_2 = k_2(e_i - e_2)$, for $e_i > e_2$;

$I_3 = k_3(e_i - e_3)$, for $e_i > e_3$; etc.

In the circuit shown in Figure 5.3, the break points, e_1, e_2, e_3, . . ., etc. are each set by a diode, a resistive divider, and a reference voltage supply. The input voltage is connected to these networks. The reference voltage polarity and the diode orientations shown are appropriate to a positive input signal.

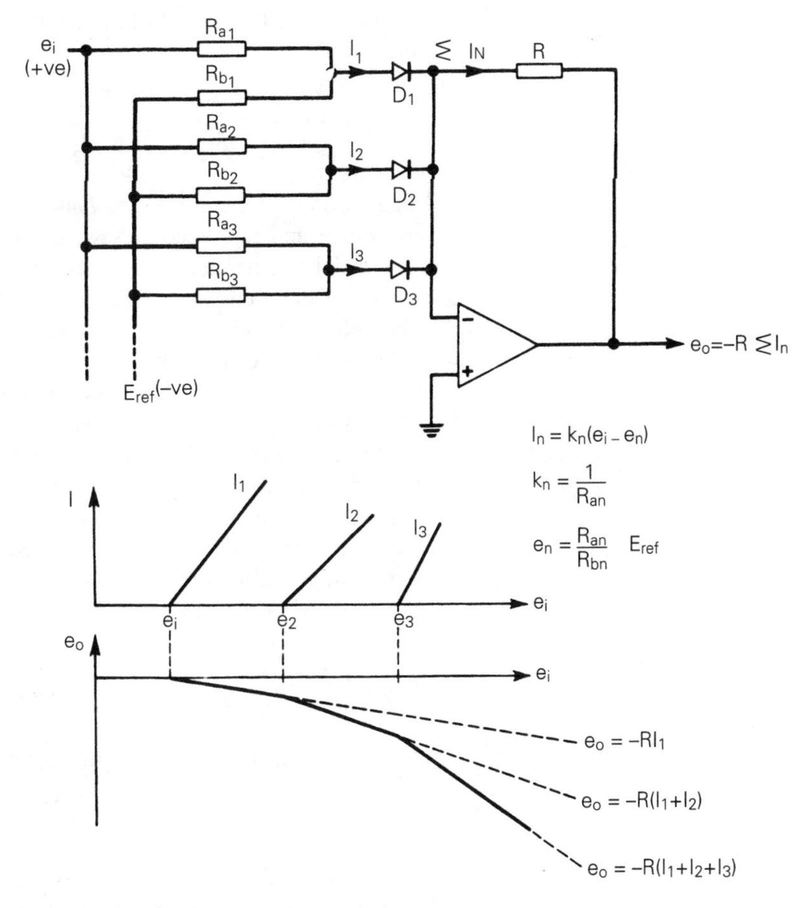

Figure 5.3 *Synthesized non-linear response*

Diode D_1 becomes forward biased when the input voltage exceeds the first break point $e_1 = E_{ref}(R_{a_1}/R_{b_1})$. Feedback from the amplifier output, through the resistor R, holds the inverting input terminal of the amplifier at earth potential. Neglecting the diode voltage drop, the current through diode D_1 for values of the input voltage greater than the first break voltage e_1 is thus

$$I_1 = \frac{1}{R_{a_1}}(e_i - e_1)$$

Similar reasoning gives the values of the currents through D_2, D_3, . . ., D_n, as

$$I_n = \frac{1}{R_{a_n}}(e_i - e_n)$$

The values for the break voltages are given by

$$e_n = E_{ref}\frac{R_{a_n}}{R_{b_n}}$$

The slopes of the straight-line segments used to approximate the desired function are

$$S_n = R\left[\frac{1}{R_{a_1}} + \frac{1}{R_{a_2}} + \ldots + \frac{1}{R_{a_n}}\right]$$

Negative input signals may be handled by using additional input networks, with diode and reference voltage polarities reversed. Non-monotonic functions can be generated by the use of an additional op-amp to invert the polarity of the input signal. Input networks with appropriate diode and reference voltage polarities must follow the op-amp.

This simple treatment has neglected diode voltage drops. Practical diodes exhibit a non-zero forward voltage drop with the added complication of temperature dependence. By using additional op-amps, diode effects can be reduced to negligible proportions and give break point voltages that change insignificantly with temperature. The circuit shown in Figure 5.4 illustrates a method of reducing diode effects.

The op-amp diode combinations used in the input network act essentially as precision rectifiers (see Chapter 8). Break point voltages are determined by E_{ref}, resistors R_c and resistors R_{b_1}, R_{b_2}, . . ., R_{b_n}, and

$$e_n = E_{ref}\frac{R_c}{R_{b_n}}$$

The slopes of the line segments are determined by resistors R_{e_1}, R_{e_2}, . . . , R_{a_n}, and

$$S_n = R\left[\frac{1}{R_{a_2}} + \frac{1}{R_{a_2}} + \ldots + \frac{1}{R_{a_n}}\right]$$

The circuit shown in Figure 5.5 illustrates another method of producing temperature stable break points. The external transistors should all be of the same type and have a high current gain.

In the circuit of Figure 5.5, the gain for small output signals is R_2/R_1. Transistors *Tr*2 and *Tr*3 are conducting, but feed back very little current to the amplifier summing point. When the output voltage rises to a certain level (set by R_3, R_4 and V_s), transistor *Tr*2 saturates and effectively connects R_3 in parallel R_2. This makes the gain of the circuit reduce to:

$$\text{Gain} = \frac{R_2 \,//\, R_3}{R_1}$$

When the output voltage rises further, to a level set by R_5, R_6 and V_s, saturation of transistor *Tr*3 occurs and connects R_5 in parallel with R_3 and R_2. The gain is thus reduced further to:

$$\text{Gain} = \frac{R_2 \,//\, R_3 \,//\, R_5}{R_1}$$

Temperature compensation is achieved in the circuit by the inclusion of transistors *Tr*1 and *Tr*4. Transistor *Tr*1 is used to temperature compensate the base-emitter voltages of *Tr*2 and *Tr*3. This arrangement keeps the voltage

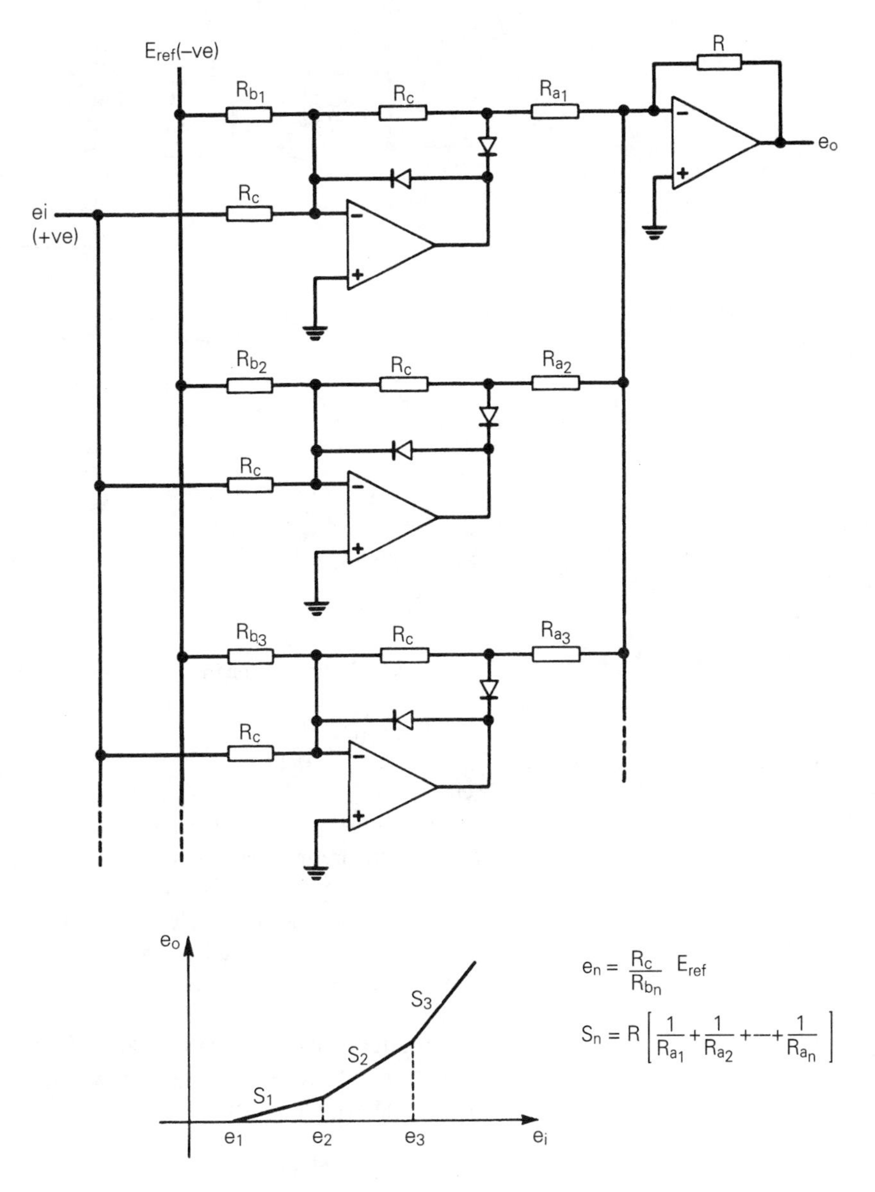

Figure 5.4 *Non-linear amplifier, break points stabilized*

across the feedback resistors R_3 and R_5 equal to the output voltage across the feedback resistor R_2. Transistor *Tr*4 is used to provide temperature compensation for change in saturation voltage of the transistors *Tr*3 and *Tr*2.

5.3 Logarithmic conversion with an inherently logarithmic device

The non-linear effects of diodes and transistors are often used to obtain logarithmic amplification. The logarithmic performance obtained using an op-amp with non-linear components is influenced by the characteristics of both the amplifier and the non-linear component. Therefore, an understanding of accuracy limitations requires some knowledge of the non-linear component.

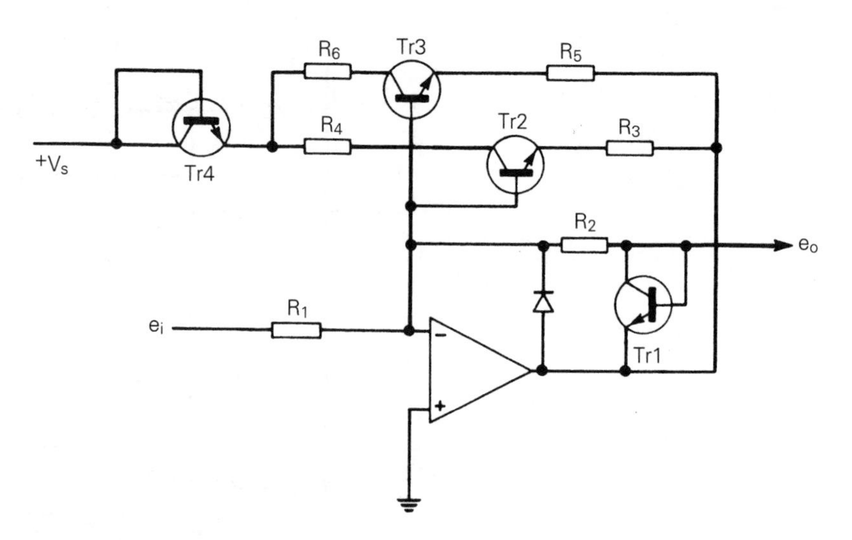

Figure 5.5 *Non-linear amplifier with temperature compensated break points*

Shockley's first order theory for a single pn junction gives the relationship

$$I = I_o\,(e^{qV/kT} - 1) \tag{5.1}$$

where:
I is the current through the junction (A),
I_o is the theoretical reverse saturation current (A),
V is the voltage across the junction,
q is the magnitude of the electronic charge (1.6×10^{-19} C),
k is Boltzmann's constant (1.38×10^{-23} J/K) and
T is the temperature in Kelvin.

Substituting values of constants gives $kT/q \sim 26$ mV at 27°C; thus for values of V greater than say 100 mV the exponential term in equation 5.1 predominates and we may write:

$$I = I_o\, e^{qV/kT} \quad (V > 100\ \text{mV})$$

Now, by taking natural logarithms:

$$\ln\left(\frac{I}{I_o}\right) = \frac{qV}{kT}$$

$$\text{hence } V = \frac{kT}{q}\ln\left(\frac{I}{I_o}\right)$$

To give this result in terms of logarithms to base 10, i.e. $\log(I/I_o)$, we use the mathematical relationship $\log(x) = \ln(x)/\ln(10)$, where $\log(x)$ is to the base 10.

$$\text{Thus, } \log_{10}\left(\frac{I}{I_o}\right) = \frac{\ln\left(\frac{I}{I_o}\right)}{\ln(10)} \text{ and } \ln(10)\log_{10}\left(\frac{I}{I_o}\right) = \ln\left(\frac{I}{I_o}\right)$$

Note that $\ln(x) = 2.3 \log_{10}(x)$, because $\ln(10) = 2.3$.

In terms of diode junction voltage, we have:

$$V = 2.3\frac{kT}{q}\log_{10}\left(\frac{I}{I_{10}}\right) \tag{5.2}$$

According to equation 5.2, a plot of log(I) against V gives a straight line of slope 2.3 kT/q volts per decade of current change. (Note the factor 2.3 $kT/q \sim 60$ mV at 27°C.)

A diode, which is assumed to obey equation 5.2, is shown connected as the feedback element in the circuit illustrated in Figure 5.6.

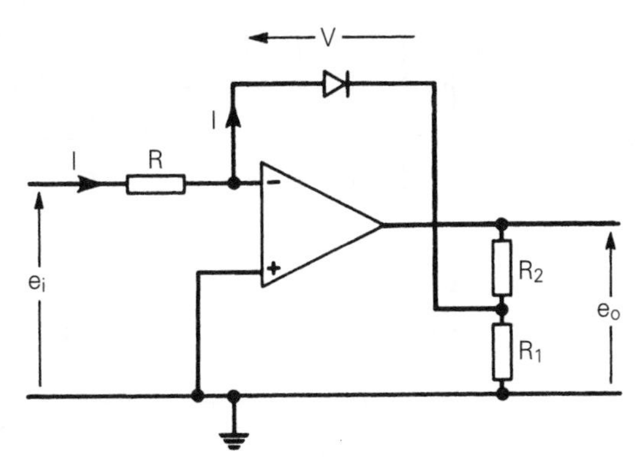

Figure 5.6 *Log amplifier with a diode as log element*

Referring to the circuit in Figure 5.6, and assuming ideal op-amp performance:

$$V = \frac{-R_1 e_o}{R_1 + R_2} = 2.3\frac{kT}{q}\log_{10}\left(\frac{I}{I_o}\right)$$

$$\text{or } e_o = -2.3\frac{kT}{q}\log_{10}\left(\frac{I}{I_o}\right)\left(\frac{R_1 + R_2}{R_1}\right) \tag{5.3}$$

The input current in the circuit shown is $I = e_i/R$. In the derivation of equation 5.3 we neglect the loading effect of the current, I, on the resistive divider R_1 and R_2. This divider is used to set a convenient scaling factor. The 60 mV/decade current change is a somewhat inconvenient factor, and a 1 V/decade scaling factor is usually preferred.

The circuit given in Figure 5.6 is attractively simple but is unfortunately rather limited in performance. Even assuming the availability of diodes that accurately obey equation 5.1, there remains the problem of temperature dependence. The scaling factor of $2.3kT/q$ is linearly dependent on temperature, with a positive coefficient of 0.3 per cent/K. This temperature dependence can be compensated by replacing resistor R_1 with a temperature sensitive resistor, having a temperature coefficient closely matched to the scaling factor.

Most diodes do not accurately obey equation 5.1. The derivation of this equation is based upon a single diffusion mechanism of current flow. There are actually several mechanisms operating and diode current is more accurately represented as the sum of several (N) components. The current components each have the form

$$I_j = I_{oj}\,(e^{qV/m_jT} - 1) \quad j = 1, 2, \ldots, N$$

where m_j can take values between 1 and 4.

A typical example of $V/\log(I)$ plots for general-purpose silicon diodes is shown in Figure 5.7. The two straight lines in this case have slopes corresponding to values of m equal to 1.78 and 1.55.

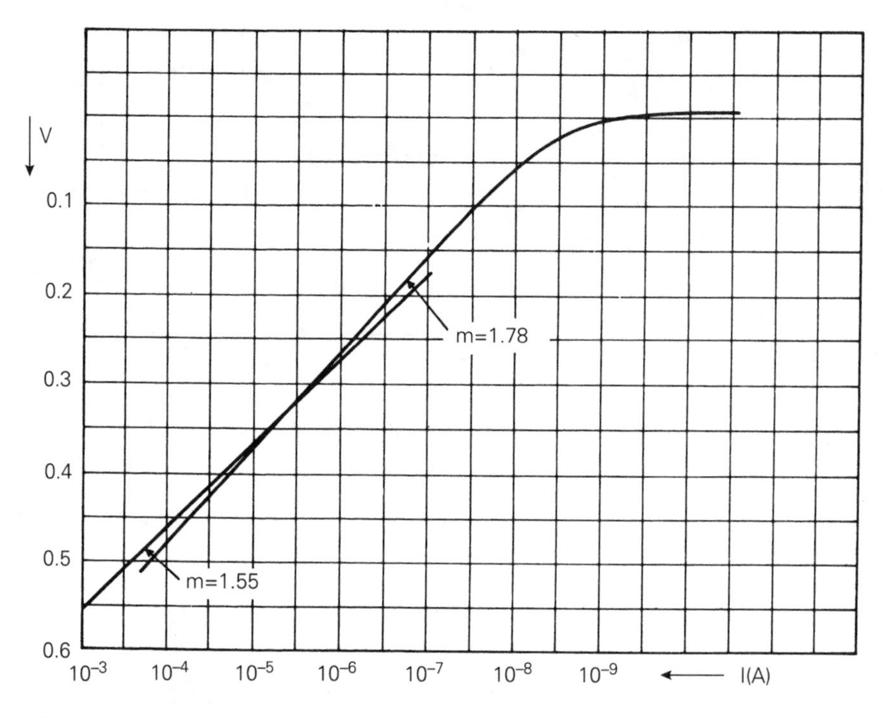

Figure 5.7 *Typical V/log(I) plot for a general-purpose diode*

The resistance of bulk semiconductor material causes errors in the logarithmic relationship. The voltage across a diode is that across the junction and the internal resistance. At higher currents, the voltage drop across this resistance becomes significant, hence only a fraction of the total diode voltage appears across the junction.

The above factors make general-purpose diodes unsuitable for accurate logarithmic conversion, except over a restricted range (three decades of current at the most). Temperature compensation requires the selection of matched diodes (matched m factors), and this presents an added difficulty.

So-called 'log diodes' are available which are said to exhibit a 7-decade current logarithmic range, but they are expensive. Transistors, which we will now consider, appear to be the most convenient elements for accurate logarithmic conversion.

A bipolar transistor consists essentially of two interacting pn junctions; the circuit symbol and a simple model for an npn transistor are illustrated in Figure 5.8. The collector current of a transistor can be accurately represented by the equation

$$I_c = \alpha_F I_{ES} (e^{-qV_E/kT} - 1) - I_{CS} (e^{-qV_C/kT} - 1) - \sum I_{CS_j} (e^{-qV_C/m_jT} - 1) \quad (5.5)$$

where:

α_F is the current transfer ratio between emitter and collector; it is very nearly unity,

I_{CS} is the collector reverse saturation current with the emitter shorted to the base,

I_{ES} is the emitter reverse saturation current with the collector shorted to the base,

m_j is the ideality factor that takes on values between 1 and 2 for silicon transistors and up to 10 for III–V materials, and

j is the number of the current path (1, 2, . . ., N).

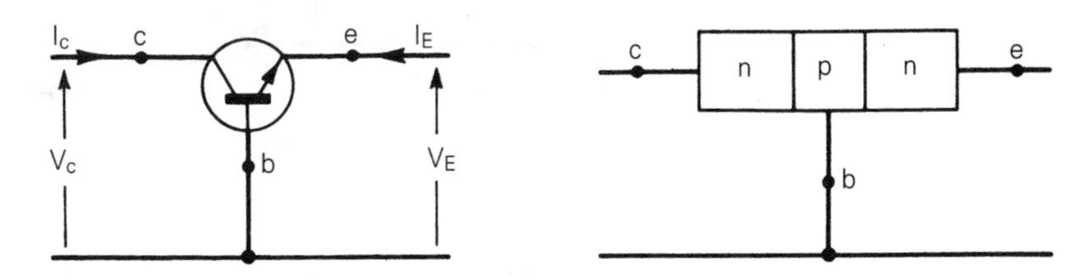

Figure 5.8 *Simple transistor model and sign conventions*

The sign convention adopted is shown in Figure 5.8. Equation 5.5 is appropriate for an npn transistor, the senses of V_S and I_S being reversed for a pnp device.

The first term in equation 5.5 represents that part of the emitter current, comprised of minority carriers in the base, which diffuses to the collector. The second and third terms are analogous to the diode current equations (equations 5.1 and 5.4); they give the collector current for the emitter shorted to the base.

The adoption of a circuit configuration which makes $V_C = 0$ causes all but the first term in equation 5.5 to become zero and the collector current is then given by the equation

$$I_c = \alpha_F I_{ES} (e^{-qVE/kT} - 1) \quad (5.6)$$

This is analogous to the 'ideal' diode relationship of equation 5.1.

Note that the $m \neq 1$ components of collector current become zero. The emitter ($m \neq 1$) current components behave largely as majority carriers in the base and as such do not diffuse to the collector. I_{ES} is typically of the order 10^{-13} A and α_F is very nearly unity. For values of collector current $I_C >> I_{ES}$ the exponential term in equation 5.6 predominates. Under these conditions, the following relationship holds:

$$-V_E = 2.3\frac{kT}{q}\log\left(\frac{I_C}{I_o}\right) \tag{5.7}$$

where $I_o = \alpha_F I_{ES}$.

Note that α_F should not be confused with the commonly used grounded base current gain $\alpha = I_C/I_E$. The value of α_F remains essentially constant over the range of collector currents for which equation 5.7 is valid.

The $V_C = 0$ condition may be obtained by connecting the collector of the transistor to the summing point of an op-amp, and the transistor base is connected to earth. This connection is made in the circuit shown in Figure 5.9. The circuit illustrates the so-called transdiode (Patterson diode) logarithmic configuration. The amplifier output terminal is connected to the emitter and provides the driving voltage ($e_o = V_E$).

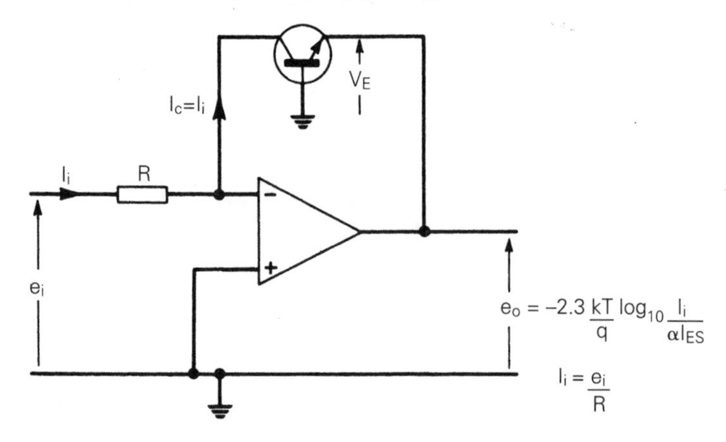

Figure 5.9 *Logarithmic amplifier, transdiode configuration*

The transdiode configuration of Figure 5.9 is capable of the widest range of logarithmic (log) conversion of input current. Accurate log conversion requires that α_F remain constant over a wide range of current values. Silicon planar transistors have this characteristic and can have a range of up to 10 decades. The upper end of the useful current range is determined by semiconductor bulk resistance effects and is usually between 1 mA and 10 mA.

The earthed base used in the transdiode configuration has two disadvantages. It allows only single polarity input signals; the reverse polarity requires the use of a complementary transistor type. Also, the transistor has a frequency dependent gain and, since it is connected inside the feedback loop, this introduces closed-loop stability problems.

An alternative arrangement is illustrated in Figure 5.10. In this circuit the collector and base are connected together and the transistor acts as a diode.

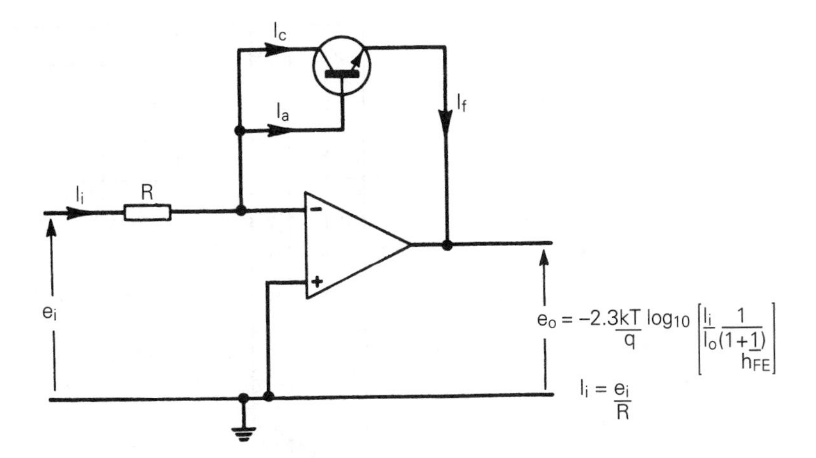

Figure 5.10 *Log amplifier, diode connected transistor*

The circuit in Figure 5.10 is not capable of such a wide range as the trans-diode circuit, but in many respects it is more versatile. Since it is a two-terminal device, its polarity can be reversed to allow a reversed input polarity. Several diodes may, if required, be connected in series for greater output voltages. Since the transistor produces no gain when connected as a diode, closed-loop stability is achieved.

In the diode configuration the feedback current (I_f) is not exactly equal to the collector current (I_C). But,

$$I_f = I_C + I_B$$

$$= I_C(1 + 1/h_{FE})$$

where $I_B = I_C/h_{FE}$ is the base current drawn by the transistor and h_{FE} is the common emitter DC current gain. Equation 5.7 becomes

$$-e_o = -V_E = 2.3\frac{kT}{q}\log\left(\frac{I_i}{\alpha_F I_{ES}}\frac{1}{1+\dfrac{1}{h_{FE}}}\right) \tag{5.8}$$

Transistors with a large value of h_{FE} should be used in order to reduce the error term. The fall in h_{FE}, which occurs at low current levels, sets the lower level of the input current at which the configuration departs significantly from logarithmic accuracy. The logarithmic range obtainable is typically within the range 10^{-3} A to 10^{-9} A.

The curves illustrated in Figure 5.11 show the logarithmic characteristics of diode connected type 2N3707 transistors. The curves show the typical upper and lower limits of logarithmic range.

A third transistor logarithmic configuration, which is sometimes used, is illustrated in Figure 5.12. The most useful feature of this connection is the reduced loading on the op-amp output (only a small base current is required). Disadvantages of the circuit are the lack of reversibility, the separate supply for the collector, and a reduced logarithmic range.

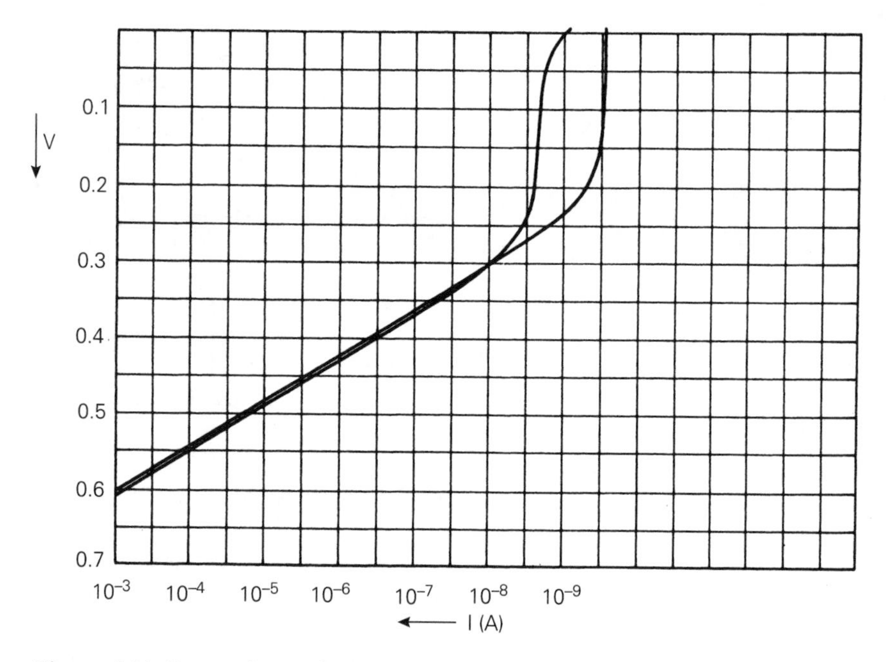

Figure 5.11 *Logarithmic characteristics of diode connected transistors (type 2N3707)*

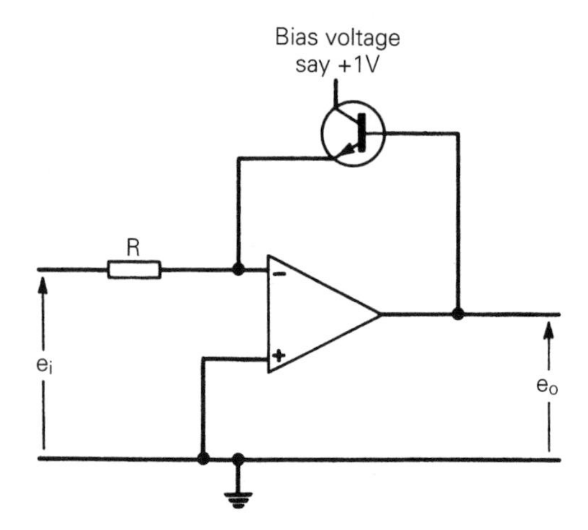

Figure 5.12 *Log amplifier, transistor connection*

The current fed back to the op-amp summing point is the emitter current of the transistor (npn) which is given by equation 5.9.

$$I_E = \alpha_R I_{CS}(e^{-qV_C/kT} - 1) - I_{ES}(e^{-qV_E/kT} - 1) - \sum I_{ESj}(e^{-qV_E/mkT} - 1) \quad (5.9)$$

Where α_R is the reverse current gain of the transistor ($\alpha_R \sim 0.2$).

The collector is usually taken to a reverse bias of order 1 V. This gives $V_c \neq 0$ and the first term in the equation contributes a small error. A more significant error is contributed by the $m_j \neq 1$ components of current represented by the third term of the equation. The useful logarithmic range with this configuration is typically within the range 10^{-5} A to 10^{-8} A.

5.4 Logarithmic amplifiers: practical design considerations

We will now look at practical considerations for logarithmic amplifiers. A more general treatment of practical considerations for op-amp circuits is given in Chapter 9.

The following are some of the more important points requiring attention in a practical logarithmic converter:

(1) The designer must ensure closed-loop stability. The method used to achieve this may affect the output slew rate, so this must be considered.
(2) Offsets must be balanced out if the full capability of the op-amp is to be exploited. The logarithmic range is usually determined by op-amp offsets, rather than by the logarithmic range of the transistor. The relative importance of voltage and current offset is determined by the magnitude of the source resistance (see Chapter 2).
(3) The transistor must be protected against possible damage caused by accidentally applying a reverse polarity voltage.
(4) A means of temperature compensating the logarithmic transistor must be employed, unless the circuit is going to be used in a temperature controlled environment.

5.4.1 Closed-loop stability

Chapter 2 discussed amplifier stability. Stable (non-oscillatory) closed-loop operation requires that the loop gain (βA_{VOL}) should be less than unity at frequencies where the phase shift around the loop reaches 180°. The condition implies that, on a Bode plot, the intersection of $1/\beta$ and A_{VOL} should occur with a rate of closure of less than 40 dB/decade.

In the feedback circuits considered so far we have assumed the feedback fraction β to be determined by purely resistive components. This makes $1/\beta$ real at all frequencies and never less than unity. Under these conditions, an op-amp open-loop response characterized by a 20 dB/decade roll-off, down to unity gain, ensures closed-loop stability for all values of input and feedback resistors.

In practical circuits, the 20 dB/decade roll-off does not always ensure closed-loop stability. Stray capacitance between the op-amp's summing point and earth causes a phase lag in the feedback fraction β at the higher frequencies. This produces a corresponding phase lead in $1/\beta$. Capacitance at the op-amp output can cause an additional phase lag. Both effects can lead to instability.

The problem of stability in logarithmic amplifiers is further complicated by the non-linear nature of the feedback. The feedback is greater, and therefore $1/\beta$ is smaller, at the higher input currents. In examining stability criteria, it is convenient to assume an op-amp with a finite open-loop gain with a 20 dB/decade roll-off down to unity gain. The effects of other departures from the ideal op-amp are initially neglected.

Since the feedback fraction β is dependent on the operating current, we examine stability in terms of a small-signal feedback ratio. The small-signal feedback ratio $\Delta e_f/\Delta e_o$ is assumed to be defined about some DC operating current I_C. Referring to the circuit shown in Figure 5.13, the current fed back to the op-amp summing point (I_f) is equal to the collector current of the transistor (I_C).

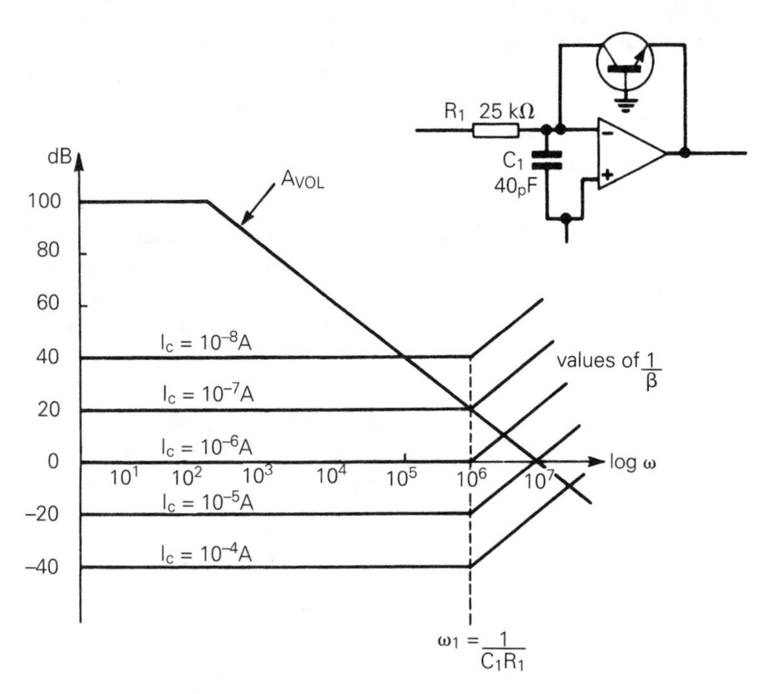

Figure 5.13 *Bode plot for transdiode configuration*

Assuming a predominance of the exponential term in equation 5.6 we may write this equation as

$$I_f = \alpha_F I_{ES} e^{-qe_o/kT} \tag{5.10}$$

Differentiating equation 5.10 with respect to e_o gives the small signal feedback resistance r_E.

$$\text{Thus} \quad \frac{\delta I_f}{\delta e_o} = \frac{-qI_f}{kT} = -\frac{1}{r_E}$$

$$\text{And} \quad r_E = \frac{kT}{qI_f} \approx \frac{1}{40I_f} \tag{5.11}$$

Note that for an operating current of 1 mA, the transistor's intrinsic emitter resistance $r_E = 25\ \Omega$, but when the operating current is, say, 1 nA, $r_E = 25\ \text{M}\Omega$. A change in the output voltage Δe_o results in a change in the feedback current $\Delta I_f = -\Delta e_o/r_E$. This in turn causes a change $-\Delta I_f Z_1$ in the voltage fed back to the op-amp summing point. We may thus write the value of the small-signal feedback ratio as

$$\beta = Z_1/r_E \tag{5.12}$$

Z_1 is the impedance between op-amp summing point and earth. In Figure 5.13, $Z_1 = R_1/(1 + j\omega C_1 R_1)$ where C_1 is the total capacitance between op-amp summing point and earth. C_1 is taken to include the capacitance between the collector and base of the transistor. The shunting effect of the collector output resistance is neglected.

Substituting for Z_1 gives

$$1/\beta = r_E(1 + j\omega C_1 R_1)/R_1 \tag{5.13}$$

Note that, at the higher operating currents, it is possible for $1/\beta$ to be considerably less than unity ($r_E < R_1$). This feature is peculiar to the transdiode configuration; in other feedback circuits the lower limit of $1/\beta$ is unity. Remember that in the transdiode configuration the transistor acts as a common base amplifier for feedback signals and, as such, it can provide a voltage gain which is greater than unity.

Values of $1/\beta$ for different operating currents are shown in Figure 5.13. For the purpose of the discussion, component values are chosen to simplify the arithmetic. The op-amp is assumed to have a unity-gain bandwidth product of $10^7/(2\pi)$ Hz. We see immediately that the circuit fails to satisfy the closed-loop stability criterion for operating currents greater than 1 μA.

One solution to the stability problem is to connect a capacitor C_2 between the op-amp output terminal and the summing point. This capacitor and r_E cause a break in the Bode plot at an angular frequency $\omega_2 = 1/(C_2 r_E)$. This causes attenuation in the value of $1/\beta$. But remember that the value of r_E depends on the level of the operating current.

The magnitude of C_2 required to ensure closed-loop stability at the higher operating currents places a severe restriction on the bandwidth and output slew rate at the lower levels of operating current. For example, to make $\omega_2 = 10^6$ rad/s at an operating current of 1 mA requires a value of C_2 equal to 0.04 μF. This value of C_2 makes $\omega_2 = 1$ rad/s at an operating current of 1 nA, this has a time constant of 1 s.

Another practical difficulty arises because of the finite open-loop output impedance of the op-amp. This inevitably causes a reduction in open-loop gain when the amplifier is used to supply a low value load resistor. At an operating current of 1 mA, the transistor's intrinsic emitter resistance is $r_E = 25\ \Omega$; such a small value is likely to have a marked effect on the gain-bandwidth product of the amplifier.

A remedy is to connect a resistor R_E in series with the emitter of the transistor. In addition to reducing the loading on the op-amp's output, the introduction of R_E allows the use of smaller values for C_2. This in turn gives the system a wider bandwidth and an increased slew rate at the lower levels of input current. The arrangement is illustrated in Figure 5.14.

Closed-loop stability is again conveniently examined in terms of a small-signal value of the feedback fraction β. Referring to Figure 5.14,

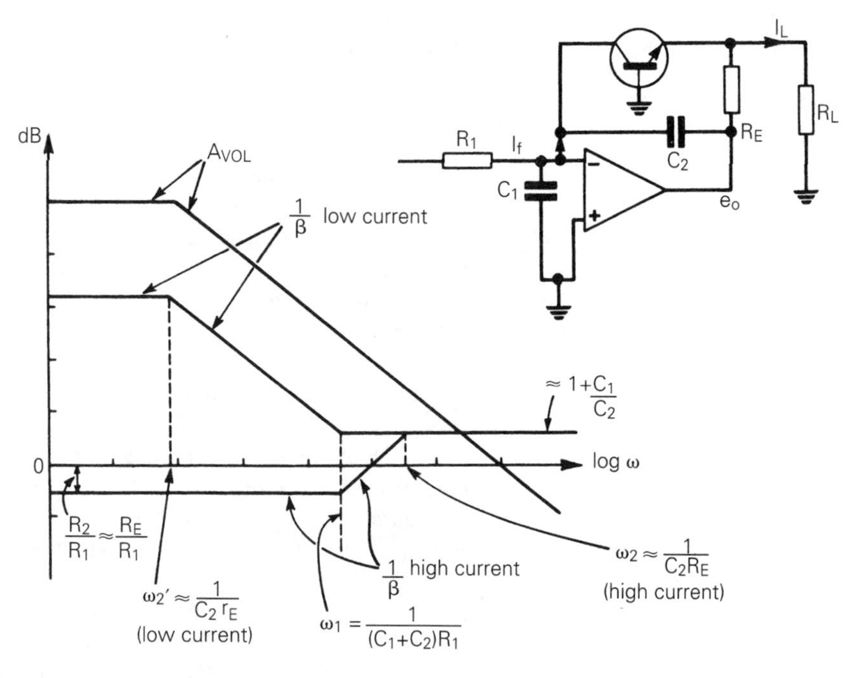

Figure 5.14 *Bode plots for stable closed-loop operation*

$$-\Delta I_f = \frac{\Delta e_o}{R_E + r_E} + j\omega C_2 (\Delta e_o - \Delta e_f)$$

e_f is the feedback voltage developed between the amplifier summing point and earth.

$$\Delta e_f = -\Delta I_f R_1/(1 + j\omega C_1 R_1)$$

The small-signal feedback ratio

$$\beta = \Delta e_f/\Delta e_o$$

Manipulation of the above equations gives

$$\frac{1}{\beta} = \frac{R_2}{R_1}\frac{1 + j\omega (C_1 + C_2) R_1}{1 + j\omega C_2 R_2} \tag{5.14}$$

where $R_2 = R_E + r_E$.

The larger the value used for R_E, the smaller is the value of C_2 required to ensure closed-loop stability at the higher operating currents. The Bode plot breakout frequency for $1/\beta$, $\omega_2 = 1/(C_2R_2)$ (at the higher operating currents) should be made to occur at least an octave before the intercept of $1/\beta$ with A_{VOL}. The maximum value which may be used for R_E is limited by the maximum output voltage swing of the op-amp, bearing in mind that the maximum output voltage across the logarithmic transistor is approximately 0.6 V. Thus R_E should be chosen so that

$$V_{o\,max} - 0.6\text{ V} > (I_{L\,max} + I_{C\,max})R_E \tag{5.15}$$

The Bode plots in Figure 5.14 show values of $1/\beta$ obtained from equation 5.14.

If the transdiode configuration is used for logarithmic scaling of current, $R_1 > \infty$ and equation 5.14 becomes

$$1/\beta \approx R_2 j\omega(C_1 + C_2)/(1 + j\omega C_2 R_2) \tag{5.16}$$

The Bode plots in Figure 5.15 show values of $1/\beta$, given by equation 5.16.

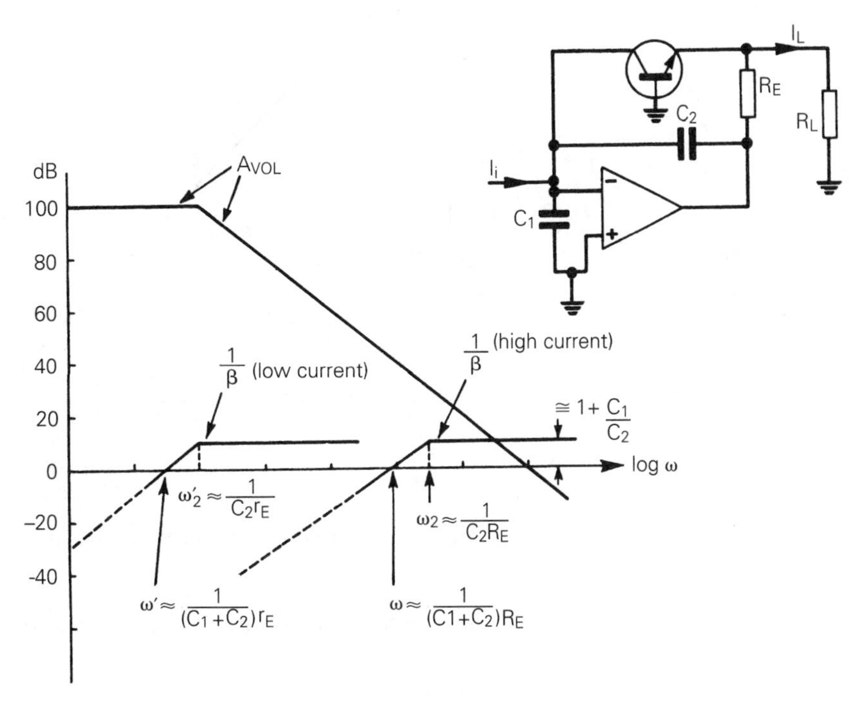

Figure 5.15 *Bode plots, logarithmic current scaling*

In Figures 5.14 and 5.15, the small-signal value of $1/\beta$ tends to the value $1 + C_1/C_2$ at high frequencies. In both cases the breakout frequency for $1/\beta$ at the higher operating currents is

$$\omega_2 \approx 1/(C_2 R_E) \text{ when } R_E >> r_E$$

A suitable choice for C_2R_E ensures closed-loop stability at the higher operating currents. The use of the maximum value of R_E allowed by equation 5.15 permits the smallest value of C_2, and hence gives the fastest slew rate at the low current levels. The low current value of ω_2 is $\omega_2' \approx 1/(C_2 r_E)$.

In the diode configuration shown in Figure 5.16, the gain of the transistor is shorted out which means that $1/\beta$ cannot be less than unity. The circuit used with an op-amp having a 20 dB/decade roll-off may be closed-loop stable without the addition of a stabilizing network. If the simple circuit is not closed-loop stable, it may be stabilized in the same way as the trans-diode circuit. If a transistor's operating current makes its value of r_E less than the rated load of the op-amp, a resistor R_E connected in series with the emitter will be needed.

The diode configuration Bode plots are illustrated in Figure 5.16.

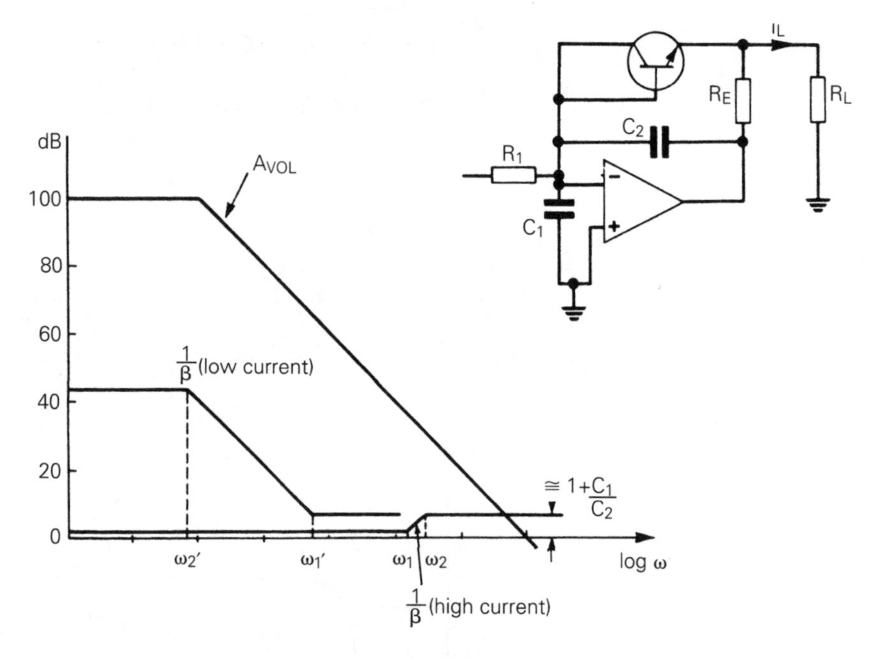

Figure 5.16 *Bode plots, diode configuration*

In the circuit shown the small-signal value of $1/\beta$ is given by the relationship

$$1/\beta = 1 + Z_2/Z_1$$

where $Z_2 = R_2/(1 + j\omega C_2R_2)$, $Z_1 = R_1/(1 + j\omega C_1R_1)$ and $R_2 = R_E + r_E$. Substituting for Z_2 and Z_1 gives

$$\frac{1}{\beta} = \frac{1 + \dfrac{R_2}{R_1} + j\omega\,(C_1 + C_2)\,R_1}{1 + j\omega\,C_2\,R_2} \tag{5.17}$$

At high currents, ($r_E << R_E$), the break frequency in $1/\beta$ is

$$\omega_1 \cong \frac{1 + \dfrac{R_E}{R_1}}{(C_1 + C_2)R_E}$$

and the break out frequency is

$$\omega_2 \approx 1/(C_2R_E)$$

At low current

$$\omega_1' \approx 1/[(C_1 + C_2)R_1] \text{ and } \omega_2' \approx 1/(C_2 r_E)$$

Stability considerations in the transistor configuration of Figure 5.12 are similar to those encountered in the diode configuration, although the problem of the loading of the op-amp output does not arise. We will not consider this connection in any detail.

5.4.2 Offset errors

The lower limit to the range of a logarithmic converter is, in many DC applications, determined by op-amp offsets rather than by the logarithmic range of the transistor. In the circuit for the transdiode configuration illustrated in Figure 5.17, we represent op-amp offsets in terms of equivalent input generators. The op-amp is assumed to have infinite open-loop gain.

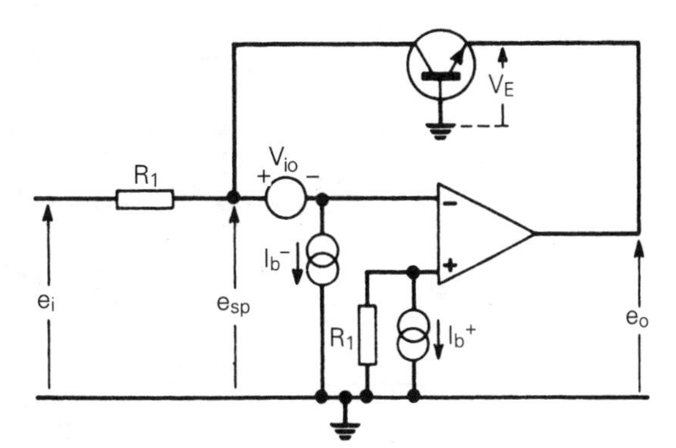

Figure 5.17 *Offset errors, transdiode configuration*

Referring to Figure 5.17 we see that the voltage at the summing point of the amplifier

$$e_{sp} = V_{io} - I_b^+ R_1$$

A finite voltage at the summing point makes $V_C \neq 0$ giving the possibility of a logarithmic error through the I_{CS} terms of equation 5.5. Clearly any appreciable forward collector bias (V_C negative in the case of an npn transistor) must be avoided. Dependent on the magnitude of the bias current I_b^+, it may be advisable to omit the bias current compensating resistor and return the non-inverting input of the op-amp directly to earth. Reverse collector bias (V_C positive in the case of an npn transistor) can only contribute a small error since $I_{CS} < 1$ pA.

Assuming that the I_{CS} terms in equation 5.5 can be neglected, we may use equation 5.7. Thus

$$V_E = e_o = -E_o \log\left(\frac{I_C}{I_o}\right)$$

Where we write

$$E_o = 2.3kT/q \text{ and } I_o = \alpha_F I_{ES}$$

In Figure 5.17,

$$I_C = \frac{e_t - e_{sp}}{R_1} - I_b^- = \frac{e_i - V_{io}}{R_1} - I_{io}$$

$$\text{Therefore, } e_o = -E_o \log\left(\frac{\frac{e_i - V_{io}}{R_1} - I_{io}}{I_o}\right) \tag{5.18}$$

We may use equation 5.18 to estimate the offset error for the transdiode configuration. If the bias current compensating resistor is omitted from the circuit we replace I_{io} in the equation by I_b.

A similar analysis may be carried out for the diode configuration illustrated in Figure 5.18.

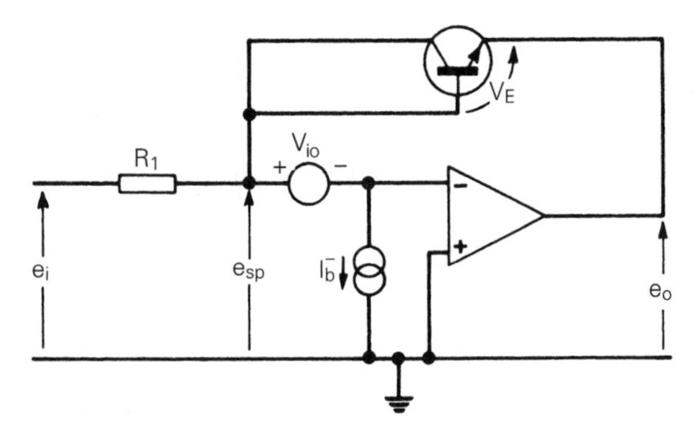

Figure 5.18 *Offset errors, diode configuration*

In this circuit,

$$V_E = e_o - e_{sp} = e_o - V_{io}$$

A bias current compensating resistor does not balance out bias currents in this configuration (because of its effect on e_{sp}) and we return the non-inverting input of the op-amp directly to earth.

$$\text{Now } I_C = \frac{e_i - V_{io}}{R_1} - I_b^-$$

Making use of equation 5.8 and neglecting the $1/h_{FE}$ term gives

$$e_o = V_{io} - E_o \log\left(\frac{\frac{e_i - V_{io}}{R_1} - I_b^-}{I_o}\right) \tag{5.19}$$

5.4.3 Balancing offsets

Initial op-amp offsets may be balanced out; errors are then due to offset drifts. Separate biasing of voltage and current offsets reduces errors and gives maximum logarithmic range. Voltage bias should be applied to the non-inverting input of the op-amp. Current bias should be applied to the inverting input of the op-amp. Figure 5.19 illustrates a typical biasing arrangement.

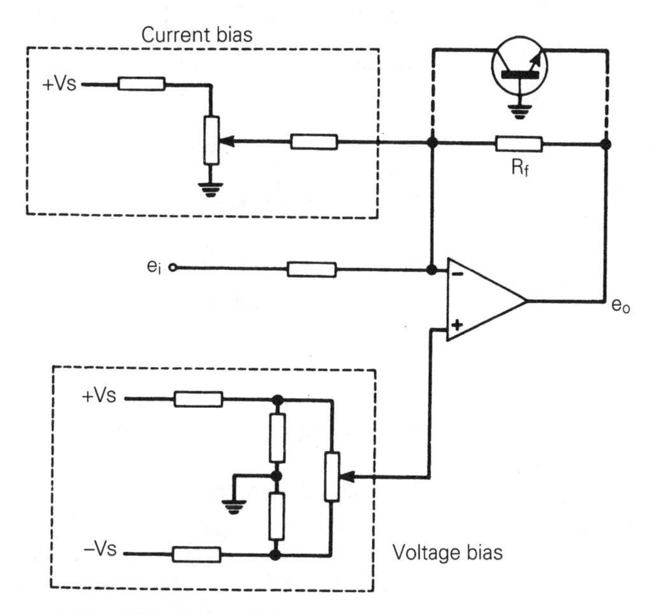

Figure 5.19 *Offset balancing*

When balancing offsets, a high value resistor (R_f) in the feedback path should replace the logarithmic element. A resistor of the same order of magnitude as the highest value of r_E to be encountered should be used. Input offset voltage is balanced by first shorting the inverting input of the op-amp to earth and then, after removing the short, adjusting the bias current to zero the amplifier output.

In practice it is advisable to inject a bias current slightly larger than I_b^- to ensure that the collector current is not zero when the input is zero (see slew rate considerations in previous section). A collector current equal to say 1 per cent of the smallest input current to be measured should be suitable. Component values used in the biasing networks should be chosen to allow balancing of the maximum specified values of V_{io} and I_b^-. Once the adjustments have been made the feedback resistor R_f is replaced by the logarithmic element.

5.4.4 Circuit protection

A small input signal of the wrong polarity applied to a logarithmic circuit can cause a large reverse emitter bias, with possible destruction of the transistor.

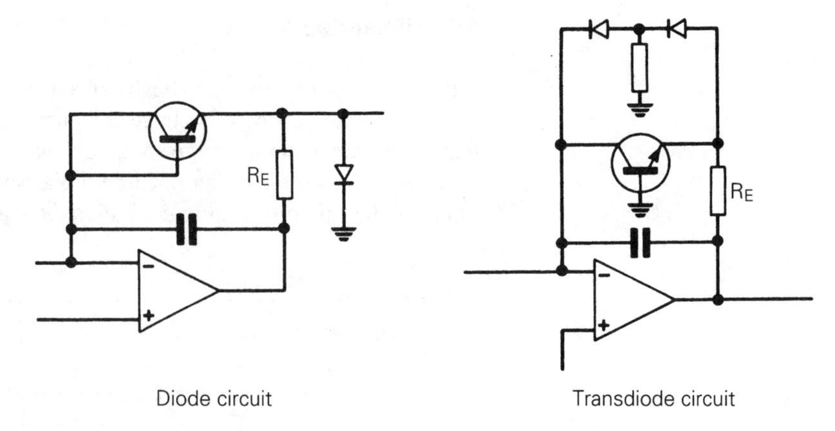

Figure 5.20 *Protection against inverse polarity*

It is advisable to provide logarithmic elements with protection against excessive inverse voltage. Examples of such protective circuits are illustrated in Figure 5.20.

5.4.5 Temperature compensation

Transistor logarithmic elements have an inherent temperature dependence, which makes a single transistor log converter inaccurate unless the temperature is kept very constant. The main effect is due to the variation with temperature of the term $I_o = \alpha_F I_{ES}$, this approximately doubles for every 10 degrees Celsius change in temperature. A less significant effect is due to the linear temperature dependence of the multiplying factor

$$E_o = 2.3kT/q$$

The slope of the logarithmic characteristic changes with temperature, by 0.3 per cent per degree Celsius in the vicinity of 27°C.

The use of matched transistors enables cancellation of the I_o terms. Consider two transistors with saturation currents I_{o_1} and I_{o_2}. We may write

$$V_{E1} = -E_o \log\left(\frac{I_{C_1}}{I_{o_1}}\right)$$

And $V_{E_2} = -E_o \log\left(\frac{I_{C_2}}{I_{o_2}}\right)$

This gives $V_{E_2} - V_{E_1} = E_o \log\left(\frac{I_{C_1}}{I_{o_1}}\frac{I_{o_2}}{I_{C_2}}\right)$

Matched transistors make $I_{o_1} = I_{o_2}$ and

$$V_{E_2} - V_{E_1} = E_o \log\left(\frac{I_{C_1}}{I_{C_2}}\right) \qquad (5.20)$$

Thus, a circuit using a matched pair of transistors performs a subtraction operation (by effectively taking the logarithm of a current ratio). This replaces the uncontrollable I_o term with a fixed adjustable reference current I_{C2}. Even if the transistors are not perfectly matched it is generally found that, for transistors of the same type, the ratio I_{o_1}/I_{o_2} remains fairly constant with change in temperature.

The linear temperature dependence of the scaling factor E_o can be compensated by using an op-amp with a temperature sensitive feedback divider (see Figure 5.6 and equation 5.3). The scaling factor of $E_o = 60$ mV at 27°C is somewhat inconvenient. This system has gain that can be used to give a more convenient scaling factor; 1 V per decade of current change is normally preferred.

5.5 Some practical log and antilog circuit configurations

Circuits suitable as a basis for implementing practical log and antilog applications are now discussed. Because of the temperature dependence of transistors, circuits that use a single transistor are only suitable for non-critical applications and then only when ambient temperature variations are small. Most practical circuits employ a pair of matched transistors and achieve temperature compensation using the method outlined in the previous section.

Silicon planar transistors exhibit a logarithmic characteristic, but those designed for use at low values of collector current should be chosen if a wide logarithmic range is required. Care should be taken to ensure that the transistors are maintained at the same temperature; the use of dual transistors ensures matching and thermal tracking.

In some circuits, a temperature sensitive resistor is used to compensate for the temperature dependence of the scaling factor. This needs to be maintained at the same temperature as the transistor.

5.5.1 Temperature compensated log of voltage and current

The circuit for a temperature compensated log converter is shown in Figure 5.21. With care in design and construction, the circuit may be expected to provide logging of positive input voltages (use pnp transistors for negative input signals) in the range 10 mV to 10 V. The accuracy over the whole range may be in the order of 3 per cent, referred to the input, for temperature changes of ±10°C. Errors are greatest at the lower end of the logarithmic range. The use of low drift op-amps improves accuracy and extends the lower limit of the useful logarithmic range.

The circuit uses two op-amps and two transistors. The output of amplifier A_1, attenuated by the resistive divider R_3 and R_4, provides the emitter base differential voltage between transistors T_1 and T_2. Neglecting the base current loading imposed by transistor T_2, the following relationship holds:

$$V_{E_1} - V_{E_2} = e_o \frac{R_3}{R_3 + R_4} \tag{5.21}$$

Negative feedback around amplifier A_2 forces V_{E_2} to take on that value which causes the collector current $I_{C_2} = I_{ref}$ to flow in transistor T_2. The collector

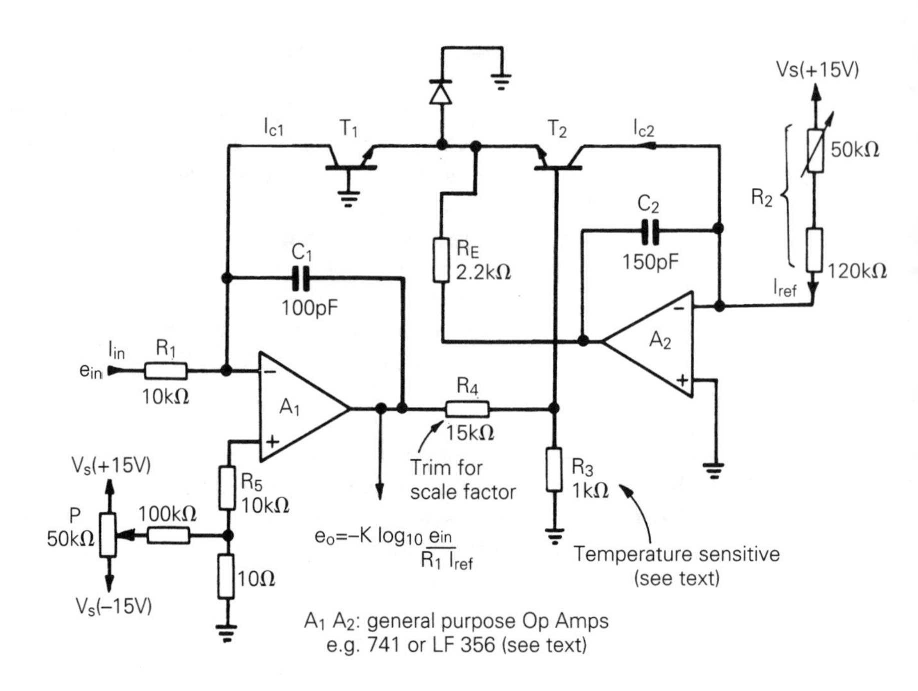

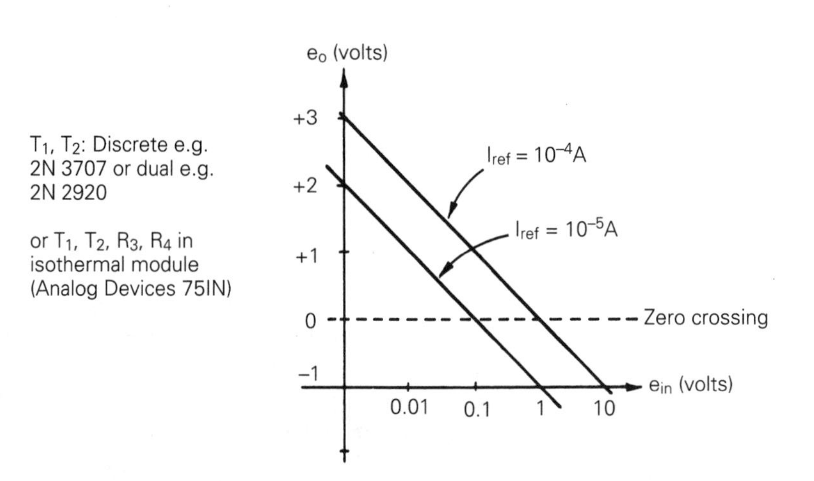

Figure 5.21 *Temperature compensated log of voltage converter*

current $I_{C_1} = I_{in}$ flows through transistor T_1 because of V_{E_1} imposed by negative feedback around amplifier A_1.

Substituting V_E values from equation 5.7 into equation 5.21 and rearranging gives the following circuit performance equation

$$e_o = -\frac{R_3 + R_4}{R_3} 2.3 \frac{kT}{q} \log \left(\frac{I_{in}}{I_{ref}} \frac{I_{o_2}}{I_{o_1}} \right) \tag{5.22}$$

where $I_{ref} = V_s/R_2$ and $I_{in} = e_{in}/R_1$ in a logarithmic voltage circuit.

In a logarithmic current circuit I_{in} is supplied directly to the inverting input of amplifier A_1. There are different considerations regarding the offset balancing of voltage and current circuits, which will be discussed shortly.

Output zero crossing

Readers who are unfamiliar with log amplifiers should carefully consider the significance of equation 5.22. Zero input signal does not give a zero output (log(0) is not defined). The output signal given by the circuit of Figure 5.21 is proportional to the log of the current ratio $I_{in}/I_{ref} \times I_{o_2}/I_{o_1}$. Since $\log(1) = 0$, zero output occurs if $I_{o_1} = I_{o_2}$ and $I_{in} = I_{ref}$.

The point at which the input voltage causes zero crossing at the output is under the control of I_{ref}. In Figure 5.21 $I_{ref} = V_s/R_2$ and zero crossing of the output may be adjusted by choice of R_2.

Wide range logarithmic applications require a very low drift op-amp for A_1. An amplifier of more modest drift performance can be used for A_2 by making I_{ref} greater than the smallest value of I_{in}. If $I_{in} << I_{ref}$ the base drive to T_2 forward biases its collector base junction, causing an effective change in the reference current. The extent to which this contributes an appreciable error is dependent upon the I_{CS} terms in equation 5.5 and upon the value of I_{ref}.

For example: Assume $I_{in} = 1$ nA, $I_{ref} = 100$ μA, $I_{CS} = 1$ pA and $R_3/(R_3 + R_4) = 1/16$. There are five decades difference in I_{in} and I_{ref}, the output voltage is +5 V and a bias 5/16 V is applied to the collector base junction of T_2. If we take only the $m = 1$ term of the collector current equation (equation 5.5)

$$I_{CS}\,[e^{-qV_C/kT} - 1] \cong 10^{-12}\, e^{5/16 \times 0.025}$$

$$\approx 10^{-12}\; 2.6 \times 10^{5}$$

Expressed as a percentage of I_{ref} this represents a 0.26 per cent error. If the expected operating conditions are more extreme, the collector base voltage of T_2 can be held near zero by returning the non-inverting input of op-amp A_2 to the base of transistor T_2, instead of to earth. A current source is then required to supply the reference current I_{ref} instead of the resistor R_2.

Scaling factor

The scaling factor for the circuit

$$K = 2.3\,\frac{kT}{q}\,\frac{R_3 + R_4}{R_3} \tag{5.23}$$

may be set at any convenient value by choice of resistor values R_3 and R_4. A scaling factor of unity is often convenient; it corresponds to a 1 V change in output for each decade change in input current. Substitution of constants gives

$$K = 59 \times 10^{-3}\,\frac{R_3 + R_4}{R_3} \quad \text{at } 25°\text{C}$$

If $R_3 = 1\ \text{k}\Omega$ a value of $R_4 = 15.9\ \text{k}\Omega$ is required to make the scaling factor unity. Trimming the value of R_4 provides a convenient method for adjusting the scaling factor. The scaling factor varies by 0.33 per cent per degree Celsius. If such a variation is not tolerable the scale factor may be compensated by using a temperature sensitive resistor for R_3 with resistance varying directly with temperature. This resistor must be kept at the same temperature as the transistors.

Logarithmic voltage

Amplifier input offset voltage and bias current give rise to an equivalent input offset voltage:

$$E_{OS} = \pm V_{io} + I_b^- R_S^- - I_b^+ R_S^+$$

In a logarithmic voltage circuit, bias current can be compensated for temperature by making $R_S^- = R_S^+$ (the function of resistor R_5 in Figure 5.21), where R_S^- is effectively equal to R_1 plus the resistance of the signal source. E_{OS} acts as a signal in series with the input signal source and the output voltage given by the circuit may be expressed as

$$e_o = -K \log \left(\frac{e_{in} \pm E_{OS}}{I_{ref} R_1} \right) \tag{5.24}$$

Consideration of the implications of equation 5.24 may be used to arrive at a convenient adjustment procedure for practically trimming offsets. E_{OS} could quite easily be trimmed to within 100 μV of zero (or closer for low drift op-amps). In equation 5.24 substitute $e_{in} = 0$, $E_{OS} = 10^{-4}$, $R_1 = 10\ \text{k}\Omega$, $I_{ref} = 10^{-4}$, $K = 1$. This gives

$$e_o = -1 \log(10^{-4}) = +4\ \text{V}$$

The initial offset of amplifier A_1 can thus be adjusted to within 100 μV of zero by setting e_{in} to zero and adjusting the trim potentiometer P for an output of +4 V. Subsequent temperature drift of E_{OS} may degrade accuracy.

Logarithmic current

In logarithmic current circuits, input current is supplied directly to the inverting input of op-amp A_1. It is inadvisable to use a bias current compensating resistor at the non-inverting input of the op-amp, because current sources usually have high impedance. The equivalent input offset voltage is

$$E_{OS} = \pm V_{io} + I_b^- R_S$$

where R_S is the output resistance of the signal source.

The input error is more usefully expressed as an equivalent input offset current by dividing the equation by R_S. Thus

$$I_{OS} = E_{OS}/R_S = \pm V_{io}/R_S + I_b^-$$

The effect of op-amp input offset voltage V_{io} is negligible for large values of R_S. Accuracy at the lower levels of the input current range depends upon the op-amp bias current.

In logarithmic current circuits, which have low source impedance, it may be necessary to balance out the initial value of V_{io}. An adjustable voltage bias may be applied to the non-inverting input of the op-amp. When making the adjustment, the inverting input should be connected to earth through a 10 kΩ resistor and the procedure for offset trimming outlined in the previous section for a logarithmic voltage circuit should be carried out.

The use of low offset drift FET input op-amps in the circuit of Figure 5.21 makes it suitable for both logarithmic voltage and current conversion. A circuit to produce the logarithm of very small input currents requires low bias current op-amps.

Closed loop stability and dynamic response

The dynamic response of a log amplifier is directly determined by the components used to achieve closed-loop stability (C_1, C_2 and R_E). Log amplifiers have a non-linear feedback path; the small-signal feedback fraction varies with the level of the input signal. This makes the transient response and small-signal bandwidth dependent upon the signal level. The effect can be investigated in the circuit of Figure 5.21 by superimposing an input signal variation on top of a DC bias, using the test arrangement shown in Figure 5.22.

The results shown in this figure were obtained by adjusting the amplitude of the input square wave and the value of the DC bias. The adjustments produced an output step covering the various levels of its full range. Note that the response time for a 1 V step at the output depends upon the input signal level. The response time for increasing input signals is less than that for decreasing signals (note that the log converter is inverting, so the 10 V input produces −1 V output).

The time taken for the output to slew through its full output range is dominated by the time taken to slew through the range corresponding to the smallest decade of the input signal (1 mV to 10 mV). Measured response times for the circuit obtained by observing the output steps with expanded time scale are as follows:

e_{in}	e_o	Input increasing	Input decreasing
1 V to 10 V	0 to −1 V	lightly damped	
100 mV to 1 V	0 to +1 V	22 μs	92 μs
10 mV to 100 mV	+1 V to +2 V	230 μs	1 ms
1 mV to 10 mV	+2 V to +3 V	2.2 ms	9.6 ms

The test arrangement of Figure 5.22 can be used to measure the small-signal 3 dB bandwidth of the circuit for sinusoidal signals. Log amplifiers accept only single polarity input signals; the sinusoidal signal must be superimposed upon a steady DC bias. The small-signal bandwidth, like the transient

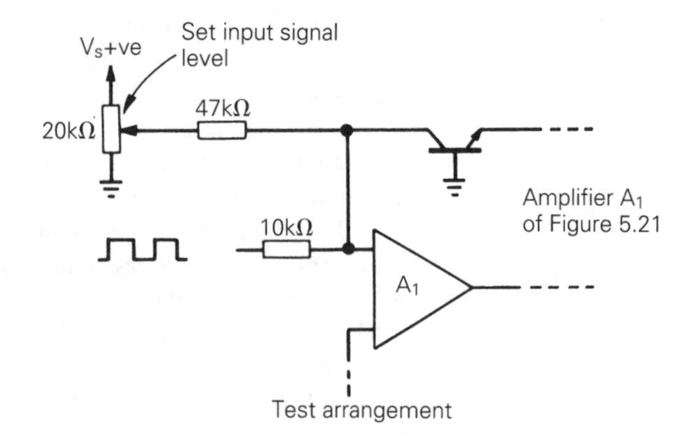

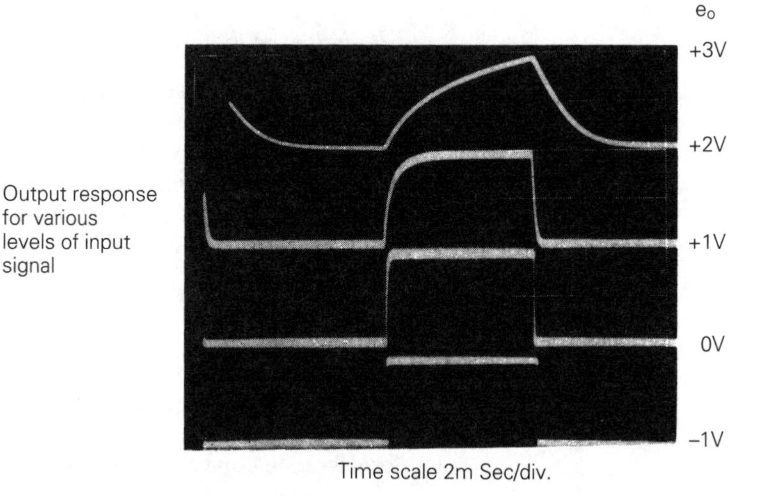

Figure 5.22 *Testing the dynamic response of a log converter*

response time, depends upon input signal level. The output signal is of course non-sinusoidal.

Adequate phase margin for op-amp A_2 is ensured by choosing C_2 and R_E so that the frequency f_{C_2} occurs at least an octave before the loop gain becomes unity. The frequency f_{C_2} is given by:

$$f_{C_2} = \frac{1}{2\pi C_2 [R_E + r_{E_2}]}$$

The value of r_{E_2} depends upon the value of the fixed reference current I_{ref} (see Section 5.4.1). The closed-loop small-signal 3 dB bandwidth of op-amp A_2 is equal to the frequency f_{C_2}.

Now consider the feedback loop around op-amp A_1. At frequencies much less than f_{C_2}, op-amp A_2 holds V_{E_2} constant at a value determined by the reference current. Changes in the output signal of op-amp A_1, attenuated by the dividers R_3 and R_4, are in effect applied directly to the emitter of T_1. The effective feedback resistance around op-amp A_1 is thus:

$$r_{eff} = \frac{R_3 + R_4}{R_3} r_{E_1}$$

But at frequencies approaching f_{C_2}, reduction in the gain of op-amp A_2 causes an effective increase in the feedback path impedance and introduces a phase lag. Critical damping of the response of op-amp A_1, for input signals at the upper end of the range, requires careful choice of the lead capacitor, C_1. Capacitor C_1 should be chosen so that the break frequency $f_{C_1} = 1/(2\pi C_1 r_{ref})$ occurs well below f_{C_2}.

In practice C_1 is often made less than that required for critical damping at the upper end of the input range. The lightly damped response at the higher input signal levels is accepted, in order that the response time at the lower levels of the input signal should not be excessive. Using amplifiers with higher unity-gain bandwidth can decrease response times; this allows the use of smaller frequency compensating capacitors.

5.5.2 Temperature compensated antilog converter

Circuitry of the type used in the temperature compensated log converter of Figure 5.21 can be rearranged to give a circuit that will perform the antilog conversion. Such a circuit is shown in Figure 5.23.

The input signal to the circuit, attenuated by the resistive dividers R_3 and R_4, provides the emitter base differential voltage between transistors T_1 and T_2 and

$$e_{in} \frac{R_3}{R_3 + R_4} = V_{E_2} - V_{E_1}$$

Negative feedback around op-amp A_1 forces V_{E_1} to take on that value which will cause the current $I_1 = I_{ref}$ to flow as a collector current in transistor T_1. If I_1 is held constant, V_{E_1} is constant and V_{E_2} varies directly with the input signal. The voltage V_{E_1} determines the collector current that flows in transistor T_2. Negative feedback around op-amp A_2 forces this current to flow through resistor R_2. Op-amp A_2 develops an output voltage

$$e_o = I_2 R_2$$

Substitution of V_E values from equation 5.7 gives

$$e_{in} \frac{R_3}{R_3 + R_4} = \frac{kT}{q} \ln\left(\frac{I_1}{I_2} \frac{I_{o_2}}{I_{o_1}}\right) \tag{5.25}$$

where $I_1 = I_{ref} = V_S/R_1$ and $I_2 = e_o/R_2$.

Antilog and rearrangement of equation 5.25 gives

$$e_o = R_2 I_{ref} \frac{I_{o_2}}{I_{o_1}} e^{-e_{in} R_3/R_3 + R_4 \, q/kT} \tag{5.26}$$

The equation can be interpreted in terms of any base other than the exponential by the use of the mathematical identity

$$a^x = b^{x \log_b a}$$

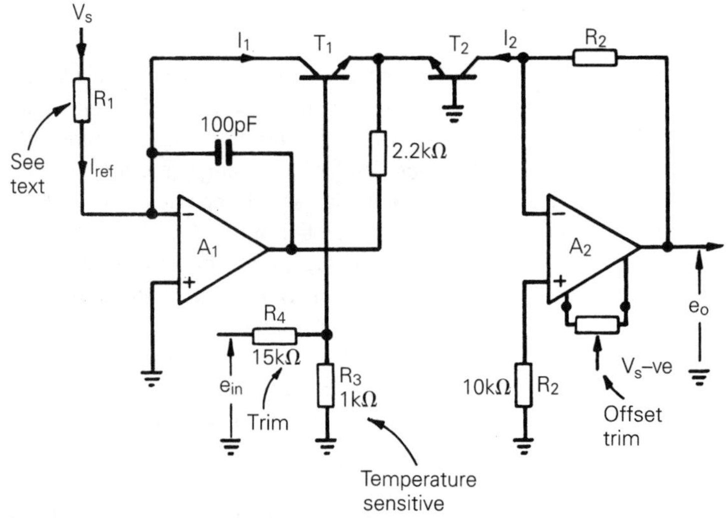

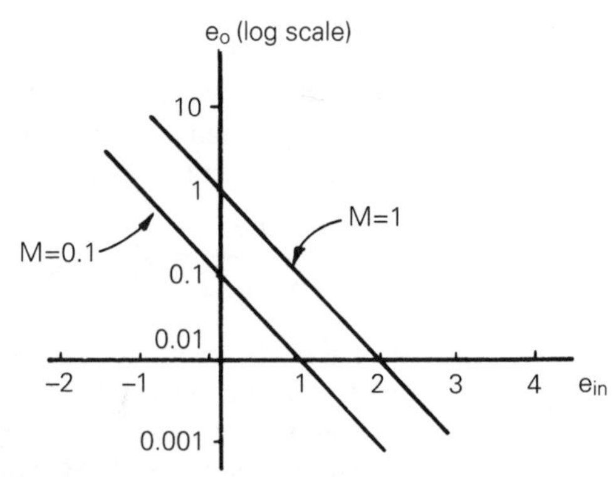

Figure 5.23 *Temperature compensated antilog converter*

Expressed in terms of the normal base 10 the circuit performance equation becomes

$$e_o = R_2 I_{ref} \frac{I_{o_2}}{I_{o_1}} 10^{-e_{in}/K} \tag{5.27}$$

$$\text{where } K = 2.3 \frac{kT}{q} \frac{R_3 + R_4}{R_3}$$

Adjustment procedure

1. Balance A_2 offset. Set e_{in} sufficiently positive to cut off transistor T_2 completely (say $e_{in} = +5$ V) and adjust the offset trim on A_2 for zero output.
2. Trim multiplying constant. The multiplying constant, $M = I_{ref}R_2(I_{o_2}/I_{o_1})$, may be set at any convenient value which allows output signals within the capability of the op-amp A_2. Set e_{in} at zero and adjust I_{ref} (by adjusting

the value of resistor R_1) to make the output voltage of A_2 exactly equal to the value of the desired multiplier factor M.

3. Trim value of base. Apply an input signal of -1 V and trim R_4 to make the output of op-amp A_2 exactly bM volts ($10M$ for base 10). Response curves for the antilog converter are shown in Figure 5.25 plotted in terms of log e_{out}/e_{in}; note that e_{in} can be positive or negative but the output is always single polarity. If negative output signals are desired, pnp transistors should be used as the logarithmic elements.

5.6 Log–antilog circuits for computation

Log and antilog converters can be combined to generate a variety of both linear and non-linear functions. The circuits are interconnected in such a way that they perform operations normally involved in logarithmic computations. Examples of such computations are:

$$\text{antilog}[n \log(x)] = x^n \tag{5.28}$$

$$\text{antilog}[\log(x) + \log(y) - \log(z)] = xy/z \tag{5.29}$$

5.6.1 A log–antilog true RMS-to-DC converter

The circuit configuration shown in Figure 5.24 gives a DC output signal proportional to the true RMS value of an input signal. This signal may have a complex alternating waveform or an alternating wave superimposed upon a DC level. The circuit consists of a precise rectifier (see Section 8.10) that is used to provide unidirectional signals to a following log–antilog computing circuit.

The performance equation can be derived, by summing emitter voltages and using the basic transistor log relationship. Thus, starting at the base of T_1 and ending at the base of T_4 we have

$$V_{E_1} + V_{E_2} - V_{E_3} - V_{E_4} = 0$$

Substituting for V_E values and cancelling out the temperature dependent scaling factor after antilog conversion gives:

$$\frac{I_{C_1} I_{C_2} I_{o_3} I_{o_4}}{I_{C_3} I_{C_4} I_{o_1} I_{o_2}} = 1 \tag{5.30}$$

Neglecting transistor base currents

$$I_{C_1} = I_{C_2} = e_{in}/R_1$$

Amplifier A_3 performs a running averaging of the current I_{C_3}. Provided that the averaging time constant ($C \times R_2$) is considerably longer than the period of any alternating input signal, the output of A_3 is a steady voltage proportional to the average value of I_{C_3}. We write:

$e_o = \overline{I_{C_3}} R_2$, where $\overline{I_{C_3}}$ represents the average value of the current I_{C_3}.

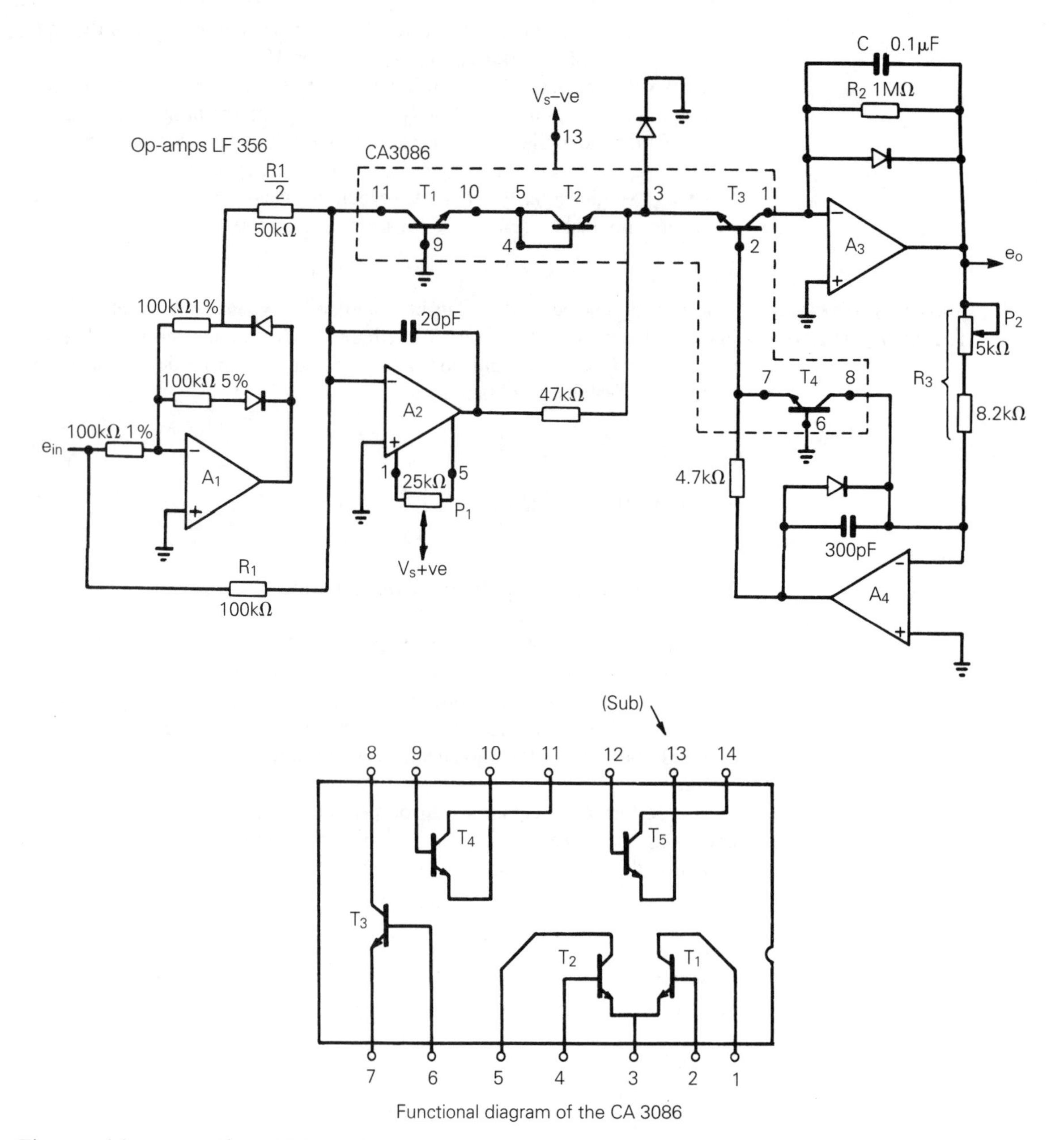

Figure 5.24 *Log–antilog RMS-to-DC converter*

Op-amp A_4 forces the relationship:

$$I_{C_4} = e_o/R_3 = I_{C_3}R_2/R_3$$

Substitution of current values in equation 5.30 gives

$$I_{C_3}I_{C_3} = \frac{R_3}{R_1^2 R_2} \frac{I_{o_3} I_{o_4}}{I_{o_1} I_{o_2}} e_{in}^2$$

$$\text{or } I_{C_3}^2 = \frac{R_3}{R_1^2 R_2} \frac{I_{o_3} I_{o_4}}{I_{o_1} I_{o_2}} \overline{e_{in}^2}$$

$$\text{Thus: } I_{C_3}R_2 = e_o = \sqrt{\frac{R_2 R_3}{R_1{}^2} \frac{I_{o_3} I_{o_4}}{I_{o_1} I_{o_2}} \overline{e_{in}^2}} \tag{5.31}$$

Mismatch in the I_o terms can be balanced, and the scaling factor can be set to unity, by adjustment of the resistor R_3.

Practical points

In the circuit shown, the four transistors are conveniently provided by a transistor array (such as CA3086). Note that one of the transistors in the array is unused and the substrate pin 13 is connected to the −15 V supply rail. Four separate transistors or two dual transistors could be used as an alternative to the transistor array provided that they are maintained at the same temperature. FET input op-amps with offset adjustment points could be used to allow the use of an offset balancing potentiometer.

Setting-up procedure

1. Set e_{in} = 0 V and adjust potentiometer P_1 to make e_o approximately 10 mV.
2. Make e_{in} = +10 V; adjust potentiometer P_2 to make the output read +10 V.

Errors no greater than 1 per cent of full scale (full-scale input equals 10 V peak) are achievable. Accuracy at the lower level of input signals could be improved by separately balancing the offsets of all op-amps.

5.7 A variable transconductance four quadrant multiplier

Log–antilog multiplier circuits allow only single polarity signals. Multipliers designed for four quadrant operation make use of alternative methods to obtain the multiplier operation. A commonly used technique is variable transconductance. Four quadrant multiplier integrated circuits are used in many signal processing applications.

Four quadrant variable transconductance multipliers do not provide the accuracy at low signal levels that log–antilog multipliers give, but they allow operation with alternating signals and provide greater speed and bandwidth than log multipliers.

As an alternative to buying a ready built transconductance multiplier it is possible to build a useful general-purpose four quadrant multiplier out of an op-amp, a five transistor array and a few resistors. The circuit shown in Figure 5.25 is based upon an offset-linearized, two quadrant multiplier cell.

The action of the circuit in Figure 5.25 can be understood in terms of the basic model shown in Figure 5.26(a) and (b).

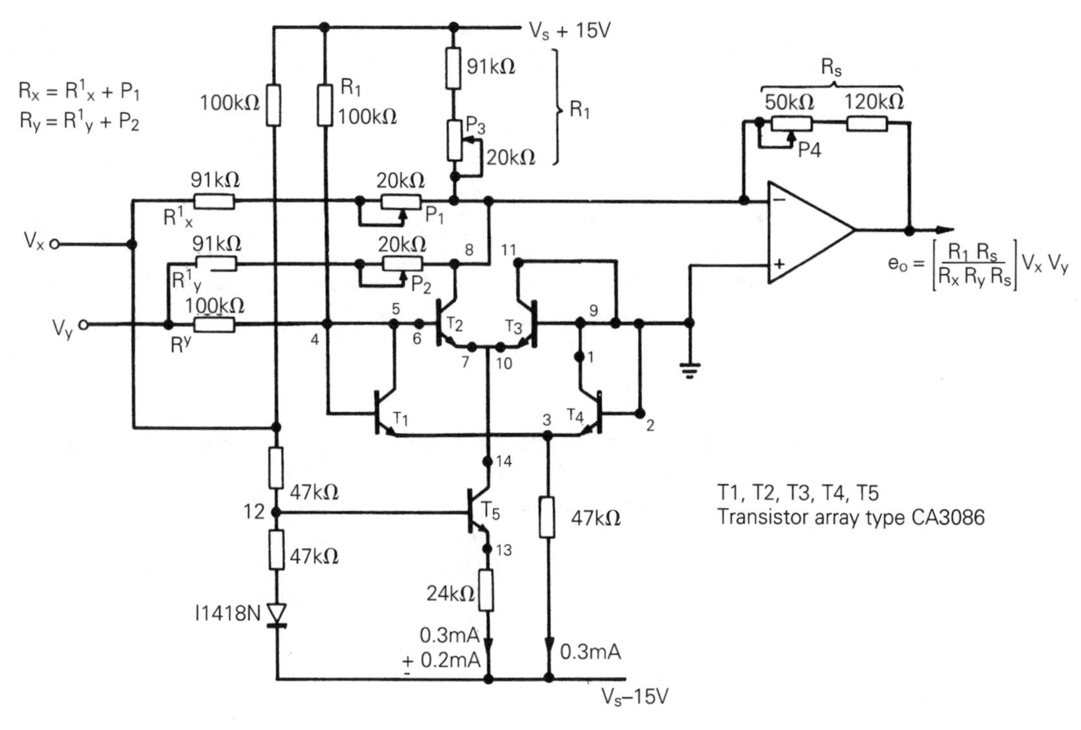

Figure 5.25 *Variable transconductance four-quadrant multiplier*

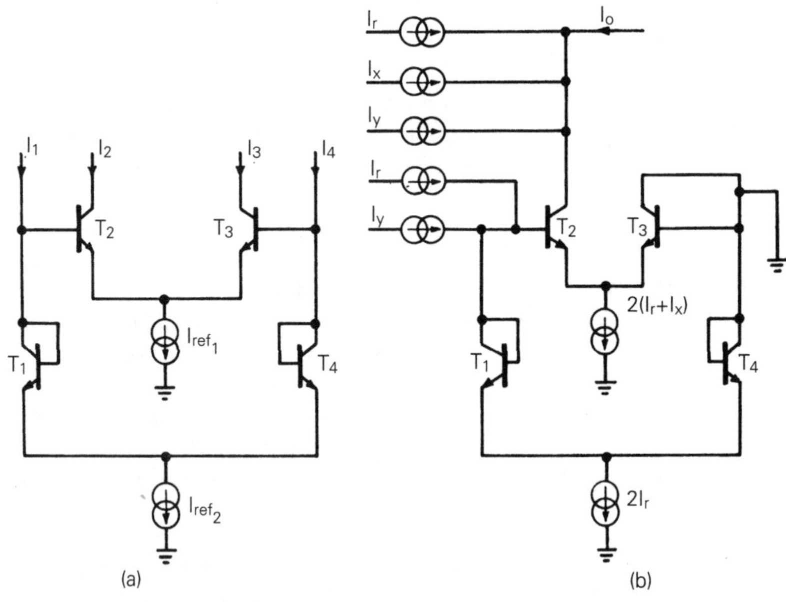

Figure 5.26 *Models of transconductance multiplier cell. (a) Basic linearized transconductance cell. (b) Model of offset transconductance cell*

Current relationships which must hold for the circuit in Figure 5.26(a) can be derived by summing emitter voltages. Starting at the emitter of T_1 and ending at the emitter of T_4, the emitter base voltages must sum to zero, thus:

$$V_{E_1} - V_{E_2} + V_{E_3} - V_{E_4} = 0$$

Analysis requires a few assumptions. Transistors T_2 and T_3 can be assumed matched, with negligible base currents. The transistor's collector and base voltages can be assumed to be zero. Now we can make use of the basic log relationship between collector current and emitter voltage (equation 5.7). After cancelling out the temperature dependent scaling factor and antilogging the relationship between the currents in the multiplier cell we obtain the relationship:

$$I_1 I_3 = I_2 I_4 \tag{5.32}$$

In Figure 5.25 currents are supplied by input signal voltages V_x and V_y as shown by the simplified model in Figure 5.26(b). The tail reference current of the pair T_2 and T_3 is made to vary with V_x. Note that $I_1 = I_r + I_y$, $I_4 = I_r - I_y$ and $I_3 = 2(I_r + I_x) - I_2$. Substituting values in equation 5.31 and rearranging gives

$$I_2 = I_r + I_x + I_y + I_x I_y / I_r$$

Also we have

$$I_2 = I_r + I_x + I_y + I_o$$

The output current is converted to an output voltage by the op-amp:

$$I_o = I_x I_y / I_r$$

$$\text{Now } I_x = V_x/R_x,\ I_y = V_y/R_y,\ I_r = V_S/R_1$$

and the output voltage is determined by the relationship

$$I_o R_S = e_o = \left[\frac{R_1 R_S}{R_X R_Y V_S}\right] V_X V_Y$$

Component values are selected to give a scaling factor of 1/10. Input signals V_x and V_y may be of either polarity; with $V_x = V_y = 10$ V the multiplier gives its full-scale output of 10 V. The bandwidth and the output slew rate of the multiplier are determined by the dynamic response characteristics of the op-amp that are used as the output current-to-voltage converter.

Setting-up procedure

1. *x* and *y* offsets. Earth V_y making $V_y = 0$. Apply a 20 V peak-to-peak 100 Hz input to V_x and adjust P_1 for minimum AC output. Make $V_x = 0$, apply a 20 V peak-to-peak 100 Hz input to V_y and adjust P_2 for minimum AC output.

2. Output offset. Make $V_x = V_y = 0$, adjust P_3 for zero output.
3 Scale factor. Make $V_x = V_y = 10$ V (DC), adjust P_4 for 10 V output.

Exercises

5.1 The non-linear amplifier shown in Figure 5.5 uses the following component values: $R_1 = 10\ \text{k}\Omega$, $R_2 = 10\ \text{k}\Omega$, $R_3 = 10\ \text{k}\Omega$, $R_4 = 150\ \text{k}\Omega$, $R_5 = 4.7\ \text{k}\Omega$, $R_6 = 22\ \text{k}\Omega$, $V_S = +15$ V. Deduce the values of the break point voltages and the slopes of the straight-line segments in the amplifier response. Sketch the relationship between amplifier input and output voltage. Neglect the voltage drop across a saturated transistor.

5.2 Calculate the output voltage of the simple log amplifier shown in Figure 5.9 if the input current is 1 nA and the ambient temperature is 27°C. Assume that $\alpha_F I_{ES} = 1$ pA at 27°C and that the op-amp behaves ideally. What does the output voltage become if the temperature rises by 10°C?

5.3 In the circuit shown in Figure 5.14, $C_1 = 100$ pF, $C_2 = 100$ pF, $R_1 = 47\ \text{k}\Omega$, $R_E = 10\ \text{k}\Omega$. Sketch the Bode plots for $1/\beta$ for input currents of 100 μA, 1 μA and 10 nA.

5.4 The basic transdiode log configuration, Figure 5.9, is used with an input resistor $R = 100\ \text{k}\Omega$. The op-amp has an input offset $V_{io} = 1$ mV and a bias current $I_b^- = 20$ nA. If no offset balancing is employed, what is the smallest input signal voltage that can be logarithmically converted with an error no greater than 2 per cent? Assume that the temperature remains constant.

5.5 What values are required for resistors R_4 and R_2, in Figure 5.21, in order that an input voltage change from +10 mV to +10 V will cause an output voltage change from 0 to −10 V? Assume that the transistors are perfectly matched and that all other circuit parameters are as shown.

5.6 In the circuit of Figure 5.21 resistor R_2 is 1.5 MΩ; other circuit parameters are as shown. There is a mismatch in the transistors such that $I_{o_1} = 0.7 I_{o_2}$. Find the value of the input signal for which the output is zero. Neglect amplifier offsets.

5.7 The circuit of Figure 5.21 is used as a logarithm of current converter. The non-inverting input of A_1 is connected to earth. The input current is supplied directly to the inverting input of A_1. Let A_1 be a FET input op-amp, with bias current $I_B = 50$ pA and input offset voltage $V_{io} = 2$ mV. What is the smallest input current which can be converted with no more than 2 per cent error if it is supplied by (a) a true current source; (b) a current source of internal resistance 1 MΩ?

5.8 In the circuit of Figure 5.23, $I_{ref} = 0.1$ mA, $R_3 = 1\ \text{k}\Omega$. What values are required for resistors R_4 and R_2 in order that the circuit should generate an output signal $e_o = 2^{-e_{in}}$? Assume that the temperature is

300 K and that transistors are perfectly matched. (Boltzmann's constant $k = 1.38 \times 10^{23}$ J/K, electronic charge $q = 1.6 \times 10^{-19}$ C.)

5.9 Assume ideal op-amp action and that transistors are matched and follow the log relationship (equation 5.7). Find the relationships between output and input signals for the circuits shown in Figure 5.27. Discuss problems likely to be encountered in practical realizations of the circuits.

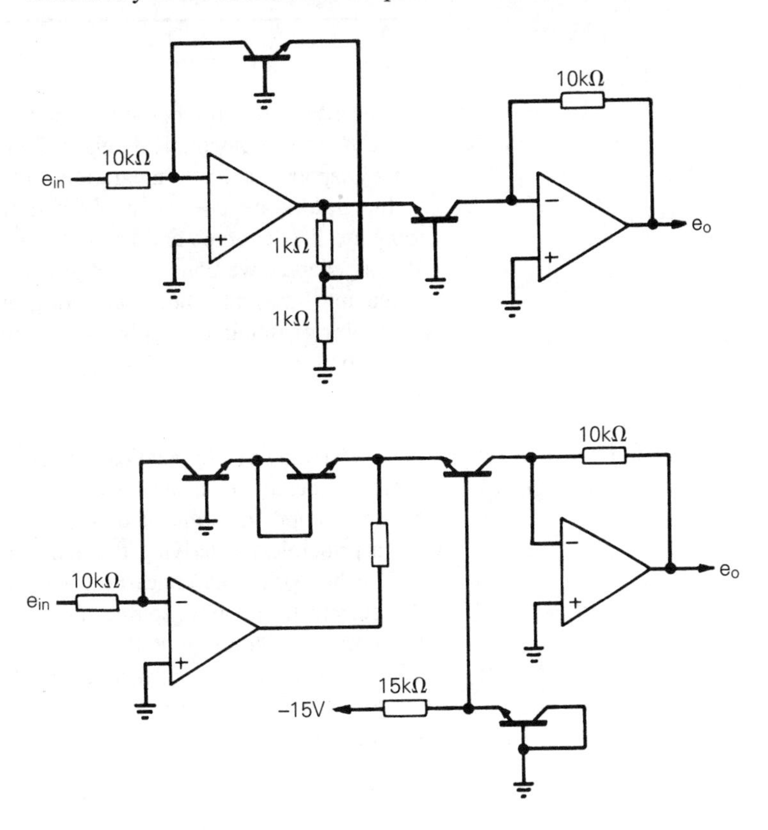

Figure 5.27 *Circuits for Exercise 5.9*

6 Integrators and differentiators

At the heart of most op-amp applications lies the ability of the circuit to force precise mathematical relationships between input and output signals. This chapter examines integration and differentiation circuits.

Integrators are used to perform timing functions, to measure charge, to generate linear ramps and triangular waves and in many other applications. In this chapter we consider integrator action and the factors which must be taken into account when connecting and using practical integrators. Many examples of integrator applications will be found in subsequent chapters of the book.

6.1 The basic integrator

An understanding of practical integrator circuits is helped by first considering the behaviour of an ideal circuit. Errors in practical circuits may then be understood in terms of departures from ideal behaviour. There are two main principles underlying the action of an ideal integrator.

The first principle concerns the ideal amplifier summing point restraints. All current from signal sources arriving at the inverting input terminal of an ideal amplifier must exit through the negative feedback path. The output voltage of the amplifier takes on just that value needed to keep the inverting input terminal at the same potential as the non-inverting input terminal. This prevents accumulation of charge at the inverting input terminal.

The second principle concerns the relationship between the voltage across a capacitor and the charge upon its plates, $V_c = (q/C)$. For charge to exist on the plates of a capacitor, charge must flow on to its plates. This charge flow represents a current and we may write:

$$i = \frac{dq}{dt}$$

$$\text{thus} \quad V_c = \frac{\int I_{in}\, dt}{C}$$

The voltage across a capacitor is proportional to the integral of capacitor charging current with respect to time.

The two principles applied to the basic integrator circuit of Figure 6.1 lead directly to the ideal performance equation. Thus, an input current $I_{in} = e_{in}/R$ arrives at the op-amp's inverting input. Since the op-amp takes no current into its inverting input, all the current that flows through the resistor must also flow through the capacitor.

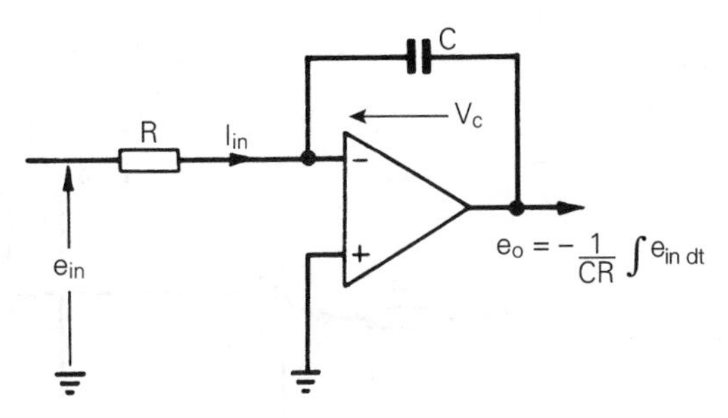

Figure 6.1 *The basic integrator*

Negative feedback forces the op-amp to produce an output voltage that maintains a virtual earth at the op-amp's inverting input. This means that the output voltage will be such that the current through the capacitor will be equal to I_{in}. For positive e_{in}, the output of the op-amp will be negative, relative to earth. Since the capacitor is connected between the op-amp's inverting input (which is at earth potential) and the op-amp's output (which is negative), the potential developed across the capacitor V_C will be positive at the end connected to the op-amp's inverting input.

$$e_o = -V_C = -\frac{\int I_{in}\,dt}{C}$$

$$e_o = -\frac{1}{CR}\int e_{in}\,dt \tag{6.1}$$

The input impedance of the integrator circuit is equal to the resistance R. The output impedance is low because of the negative feedback that is inherent in the circuit. CR gives the characteristic time of the integrator. It is sometimes useful to think of $1/CR$ as the integrator 'gain' in terms of V/s output for each volt of input signal.

For example, if $C = 1\ \mu$F and $R = 1\ \text{M}\Omega$, an input signal of +1 V would cause a current of 1 μA to flow towards the amplifier summing point. To maintain this charging current through capacitor C, the output would have to decrease linearly with time at a rate of −1 V/s.

If during the integration process the input signal were switched to zero, the input current would become zero and the output voltage of the ideal integrator would remain constant (hold) at any value it happened to have reached. If the input polarity were reversed ($V_{in} = -1$ V) this would require the output voltage to increase linearly at a rate of 1 V/s to maintain the 1 μA current flow away from the amplifier summing point.

6.2 Integrator run, set and hold modes

In a practical integrator circuit it is necessary to provide some means of setting a desired initial value of the integrator output voltage at the start of the integration period. In some systems, it is also desirable to be able to stop the integrator at any time and for the integrator output then to remain constant at the value it has reached at that time. The principles underlying

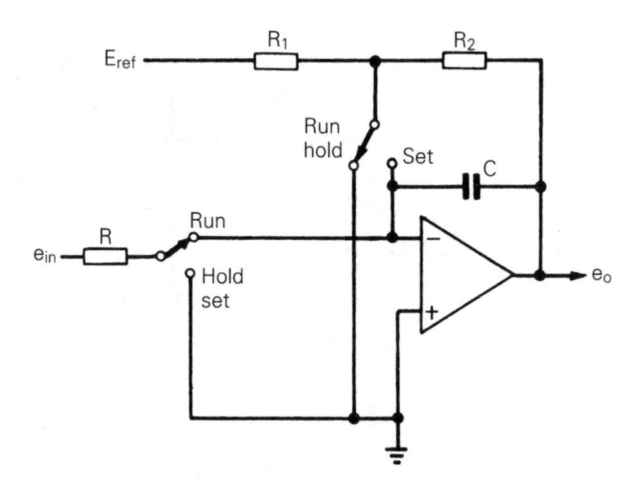

Figure 6.2 *Integrator run, set and hold modes*

the switching of an integrator between its various modes of operation are shown in Figure 6.2. Manual switching, relay switching or some form of solid state switching can be used.

The switches put in the 'set' position allow the initial value of the integrator output to be set at any desired value within the output capability of the amplifier.

$$e_{o(t=0)} = \frac{-R_2}{R_1} E_{ref}$$

The integrator output does not immediately take on this value when switched to the set mode. It approaches the value exponentially in accordance with the relationship

$$e_o = e_{o(t=0)} + (e'_o - e_{o(t=0)}) \exp\left(\frac{-t}{R_2 C}\right)$$

where e'_o is the value of e_o at the instant of switching to the set mode. Note the period of the set mode must be long enough for the exponential to decay.

When switched to the 'run' mode the circuit integrates the input voltage and

$$e_o = e_{o(t=0)} - \frac{1}{CR}\int_0^t e_{in}\, dt$$

If the integrator is switched to the 'hold' mode integration is stopped and ideally the output of the integrator then remains constant at any value it may have reached. In practice, in both the 'run' and 'hold' modes, drift causes an integrator error.

6.3 Integrator errors

The deviations from ideal behaviour that are exhibited by a practical integrator circuit are conveniently treated as errors. A firm understanding of the sources of error enables the designer to choose an amplifier and associated circuit to minimize errors.

6.3.1 Offset and drift errors in practical integrators

The greatest source of error in practical integrators is usually due to offset and drift of the op-amp. Even with zero applied input signal, the op-amp's input offset voltage and bias current cause a continuous charging of the feedback capacitor. Consequently, the output voltage of a practical free running integrator will change continuously. Eventually, the op-amp's output will drift into either positive or negative saturation.

Integrator output voltage drift with time can be adjusted to zero by cancelling the effects of the amplifier offsets with a suitable balance control (see Section 9.6). However, amplifier offsets are temperature dependent, supply voltage dependent, and they show long-term time dependence. This means that a zero output drift condition established with a balance control is not maintained and a free running integrator therefore always ends up in one of its saturated states. Integrator error due to amplifier input offset voltage and bias current is readily deduced from the equivalent circuit shown in Figure 6.3.

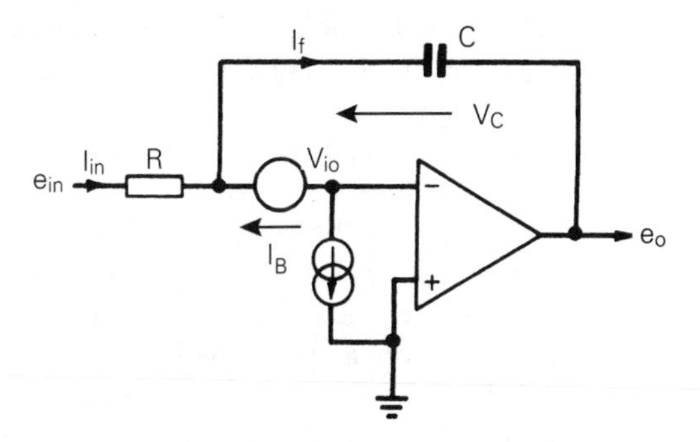

Figure 6.3 *Equivalent circuit used for estimating error due to input offset voltage and bias current*

For the moment we assume that the open-loop gain and open-loop input impedance of the amplifier are infinite. We may write

$$I_f = I_{in} - I_B^-$$

where $I_{in} = (e_{in} - V_{io})/R$.

$$\text{Now, } V_c = V_{io} - e_o = \frac{\int I_f \, dt}{C}$$

Thus e_o = ideal performance equation − error due to offset and bias current.

$$e_o = -\frac{1}{CR}\int e_{in}\,dt + \frac{1}{CR}\int V_{io}\,dt + \frac{1}{C}\int I_B\,dt + V_{io} \tag{6.2}$$

The percentage error after a particular integration time may be written

$$\frac{V_{io} + I_B^- R}{e_{in}} \times 100\% \tag{6.3}$$

where $\overline{e_{in}}$ is the time average of the input signal over the integration period.

In some applications it is more useful to refer integrator offset errors to the output. Offsets cause the output of an integrator to have an error in the form of an output drift rate (a ramp) determined by the relationship:

$$\frac{de_o}{dt_{(\text{due to offset})}} = \frac{V_{io}}{CR} + \frac{I_B^-}{C} \tag{6.4}$$

It should be noted that V_{io} and I_B^- are initial values plus accumulative drift. If initial values are balanced this leaves temperature drift values only.

The error component due to amplifier bias current can be reduced by connecting a resistor equal in value to that of the integrating resistor between the non-inverting input terminal of the amplifier and earth. With this resistor in circuit, values of the amplifier input difference current I_{io} should be substituted for I_B^- in the drift error equations.

When switched to the hold mode, an integrator normally has R open circuit so that, according to the error equations, it is bias current alone that accounts for drift in the hold mode. However, as will be shown later, finite open-loop gain and finite amplifier input impedance give rise to an additional source of error in the hold mode.

Examination of equation 6.4 suggests that for a particular integrator characteristic time drift error is minimized by using a large value capacitor. The value needs to be large enough to make the bias current contribution to the drift negligibly small, compared with the input offset value contribution.

There are practical considerations that limit the capacitor value. A large value of C requires a correspondingly small value of R for a particular CR value. The input impedance of the integrator is set by the value of the resistor R; the minimum value that can be used is dependent upon signal loading error. Capacitor leakage represents an additional source of integrator drift. Large value high performance capacitors are expensive, and should have dielectric leakage current that is less than the amplifier bias current. The dielectric absorption of a capacitor will also cause drift. Polypropylene and polystyrene dielectrics have lowest absorption, whilst electrolytic and tantalum have the highest.

For low drift integrators with long-term stability it is best to use low current FET input op-amps. These allow the bias current contribution to integrator drift to be made negligible without the use of excessively large capacitor values. When such op-amps are used, it is important to prevent leakage paths to the summing point from degrading performance (see Section 9.4.3).

6.3.2 Integrator errors due to finite open-loop gain, finite input impedance and finite bandwidth

The ideal performance equation for an integrator (equation 6.1) was obtained from the assumption that the op-amp used in the circuit had infinite open-loop gain and bandwidth. In all op-amp circuits using negative feedback, the extent to which a practical circuit performance departs from the ideal is governed by the loop gain βA_{OL} (see Section 2.2). The larger the loop gain the closer the practical circuit conforms to the ideal.

In a practical integrator, finite open-loop gain causes integrator performance errors for very low frequency input signals, and finite bandwidth causes errors

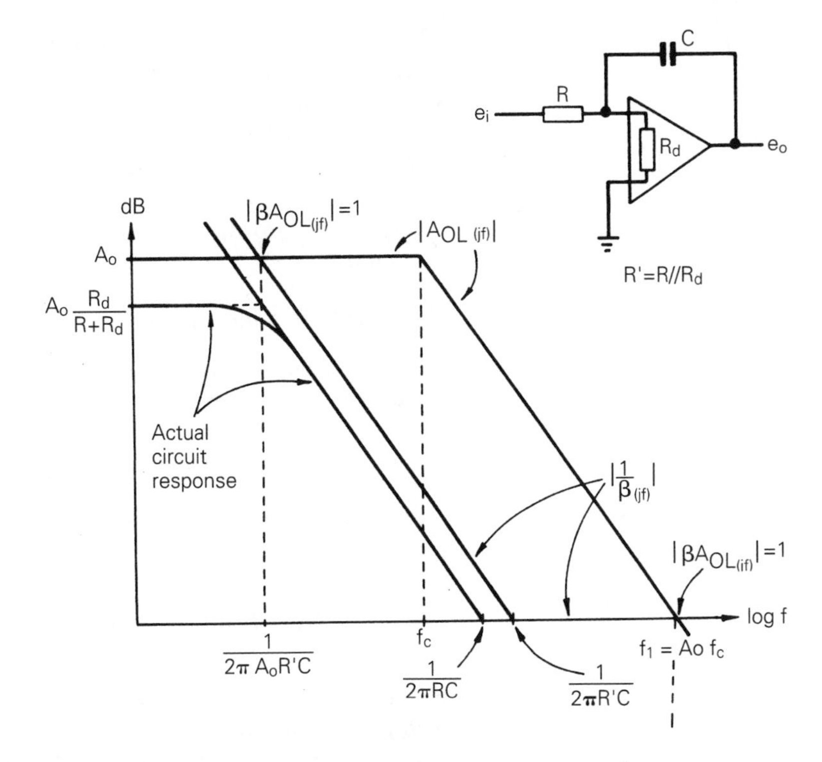

Figure 6.4 *Effect of finite open loop gain and input impedance*

for high frequency input signals. Integrator errors due to inadequate loop gain are discussed in terms of the circuit and Bode plots shown in Figure 6.4.

The op-amp used in the circuit of Figure 6.4 is assumed to have a finite differential input resistance R_d, finite open-loop gain and a first order frequency response described by the relationship:

$$A_{OL(jf)} = \frac{A_O}{1 + j\dfrac{f}{f_c}} \quad \text{(see Section 2.4)}$$

The closed-loop performance equation for the circuit expressed in the form: actual performance equation = ideal performance equation × gain error factor (see Section 2.3) is

$$\frac{e_{o(jf)}}{e_{i(jf)}} = -\frac{1}{j2\pi fCR}\left[\frac{1}{1 + \dfrac{1}{\beta A_{OL(jf)}}}\right] \tag{6.5}$$

Provided that the magnitude of the loop gain is large the integrator performance closely approximates the ideal. The feedback fraction for the circuit is

$$\beta = \frac{R'}{R' + \dfrac{1}{j2\pi fC}}$$

where $R' = RR_d/(R + R_d)$ and $1/\beta = 1 + 1/(j2\pi fCR')$.

The intersection of the Bode plots for $1/\beta$ with the open-loop response shows that the magnitude of the loop gain becomes unity at the frequencies $1/(2\pi A_o CR')$ and f_1. The integrator must thus be expected to perform near ideally at frequencies such that:

$$\frac{1}{2\pi A_o CR'} \ll f \ll f_1$$

Errors at high frequencies due to finite open-loop bandwidth

At frequencies approaching and exceeding the amplifier unity-gain frequency f_1

$$\beta \rightarrow 1,\ A_{OL(jf)} \rightarrow -j(f_1/f)$$

and, equation 6.5 approximates to

$$\frac{e_{o(jf)}}{e_{i(jf)}} = -\frac{1}{j2\pi fCR}\left[\frac{1}{1 + j\dfrac{f}{f_1}}\right] \tag{6.6}$$

Equation 6.6 represents the equation for an ideal integrator when cascaded with a first order low-pass function having a break frequency equal to the open-loop unity-gain frequency of the op-amp. The attenuation and phase shift produced by this first order function represents the errors in the steady state sinusoidal response of the integrator at frequencies approaching f_1. Associated with this low-pass function, the output of a practical integrator exhibits a time lag in response to an input step signal as shown in Figure 6.5. The time lag is inversely proportional to the open-loop unity-gain frequency f_1 of the op-amp.

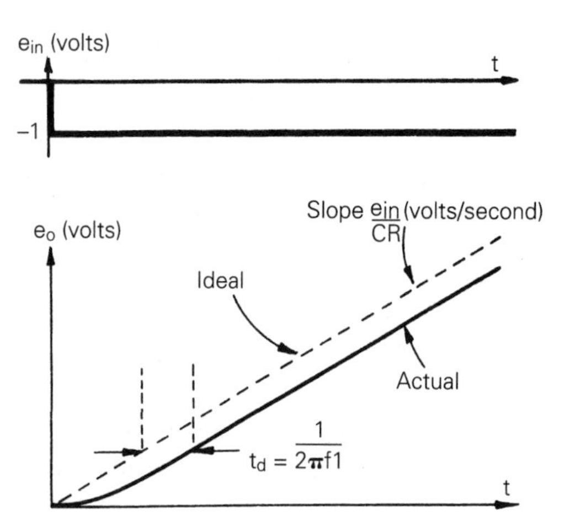

Figure 6.5 *Time lag in integrator step response due to finite open-loop bandwidth*

Errors at low frequencies due to finite open-loop gain

The gain of an ideal integrator circuit continues to increase as the signal frequency is decreased, but clearly in a practical integrator circuit the gain cannot be greater than the open-loop gain of the amplifier A_o ($A_oR_d/(R + R_d)$, if we allow for finite input resistance).

At frequencies less than f_1/A_o, $A_{OL(jf)} \rightarrow A_o$ and

$$\frac{1}{\beta A_{OL(jf)}} = \frac{1}{A_o} + \frac{1}{j2\pi fA_oCR'} \cong \frac{1}{j2\pi fA_oCR'}$$

Equation 6.5 approximates to

$$\frac{e_{o(jf)}}{e_{1(jf)}} = -\frac{1}{j2\pi fCR}\left[1 + \left(\frac{1}{\dfrac{1}{j2\pi fA_oCR'}}\right)\right] = -\frac{A_o\dfrac{R'}{R}}{1 + j2\pi fA_oCR'} \qquad (6.7)$$

for $f < f_c$.

From equation 6.7 we can gain some insight into the operation of integrators at low frequencies. Equation 6.7 is equivalent to the response of an ideal op-amp with infinite gain and infinite input resistance. However, it has a feedback impedance consisting of a capacitor C in parallel with a resistor A_oR' as shown by the equivalent circuit in Figure 6.6.

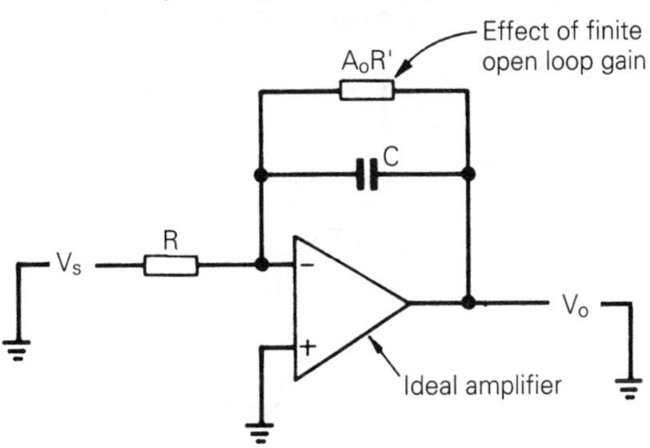

Figure 6.6 *Integrator low frequency equivalent circuit representing effect of finite open-loop gain*

Connecting the input resistor R to a constant DC voltage produces a linear ramp generator. In such applications, low frequency errors due to finite open-loop gain cause departures from the ramp linearity. The output response of the low frequency equivalent circuit of Figure 6.6 to an input step voltage V_S is determined by the relationship

$$V_{o(t)} = -A_o\frac{R'}{R} \cdot V_S\left[1 - \exp\left(\frac{-t}{A_oCR'}\right)\right] \qquad (6.8)$$

Expanding the exponential in equation 6.8 as a power series gives

$$V_{o(t)} = -V_S\left[\frac{t}{CR} - \frac{t^2}{2A_o(CR')\,CR} + \ldots\right] \tag{6.9}$$

The first term in equation 6.9 represents the ideal response (a linear ramp). The second and subsequent terms represent the error due to finite gain, which causes a departure from linearity. The departure from linearity is governed principally by the second term; expressing this as a percentage of the ideal linear term gives

$$\text{Ramp non-linearity error} = \frac{-t}{2A_oCR'} \times 100\% \quad \Big| \quad t << CR'A. \tag{6.10}$$

Finite open-loop gain causes errors in hold mode

If during an integration process the input voltage to the integrator is switched to zero, the output of the integrator should ideally remain constant (hold) at any value it may have reached. Finite open-loop gain and finite input resistance, in addition to amplifier offsets, contribute errors in the hold mode.

Minimum error in the hold mode is obtained by open circuiting the input resistor R. This makes the effective leakage resistance A_oR' equal to A_oR_d, due to the amplifier's finite open-loop gain and finite input resistance. This effective leakage resistance tends to discharge any fixed voltage stored across the integrating capacitor and the equivalent circuit shown in Figure 6.7 may be used to compute drift error in the hold mode.

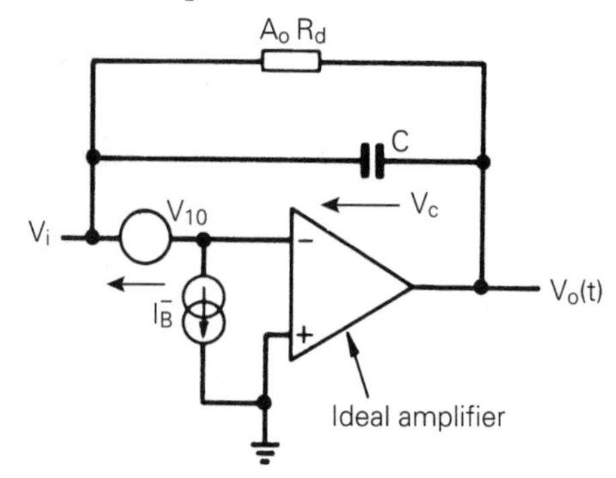

Figure 6.7 *Equivalent circuit used to find error in hold mode*

An expression for the output drift error in the hold mode can be obtained by assuming that when the integrator is switched to the hold mode the capacitor in the equivalent circuit of Figure 6.7 has an initial voltage V_i. The subsequent time variation of the amplifier output voltage is determined by the relationship:

$$V_{o(t)} = V_i \exp\left(\frac{-t}{A_oCR_d}\right) + I_B^- A_oR_d\left[1 - \exp\left(1 - \frac{-t}{A_oCR_d}\right)\right] + V_{io} \tag{6.11}$$

If we write

$$\exp\left(\frac{-t}{A_o CR_d}\right) = 1 - \frac{-t}{A_o CR_d} \quad \text{for } t < CA_o R_d$$

The output drift value in the hold mode may be written as

$$\frac{dV_o}{dt_{\text{(hold mode)}}} = \frac{-V_i}{A_o CR_d} + \frac{I_B}{C} \tag{6.12}$$

In a practical integrator, the drift error in the hold mode is normally dominated by the effect of amplifier bias current and stray leakage currents. However, equation 6.12 shows that even if all leakage paths were eliminated and amplifier bias current compensated, an error would still remain because of the amplifier's finite open-loop gain and finite input resistance.

6.3.3 Slew rate errors

Fast integrators require an output voltage that changes rapidly. Slew rate limitations, which set the maximum rate at which the op-amp's output can change, can cause performance errors.

Slew rate limitations are inherent and arise from the basic mechanism of capacitor charging. Amplifier slew rates are normally specified for an op-amp when it is used to drive a resistive load. The slew rate is determined by the charging of the op-amp's frequency compensating capacitor (see Section 2.7.1).

In integrator applications, the output current of the amplifier charges the feedback capacitor. The output current limit of the amplifier may impose an output slew rate limit that is less than the published amplifier slew rate. Remember that the output current of the amplifier must supply any external load as well as the current taken by the feedback capacitor. For example, consider an amplifier with an output current limit of ±5 mA used in an integrator circuit with a feedback capacitor of 0.01 μF. Even if all the output current were available to supply the feedback capacitor, the maximum rate of change of output voltage would be:

$$\frac{I_o}{C} = \frac{5 \times 10^{-3}}{10^{-8}} = 0.5\ V/\mu s$$

6.4 Extensions to a basic integrator

There are a variety of external circuit modifications that can be made to the basic integrator circuit in order to change its response characteristics and extend its usefulness.

6.4.1 Summing integrator

The current summing property of the inverting input terminal of a differential input op-amp could be exploited, to allow a single amplifier to perform both

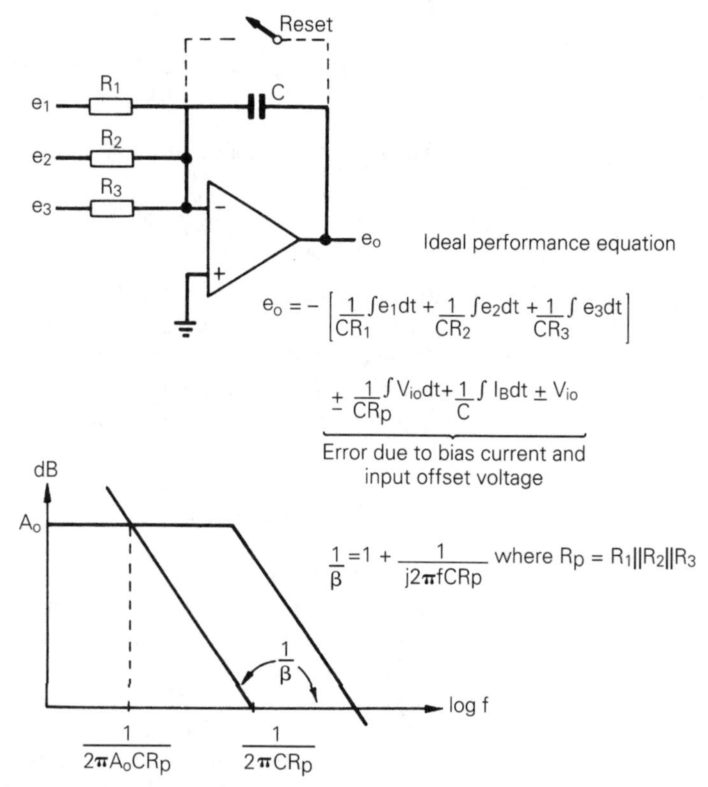

Figure 6.8 *Summing integrator*

summation and integration at the same time. The circuit shown in Figure 6.8 illustrates the principle.

Note that by using different input resistor values the contributions to the output of the several inputs is weighted in inverse proportion to the resistor values.

Considerations involved in determining performance errors are much the same as those outlined for the basic integrator. The error equations of the basic integrator included a single input resistor R. However, in the summing integrator, a resistor R_p that is equal to the parallel sum of all input resistors must be substituted in place of R.

Note that low frequency errors due to finite gain occur at frequencies approaching and below the frequency at which the Bode plots for $1/\beta$ and the open-loop gain intersect. For the summing integrator $1/\beta$ is determined by the parallel sum of all input resistors R_p:

$$\frac{1}{\beta} = 1 + \frac{1}{j2\pi f C R_p}$$

6.4.2 Augmenting integrator

A resistor connected in series with the feedback capacitor of a basic integrator (Figure 6.9) makes the circuit produce a composite output consisting

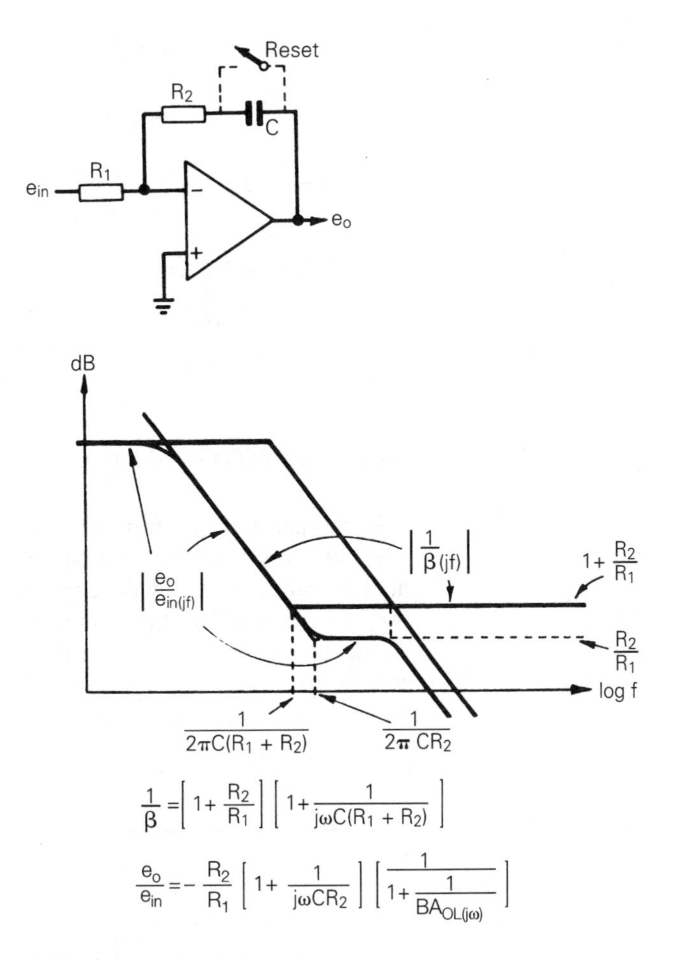

Figure 6.9 *Augmenting integrator*

of a component proportional to the input signal added to a component proportional to the time integral of the input signal. The principle may also be adapted to the summing integrator of Figure 6.8 by connecting a resistor in series with the feedback capacitor in that circuit.

6.4.3 Differential integrator

The subtraction principle of Section 4.4 can be applied to give a circuit in which a single differential input op-amp produces an output signal proportional to the time integral of the difference between two input signals. A circuit for this purpose is shown in Figure 6.10.

This circuit may be used to integrate the output of a floating source whilst rejecting common mode input signals. The ability of the circuit to reject common mode signals depends on the CMRR of the amplifier but, in addition, it is also very much dependent upon an accurate matching of the time constants of the networks connected to the two input terminals.

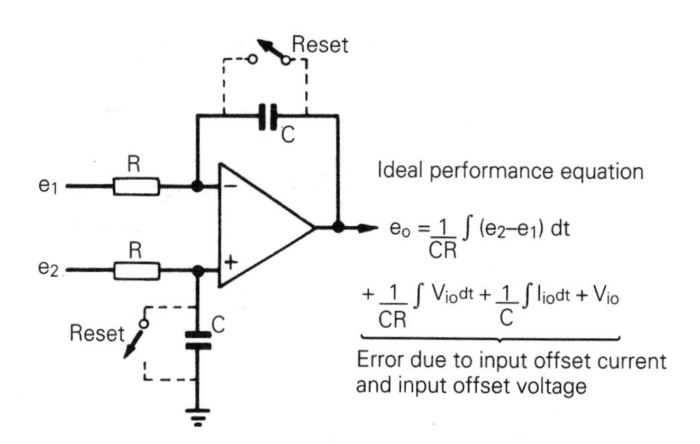

Figure 6.10 *Differential integrator*

Performance errors in a practical circuit are similar to those of a basic integrator. The errors are determined by amplifier input offset voltage, amplifier bias current and by finite amplifier open-loop gain and bandwidth. Using two capacitors with a single amplifier increases the problems associated with providing a practical circuit with reset and hold modes. It may be found more convenient to use the two-amplifier one-capacitor circuit shown in Figure 6.11 to perform the differential integrator operation.

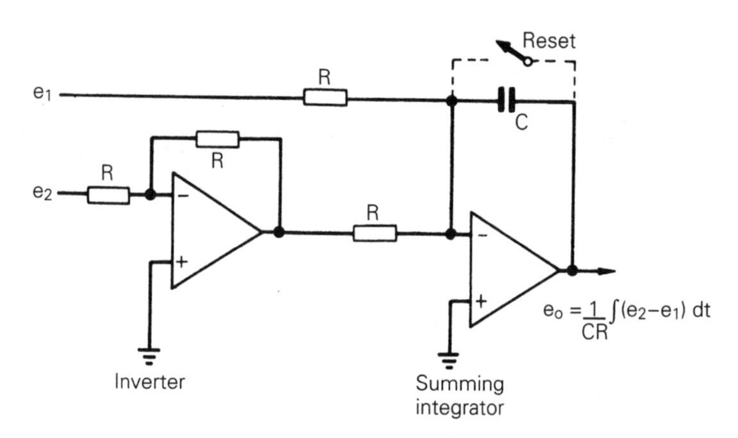

Figure 6.11 *Differential integrator using two op-amps*

In this circuit one amplifier acts as a simple inverter and the other acts as a summing integrator. The CMRR of the circuit does not depend upon the CMRR of the amplifiers but it is still dependent upon accurate resistor matching.

6.4.4 Current integrator

It is sometimes desirable to form an output signal voltage proportional to the integral with respect to time of input current rather than input voltage

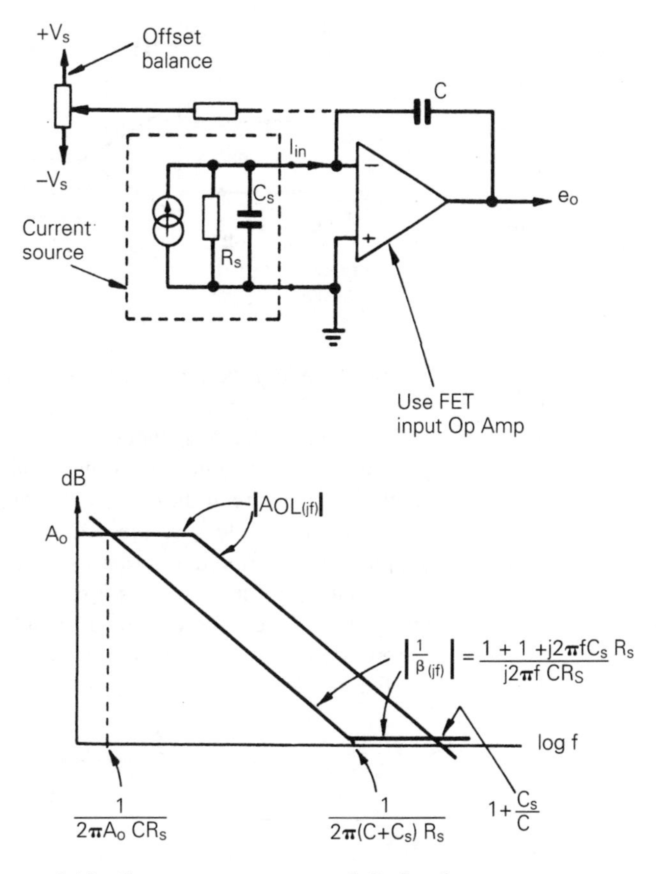

Figure 6.12 *Current integrator and Bode plots*

(see also Section 8.1.2). The basic integrator is readily adapted to current integration by simply omitting the input resistor *R*. A practical current integrator circuit is outlined in Figure 6.12.

The circuit is suitable for integrating small high impedance currents to earth. It produces a negligible voltage intrusion into the measurement circuit. Assuming that offsets are nulled and suitable precautions taken to avoid leakage, the circuit may be expected to provide accurate integration for very small input currents. If external leakage is reduced to negligible proportions, the accuracy limitations are set by amplifier bias current drift. A FET input op-amp has low bias current drift and should be used.

6.4.5 Integral of current sum and current differences

The current integrator circuit (Figure 6.12) is readily adapted to summation, all that is necessary being to supply the extra input currents to the amplifier summing point. A one-amplifier circuit can also be used to generate an output voltage proportional to the integral of a current difference. The circuit shown in Figure 6.13 illustrates the principle.

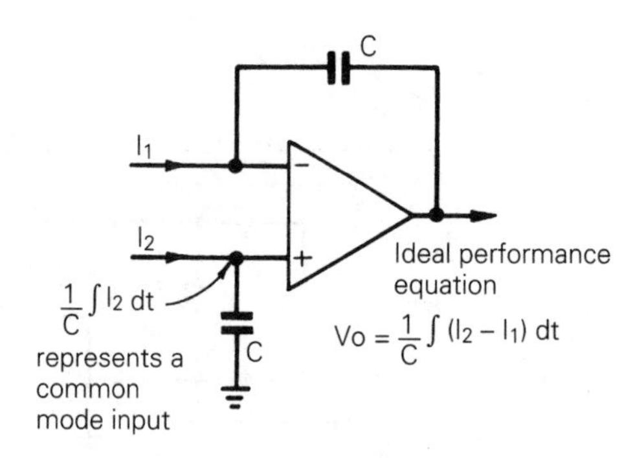

Figure 6.13 *Simple one-amplifier circuit for integral of current difference*

However, with two capacitors, practical problems are involved in the provision of reset. Capacitors must be accurately matched if CMRR is not to be degraded. The circuit introduces a voltage drop, $1/(C\int I_2\,dt)$, into the measurement circuit. Because of these difficulties it is usually more convenient to employ two amplifiers in order to perform the integration of a current difference, one amplifier acting as a current inverter and the other as a summing integrator. Offsets and drift of the amplifier types used in the practical circuit determine performance accuracy in much the same way as before. A circuit is illustrated in Figure 6.14.

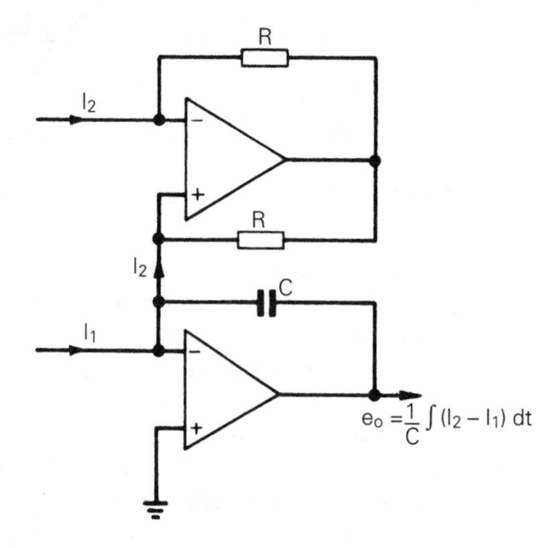

Figure 6.14 *Two-amplifier circuit for integral of current difference*

6.5 Integrator reset

Unlike a normal amplifier circuit, the output of an integrator does not return to zero when the input signal is made zero. Therefore, a practical integrator must always be provided with some means of resetting its output voltage to zero (or some desired initial value). Switching of an integrator between its various modes of operation can be performed by mechanical switches or relays (see Section 6.2) but it is sometimes desirable to provide a solid state reset switch or to arrange that the integrator automatically resets when its output reaches some predetermined level.

Reset switches are connected in parallel with the integrator feedback capacitor. In the run mode, any leakage across the open switch adds to the error due to amplifier bias current. The extent to which switch leakage becomes a design consideration is dependent upon the magnitude of the integrator's input signal current and the desired accuracy. If input currents are large compared to the leakage of a simple solid state switch then a simple switch will suffice. If not, then a leakage reduction switch configuration must be sought.

A low leakage reset switch can be implemented using two MOSFETs as shown in Figure 6.15. When using p-channel MOS switches, the source substrate junction must not become forward biased. The substrate must therefore never be allowed to become negative with respect to the input signal. The leakage current of a MOS switch in the 'off' state occurs mainly across the substrate to drain junction.

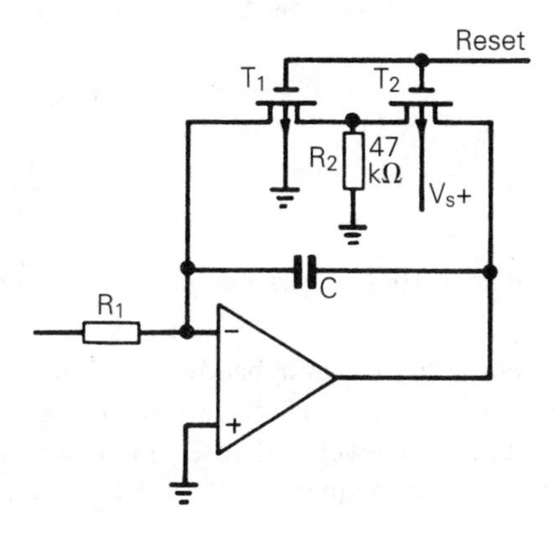

Figure 6.15 *Low leakage integrator reset*

In Figure 6.15 a negative going reset pulse turns on T_1 and T_2 shorting the integrator capacitor and setting the output voltage to near zero (to a voltage $\cong -V_{in}2R_{ON}/R_1$, where R_{ON} is the low 'on' existence of the MOS switch, $R_{ON} << R_2$). When the switches turn off the leakage current of T_2 passes through resistor R_2. The small voltage across R_2 is blocked from the amplifier's summing junction by T_1. T_1 has practically no voltage across its junctions because its substrate is earthed and hence leakage currents are negligibly small.

The reset switch of Figure 6.15 can be made to provide an automatic reset by the addition of a comparator (see Section 7.1) used to sense the integrator output and to provide the reset drive.

The circuit shown in Figure 6.16 allows independent adjustment of both the integrator-reset voltage and the output level to which the integrator is reset.

Reset in this circuit occurs when the output voltage of the integrator reaches the comparator trip point. During reset, the capacitor discharges until the integrator output voltage reaches the lower comparator trip point, as determined by comparator hysteresis. Integrator errors are like those of the basic integrator, and are governed by amplifier offsets and bias currents, finite

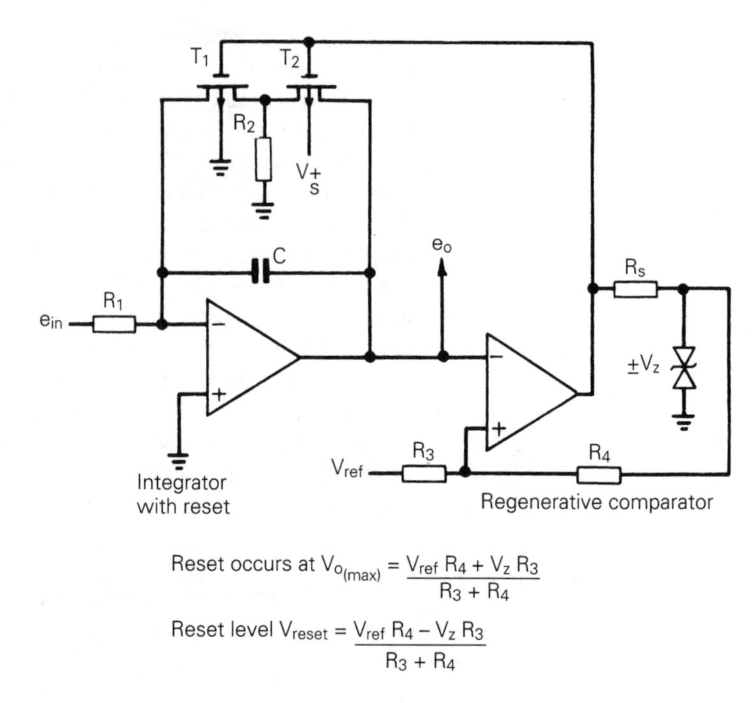

Figure 6.16 *Comparator provides automatic integrator reset*

open-loop gain and bandwidth. Finite comparator switching time and the bias current and offset voltage of the comparator amplifier introduce errors in the reset level and reset point but these can be compensated practically by choosing values of V_{ref} and V_Z to give desired values of $V_{o(max)}$ and V_{reset}.

6.6 AC integrators

In integrator applications not requiring a response down to DC it is possible to avoid having an output reset. To do so means that the DC closed-loop gain must be limited. The circuit shown in Figure 6.17 has a steady state sinusoidal response governed by the relationship:

$$\frac{e_o}{e_{in}} = -\frac{R_2}{\dfrac{1 + j2\pi fCR_2}{R_1}}\left[\frac{1}{1 + \dfrac{1}{\beta A_{OL}}}\right] \qquad (6.13)$$

For $\beta A_{OL} >> 1$ and for frequencies greater than $1/(2\pi CR_2)$ the response approximates that of the ideal integrator

$$\frac{e_o}{e_{in}} \cong -\frac{1}{j2\pi fCR_1}$$

At a frequency a decade away from $1/(2\pi CR_2)$ the magnitude error is only 0.5 per cent.

The presence of R_2 prevents integrator drift due to amplifier bias current and offset voltage from causing the amplifier to drift into saturation. Instead the output assumes a DC value of

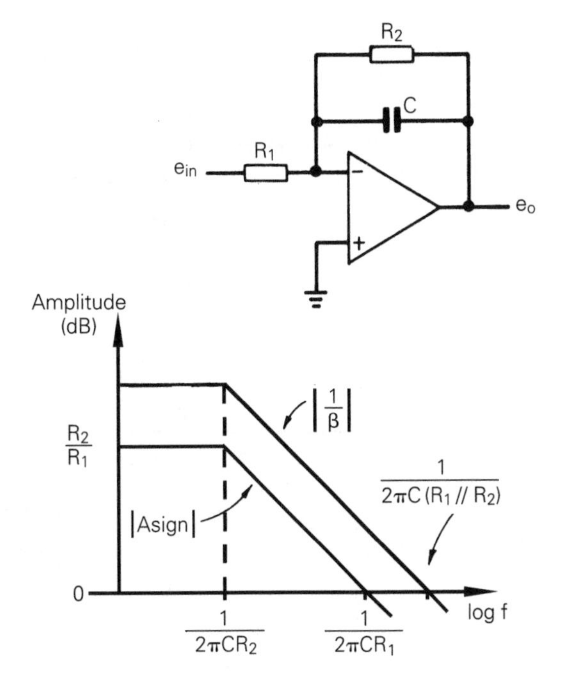

Figure 6.17 *AC Integrator*

$$V_{\text{o(offset)}} = \left[1 + \frac{R_2}{R_1}\right]\left[\pm V_{\text{io}} - I_{\text{B}}\frac{R_1R_2}{R_1 + R_2}\right]$$

This output offset limits the dynamic range for AC output signals. As in all applications an offset balance can be used to cancel initial values of amplifier input offset voltage and bias current; output offset is then due to amplifier drift.

6.7 Differentiators

The differentiator is not as widely employed as the integrator operation, but is nevertheless sometimes useful in signal processing applications. The reasons for this are: (1) differentiation, unlike integration, is a noise amplifying process – noise problems are inherent in differentiators and are not just a defect of practical circuits; (2) differences between the ideal differentiator circuit and the practical circuit are more marked than between the ideal and real integrator.

6.7.1 The basic differentiator

A simple differentiator circuit (Figure 6.18) is obtained by interchanging the position of the resistor and capacitor in the basic integrator circuit. The ideal performance equation for the simple differentiator is readily derived from the usual ideal amplifier assumptions. Since the input signal is applied through a capacitor there is current flow to the amplifier summing point and a non-zero output voltage only when the input voltage changes.

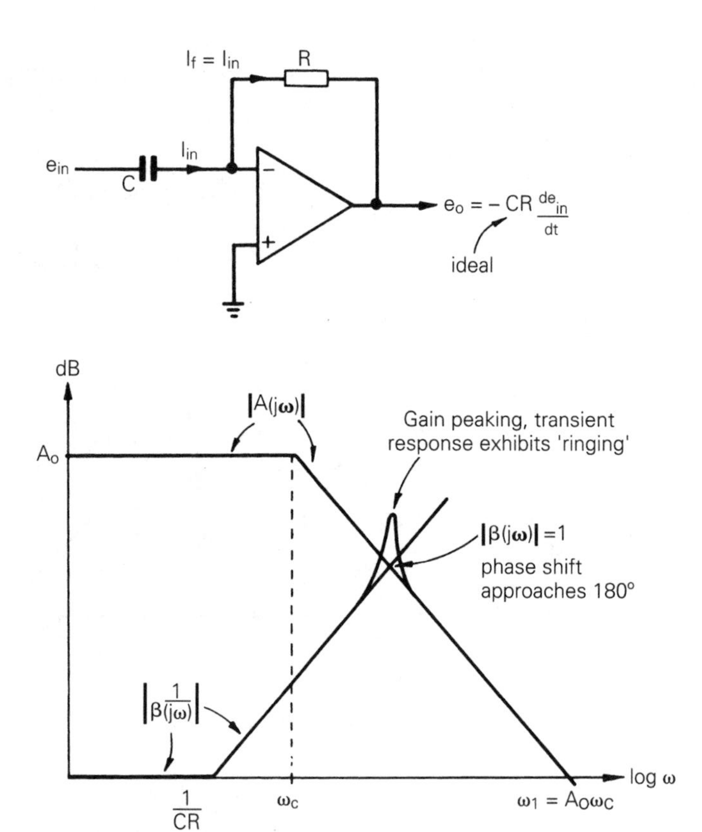

Figure 6.18 *Basic differentiation and Bode plots*

The current to the amplifier summing point is

$$I_{\text{in}} = C\frac{\mathrm{d}e_{\text{in}}}{\mathrm{d}t}$$

In the ideal case this current must be equal to the current through the feedback resistor R, thus:

$$e_{\text{o}} = -I_{\text{f}}R$$

$$\text{or } e_o = -CR\frac{\mathrm{d}e_{\text{in}}}{\mathrm{d}t} \tag{6.14}$$

The basic differentiator is unstable because of amplifier finite open-loop bandwidth.

The feedback fraction β for the simple differentiator circuit of Figure 6.18 is

$$\beta = \frac{\dfrac{1}{j\omega C}}{R + \dfrac{1}{j\omega C}} = \frac{1}{1 + j\omega CR}$$

At angular frequencies greater than $1/CR$ the feedback signal lags behind the amplifier output signal by a phase angle approaching 90°. Because of the finite open-loop bandwidth of practical op-amps signals with frequency above the open-loop bandwidth undergo an additional phase lag in their passage through the amplifier. The two phase lags can readily add up to 180° making the overall feedback positive rather than negative and resulting in oscillation.

Bode plots of $1/\beta$ and A_{OL} for the simple differentiator are shown in Figure 6.18. The two plots intersect with a rate of closure of 40 dB/decade, indicating a near zero phase margin. The simple differentiator produces gain peaking for signal frequencies approaching the frequency at which the magnitude of the loop gain is unity. Any transient disturbance in the circuit gives rise to output ringing. This very lightly damped response means that any additional phase shift in the feedback loop, say due to capacitive loading at the output, can cause sustained oscillations.

6.8 Practical considerations in differentiator design

6.8.1 Bandwidth limits

Practical differentiator circuits often use some means of limiting the bandwidth in order to achieve closed-loop stability. One method shown in Figure 6.19 is to connect a resistor R_1 in series with the input capacitor C. Bode plots for $1/\beta$ and A_{OL} intersect with a rate of closure of 20 dB/decade, thus ensuring adequate stability phase margin. Resistor R_1 also serves to increase the differentiator's effective input impedance and to reduce its high frequency gain.

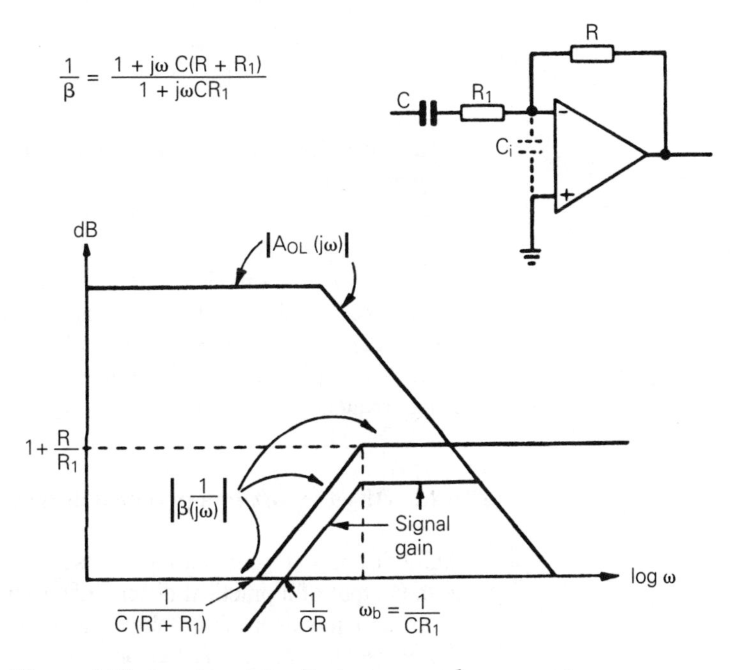

Figure 6.19 *Input resistor R_1 increases phase margin*

Note that the gain of a differentiator increases with frequency. This means that high frequency noise is amplified, which may obscure the wanted signal at the output. Adding the resistor R_1 introduces a break in the differentiator 20 dB/decade rise at an angular frequency

$$\omega_b = \frac{1}{CR_1}$$

In order to obtain additional attenuation of frequencies above those of interest, a capacitor C_1 may be connected in parallel with the differentiator feedback resistor as shown in Figure 6.20.

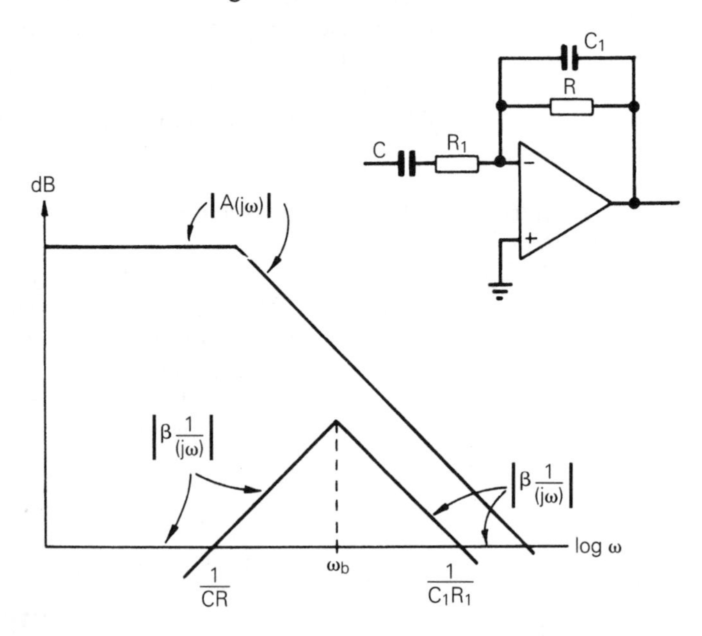

Figure 6.20 *Noise reduction in differentiation by restricting high frequency gain*

With C_1 chosen so that $C_1R = CR_1$, Bode plots are as shown in Figure 6.20. Note that the circuit acts as an integrator for frequencies greater than $1/(2\pi C_1R)$. In the circuits of Figures 6.19 and 6.20, component values should be chosen to place the break point ω_b well enough above the maximum operating frequency, to ensure the required accuracy.

6.8.2 Offset errors in a practical differentiator

Amplifier bias current and input offset voltage give rise to a DC offset error at the output of a practical differentiator circuit. The error is readily deduced from the equivalent circuit shown in Figure 6.21.

For the purpose of offset error evaluation the open-loop gain of the amplifier is assumed to be infinite. We may write

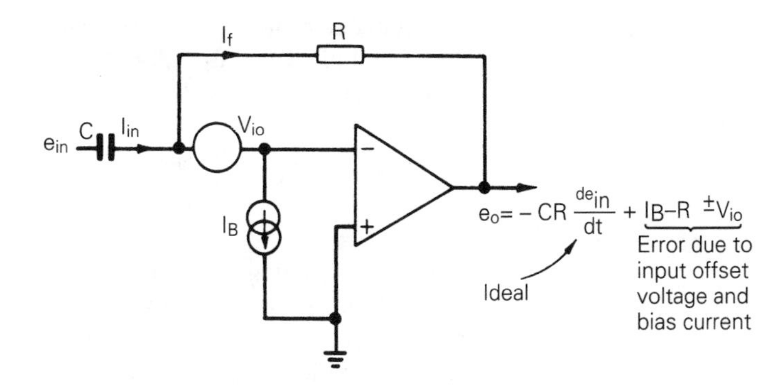

Figure 6.21 *Equivalent circuit for evaluation of differentiator offset error*

$$I_{\text{in}} = C\frac{\mathrm{d}e_{\text{in}}}{\mathrm{d}t}$$

$$\text{and} \quad I_{\text{f}} = I_{\text{in}} - I_{\text{B}} = \pm\frac{V_{\text{io}} - e_{\text{o}}}{R}$$

$$\text{Thus} \quad e_{\text{o}} = CR\frac{\mathrm{d}e_{\text{in}}}{\mathrm{d}t} + I_{\text{B}}R \pm V_{\text{io}} \tag{6.15}$$

e_{o} = ideal performance equation − error due to input offset voltage and bias current.

The output offset error can be referred to the input of the differentiator where it represents an equivalent input error:

$$\frac{\mathrm{d}V_{\text{in}}}{\mathrm{d}t_{(\text{error})}} = -\frac{I_{\text{B}}}{C} \pm \frac{V_{\text{io}}}{CR} \tag{6.16}$$

Amplifier input offset voltage and bias current are initial values plus drift. If initial values are balanced this leaves drift values only. The error component due to amplifier bias current can be reduced by connecting a resistor, equal in magnitude to the feedback resistor, between the non-inverting input terminal of the amplifier and earth. With this resistor in circuit, values of the amplifier input difference current I_{io} should be substituted for I_{B} in the error equations.

Equation 6.16 suggests that for a particular *CR* value the equivalent input error is reduced by making *C* as large as possible. But note that the current through the feedback resistor *R* is supplied by the amplifier; *R* must therefore not be so small as seriously to load the amplifier output. Also there are problems involved in large capacitor values (say *C* greater than 10 μF) as discussed in Section 6.3.1.

6.8.3 Choice of differentiator component values

The component values and the amplifier type to be used in a practical differentiator are dictated by the accuracy requirements assessed in terms of the expected frequency content and magnitude of the input signal. The starting point in a differentiator design is normally the choice of characteristic time *CR*.

It is convenient to select a characteristic time such that the amplifier will give near full-scale output for the maximum expected rate of change of the input signal.

$$\text{Choose} \quad CR = \frac{|V_o|_{max}}{\left(\dfrac{de_{in}}{dt}\right)_{max}} \tag{6.17}$$

An output bounding circuit can be added to prevent unexpectedly fast input signals from driving the output into saturation limits. Select resistor value say in the range 10 kΩ–100 kΩ, which is not too low as to draw seriously on the amplifier output current, and then calculate the necessary value of capacitor. If the calculation calls for a value of C greater than 10 μF, it is usually better to increase the value of R and use an amplifier type with smaller bias current. This is especially true if the equivalent input offset error (equation 6.16) is such as to reduce the accuracy below the design limit.

Ensure adequate closed-loop phase margin and reduce high frequency noise by selecting a value for R_1 (Figure 6.19) or C_1R_1 (Figure 6.20). To minimize noise, the break in the 20 dB/decade increasing gain should be set to the highest expected input signal frequency.

6.9 Modifications to the basic differentiator

Modifications comparable to those discussed in Section 6.4 can be made in order to vary the response characteristics of a differentiator circuit.

6.9.1 Summing differentiator

The derivative of several input signals may be combined in a summing differentiator. Introducing additional capacitive input paths to the amplifier summing point forms a summing differentiator, shown in Figure 6.22.

6.9.2 Differential differentiator

A circuit that produces an output proportional to the difference between the derivatives of two input signals is shown in Figure 6.23.

Exercises

6.1 A simple integrator circuit (Figure 6.l) uses $C = 0.1$ μF, $R = 100$ kΩ. The input point of the circuit is connected to earth and the output drift rate is adjusted to zero by means of a suitable offset balance (Section 10.6). Find the output drift rate if the temperature changes by 10°C assuming that the op-amp has $\Delta V_{io}/\Delta T = 20$ μV/°C and $\Delta I_B/\Delta T = 0.5$ nA/°C (see Section 6.3.1).

6.2 Show how you would use a single op-amp to generate the relationship

$$e_o = -\int_0^t (e_1 + 2e_2 + 10e_3)\, dt$$

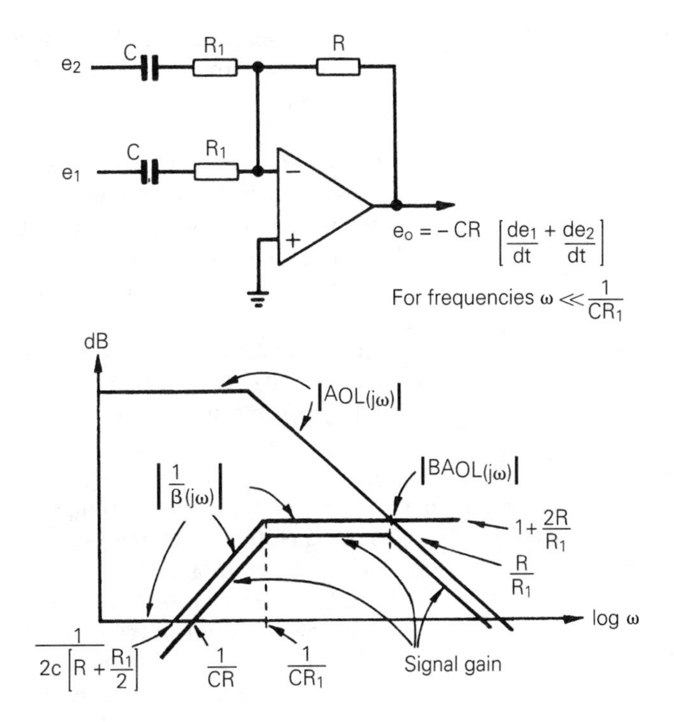

Figure 6.22 *Summing differentiator*

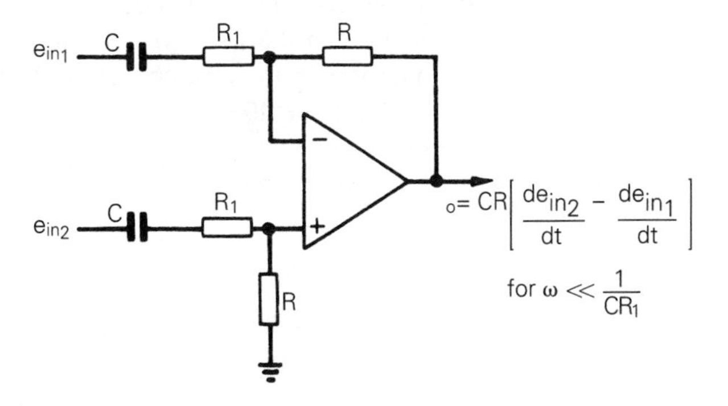

Figure 6.23 *Differential differentiators*

Find component values if the integrating capacitor has a value 1 μF. Assume ideal operational amplifier action. If the op-amp has input offset voltage V_{io} = 10 μV and bias current I_B = 2 nA what is the integrator output drift rate? (see Section 6.3.1 and Figure 6.8).

6.3 In the simple integrator circuit of Figure 6.1, C = 0.1 μF and R = 100 kΩ. If the open-loop gain of the operational amplifier at zero frequency is 80 dB what is the lowest sinusoidal signal frequency for which the phase error will be no more than 5 degrees?

What is the amplitude error at this frequency?

What happens to the phase and amplitude errors if an additional input path in the form of a second 100 kΩ resistor is added to the circuit? (see Section 6.3.2).

6.4 A simple operational integrator with $C = 0.1\ \mu F$ and $R = 100\ k\Omega$ is to be used to produce a linear ramp of amplitude 10 V and duration 100 ms. What is the minimum value of the open-loop gain of the amplifier to ensure that the departure from linearity of the ramp is less than 0.1 per cent? (see equation 6.10).

6.5 An AC integrator (Figure 6.17) uses the component values $R_1 = 10\ k\Omega$, $R_2 = 1\ M\Omega$, $C = 0.1\ \mu F$. What is the lowest input signal frequency for which the magnitude of the output is in error by no more than 0.5 per cent? (see Section 6.6).

6.6 An internally frequency compensated operational amplifier with open-loop gain 100 dB and unity-gain frequency 10 MHz is used as a simple differentiator (Figure 6.18) with $C = 0.1\ \mu F$ and $R = 100\ k\Omega$. The circuit is found to be very lightly damped; explain this fact and estimate the approximate value of the frequency at which the output 'rings' when the circuit is subjected to a transient disturbance. In order to overcome the stability problem, a resistor $R_1 = 1\ k\Omega$ is connected in series with the input capacitor C.

Explain the action of this resistor and estimate the highest frequency for which the magnitude of the output signal will be in error by no more than 0.5 per cent with R_1 in circuit. Illustrate your answer with appropriate Bode plots (see Sections 6.7.1 and 6.8.1).

If the operational amplifier has a bias current $I_B = 0.1\ \mu A$ and input offset voltage $V_{io} = 4\ mV$ what will be the output offset in the circuit? (see Section 6.8.2).

7 Comparator, monostable and oscillator circuits

Negative feedback has been used in the op-amp circuits described so far. In such circuits, the op-amp's output voltage usually lies between its positive and negative saturation limits. The differential input is very small at all times, due to the use of negative feedback and because the op-amp has a high gain.

When an op-amp is used without feedback (open-loop operation) its output will usually be in one of its saturated states. The application of a small differential input signal, of the appropriate polarity, will cause the output to switch to its other saturated condition.

In this chapter we will consider circuits in which op-amps, and related analogue integrated circuits, have positive feedback applied to them. Positive feedback can be used to produce a comparator action, or a controlled oscillation of definite frequency. Integrated circuits (ICs) that are designed specifically to operate as comparators will be described, in addition to the ubiquitous 555 timer IC and the 8038 function generator.

7.1 Comparators

A comparator is a circuit used to sense when a varying signal reaches some threshold value. Comparators find application in many electronic systems. They may be used to sense when an electrical signal reaches or exceeds some defined voltage level.

A comparator may be built using dedicated comparator ICs. These dedicated devices are like op-amps except they usually have fast switching outputs designed for driving digital circuits. Alternatively, op-amps may be used as comparators, and are able to drive digital circuits if suitable output voltage limiting is provided. However, not all op-amp types are suitable. In addition to low offset and drift, rapid switching times are normally essential.

Comparators have a differential amplifier at their input. The output is a switching driver stage. The simplest comparator circuit has the signal voltage directly to one of the input terminals and a reference voltage to the other. The principle for an op-amp is illustrated in Figure 7.1; the same principle applies for a comparator IC, except that the minimum output voltage is usually 0 V (earth).

The op-amp is used open loop. Its output makes a transition between saturated states as the input signal passes through a value equal to E_{ref}. E_{ref} must not exceed the maximum common mode voltage if the circuit uses an op-amp. Swapping the connections of the signal and reference input changes the polarity of the output.

The circuit in Figure 7.2 illustrates an alternative arrangement for comparator operation.

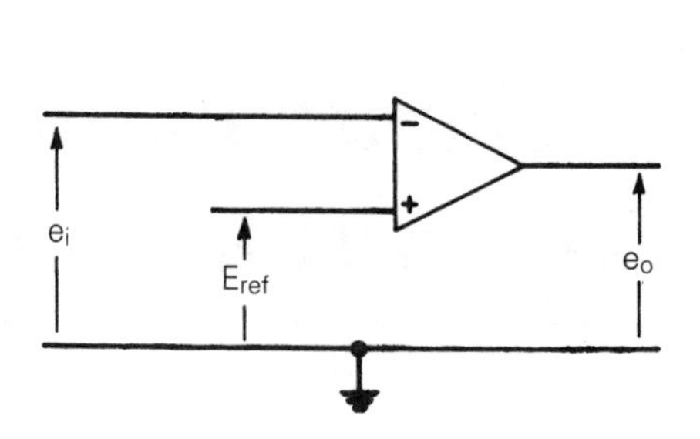

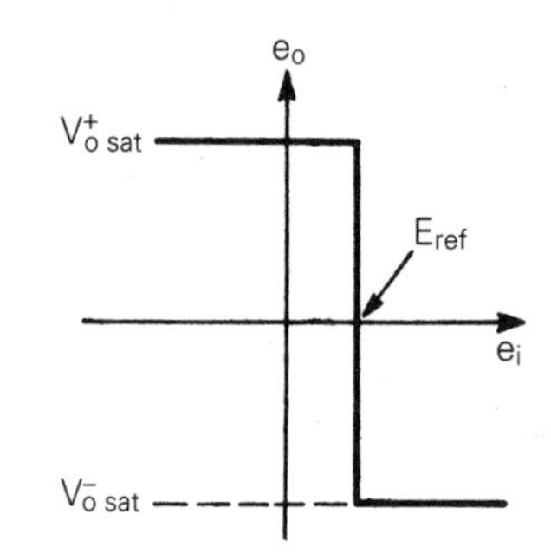

Figure 7.1 *Simple op-amp comparator*

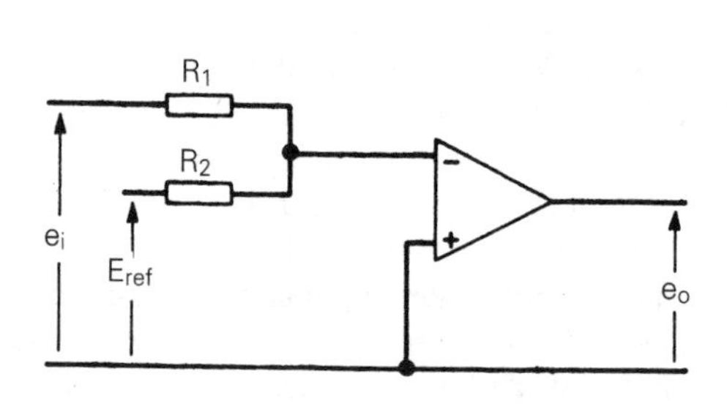

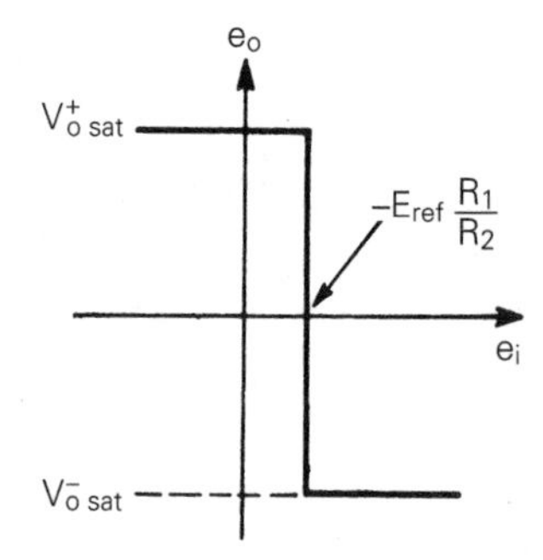

Figure 7.2 *Single ended input op-amp comparator*

In Figure 7.2, signal and reference voltages are applied to the same input terminal through appropriate resistors. The other input terminal is earthed, which means that the circuit is not subject to common mode voltage limitations and that an op-amp could be used. The output transition occurs when

$$e_i = e_t = -E_{ref}\frac{R_1}{R_2}$$

The threshold voltage e_t can be set by choice of input resistors. The reference voltage E_{ref} may be any convenient voltage of opposite polarity to the input signal.

In both the comparators shown, for the full output transition to take place the input voltage must swing past the threshold voltage by an amount

$$\frac{V_{o\ sat}^{+} - V_{o\ sat}^{-}}{A_{VOL}}$$

In the case of rapidly changing input signals the output transition time is dependent on amplifier characteristics. The switching time with slowly varying input signals depends on the rate of change of the input voltage. In such cases it is often advantageous to speed up the output transition by using some form of positive feedback. A comparator circuit with positive feedback is regenerative, in that the feedback adds to the differential input signal. A regenerative comparator (also known as a Schmitt trigger) using an op-amp is shown in Figure 7.3.

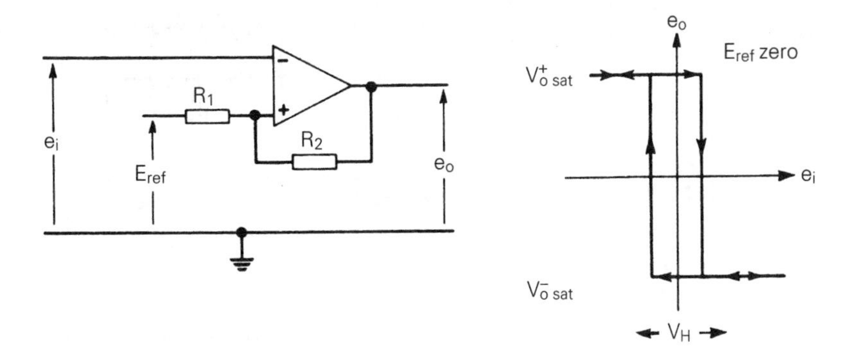

Figure 7.3 *Regenerative comparator (Schmitt trigger)*

Positive feedback is applied via resistor R_2. This is connected between the amplifier output and the non-inverting input terminal. When e_i reaches the threshold voltage the amplifier switches between saturated states. Positive feedback causes the voltage on the non-inverting terminal to rise or fall (depending on the reference voltage polarity) such that the differential input voltage is greater. This increases the drive to the output stage and the output transition time is made virtually independent of the rate of change of input voltage.

The circuit in Figure 7.3 exhibits hysteresis; that is, the transition takes place for different values of e_i dependent on whether e_i is increasing or decreasing towards the threshold value. The transfer curve for the comparator is illustrated for an op-amp with a value of E_{ref} equal to zero. The threshold value for e_i at which the transition takes place has a value, neglecting offsets, equal to

$$V_{o\,sat} \frac{R_1}{R_1 + R_2}$$

$V_{o\ sat}$ can take on both its positive and negative saturation values and the amount of hysteresis is thus

$$V_H \approx (V^+_{o\,sat} - V^-_{o\,sat}) \frac{R_1}{R_1 + R_2} \tag{7.1}$$

The amount of hysteresis is directly dependent on the magnitude of the positive feedback fraction,

$$\beta = \frac{R_1}{R_1 + R_2}$$

Avoid using a regenerative comparator (Schmitt trigger) with a very small amount of hysteresis (small β) because this usually results in high frequency oscillations at switching.

In all comparator circuits outputs may be clamped to desired values rather than using saturation limiting. It must be stressed that care should be taken to ensure that reference and input voltages do not exceed allowable limits for common mode and differential input signals. A modification of the simple comparator of Figure 7.2 is illustrated in Figure 7.4.

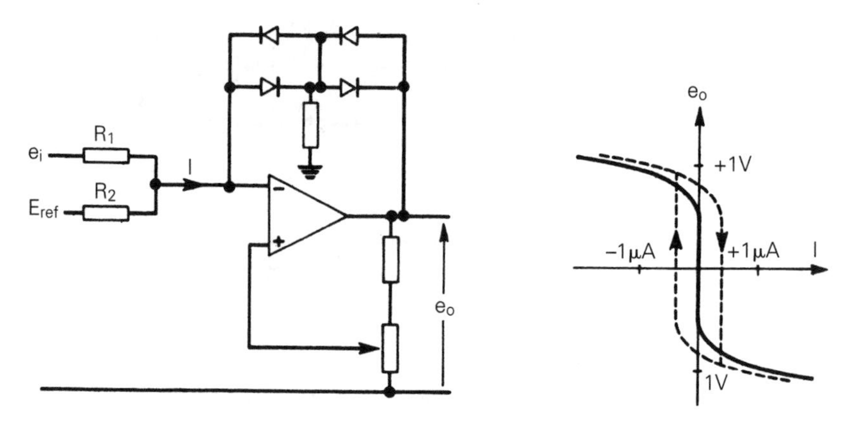

Figure 7.4 *Comparator with diode bounding*

Diodes are used to impose output bounds; the output voltage varies approximately logarithmically with the current into the amplifier summing point. The circuit incorporates a variable amount of hysteresis, which may be used to speed up the output transition for slowly varying input signals.

In both the circuits of Figures 7.2 and 7.4 the state of the output depends essentially on the direction of the current flowing towards the amplifier summing point. The circuits may thus be used to compare the sum of several voltages against a reference merely by adding appropriate resistors to the amplifier summing point. The principle is illustrated by the circuit shown in Figure 7.5.

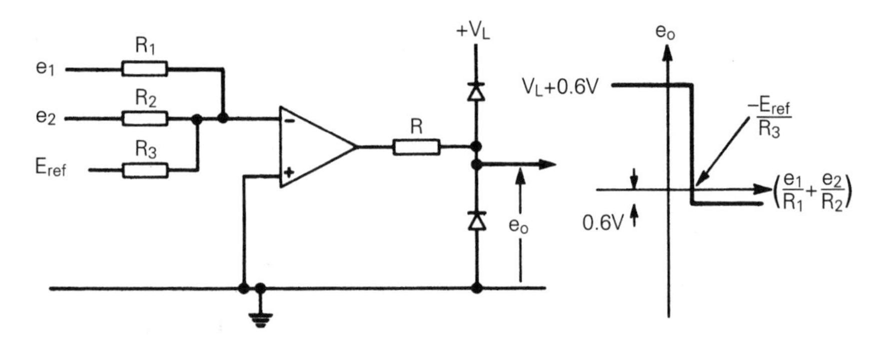

Figure 7.5 *Comparison of sum of input signals*

The output transition in this circuit occurs when

$$\frac{e_1}{R_1} + \frac{e_2}{R_2} = -\frac{E_{ref}}{R_3}$$

The circuit includes a method of restricting the comparator output for compatibility with digital integrated circuits. Resistor R is included to limit the amplifier output current.

7.2 Multivibrators

Multivibrators are a group of circuits that have two states; they are used extensively in pulse systems. There are two types described here: astable multivibrators (free running) and monostable multivibrators (one shot). Generally, circuits to perform these functions are available as digital integrated circuits. However, sometimes it is necessary to use analogue devices. Op-amps with positive feedback can be made to operate as multivibrators.

7.2.1 Astable multivibrators

In an astable multivibrator the two states of the circuit are momentarily stable and the circuit switches repetitively between these two states.

The circuit illustrated in Figure 7.6 shows a differential input op-amp acting as a free running symmetrical multivibrator.

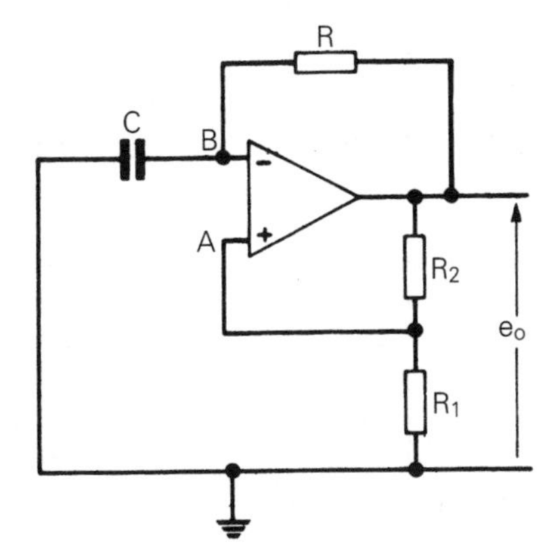

Figure 7.6 *Free running symmetrical multivibrator*

The two states of the circuit between which it switches are those in which the amplifier output is at positive and negative saturation. The amplifier output is thus a square wave. The period of the square wave is determined by the time constant CR and the feedback ratio established by the potential divider R_1, R_2.

The action of the circuit is conveniently described by reference to the waveforms illustrated in Figure 7.7. Starting at the time t' when the amplifier is in negative saturation, the voltage at terminal A is

$$V_A = \beta V_{o\ sat}^{-}$$

where $$\beta = \frac{R_1}{R_1 + R_2}$$

Terminal B is positive with respect to terminal A and its potential is decreasing as C charges down through R. When the potential difference between the two input terminals approaches zero the amplifier comes out of saturation. The positive feedback from the output to terminal A causes a regenerative switching which drives the amplifier to positive saturation. The voltage across a

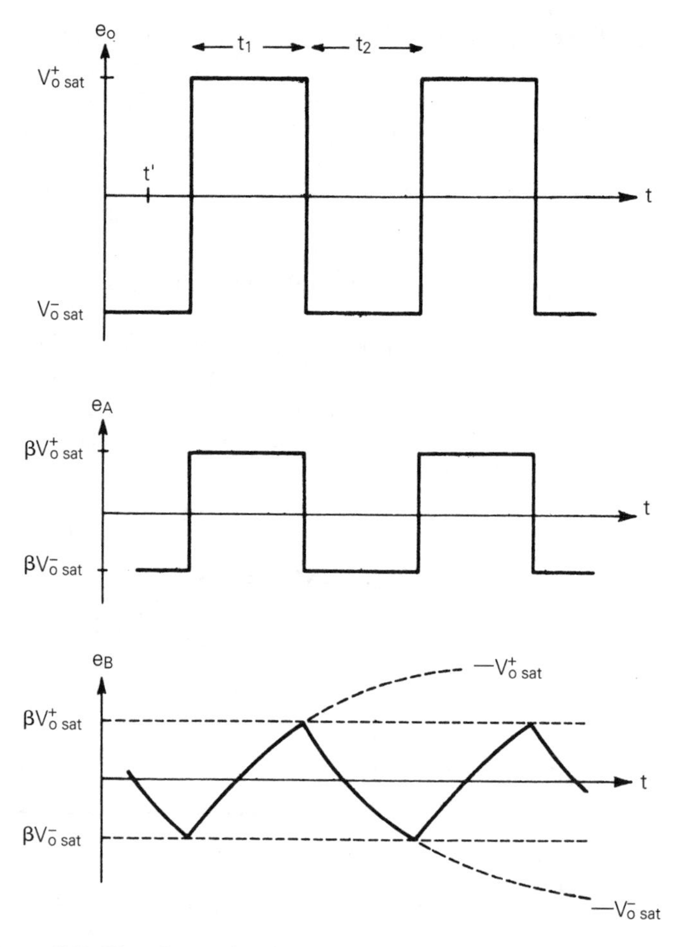

Figure 7.7 *Waveforms for free running multivibrator*

capacitor in series with a resistor cannot change instantaneously, and the potential at the terminal *B* therefore remains substantially constant during this rapid transition. Capacitor *C* now charges up through *R* and the potential at *B* rises exponentially; when it reaches $V_B = \beta V_{o\ sat}^{+}$ the circuit switches back to the state in which the amplifier is in negative saturation.

The period of the oscillations may be obtained by making use of the general equation for capacitor charging. A capacitor *C* with an initial voltage V_i, charged through resistor *R* by a voltage V_f reaches voltage V_b in time given by:

$$t = CR \ln\left(\frac{V_f - V_i}{V_f - V_b}\right) \tag{7.2}$$

Substitution of appropriate voltages from Figure 7.7 gives the timing periods:

$$t_1 = CR \ln\left(\frac{V_{o\ sat}^{+} - \beta V_{o\ sat}^{-}}{V_{o\ sat}^{+} - \beta V_{o\ sat}^{+}}\right)$$

$$t_1 = CR \ln \left(\frac{V^+_{o\,sat} - \beta V^-_{o\,sat}}{V^+_{o\,sat}(1 - \beta)} \right) \tag{7.3}$$

$$\text{and } t_1 = CR \ln \left(\frac{V^-_{o\,sat} - \beta V^+_{o\,sat}}{V^-_{o\,sat} - \beta V^-_{o\,sat}} \right)$$

$$t_1 = CR \ln \left(\frac{V^-_{o\,sat} - \beta V^+_{o\,sat}}{V^-_{o\,sat}(1 - \beta)} \right) \tag{7.4}$$

If the positive and negative values of the amplifier saturation voltage have the same magnitude, $t_1 = t_2$ and the expression to give the period of oscillation becomes:

$$t = t_1 + t_2 = 2CR \ln \left(\frac{1 + \beta}{1 - \beta} \right)$$

$$\text{which simplifies to: } T = 2CR \ln \left(1 + 2\frac{R_1}{R_2} \right) \tag{7.5}$$

Non-symmetrical multivibrator

A free running multivibrator with a non-symmetrical waveform may be obtained by using the circuit illustrated in Figure 7.8. The timing can be synchronized so that the period is an exact multiple of the period of a synchronizing signal. The synchronizing signal can be injected into the circuit at the non-inverting input terminal of the op-amp.

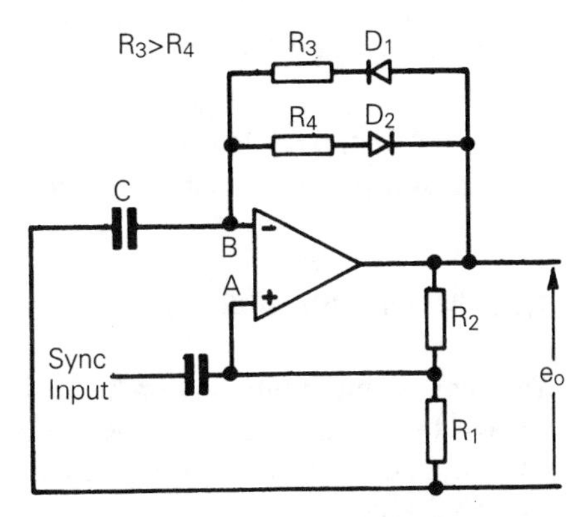

Figure 7.8 *Non-symmetrical multivibrator*

In this circuit capacitor C charges up through diode D_1 and resistor R_3. Diode D_2 is reverse biased during the timing period t_1 which is governed by the time constant CR_3. Capacitor C charges down through diode D_2 and resistor R_4; diode D_1 is reverse biased and the time constant CR_4 governs the period t_2. The waveforms are illustrated in Figure 7.9.

The timing equation for one output state of the non-symmetrical multivibrator, when the positive and negative saturation voltages have equal magnitude, is given by:

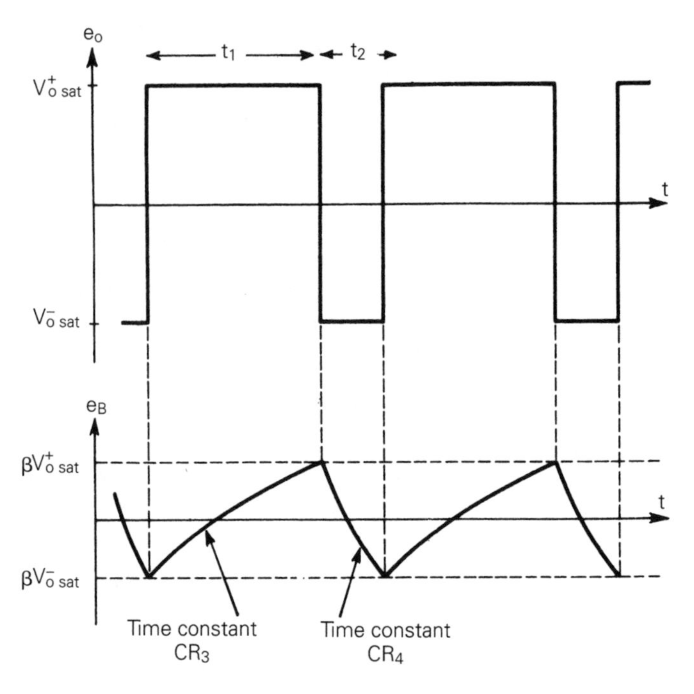

Figure 7.9 *Waveforms for non-symmetrical multivibrator*

$$t = CR \ln\left(\frac{1+\beta}{1-\beta}\right) \quad \text{and as before} \quad \beta = \frac{R_1}{R_1 + R_2}$$

Here the value of R depends on the output voltage polarity. Referring to Figure 7.8, a 'mark' is negative, so D_2 conducts and $R = R_4$. A 'space' is defined as a positive polarity output, so D_1 conducts and $R = R_3$.

7.2.2 Monostable multivibrators

The monostable multivibrator has only one stable state. It can be made to change to its other state by the application of a suitable triggering pulse. After a time interval determined by component values, the circuit returns to its stable state.

The connection of a diode in parallel with the timing capacitor in an astable circuit may be used to prevent the phase inverting input terminal of the amplifier from going positive; this gives a monostable circuit. The arrangement is illustrated in Figure 7.10.

In the permanently stable state of this circuit the amplifier output is at positive saturation, terminal B is clamped to earth by diode D_1 and terminal A is positive with respect to earth; $V_A = \beta V^+_{o\ sat}$. It is assumed that the resistor R_3 is much greater than R_1 so that its loading effect may be neglected. If the potential at the point A is brought down to earth by applying a negative pulse, the circuit switches to its temporarily stable state (in which the amplifier output is in negative saturation). Terminal A is then negative with

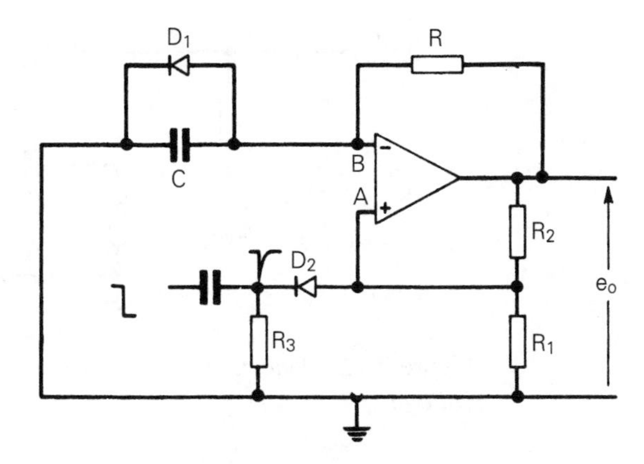

Figure 7.10 *Monostable multivibrator*

respect to earth; $V_A = \beta V_{o\ sat}^-$. The potential at B falls exponentially as C charges down through R; diode D_1 is reverse-biased. The circuit switches back to its permanently stable state when the potential at B is equal to that at A. Waveforms are illustrated in Figure 7.11.

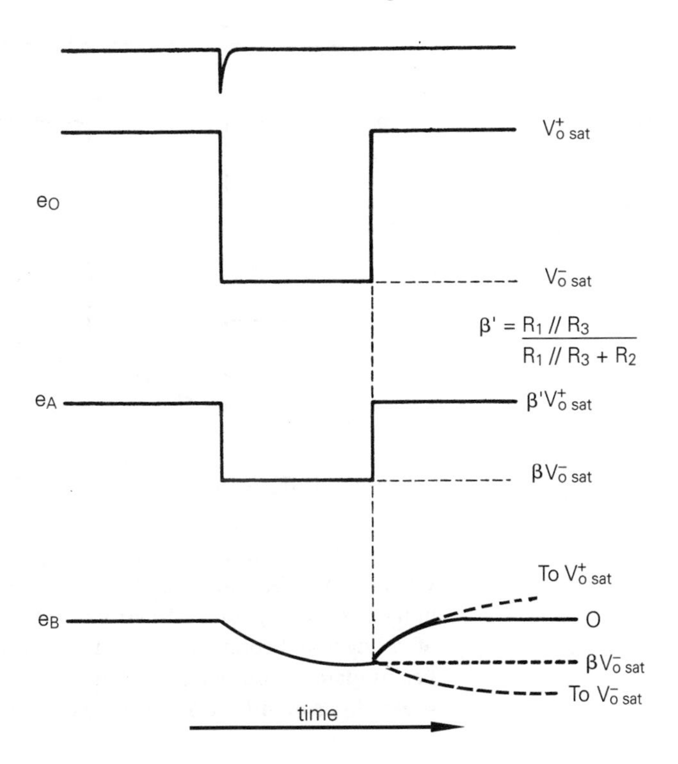

Figure 7.11 *Waveforms for monostable multivibrator*

The timing period is $T = CR \ln\left(1 + \frac{R_1}{R_2}\right)$

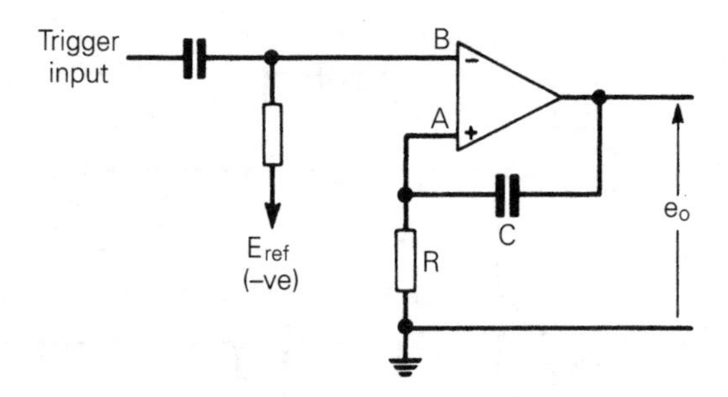

Figure 7.12 *Monostable multivibrator with period controlled by reference voltage*

The circuit illustrated in Figure 7.12 shows an alternative arrangement for a monostable circuit.

The timing period is controlled by the magnitude of a negative reference voltage that is applied to the inverting input B of the op-amp. The timing capacitor C connected between op-amp output and the non-inverting input A provides the necessary positive feedback path. Waveforms are illustrated in Figure 7.13.

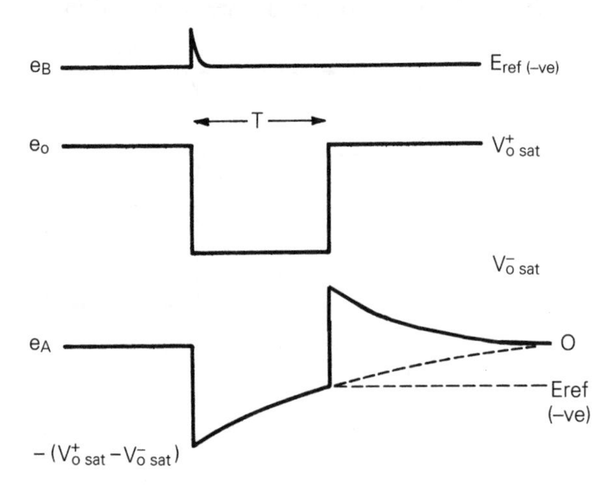

Figure 7.13 *Waveforms for circuit of Figure 7.12*

The timing periods obtained with practical astable and monostable multivibrators may be expected to show minor deviations from the values derived in the text. This is due primarily to the effect of amplifier offsets. The values of voltages and components used for practical circuits must ensure that amplifier limitations are not exceeded. The use of bounding circuits may involve some slight modification to the expressions derived for the timing periods.

7.2.3 Long delay monostable without large timing capacitor

Large value capacitors are rather expensive and bulky, and designers generally try to avoid using them. High value resistors are often used to give long time constants without having to use high value capacitors. Unfortunately, this leads to increased offset and noise in op-amp circuits. Sometimes resistor values are fixed by other circuit conditions that make large capacitance values essential.

One method to avoid the use of a large capacitor is to use a capacitance multiplier circuit. This increases the effective value of a capacitor. Capacitance multipliers can also be useful in creating an effectively variable capacitance from a fixed value capacitor.

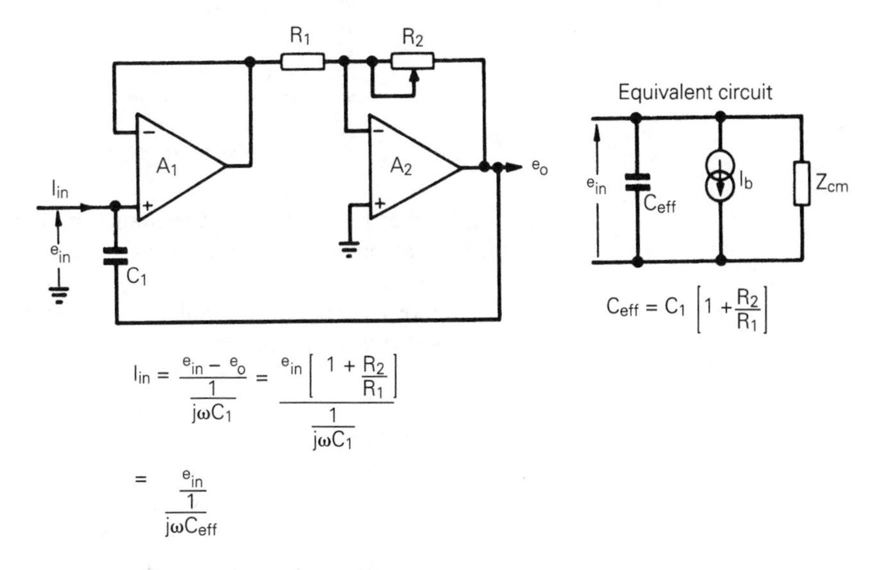

Figure 7.14 *Variable capacitance multiplier*

The circuit illustrated in Figure 7.14 allows the effective capacitance between the input terminal and earth to be adjusted by simply varying the gain of the inverting amplifier stage A_2. Amplifier A_1 acts as a unity gain follower; its function is to isolate the capacitance formed by the circuit from the loading imposed by the inverting amplifier stage.

There is a practical limit on the size of capacitance that can be created. This is determined by the fact that the capacitance multiplication achieved is almost the same as the gain of the inverting amplifier stage. Thus, the larger the capacitance multiplication, the smaller is the allowable input signal that can be tolerated without exceeding amplifier A_2 output voltage limitation.

Choice of amplifier type to be used in the practical circuit is determined largely by signal frequency requirement.

Amplifier bias currents and input offset voltage cause an offset voltage in the output of A_2, but this is not of major significance other than in its effect in limiting signal output sweep. The bias current of amplifier A_1 represents a leakage current of the synthesized capacitor, but it is not a function of the applied voltage. Bias current continues to flow even with zero applied input voltage. If the synthesized capacitor is used to perform a timing function, the bias current causes an offset error.

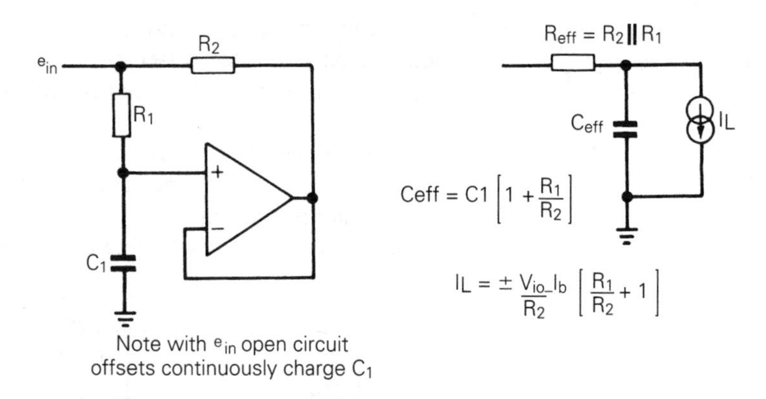

Figure 7.15 *Capacitance multiplier*

Another example of a capacitance multiplier circuit is shown in Figure 7.15. The op-amp is connected as a unity-gain follower and, neglecting offsets, its output voltage at any instant is equal to the voltage across the capacitor C_1. This output voltage is fed back via resistor R_2 to the input end of resistor R_1 in a 'bootstrap' fashion and increases the effective capacitance value between the input terminal and earth.

The performance equation and equivalent circuits given in Figure 7.15 represent circuit behaviour. Note that the multiplied capacitance has an effective resistance in series with it, so that high Q capacitors cannot be realized and the circuit cannot be used for tuned filter applications. However, it can be used in timing applications and simple RC low-pass filters where resistance is always connected in series with the capacitor.

In timing circuits, the multiplied capacitance is connected in series with an external resistor to a DC supply voltage. The voltage across the actual capacitance C_1 rises exponentially, but the time constant is determined by the multiplied capacitance value. The voltage across C_1 is available at the low impedance output terminal of the op-amp. A timing circuit using the principle is illustrated in Figure 7.16. The timing period is initiated by opening the switch, after which the voltage across capacitor C_1 rises exponentially governed by the time constant.

$$T = C_{\text{eff}}\,[R_{\text{eff}} + R_3] = C_1\left[1 + \frac{R_1}{R_2}\right][R_2 // R_1 + R_3]$$

The second amplifier in the circuit is used as a comparator, to sense when the exponentially rising voltage reaches a reference value set by the potential divider R_4 and R_5. When the reference voltage is reached, the output of A_2 switches from its positive to its negative saturation value. Long timing periods can be obtained with this circuit, without the necessity for very large *CR* values. With the component values shown in the circuit the time delay is approximately 90 s.

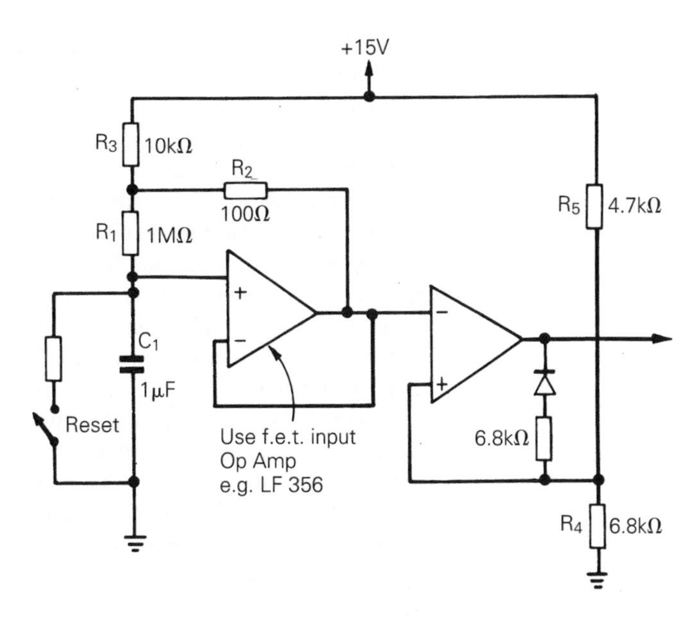

Figure 7.16 *Timing circuit not requiring excessively large CR values*

7.3 Sine wave oscillators

An oscillator continuously produces a repetitive time varying electrical signal. The most important characteristics of an oscillator are the waveform, amplitude, and frequency of the signal it produces.

Op-amps or special linear integrated circuits may be used as oscillators. The astable multivibrator discussed in the previous section uses op-amps to produce non-sinusoidal oscillations. A different arrangement of the amplifier is necessary if sinusoidal oscillations are required.

The application of sufficient positive feedback will transform any amplifier into an oscillator. In fact, when using high gain, fast roll-off amplifiers, it is easy to obtain unwanted oscillations. For this reason proper attention should be paid to decoupling and frequency compensating techniques. When an amplifier circuit is designed specifically to produce oscillations, a positive feedback loop is deliberately introduced into the circuit.

To aid understanding of feedback oscillators, consider the simple feedback loop illustrated in Figure 7.17.

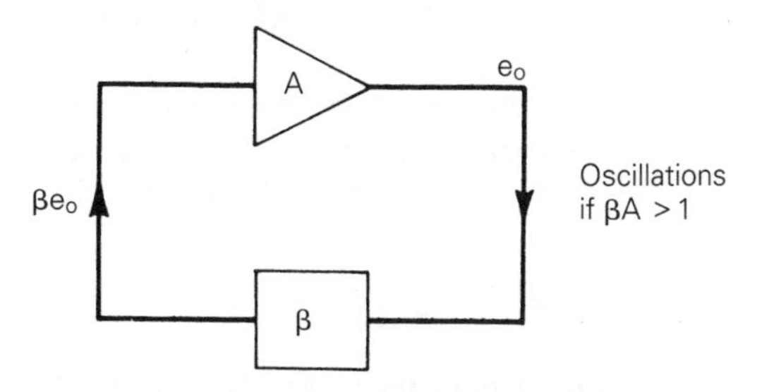

Figure 7.17 *Representation of simple feedback loop*

The diagram shows an amplifier, gain A, with a fraction β of its output signal returned to its input via a feedback network. In the general case both A and β are frequency dependent, and are represented mathematically as complex quantities. If the loop is broken at any point and a signal is injected at the amplifier input side of the break, the same signal appears at the other end of the break multiplied by the loop gain βA.

The condition that the circuit should produce continuous oscillations when the loop is closed is that the loop gain should be real, positive, and greater than unity. If this condition is satisfied, any minute disturbance (for example, noise) will trigger oscillations.

In order that the circuit in Figure 7.17 should produce a sinusoidal oscillation of defined frequency, the circuit components must be chosen so that the loop gain is greater than unity only at the desired oscillation frequency. Values of loop gain greater than unity cause a continuous growth in signal amplitude, which eventually results in waveform distortion. For stable amplitude oscillations, with undistorted waveform, it is necessary to make the effective loop gain decrease with increase in signal amplitude. Oscillations then grow to some limiting stable amplitude at which the loop gain becomes exactly unity.

7.3.1 Wien bridge oscillator

The circuit shown in Figure 7.18 illustrates the use of an op-amp in a Wien bridge oscillator.

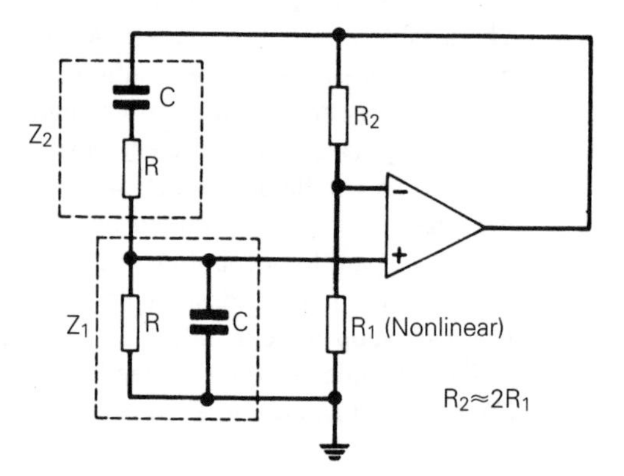

Figure 7.18 *Wien bridge oscillator*

In this circuit, feedback is applied between the output and the non-inverting input of the op-amp via the frequency dependent network Z_2, Z_1. The network produces zero phase change at a frequency

$$f_o = \frac{1}{2\pi CR}$$

Oscillations thus take place at this frequency since the feedback is positive. The output from the network

$$\left(\frac{Z_1}{Z_1 + Z_2}\right)$$

is one third that of the input at the frequency f_o. Negative feedback is applied to the amplifier via resistors R_2 and R_1 in order to reduce the loop gain to unity and so ensure a sinusoidal output waveform. If the amplifier had infinite open-loop gain, oscillations would just be maintained for values of R_2 and R_1 such that

$$\frac{R_1}{R_1 + R_2} = \frac{1}{3}$$

In a practical circuit, in order to maintain stable oscillation amplitude, a non-linear resistor is normally used for R_1. The non-linear resistor should have a positive temperature coefficient so that it increases its resistance with increasing current. This effect can be used to make the loop gain depend upon the amplitude of oscillations. An increase in the amplitude of oscillations causes an increase in the current through R_1, which results in an increase in the value of R_1. An increase in the magnitude of R_1 means a greater amount of negative feedback and a consequent reduction in loop gain and signal amplitude.

The impedances Z_2, Z_1, R_2, R_1, in fact form the arms of a bridge network (a Wien bridge). The imbalance voltage from the bridge is applied between the differential input terminals of the op-amp. Analysis of the bridge network shows that when $R_2 = 2 \times R_1$ the bridge is balanced at a frequency

$$f_o = \frac{1}{2\pi CR}$$

In practice a small imbalance must always exist but the greater the open-loop gain of the amplifier the closer is the bridge to balance and the greater is the frequency stability of the oscillator.

The circuit illustrated in Figure 7.19 shows an alternative method of ensuring amplitude stability. A field effect transistor (FET) is used in place of the non-linear resistor R_1.

For small values of drain source voltage (below 'pinch-off'), an FET behaves very much like a linear resistor. The resistance between the FET's drain and source (R_{DS}) is determined by the voltage applied between the FET's gate and source. In this circuit the oscillator output voltage is rectified by the diode D, filtered by R_5 and C_2, and applied via potentiometer R_6 to the gate of the FET. The arrangement ensures that R_{DS} takes on that value just necessary to maintain the required amplitude of oscillation. The signal amplitude applied to the bridge must be small enough to ensure that the FET is working in the linear resistance region.

7.3.2 Quadrature oscillator

The circuit illustrated in Figure 7.20 may be used to generate two sinusoidal signals in quadrature.

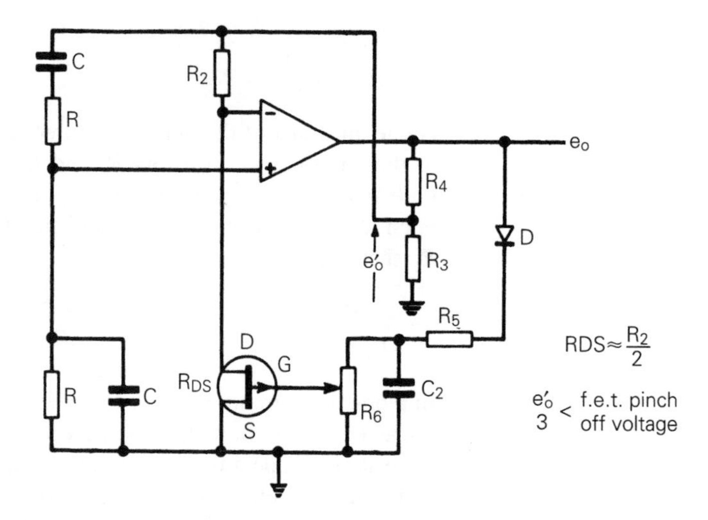

Figure 7.19 *Wien bridge oscillator with FET amplitude stabilization*

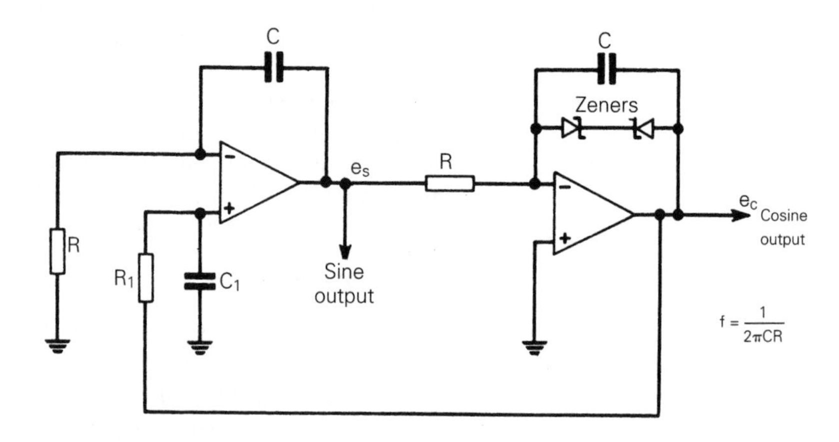

Figure 7.20 *Quadrature oscillator*

The circuit uses two amplifiers: one acts as a non-inverting integrator, the other as an inverting integrator (see Chapter 6). The two amplifiers are connected in cascade to form a feedback loop. The feedback loop is represented by the differential equations

$$RC\frac{de_S}{dt} = e_C, \quad RC\frac{de_C}{dt} = -e_S$$

The solution is represented by a sinusoidal oscillation of frequency

$$f_o = \frac{1}{2\pi CR}$$

In practice the resistor R_1 is made slightly larger than the other resistors to ensure sufficient positive feedback for oscillations. The zener diodes, used to limit the output of the inverting integrator, serve to stabilize the amplitude of oscillations.

7.4 Waveform generators

Signals with a waveshape other than sinusoidal are sometimes required. Signal generators that provide a variety of waveforms are commonly referred to as function generators. They can be built using op-amps or special linear integrated circuits.

The basic waveshapes produced by most function generators are square and triangular. These waves can be shaped by non-linear or limiting amplifiers to produce sinusoidal and other waveform shapes.

There are two basic functions performed in a waveform generator. The first is a capacitor charging, which is used to fix waveform periods and generate a triangular wave. The second is a comparator function used to sense capacitor voltage and switch between charge and discharge conditions. In the astable multivibrator circuit discussed in Section 7.2.1, both functions are performed by a single op-amp. The astable multivibrator gives a square-wave and a non-linear triangular wave.

In order to generate a linear triangular wave, the capacitor must be charged with a constant current. The astable multivibrator can be modified for constant current charging (see Appendix A1, Figure A1.7). One alternative is to use two op-amps; one providing the linear capacitor charging function and the other providing the comparator function.

7.4.1 A basic triangular square wave generator

A basic circuit for a triangular square wave generator is given in Figure 7.21. It consists of an integrator and regenerative comparator connected in a positive feedback loop.

Precise triangular waves are formed by integration of the square wave that is fed back from the comparator's output to the integrator's input. With the comparator output at its positive saturation level, the integrator output ramps down at the rate

$$-\frac{V_o^+}{CR}\ \text{V/s} \tag{7.6}$$

until it reaches the lower trip point of the comparator:

$$-V_o^+\frac{R_1}{R_2}$$

The comparator output then switches rapidly to its negative saturation level V_o^- and the integrator output then ramps up at the rate

$$-\frac{V_o^-}{CR}\ \text{V/s} \tag{7.7}$$

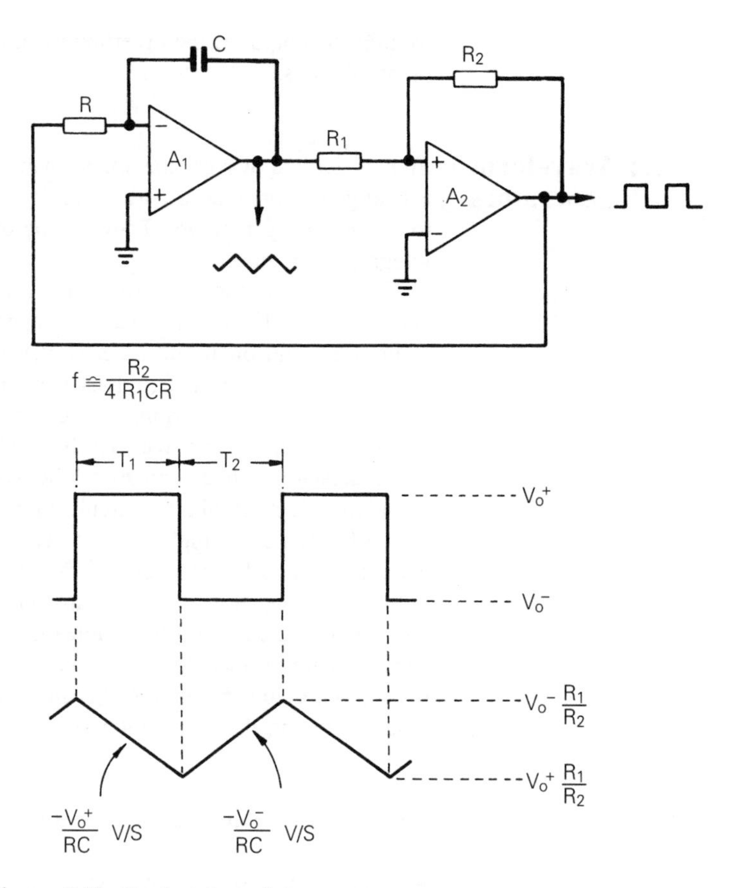

Figure 7.21 *Basic triangular square wave generator*

When the integrator output reaches the upper trip point of the comparator:

$$-V_o^- \frac{R_1}{R_2}$$

the comparator again switches states and the process repeats. The waveform periods are determined by the relationships

$$T_1 = \frac{[V_o^+ - V_o^-]\dfrac{R_1}{R_2}}{-\dfrac{V_o^+}{CR}} \text{ s} \tag{7.8}$$

$$T_2 = \frac{[V_o^+ - V_o^-]\dfrac{R_1}{R_2}}{-\dfrac{V_o^-}{CR}} \text{ s} \tag{7.9}$$

If the comparator positive and negative output limits have the same magnitude, $V_o^+ = -V_o^-$, $T_1 = T_2$ and the frequency of the oscillations is determined by the relationship

$$f = \frac{1}{T_1 + T_2} = \frac{R_2}{4R_1 CR} \tag{7.10}$$

Equation 7.10 has been derived from the assumption of ideal op-amp action. The performance limits of a practical circuit are determined by comparator slew rate and integrator bandwidth at the higher frequencies and by integrator drift at the lower frequencies.

Bias current and input offset voltage give rise to an integrator output drift. This drift increases one integration rate and decreases the other as the output of the comparator changes polarity. The effect at low frequencies is to cause a lack of symmetry in the generator waveform.

The effect of bias current and input offset voltage on the performance of A_2 is to introduce an equivalent error voltage at the non-inverting input terminal. This error voltage shifts both comparator trip points an equal amount, which in turn shifts the DC level of the triangular wave but leaves its amplitude unchanged.

7.4.2 Varying the waveform characteristics of the basic generator

The waveform characteristics of the basic function generator system of Figure 7.21 can be varied during operation by using potentiometers. The circuit shown in Figure 7.22 gives one possible arrangement. This allows adjustment of frequency, waveform symmetry, triangular wave DC offset and triangular wave amplitude. The circuit includes a zener output limiting clamp on the comparator, which sets the square wave amplitude at $\pm V_Z$.

Adjustment of the timing resistor R controls the frequency and does not alter other waveform characteristics. Potentiometer P_1 applies a voltage V_1 to the inverting input of the regenerative comparator amplifier A_2. This shifts both comparator trip points by an amount V_1/β, where β is the positive

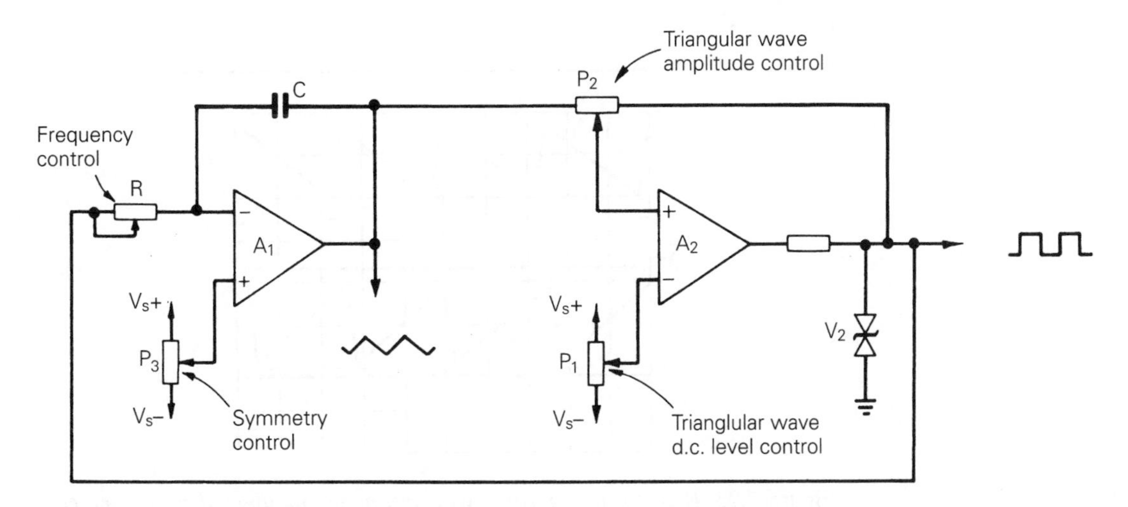

Figure 7.22 *Waveform generator with control of waveform characteristics*

feedback fraction determined by the setting of potentiometer P_2. The effect is to shift the DC level of the triangular wave by an amount V_1/β.

The setting of potentiometer P_2 determines the amount of hysteresis in the regenerative comparator. This controls the comparator trip points, and thus controls the triangular wave amplitude. Change of triangular wave amplitude is inevitably accompanied by a change in frequency. A decrease in triangular wave amplitude causes a proportional increase in frequency.

Potentiometer P_3 applies a DC offset to the integrator. This results in an increase in one timing period and a decrease in the other. This change of timing controls waveform symmetry, but it also affects the frequency.

In Figure 7.23 is an alternative circuit that allows control of waveform symmetry without altering the frequency. In the circuit, resistor values R_1 and R_2, which control the comparator trip points, are chosen so as to give a triangular wave of amplitude approximately 10 V peak-to-peak. This allows a single polarity triangular wave or ramp to be generated by adjustment of the triangular wave offset control potentiometer P_1. The traces as given in Figure 7.23 show the control of waveform symmetry obtained by adjusting the symmetry control potentiometer P_3.

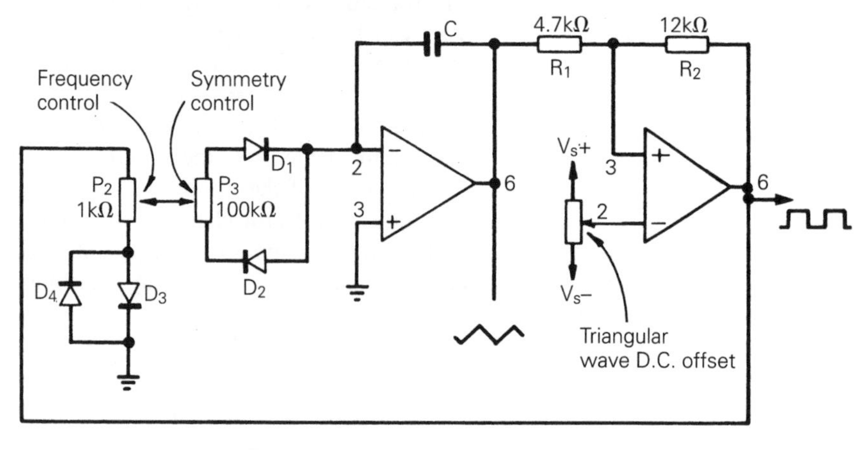

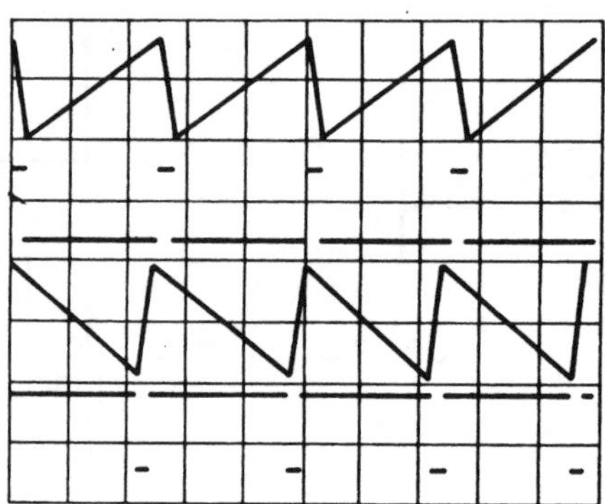

Figure 7.23 *Waveform generator with frequency unaffected by symmetry control*

Note that there is some interaction between the symmetry control potentiometer and the frequency control potentiometer, at the extreme settings of the symmetry control. This is due to unequal loading of the frequency control potentiometer on the run up and run down portion of the triangular wave. If this interaction is not tolerable, a follower can be used to buffer the output of the frequency control potentiometer.

Temperature dependence of the forward voltage drops of diodes D_1 and D_2, which can be expected to cause frequency instability at the lower levels of frequency, are compensated by diodes D_3 and D_4.

7.4.3 Waveform generator with voltage control of frequency

It is often convenient to be able to control, or modulate, the frequency of a waveform generator with a control voltage. To achieve this in the circuit of Figure 7.21, the magnitude of the current to the integrator must be varied in response to an externally applied control voltage. The sign of the integrator current must change during operation, to allow the integrator output to ramp both up and down.

A four-quadrant multiplier (see Chapter 5) could be connected between the comparator and the integrator. This is shown in Figure 7.24. The multiplier can be used for voltage control of the waveform generator's frequency. The multiplier may be thought of as acting as a voltage controlled potentiometer.

Assuming that the scaling factor of the multiplier is the normal 1/10, the square wave is multiplied by $V_c/10$ before being applied to the integrator. The equations for the waveform periods (equations 7.8 and 7.9) are in effect multiplied by $10/V_c$ and the expression for frequency (equation 7.10) is multiplied by $V_c/10$. If the comparator positive and negative output limits are equal in magnitude the frequency of oscillations given by the circuit in Figure 7.22 is thus:

$$f = \frac{V_c}{10} \frac{R_2}{4R_1CR} \tag{7.11}$$

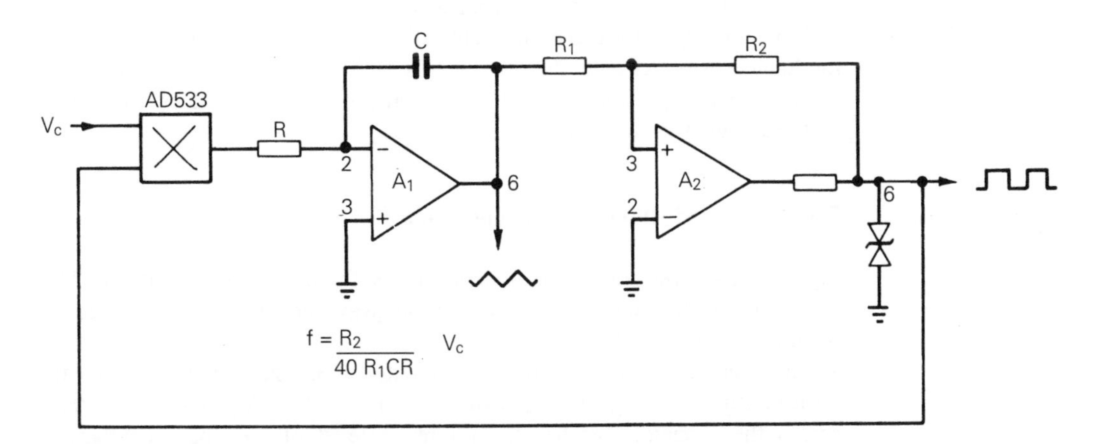

Figure 7.24 *Four quadrant multiplier allows voltage control of frequency*

7.5 The 555 timer

The 555 integrated circuit is a general-purpose timer that can be configured to give accurate time delays or oscillation frequencies. It is produced by several semiconductor manufacturers; for example, National Semiconductor (LM555), Texas Instruments and Maxim. A CMOS version of the 555 timer allows working over a wider voltage range and also draws less current. The CMOS version suffers from increased timing drift with temperature, but is otherwise identical in performance.

The 555 timer comprises two comparators, a flip-flop (set/reset bi-stable latch), a switch transistor and an output stage. The reference level of one comparator is fixed at a $1/3V_{cc}$ and the other is fixed at $2/3V_{cc}$, these levels are maintained by three equal resistors in the device which are connected across the supply voltage V_{cc}. The comparator outputs are used to set and reset the flip-flop. The flip-flop, in turn, drives both the output buffer and the switch transistor. These internal connections are shown in Figure 7.25.

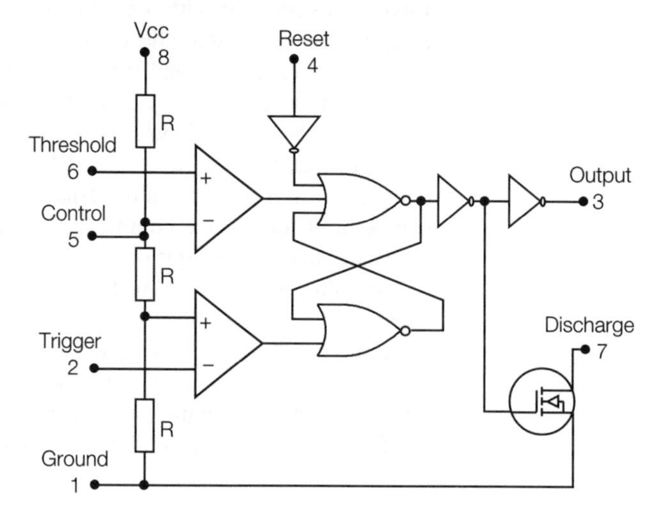

Figure 7.25 *Schematic of 555 timer*

The control voltage terminal (pin 5) allows external control of the upper and lower comparator trip points. This allows astable circuits to be frequency modulated. Most circuits do not use this facility. Instead, it is advisable to have a small capacitor between pin 5 and ground, this reduces the chance of false triggering to power supply noise.

7.5.1 The 555 timer, astable operation

Figure 7.26 gives the external connections for an astable oscillator. This circuit allows the duty cycle of the output waveform to be set by selection of timing resistor values.

An externally connected timing capacitor C is charged up towards the positive supply voltage through external resistors $R_A + R_B$. When the voltage across the capacitor reaches the reference level of the upper comparator ($2/3V_{cc}$), the comparator forces the state of the flip-flop to change, which then turns on the switch transistor. The capacitor discharges through resistor

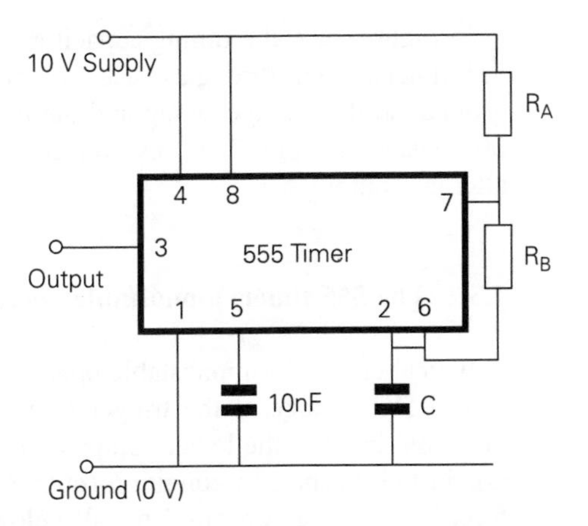

Figure 7.26 *555 timer functional schematic; external connections for free-running operation*

R_B until the voltage across it falls to the reference level of the lower comparator ($1/3V_{cc}$). This comparator then forces the state of the flip-flop to change again, which in turn turns off the switch transistor and the cycle repeats.

The oscillation frequency is given by:

$$f = \frac{1.44}{(R_A + 2R_B)\,C}$$

The duty cycle of the square wave output is determined by the ratio of R_A and R_B:

$$\text{Duty} = \frac{R_A + R_B}{R_A + 2R_B}$$

An accurate 50 per cent duty cycle can be obtained from a modified version of this circuit, as shown in Figure 7.27.

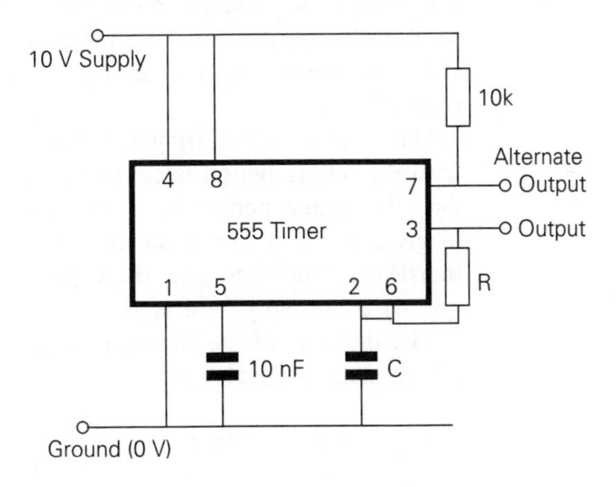

Figure 7.27 *50 per cent duty cycle oscillator using CMOS 555 timer*

In Figure 7.27, the timing capacitor C is charged and discharged from the 555 timer output, through resistor R. The CMOS output swings from V_{cc} to ground, so the voltage swing and the trip points are symmetrical about mid-rail. Discharge pin 7 is not connected to the capacitor, but provides an alternate output.

7.5.2 The 555 timer, monostable operation

A typical circuit for monostable operation of the 555 is given in Figure 7.28.

The DC voltage at the trigger terminal (pin 2) should be set above the threshold level of the lower comparator ($1/3V_{cc}$); this holds the timing capacitor in the discharged condition (output low). When a negative going pulse forces the voltage on pin 2 to fall below $1/3V_{cc}$, this triggers the flip-flop.

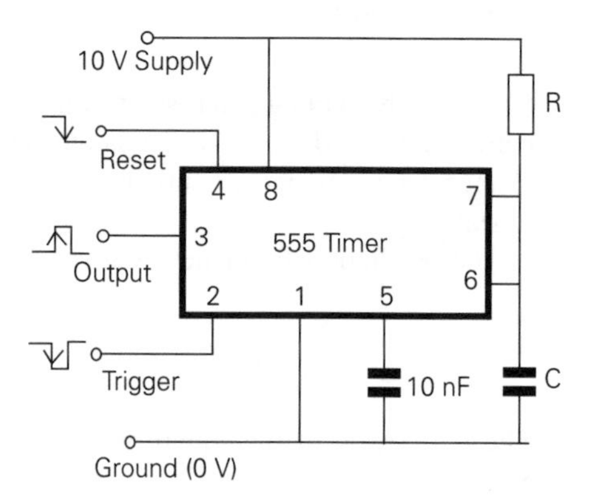

Figure 7.28 *Monostable circuit using 555 timer*

Once triggered, the flip-flop output state changes, to turn off the switch transistor that is connected across the timing capacitor. The timing capacitor then charges up exponentially through R towards V_{cc}, with the time constant CR. When the voltage across the timing capacitor reaches the threshold level of the upper comparator ($2/3V_{cc}$), the flip-flop is reset. The switch transistor is turned on, discharging the timing capacitor. The cycle is now complete.

Once the circuit is triggered it is insensitive to further triggering pulses until the timing period is complete. The triggering pulse width must be less than the timing period for proper operation. Connecting the reset terminal to ground will interrupt the timing period; this action turns on the switch transistor, which prevents the capacitor from charging. The reset terminal (pin 4) is normally held at V_{cc}.

The duration of the timing period T, during which the output level is at a high state, is given by:

$$T = -\ln(0.333)RC$$

or $T \approx 1.1\ RC$

7.6 The 8038 waveform generator

The ICL8038 waveform generator is an integrated circuit voltage controlled oscillator, which can generate frequencies up to 1 MHz. A more recent development has been the Maxim MAX038, which can generate frequencies up to 20 MHz. Both versions will be referred to as an 8038 because, apart from their operating frequency range, their functionality is identical.

The 8038 waveform generator is similar to the 555 timer, but provides additional triangular and sinusoidal output waveforms. The triangular waveform arises because the timing capacitor is charged and discharged using constant current circuits. The sinusoidal waveform is produced by internal sine shaping circuitry.

A simplified functional schematic of the 8038 devices is shown in Figure 7.29. The device contains two current sources I_1 and I_2, with a value set by external resistors R_A and R_B, respectively. A secondary control of I_1 and I_2 is through the voltage applied to IC pin 8 (FM sweep input). If pins 7 and 8 are connected together the FM sweep input voltage is set at a value $4/5V_{cc}$ by an internal potential divider, making the frequency independent of supply voltage.

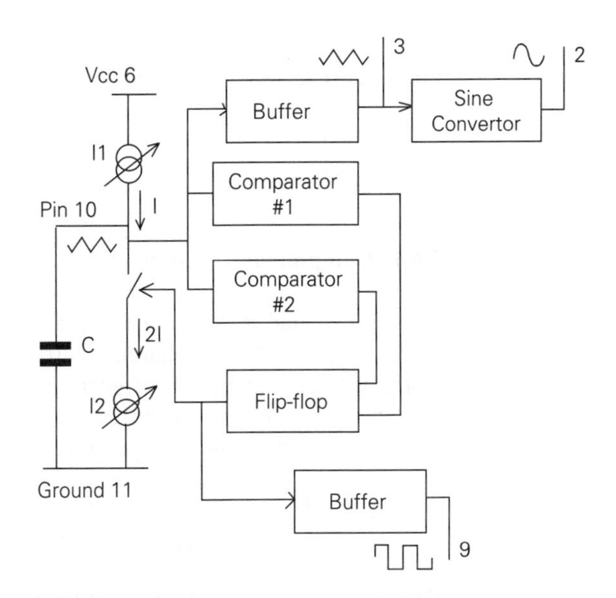

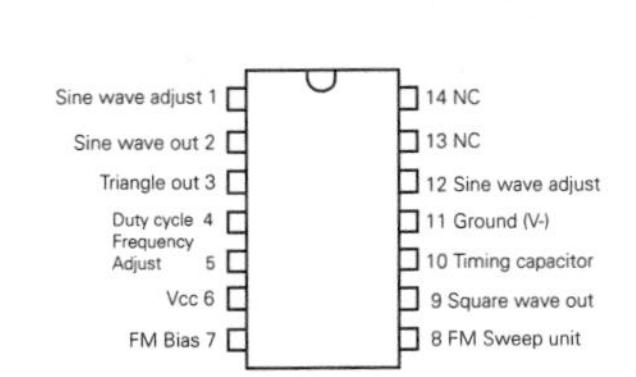

Figure 7.29 *8038 functional diagram and IC pin-out*

Current source I_1 provides a continuous charging current to timing capacitor C, connected between IC pin 10 and ground. Current source I_2 is usually set to have a current flow double that of I_1. When current source I_2 is switched on it discharges capacitor C. The net discharge current is then $I = (I_2 - I_1)$ and, if I_2 equals $2 \times I_1$, the charge and discharge currents are equal.

Like the circuit of the 555 timer, the 8038 device includes two comparators and a flip-flop. The threshold levels of the two comparators are set at $2/3V_{cc}$ and $1/3V_{cc}$ by three equal resistors connected across the supply voltage. The flip-flop is set and reset, causing I_2 to be switched on and off, by the two comparators. Thus, the capacitor is charged and discharged between the levels $1/3V_{cc}$ and $2/3V_{cc}$ giving a triangular wave of magnitude $1/3V_{cc}$.

The capacitor is charged by current I_1 alone, and this takes place in time period t_1, given by:

$$t_1 = \frac{CV}{I} = \frac{C(V_{CC}/3)\, R_A}{0.22V_{CC}} = \frac{CR_A}{0.66}$$

The capacitor's discharged current is $I_2 - I_1$, and this takes place in time period t_2, given by:

$$t_2 = \frac{CV}{I} = \frac{C(V_{CC}/3)}{2\left[0.22\,\frac{V_{CC}}{R_B}\right] - \left[0.22\,\frac{V_{CC}}{R_A}\right]} = \frac{CR_AR_B}{0.66\,(2R_A - R_B)}$$

The frequency of oscillation is given by:

$$f = \frac{1}{t_1 + t_2} = \frac{1}{\frac{R_AC}{0.66}\left(1 + \frac{R_B}{2R_A - R_B}\right)}$$

If $R_A = R_B = R$, then I_2 equals $2 \times I_1$, which gives equal charge and discharge times. This results in a 50 per cent duty cycle and a frequency given by:

$$f = \frac{0.33}{RC}$$

7.6.1 Component selection

Resistors R_A and R_B should be selected to give charging currents in the range 10 μA to 1 mA. Lower charging currents give errors due to leakage, particularly at high temperature. Higher charging currents are limited in accuracy by saturation voltages in the constant current circuits. If pins 7 and 8 are connected together, the charge current is given by:

$$I = \frac{R_1\,(V^+ - V^-)}{R_1 + R_2}\,\frac{1}{R_A} = \frac{0.22(V^+ - V^-)}{R_A}$$

Resistors R_1 and R_2 are within the 8038 device. R_1 = 11 kΩ and R_2 = 39 kΩ.

Capacitor C should be chosen, using the frequency equations, at the upper end of the required frequency range.

7.6.2 Fixed frequency operation

Circuit connections for fixed frequency operation are shown in Figure 7.30.

In Figure 7.30(a), separate resistors R_A and R_B are used to adjust the frequency and duty cycle. A fixed 82 kΩ resistor connected between pin 12 and the negative supply reduces sine wave distortion.

In Figure 7.30(b), a low-value potentiometer is used to adjust the duty cycle. A variable resistor is used between pin 12 and the negative supply, to give minimum sine wave distortion.

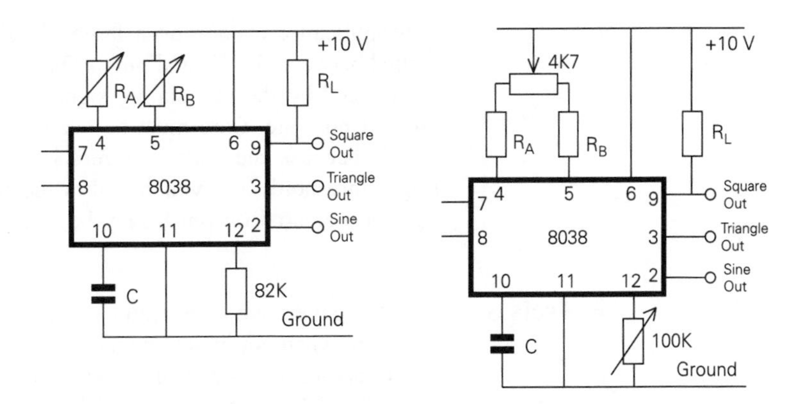

Figure 7.30 *Fixed frequency operation of the 8038*

The square wave output (pin 9) is available at the open collector of an internal transistor. This enables the output load resistor to be returned to a separate power supply. The output can thus be made TTL compatible by returning the load to +5 volts, whilst the waveform generator is powered from a higher voltage. Non-symmetrical waveforms can be produced by using different values for resistors R_A and R_B.

7.6.3 Variable frequency operation

The signal frequency produced by the 8038 is a direct function of the DC voltage at pin 8, the FM sweep input. By altering the voltage on pin 8, frequency modulation is achieved. For small deviations (e.g. up to ±10 per cent) the modulating signal can be applied to pin 8 via a decoupling capacitor. The connections are shown in Figure 7.31.

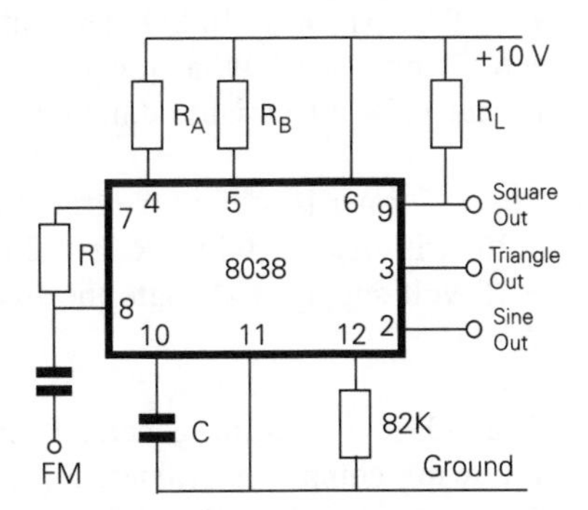

Figure 7.31 *Frequency modulated (FM) oscillator*

The external resistor between pins 7 and 8 is not necessary. Without it, the input impedance is 8 kΩ. Adding the external resistor increases the input impedance, by an amount equal to its resistance:

$$Z = R + 8\,\text{k}\Omega.$$

In applications requiring a large FM deviation, the modulating signal is applied between the V_{cc} and pin 8. Thus the entire bias for the current sources is produced by the modulating signal and a very large (e.g. 1000:1) sweep range is possible. Care must be taken to regulate V_{cc} in this configuration. This is because the charge current is no longer a function of V_{cc}, but the trigger thresholds are, which makes the oscillation frequency dependent on V_{cc}. The potential on pin 8 may be swept from V_{cc} to about $2/3V_{cc}$.

Exercises

7.1 A regenerative comparator is required to give a negative output transition when an input signal rises through a value 3 V. The reverse transition is to take place when the input signal decreases through the value 2.7 V. The upper and lower limits of the comparator output are to be bound at +5 V and −1 V respectively. Sketch a suitable circuit indicating appropriate component values.

7.2 A comparator is required to indicate when the average value of three input signals, weighted in the proportions 1:2:3, is equal to 5 V. Sketch a suitable circuit.

7.3 The following component values are used in the circuit shown in Figure 7.6; $C = 0.1\ \mu F$, $R = 47\ k\Omega$, $R_1 = 10\ k\Omega$, $R_2 = 47\ k\Omega$. The output voltage is bounded to the limits +10 V and −5 V. Calculate the timing periods and sketch the circuit waveforms (analogous to Figure 7.7).

7.4 The circuit of Figure 7.8 is used to produce a non-symmetrical square wave of frequency 1 kHz and mark-space ratio 4:1. If $R_2 = 22\ k\Omega$, $R_1 = 4.7\ k\Omega$, $C = 0.01\ \mu F$, and the output limits of the amplifier are fixed at ±10 V what are the values of R_3 and R_4?

7.5 The following components are used in the circuit of Figure 7.12; $C = 0.01\ \mu F$, $R = 10\ k\Omega$. The output limits of the amplifier are set at +6 V and −1 V. What range of reference voltage is required for an output pulse of variable duration, between 50 μs and 200 μs?

7.6 The 555 timer is used in the free-running mode of operation (see Figure 7.26) with $R_A = 10\ k\Omega$, $R_B = 22\ k\Omega$ and $C = 10$ nF. The timer has a +15 volt supply. Calculate the frequency and duty cycle (see Section 7.5.1).

7.7 The 8038 waveform generator in Figure 7.30(a) is used with the following component values; $R_A = 4.7\ k\Omega$, $R_B = 5.6\ k\Omega$, $C = 22$ nF, $V_{cc} = 15$ volts. Pins 7 and 8 are connected together. What are the time periods of the rising (t_1) and falling (t_2) parts of the triangular waveform? Sketch the waveforms to be expected.

8 Sensor interface, analogue processing and digital conversion

This chapter describes sensor interface circuits, peak detector circuits, sample-and-hold circuits and both analogue-to-digital and digital-to-analogue converters.

Interface circuits are required to convert the sensor's output signal into a convenient form. A high impedance and low level sensor signal may have to be buffered and amplified before being processed by subsequent circuits.

Peak detector circuits and sample-and-hold circuits are often process sensor signals. It may be that the largest signal within a certain period is required. Or the value of a signal at a particular instance is needed.

Computers and other digital processing systems require analogue signals to be converted into digital form. An analogue-to-digital converter (ADC) performs this function. Analogue outputs from computing systems are produced by a digital-to-analogue converter (DAC).

8.1 Sensor interface circuits

The first stage in any measurement system is often a sensor or transducer. This is a device for converting the quantity under investigation such as mechanical movement, temperature, pressure and force into an electrical signal.

Op-amps are used extensively to amplify weak electrical signals from a transducer before performing any signal processing. The most suitable op-amp circuit configuration depends on the electrical output characteristics of the transducer. The expected signal level, the frequency response and the source impedance are most important.

An amplifier with high input impedance is required to interface with transducers that produce an output voltage, proportional to the measurement variable. This minimizes errors caused by loading of the transducer's output signal. A current-to-voltage converter circuit is required for transducers that produce an output current, proportional to the measurement variable (e.g. photodiodes). The bandwidth of the amplifier should be greater than the expected frequency content of the transducer output signal; otherwise there cannot be accurate reproduction of the signal waveform.

Differential transmission is used in cases where the transducer is remote from the amplifier. This technique reduces the pick-up of noise and other unwanted signals. The transducer's output is connected across a pair of wires, to produce a differential signal. At the other end of the transmission line a differential input amplifier is required. The input impedance of this amplifier is usually matched to that of the transmission line. If the end of a transmission line is terminated by high impedance, or is unbalanced, it will be susceptible to external signal pick-up.

8.1.1 Bridge amplifiers for resistive transducers

Resistive transducers are available which respond to temperature, light intensity and physical strain. When precise measurements are to be made, the transducers can be included in the arms of a balanced bridge. Changes in the physical variable to which the transducer is sensitive cause an unbalance in the bridge; the extent of the unbalance being used to measure the change in the physical variable.

There are several amplifier circuits that can be used with bridge circuits. The most suitable circuit depends on the nature of the particular application. Here are some of the points that have to be considered in choosing a particular circuit:

- earthed or floating bridge voltage supply;
- earthed or floating unknown resistor;
- output voltage linearly related to changes in the unknown resistor;
- temperature sensitivity of the bridge circuit. This last point determines whether or not the arrangement is sensitive to changes in the ambient temperature affecting *all* the arms of the bridge.

The circuit illustrated in Figure 8.1 is basically an application of the subtractor amplifier (Chapter 4). E denotes the battery voltage.

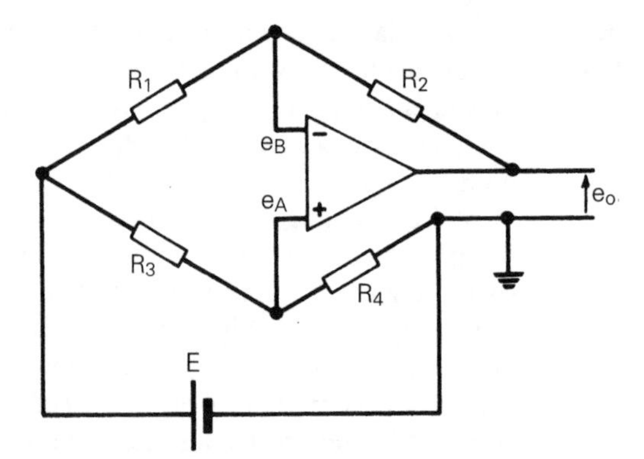

Figure 8.1 *Bridge amplifier, earthed bridge supply*

Referring to Figure 8.1 and assuming that the amplifier behaves ideally, the following analysis holds:

$$e_B = e_o + \frac{(E - e_o) R_2}{R_1 + R_2}$$

$$e_A = E\frac{R_4}{R_3 + R_4} \tag{8.1}$$

But $e_A = e_B$

Substitution and rearrangement gives

$$e_o = \left[\frac{R_4 - \dfrac{R_2}{R_1}R_3}{R_4 + R_3}\right]E \tag{8.2}$$

There are two ways in which the circuit may be used, dependent upon which arm of the bridge contains the unknown resistor (R_x). One method is to make $R_1 = R_o$, $R_2 = R_x = R_o(1 + \alpha)$, and $R_3 = R_4$. Substituting these values in equation 8.2 gives,

$$e_o = \frac{\alpha}{2}E \tag{8.3}$$

The circuit gives an output voltage which is linearly dependent upon $(R_x - R_o)$, the difference between the unknown and the standard. Linearity is maintained for large deviations from bridge balance. A possible disadvantage of this arrangement is that the unknown resistor is not earthed (it is floating).

Another way of using the circuit is to place the unknown resistor in the position occupied by R_4. We make $R_3 = R_o$, $R_4 = R_x = R_o(1 + \alpha)$ and $R_1 = R_2$. Substituting these values in equation 8.2 gives

$$e_o = \frac{\alpha}{2 + \alpha}E \tag{8.4}$$

With this arrangement the output is linear only for small deviations in the unknown ($\alpha << 2$), and is useful when one end of the unknown must be earthed. The amplifier output does not have to supply current through the unknown resistor. Thus, if this is required by the application, large currents may be passed through the unknown resistor.

An advantage of both arrangements is the earthed bridge supply. The output level from the bridge is independent of the bridge impedance. However, the circuits do not provide amplification and the measurement of small resistance changes may require an additional gain stage. Care must be taken to ensure that the maximum common mode voltage rating of the op-amp is not exceeded.

A single ended input amplifier may be used in the circuit shown in Figure 8.2. Bridge unbalance causes a voltage

$$\frac{E\alpha}{4\left(1 + \dfrac{\alpha}{2}\right)}$$

to be developed across the bridge. In order to force the amplifier input voltage (e_ε) to zero, the amplifier output voltage develops a voltage at A equal to the bridge unbalance voltage.

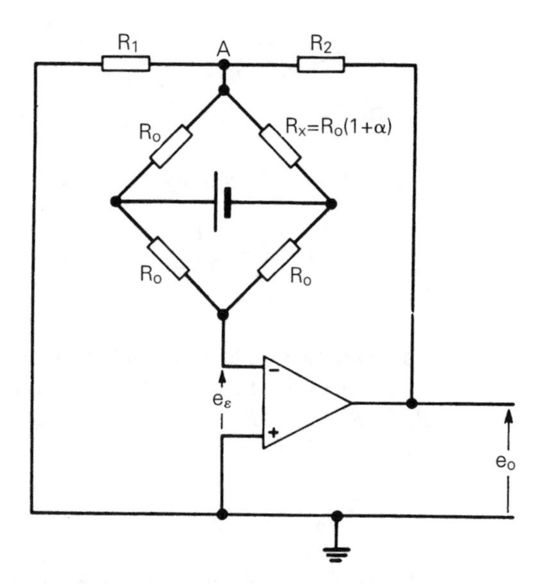

Figure 8.2 *Bridge amplifier, single ended amplifier*

$$\text{Thus, } e_o \frac{R_1}{R_1 + R_2} = \frac{E\alpha}{4\left(1 + \frac{\alpha}{2}\right)}$$

$$\text{and } e_o \frac{R_1 + R_2}{R_1} = \frac{E\alpha}{4\left(1 + \frac{\alpha}{2}\right)} \tag{8.5}$$

This is linear for $\alpha/2 << 1$.

The circuit provides amplification of the bridge unbalance voltage and is independent of bridge impedance. However, the need for a floating bridge supply may be a disadvantage.

Another bridge circuit is shown in Figure 8.3. In this circuit the feedback round the amplifier causes the opposing corners of the bridge to be at the same potential. The amplifier output voltage establishes the differential current into the bridge needed to maintain this condition.

Summing currents at B

$$\frac{E - e_B}{R_o} - \frac{e_B}{R_o} + \frac{e_o - e_B}{R} = 0$$

Summing currents at A

$$\frac{E - e_A}{R_o} - \frac{e_A}{R_o(1 + \alpha)} - \frac{e_A}{R} = 0$$

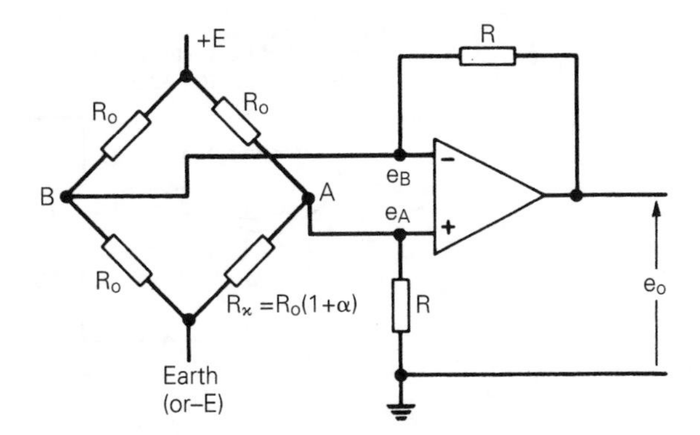

Figure 8.3 *Bridge amplifier, earthed or floating supply*

Equating $e_A = e_B$ and rearranging gives

$$e_o = \frac{R}{R_o} E\alpha \frac{1}{(1+\alpha)\left(1+\dfrac{R_o}{R}\right)+1} \tag{8.6}$$

(linear for $\alpha << 1$).

The circuit may be used with an earthed bridge supply, but it has the disadvantage of having a sensitivity that is dependent on bridge impedance. The op-amp should be chosen to be insensitive to the input common mode voltage.

8.1.2 Interfacing high impedance transducers

Some transducers have very high output impedance, which is essentially capacitive. Examples include piezoelectric accelerometers, pressure transducers and capacitive (condenser) microphones. Transducers of this kind operate by producing a charge that is proportional to the measurement variable. The charge can be converted into a voltage by using an op-amp connected as a current integrator.

The current integrator or charge amplifier arrangement has the desirable feature that the transducer is connected between the amplifier summing point and earth, and this means that the signal is unaffected by stray capacitance. The amplifier summing point is a virtual earth, so stray capacitance and cable capacitance between this point and earth has no potential across it (effectively it is earthed on both sides) and therefore has no effect on the signal.

A theoretical charge amplifier circuit is shown in Figure 8.4(a). The output of the capacitive transducer is represented by an equivalent circuit consisting of a voltage source e_t in series with a capacitance C_t. The amplifier gives an output signal $e_o = -(C_t/C_f)e_t$.

In a practical charge amplifier, it is necessary to provide a DC path for amplifier bias current. This takes the form of a feedback resistor R_f connected between the op-amp's output and its inverting input. Without this resistor, C_f continuous charges and causes the output to drift into saturation. However,

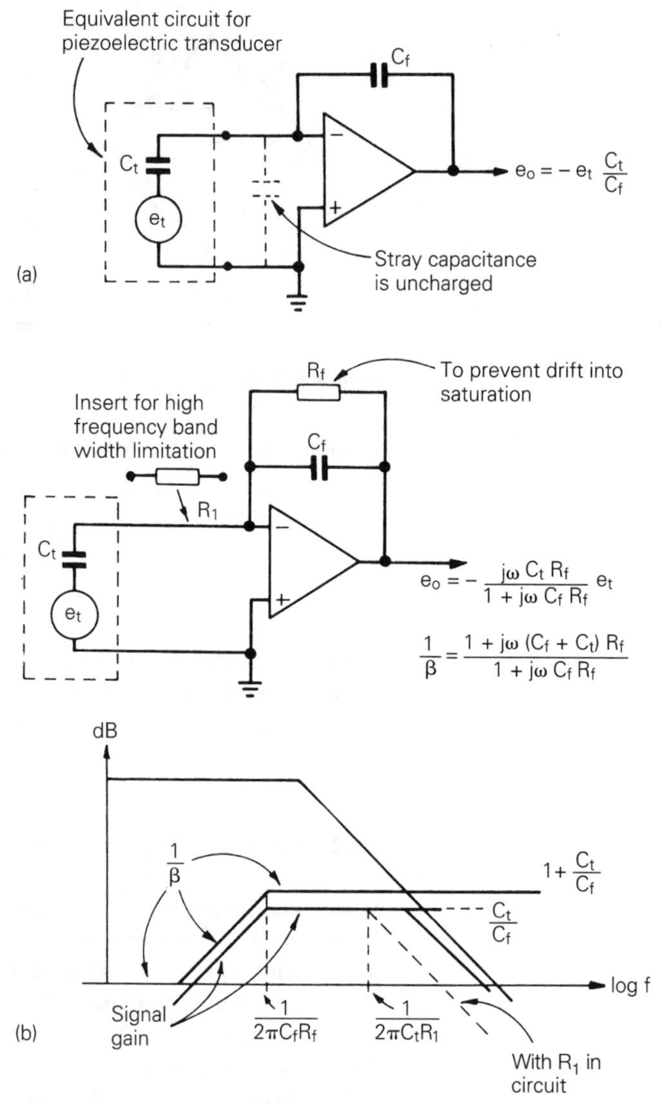

Figure 8.4 *Charge amplifier. (a) Ideal charge amplifier. (b) Practical charge amplifier and Bode plots*

the presence of the resistor R_f limits the lower bandwidth of the charge amplifier to the frequency given by:

$$f_L = \frac{1}{2\pi C_f R_f}$$

Bode plots, for the practical circuit, are illustrated in Figure 8.4(b). Note that for signal frequencies less than f_L, the output is proportional to the differential of the input signal. A very large value for R_f is required if the amplifier output is to reproduce faithfully slow changes in the measurement variable.

This normally requires the use of a low bias current op-amp (a FET input amplifier) in order that offset and drift error should not be excessive.

The closed-loop upper frequency limit is determined by the frequency at which the Bode plots for $1/\beta$ and the open-loop gain intersect. If it is required to restrict the upper frequency limit, this can be accomplished by connecting a resistor R_1 in series with the transducer as shown in Figure 8.4(b).

8.2 Hot wire anemometer with constant temperature operation

Hot wire anemometers are often used to measure air speed. The principle of operation depends upon the cooling of an electrically heated platinum filament by the movement of air around it. In the arrangement illustrated in Figure 8.5 the heated filament is included as one arm of a balanced bridge.

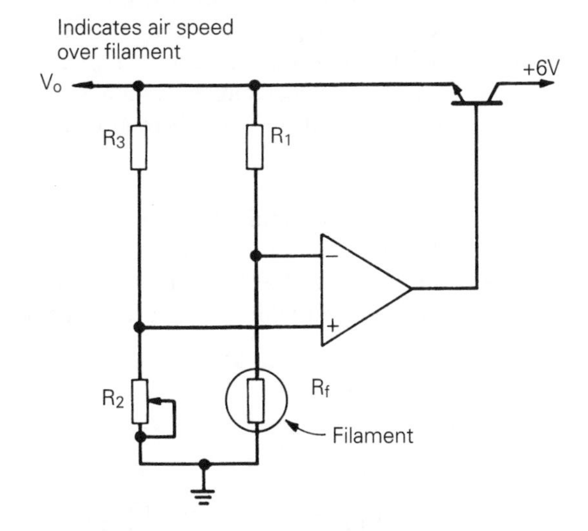

Figure 8.5 *Op-amp forces constant filament temperature in hot wire anemometer*

The output voltage of the op-amp supplies the bridge, with a simple transistor emitter follower being used to boost the amplifier output current. A Darlington connected transistor pair can be used for greater currents, if required. The amplifier output voltage changes in such a way as to force the input error voltage towards zero and in so doing it establishes the bridge balance condition

$$R_f = \frac{R_1 R_2}{R_3}$$

Platinum has a positive temperature coefficient of resistance. If the air flow over the filament increases, it loses more heat and its temperature begins to fall. The falling temperature causes the resistance of the element to reduce and this reduces the voltage on the op-amp's inverting input. The amplifier output voltage increases so as to increase the power dissipation in the filament and hold its resistance, and hence its temperature, constant.

The output voltage of the amplifier represents an amplified form of the filament voltage. This gives a measure of the air speed over the filament. However, the output varies non-linearly with air speed and sensitivity is greater at the lower speeds. Varying the value of resistor R_2 sets the operating temperature of the filament, and some experimentation is needed to find the setting that gives the best sensitivity.

It is possible for the circuit to remain inoperative when switched on. The emitter follower may not be brought into conduction because the op-amp output is in negative saturation. A positive offset applied to the non-inverting input of the op-amp ensures operation at switch-on.

Heated thermistors are sometimes used in airflow measurements. They can be operated at constant temperature, using an op-amp circuit arrangement similar to that of Figure 8.7. It is simply necessary to interchange input leads to the op-amp because most thermistors have a negative temperature coefficient of resistance. Their resistance decreases with rise in temperature. Constant temperature operation allows rapid measurements of changes in flow, since there is no thermal delay.

8.3 Temperature measurement using a thermocouple

The typical circuit shown in Figure 8.6 may be used with a K-type (see Appendix A4) thermocouple for the measurement of temperatures in the range 0°C to 100°C. If the component values are changed then this circuit may be used with other types of thermocouple at different temperatures.

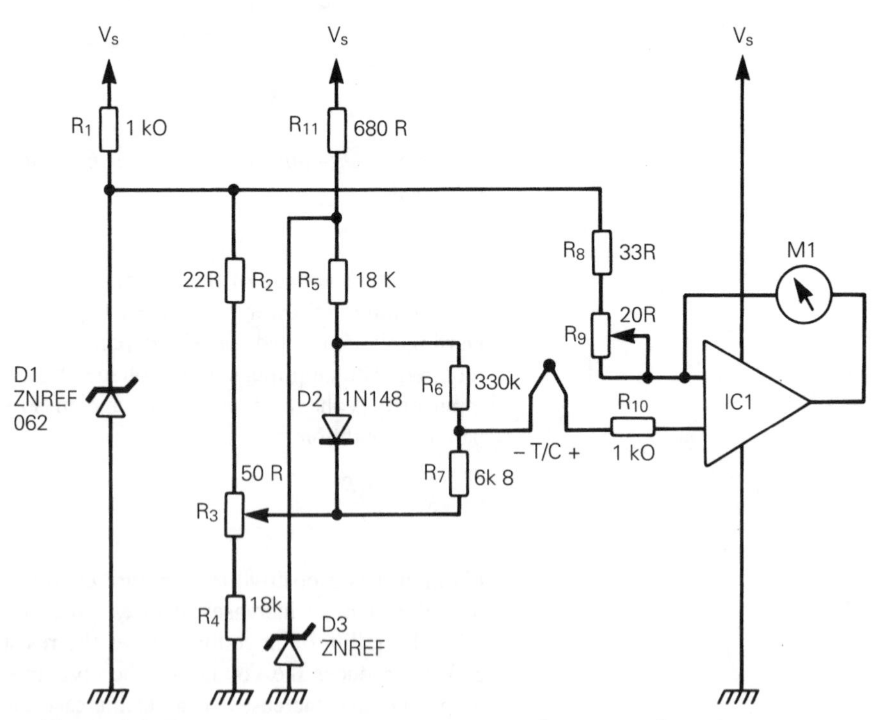

Figure 8.6 *Temperature measurement using a thermocouple and an op-amp*

The cold junction is at ambient temperature and any variation in this is compensated by the change, about 2 mV/°C, in the forward voltage drop of diode D_2. A potential divider reduces this voltage to the thermocouple output. The result, amplified by IC_1 and displayed on any 100 μA meter, is proportional to the temperature of the measurement junction of the thermocouple. The single voltage supply is intended to be from a standard 9 V battery.

The meter zero is set by adjustment of R_3; the span by R_9 and the calibration procedure is as follows:

1. Place the measurement junction of the thermocouple in freezing water to give 0°C.
2. Adjust R_3 to set a reading of zero on the meter.
3. Place the measurement junction of the thermocouple in boiling water to give 100°C.
4. Adjust R_9 to give full-scale deflection on the meter.
5. The circuit will then be calibrated to give a reading of 1 μA/°C.

8.4 Light sensitive switching

The circuit shown in Figure 8.7 is that of a very sensitive light-operated relay. The relay is energized when the light striking the light-dependent resistor, NORP12, exceeds a level determined by the setting of VR_1. If R_1 and NORP12 were interchanged, the relay would be energized in poor light conditions and would de-energize when the light intensity reached the preset level.

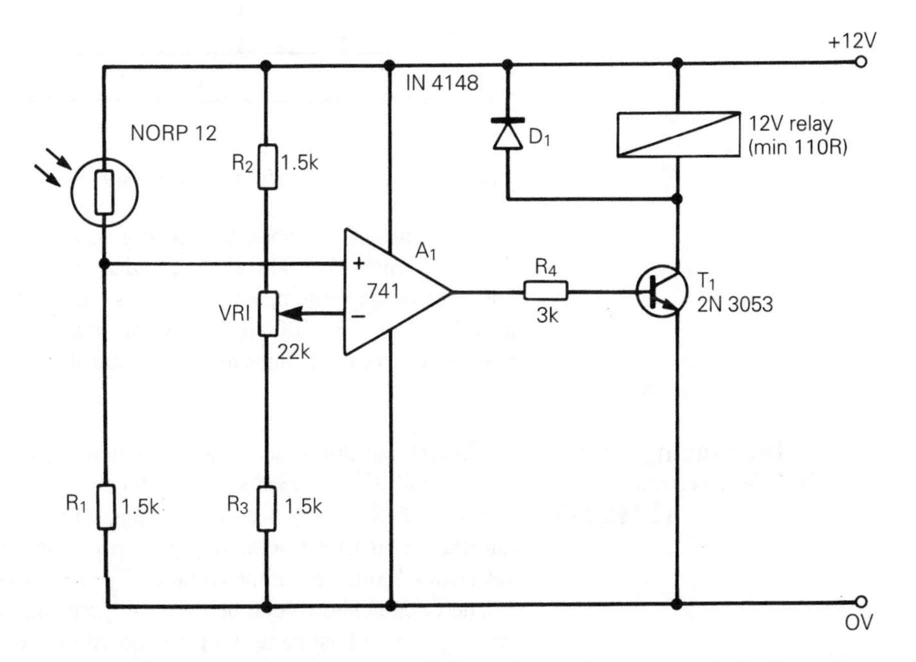

Figure 8.7 *Extremely sensitive light-operated relay using a light-dependent resistor*

8.5 Sensing analogue light levels

The circuit shown in Figure 8.8 uses a general-purpose photodiode and FET input op-amp, which has a low input bias current. The input signal to the op-amp is obtained from the photodiode. The output produced is a steady DC indication of the light level and this is particularly useful in photometry applications.

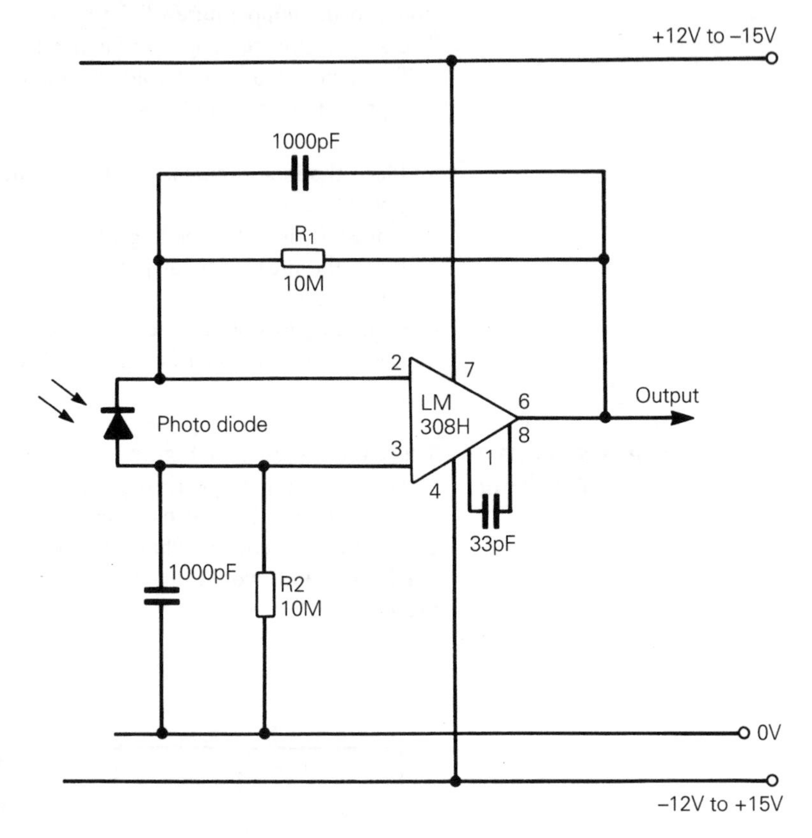

Figure 8.8 *Analogue light level sensor*

The values of components shown give sensitivity to light of approximately 14 V/mW/cm^2. The values of R_1 and R_2 may be reduced for less sensitivity but must be kept equal. For values less than 100 kΩ, a bipolar input op-amp may be used. The 1000 pF capacitors may be increased in value to reduce any ripple from AC lighting or to control the response time accordingly.

8.6 Interfacing linear Hall effect transducers (LHETs)

Hall effect transducers sense magnetic fields (see Appendix A4, Section A4.11). Linear Hall effect transducers (LHETs) output a voltage proportional to the magnetic field strength at the device. Figure 8.9 shows three methods of interfacing an LHET with single supply op-amp circuits. The op-amp characteristics limit the output voltage (V_o) equations at the high and low ends.

The circuits shown can be used with adjustable gain and adjustable offset, although the adjustments will not be completely independent. One method is to adjust the gain to the desired value with V_1 set at approximately half the op-amp supply voltage. Then adjust V_1 to give the exact offset at the V_o required for the application.

$$V_o = (V_1 - V_{IN})\frac{R_2}{R_1} + V_1$$

(a)

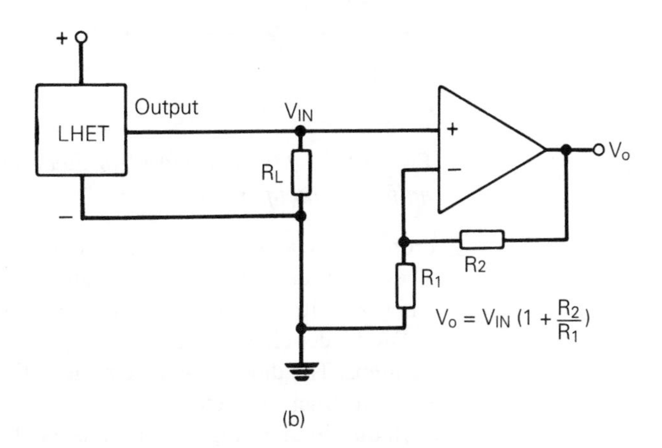

$$V_o = V_{IN}\left(1 + \frac{R_2}{R_1}\right)$$

(b)

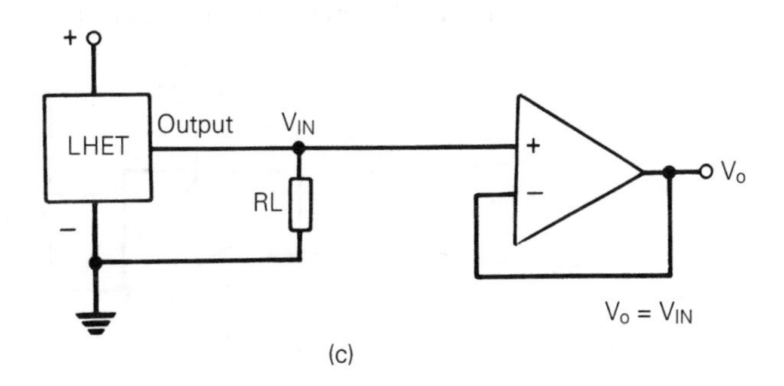

$V_o = V_{IN}$

(c)

Figure 8.9 *Interfacing LHET devices and single-supply op-amps: (a) inverting; (b) non-inverting; (c) voltage follower*

8.7 Precise diode circuits

An ideal diode is a device that exhibits zero resistance for applied voltages of one polarity and an infinite resistance to the opposite polarity. When used in a simple rectifier circuit as in Figure 8.10(b) it would completely block signals of one polarity and transmit perfectly those of the other.

The characteristics of real diodes are non-ideal, as shown in Figure 8.10. Real diodes pass no appreciable current for small voltages applied across them. They exhibit a non-linear finite resistance when conducting. The voltage

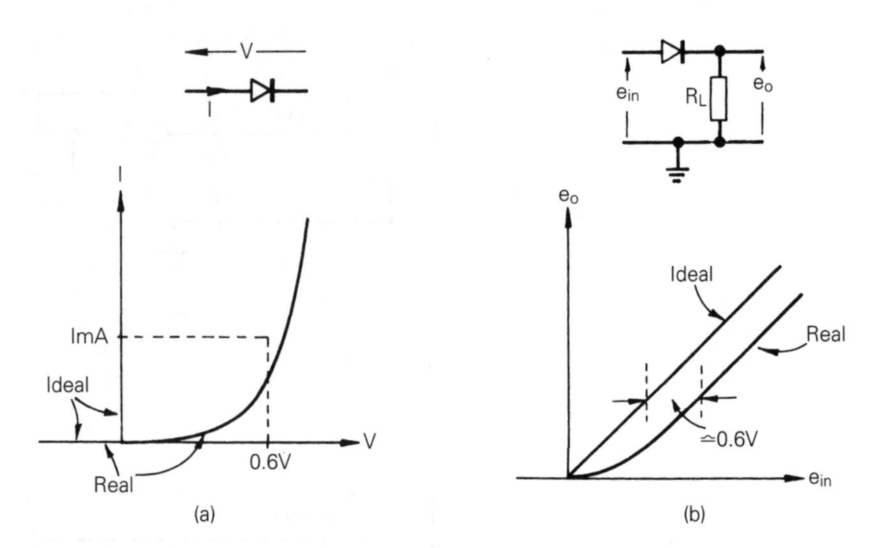

Figure 8.10 *(a) Real and ideal diode characteristics. (b) Real and ideal half-wave rectifier*

drop across a forward biased diode has marked temperature dependence. These non-ideal characteristics cause performance errors at low signal levels when a solid state diode is used in a simple rectifier circuit.

Diode deficiencies can be largely overcome by combining them with op-amps. The diode op-amp circuit of Figure 8.11 produces a near ideal half-wave rectifier characteristic.

In the circuit of Figure 8.11, diodes D_1 and D_2 are included within the feedback loop of the amplifier. If the diodes are non-conducting the amplifier is effectively acting open loop and an input signal of magnitude V_f/A_{OL} is all that

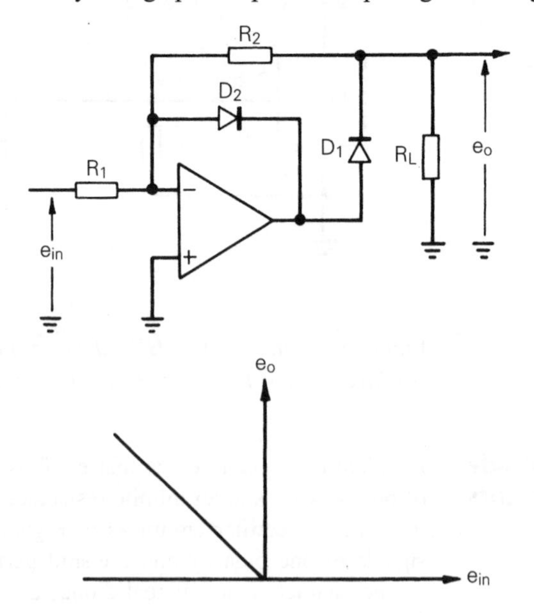

Figure 8.11 *Op-amp diode circuit performing ideal half-wave rectification*

is required to cause diode conduction (V_f is the diode forward voltage drop). Negative input signals cause diode D_1 to conduct and the output signal that appears at the cathode of D_1 is

$$e_o = -\frac{R_2}{R_1} e_{in}$$

The non-linear diode resistance has negligible effect on the output signal. This is because the diode is included within the feedback loop of the amplifier, and the resistance is effectively divided by the loop gain in the circuit (see Chapter 2).

Positive input signals cause diode D_1 to be reverse biased and cause D_2 to be forward biased. Thus D_1 is isolating and D_2 is conducting. This maintains the virtual earth at the inverting input terminal of the amplifier. The output signal is zero since it is connected directly to the op-amp's inverting input via resistor R_2.

The main performance limitation of op-amp circuits incorporating diodes arises as a result of amplifier slew rate. Because of slew rate limitations, the amplifier output voltage takes a finite time to overcome diode forward voltage drops. This restricts the frequency response of precise diode circuits.

8.8 Full-wave rectifier circuits

A full-wave rectifier circuit gives an output signal in proportion to the magnitude of its input signal. It converts bipolar signals into unipolar form. Full-wave rectifier circuits are used extensively in AC measurements. They are used to interface bipolar inputs to single quadrant (unipolar) devices, e.g. in log–antilog computation circuits.

The basic operation performed by a full-wave rectifier, using an op-amp, is to switch the amplifier's gain polarity. The circuits are arranged so that when the polarity of the input signal changes, so also does the overall gain polarity. Thus the circuit maintains a constant polarity output signal.

There are several op-amp circuit configurations possible that provide full-wave rectification. The factors to be considered when choosing a circuit configuration are: the input impedance requirements; whether or not summing is required; and the cost or performance determined by the number of close tolerance resistors required in the circuit.

Three full-wave rectifier circuits are shown in Figure 8.12. They all use an inverting amplifier configuration at their input and therefore have input impedance determined by the input resistor values.

Figure 8.12(a) is probably the most obvious approach to full-wave rectification. It comprises the precise half-wave rectifier circuit (as shown in Figure 8.11) added to a summing amplifier. Negative input signals are simply passed by the inverting summer and blocked by the half-wave rectifier. Positive signals are inverted and passed by the half-wave rectifier; they are multiplied by 2, summed with the input and inverted by amplifier A_2. The circuit requires accurate matching of two pairs of resistors plus the selection of a half-value resistor.

The circuit of Figure 8.12(b) requires only two matched resistors. Two parallel signal paths exist between input and output. Amplifier A_1 (via D_1) buffers positive input signals. Negative input signals are inverted by A_2 and passed through diode D_3. Negative signals can reach the output of A_1 through

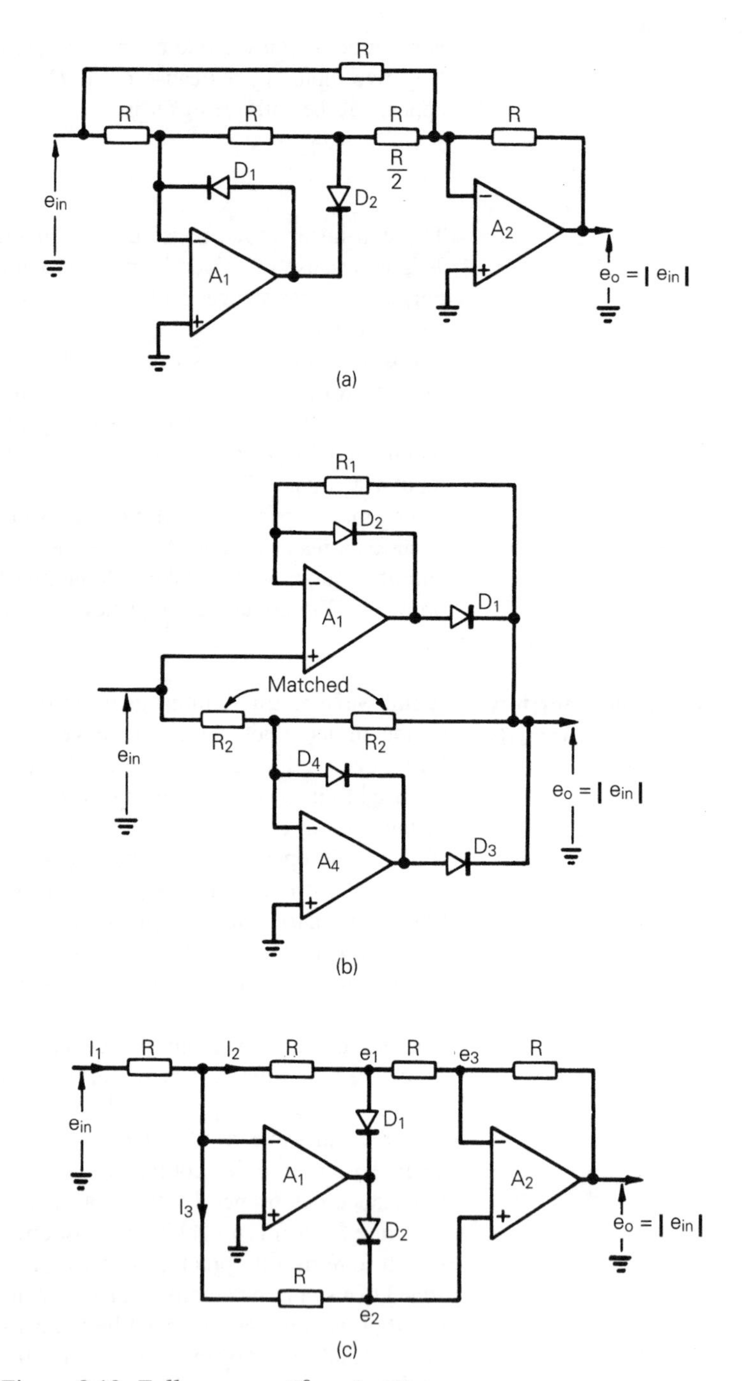

Figure 8.12 *Full-wave rectifier circuits*

R_1 and D_2. Diode D_2 is used to prevent saturation of A_1. The value of resistor R_1 must be sufficiently large to minimize this current.

The circuit of Figure 8.12(c) allows summation of signals at its input simply by adding extra input resistors. Equal value resistors are used throughout the circuit. Positive input signals produce a negative output from amplifier A_1, which reverse biases D_2 and forward biases D_1. The amplifier A_2 inverts the voltage from A_1, so a positive input produces a positive output. A negative input signal provides forward bias across D_2 but reverse bias across D_1. There are then two feedback paths to the inverting input A_1. Assuming ideal circuit performance:

$$I_1 = I_2 + I_3$$

$$\text{giving} \quad \frac{e_{in}}{R} = -\frac{e_o}{3R} - \frac{e_2}{R}$$

But $e_2 = e_3 = e_o\ 2R/3R$ and substitution gives $e_o = -e_{in}$, for e_{in} negative.

The circuit shown in Figure 8.13 is a high input impedance full-wave rectifier circuit, which uses a follower-connected op-amp at its input. The circuit requires only two closely matched resistors.

Positive input signals cause D_1 to conduct and D_2 to block. The feedback loop is connected around amplifier A_1, via A_2. No current flows through the two feedback resistors, R, so the output voltage appears at the inverting inputs of both A_1 and A_2. This results in 100 per cent feedback to both op-amps and the signal is passed to the output of A_2 at unity gain. Because two amplifiers are included within the feedback loop for positive input signals, additional phase compensation may be required. This can be obtained by a capacitor, C, connected in the position shown.

Negative input signals cause D_2 to conduct and D_1 to block. Amplifier A_1 then acts as a unity-gain follower, which passes the input signal to the point X. Amplifier A_2 acts as a unity-gain inverter on this signal at X.

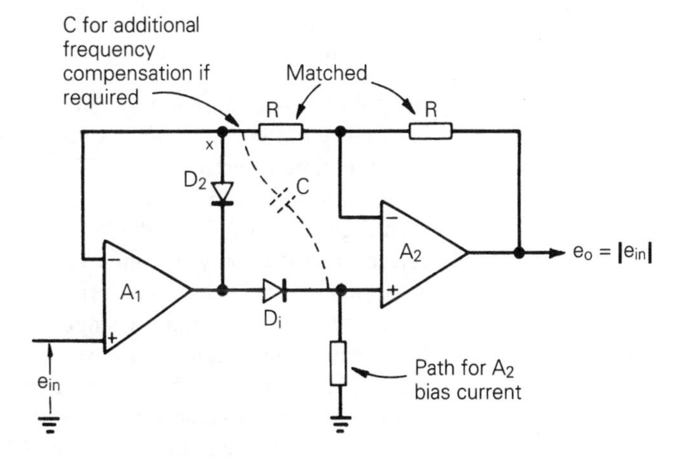

Figure 8.13 *High input impedance absolute value circuit*

8.9 Peak detectors

It is sometimes necessary to measure the maximum positive excursion (peak value) or negative excursion (valley value) of a waveform over a given time period. There may also be a requirement to capture and hold some maximum value of a positive or negative pulse. A circuit that performs this function is a peak detector.

A basic peak detector circuit consists of a diode and a capacitor connected as shown in Figure 8.14(a). The capacitor is charged by the input signal through the diode. When the input signal falls, the diode is reverse biased and the capacitor voltage retains the peak value of the input signal. The simple circuit has errors because of the diode forward voltage drop. Forward voltage drop errors can be removed by replacing the diode with a precise diode circuit as shown in Figure 8.14(b).

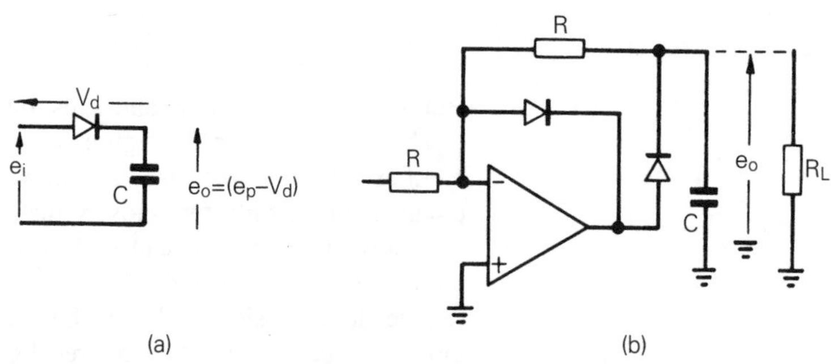

Figure 8.14 *Peak detection. (a) Simple peak detector. (b) Precise diode peak detector*

The circuit of Figure 8.14(b) is useful in applications not requiring a long hold time, for example for measuring the peak value of a repetitive signal. In the hold mode, the voltage across the capacitor decays exponentially governed by the time constant

$$t = \frac{CRR_L}{R + R_L}$$

In applications requiring an appreciable hold time some form of high input impedance buffer must read out the output voltage across the capacitor. Peak detector circuits employing FET input op-amps in the follower mode as buffers are shown in Figure 8.15.

In Figure 8.15(a) a two-diode arrangement is used to reduce diode leakage current, and it is only the input error voltage of amplifier A_2 which appears across diode D_2 in the hold period. Circuits of this type in which two op-amps are enclosed within a single feedback loop, normally require added frequency compensation; this has the effect of slowing down the rate at which the circuit responds to rapid changes in peak value.

Two separate feedback loops, one connected round each amplifier, are employed in the circuit of Figure 8.15(b). Amplifier A_1 acts both as a comparator and a unity-gain follower. The feedback loop around A_1 is open-circuit (diode D_1 blocks) when input signal levels are lower than the voltage stored on the capacitor. When the input exceeds the capacitor voltage, diode D_1 conducts. Amplifier A_1 then acts as a unity-gain follower and causes the capacitor voltage to follow the input signal.

If appreciable hold times are required, both amplifiers should be FET input types. This will minimize capacitor leakage caused by amplifier bias current. Amplifier A_1 should be a type that retains its high input impedance in the saturated overload condition and should be capable of fast recovery from

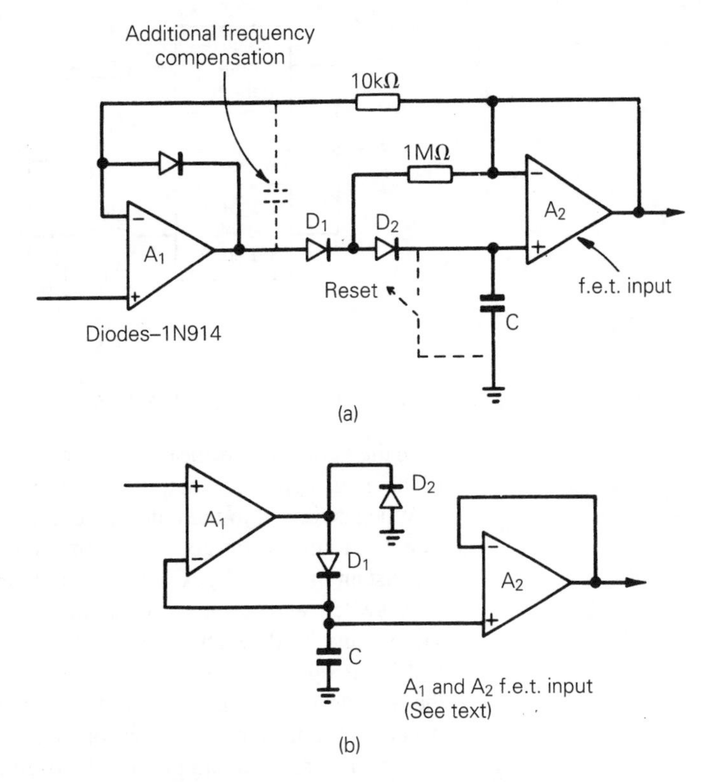

Figure 8.15 *Precise peak detector circuits. (a) Low drift peak detector. (b) Fast peak detector*

this condition. It must also be able to drive a capacitive load without serious reduction in phase margin.

The choice of capacitor values in a peak follower circuit is governed by conflicting requirements. It must be able to be charged quickly to allow a rapid acquisition of rapidly changing input peaks. It must also have a long hold time. The smaller the value of C, the more rapid is its charging rate. By the same reasoning, small values of C will also discharge rapidly due to leakage during the hold period. One way of increasing capacitor charging current, in order to obtain faster acquisition, is to use a current booster at the output of the op-amp. A simple emitter follower booster can be used since only single polarity output currents are required.

Positive peak detector circuits have been described, but they can all be modified in order to detect negative peaks (valleys) by simply reversing diode directions. Peak-to-peak detectors can be implemented by connecting the output of a positive peak detector and a negative peak detector to a subtractor connected amplifier as shown in Figure 8.16.

8.10 Sample and hold circuits

In signal processing applications, it is sometimes necessary to hold the value that a signal has at a specified instant in time. A circuit used to perform this function is called a sample and hold. As well as input and output terminals, a sample and hold circuit has control inputs to allow switching between the sample and hold mode. The phrase 'track and hold' is often used when referring to a sample and hold circuit that is in the sample mode for an appreciable time.

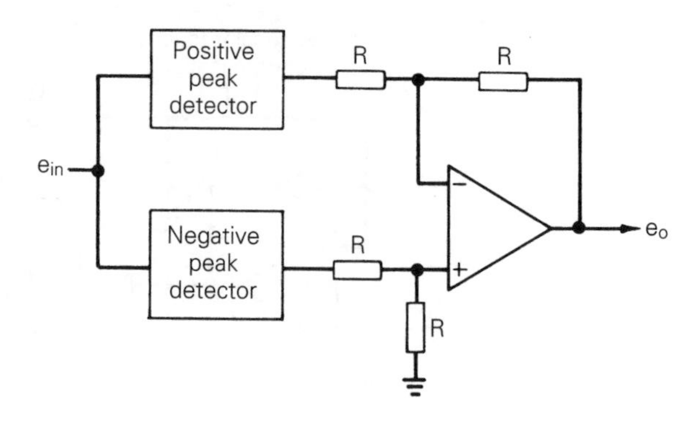

Figure 8.16 *Peak-to-peak detector*

In the sample mode, the output of the sample and hold circuit is (ideally) equal to the input signal. The output follows or tracks variations in the input signal.

When switched to the hold mode, the output of the sample and hold circuit (ideally) remains constant. The output signal is held at the value that existed at the instant of switching. Changes to the input voltage do not affect the output.

A switch and a capacitor as shown in Figure 8.17(a) can perform the sample and hold function. When the switch is closed (sample mode) output and input signals are equal and the output follows or tracks the input. When the switch is opened, the output voltage remains constant at the value that it had at the instant the switch opened.

A more practical sample and hold circuit is shown in Figure 8.17(b). It uses a FET switch and an op-amp unity-gain follower to minimize capacitor discharge in the hold mode.

Practical sample and hold circuits do not behave in an ideal manner. They usually depart from the ideal in terms of both speed and accuracy. When switched from the hold mode to the sample mode, a finite time is required for the output to become equal to the input. This time is referred to as the acquisition time.

There is a small delay from when the hold command signal is applied and the time the circuit actually goes into the hold mode. This time delay is referred to as the aperture time. With fast changing input signals, the signal held at the output is in error to an extent determined by the aperture time.

The ideal sample and hold circuit is designed to have unity gain (output signal = input signal). Sample mode accuracy is sometimes expressed in terms of the percentage gain error. In the hold mode, the output of a practical sample and hold circuit does not remain constant. Loading of the hold capacitor causes the capacitor to discharge and the output drifts towards zero.

The choice of capacitor values for use in a sample and hold circuit is normally a compromise based upon conflicting requirements. These requirements are fast acquisition time and long hold time. In the sample mode the capacitor must charge up to the value of the input signal. The larger the capacitor value, the longer it takes to charge. In the hold mode there is inevitably always some capacitor discharge current (amplifier bias current and switch leakage current). The larger the capacitor's value, the longer the time taken for the capacitor to discharge.

The capacitor dielectric also has an effect on performance. Electrolytic and tantalum capacitors have high leakage current. Dielectric absorption (or

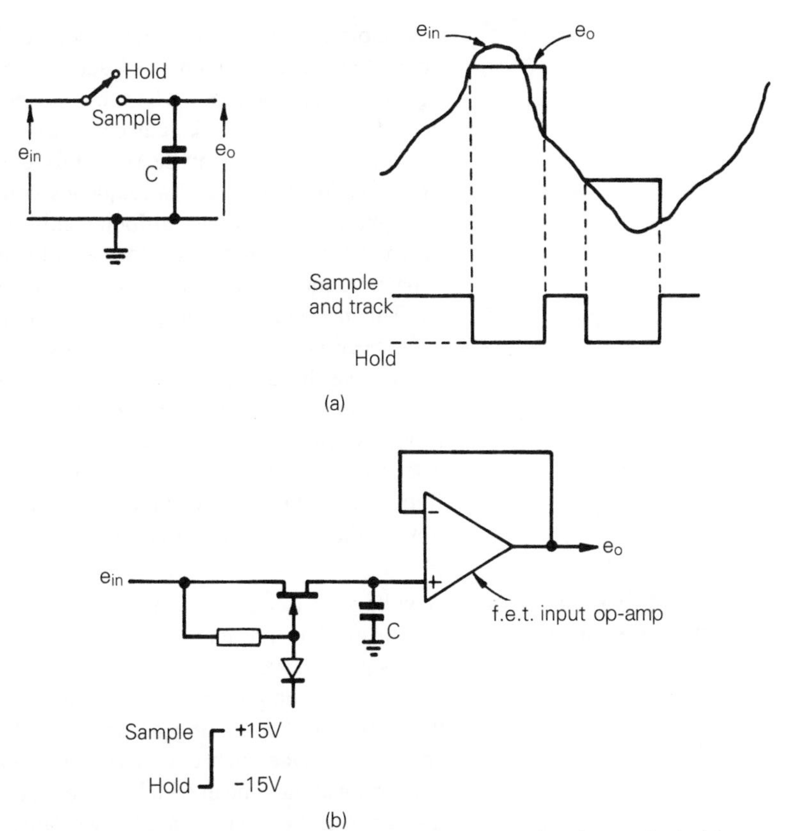

Figure 8.17 *The sample hold function. (a) Principle of sample hold circuit. (b) Simple practical sample hold circuit*

soakage) also affects these dielectrics, as it does Mylar or polyester capacitors. Polypropylene and polystyrene capacitors have lowest leakage and lowest dielectric absorption.

Sample and hold circuits are available in integrated circuit form. The user simply connects a hold capacitor externally. The choice between buying an IC, or building the circuit using op-amps and FET switches, depends on the performance requirements.

In the simple sample hold circuit of Figure 8.17(b) the storage capacitor is charged directly by the signal source through the FET switch. The capacitor loads the signal source. The capacitor charging rate, when switched from hold mode to sample mode, is determined by the time constant $C(R_s + R_t)$. Here, R_s is the source resistance and R_t is the switch-on resistance. An upper limit to the charging rate is set by the signal source's output current limit.

An alternative one-amplifier sample and hold circuit is shown in Figure 8.18(a). It is essentially an integrator that is switched between the sample mode and the hold mode. The circuit, unlike that of Figure 8.17(b), does provide some input isolation, in the form of resistor R_1. Its main deficiencies are its limited tracking bandwidth and comparatively slow acquisition time. Both tracking bandwidth and acquisition time are controlled by the time constant C_1R_1.

The circuit for a two op-amp, high accuracy, sample and hold circuit is given in Figure 8.18(b). Amplifier A_1 is connected as a voltage follower and imposes negligible holding on the signal source. In the sample mode (S_1 closed, S_2 open) the feedback loop is closed round both amplifiers and the output is forced to track the input. There may be errors due to the gain, common mode and offset errors, and the current output capability of amplifier A_1. Common mode and offset errors in the output follower A_2 are compensated by the action of the feedback loop. Extra frequency compensation in the form of C_1 and R_1 is required to ensure closed-loop stability in the sample mode, but this slows down the circuit response. In the hold mode S_1 is open, isolating the hold capacitor, and S_2 is closed so as to complete the feedback loop round amplifier A_1. This prevents A_1 from going into saturation.

If speed is more important than high accuracy, the circuit shown in Figure 8.18(c) can be used. The two amplifiers in this circuit work independently. Each amplifier has its own closed feedback loop and, in the sample mode (both switches closed), the switches are enclosed within A_1's feedback loop. The circuit is faster than that of Figure 8.18(b) because no additional frequency compensation is required. It is less accurate because of the summation of the offset and common mode errors of both amplifiers.

8.11 Voltage-to-frequency conversion

A voltage-to-frequency converter produces a periodic signal with frequency proportional to an analogue control voltage. The waveform produced may be a square wave, a pulse train, a triangular wave or a sine wave.

Pulse train output voltage-to-frequency converters could be realized using two op-amps, one acting as an integrator and the other as a regenerative comparator. One such circuit is illustrated in Figure 8.19.

Starting at the time at which the comparator switches to its positive level V^+, the action of the circuit is as follows. Diode D_1 is reverse biased and the output of the integrator falls linearly, at a rate determined by the magnitude of the positive DC voltage e_{in} and input resistor R. When the integrator output reaches a level $-V_o^+(R_1/R_2)$, the comparator switches to its negative output state. Now diode D_1 is forward biased and the integrator output voltage rises rapidly, because $R_3 << R$. When the integrator output voltage $-V_o^-(R_1/R_2)$, the comparator reverts to its positive output state and the cycle repeats.

If the integrator output rise time is made negligibly small compared to the decay time, the frequency of oscillations becomes directly proportional to the input voltage e_{in}.

$$f \cong \frac{R_2}{R_1\,(V_o^+ - V_o^-)\,CR}\,e_{in}$$

Simple voltage-to-frequency converters using the type of circuitry discussed above can be expected to provide ±1 per cent accuracy over the two to three decades at the most. Greater conversion accuracy and wider dynamic range require the use of more sophisticated circuitry, or as an alternative to building a voltage-to-frequency converter ready built modules are available.

8.12 Frequency-to-voltage conversion

Simple frequency-to-voltage conversion circuits operate by first converting the signal (whose frequency is to be measured) to a constant amplitude pulse train. The pulse train is then differentiated, rectified and averaged to give a DC indication of the frequency. A simple frequency-to-DC converter using this principle of operation is illustrated in Figure 8.20.

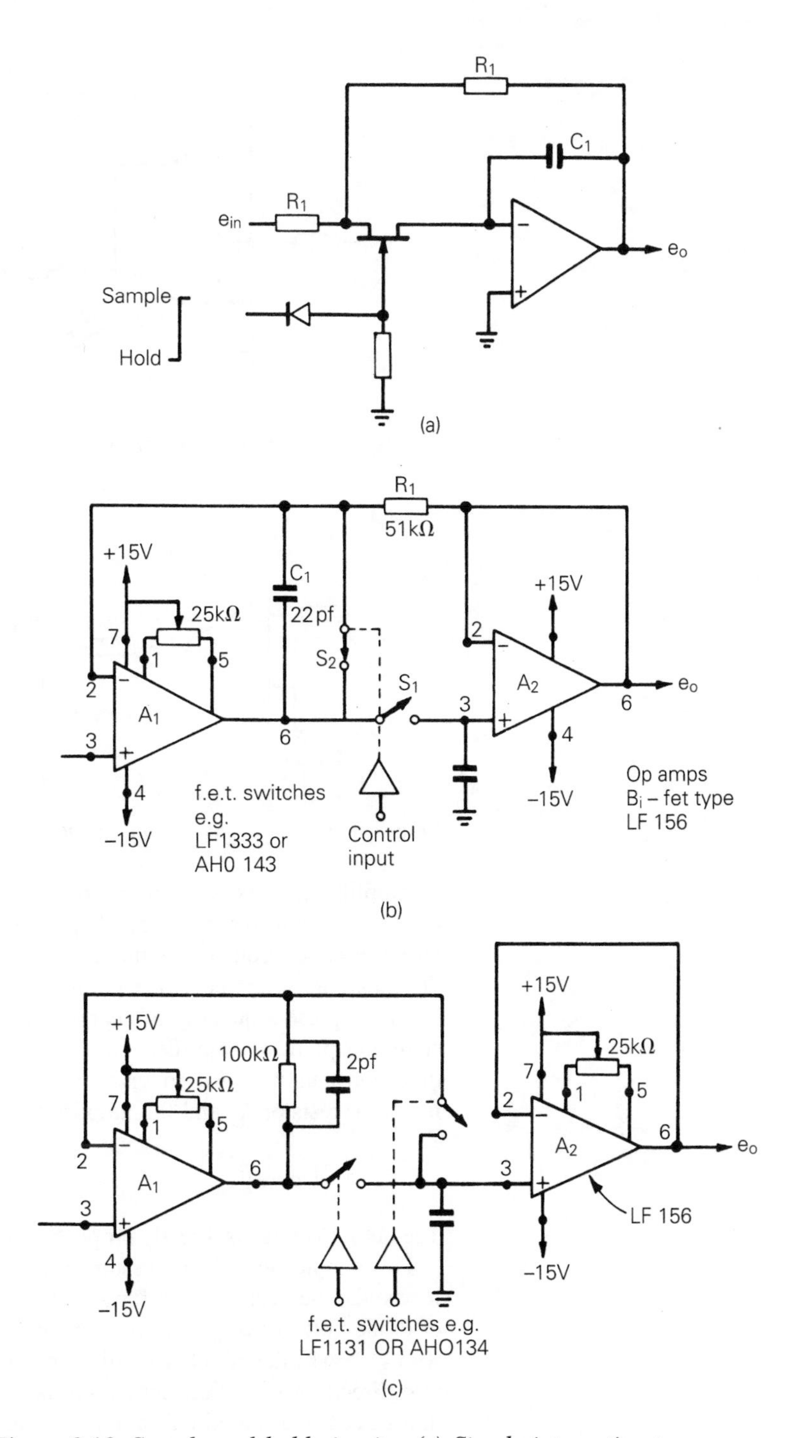

Figure 8.18 *Sample and hold circuits. (a) Simple integrating type. (b) High accuracy type. (c) Fast type*

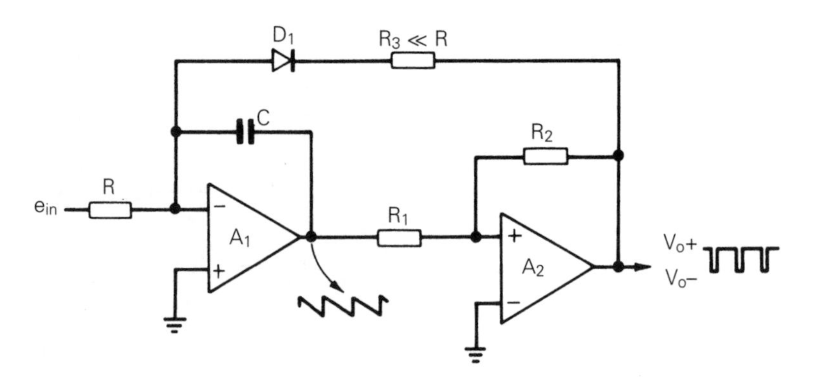

Figure 8.19 *Simple V to f converter*

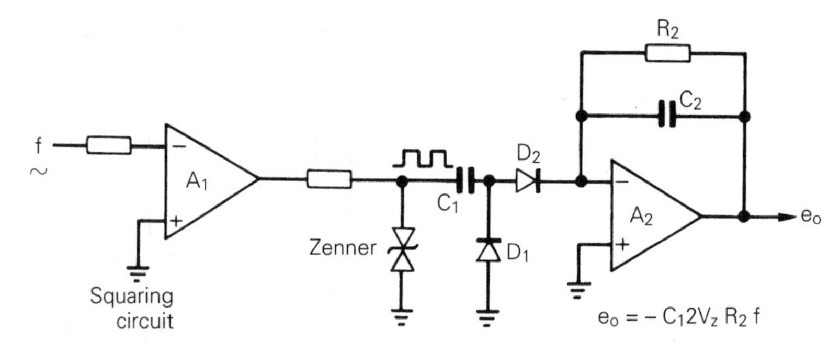

Figure 8.20 *Simple frequency-to-DC voltage conversion*

Amplifier A_1 acts as a zero reference comparator, and its output is bounded by back-to-back zener diodes. A_1 produces a constant amplitude pulse train with the same frequency as the input signal. Capacitor C_1 and diodes D_1 and D_2 constitute a simple diode pump circuit. On each positive going part of the input pulse a quantity of charge ($2V_ZC_1$) is transferred through D_2 to the summing point of amplifier A_2. The charge pulses are averaged (averaging time constant = C_2R_2) to give an average current of $2V_ZC_1f$ through the feedback resistor R_2 and the amplifier develops an output voltage

$$e_o = 2V_ZC_1fR_2$$

The output voltage is directly proportional to the frequency of the input signal.

This simple analysis has assumed that capacitor C_1 discharges completely. It has also neglected diode forward voltage drops. Temperature dependence of diode voltage, ΔV_d, may be expected to cause scaling factor changes but if $2V_Z >> \Delta V_d$ the effect is small. Temperature dependence of zener voltages also directly affects frequency-to-voltage scaling. The comparator switching times determine the circuit's upper frequency limit.

8.13 Analogue-to-digital converter (ADC)

Analogue-to-digital converter (ADC) devices are used to interface analogue circuits with microprocessors and other digital devices. There are a number of ADC architectures, the simplest is used for slowly changing input signals and is known as an integrating ADC. The integrating ADC uses a digital counter, so the time to produce a valid output depends on the voltage being measured. The conversion time is a maximum of 2^N clock cycles, where N is the number of bits.

The successive approximation ADC uses a sample and hold circuit so that rapidly changing signals can be sampled and then measured. The digital output is available after N clock cycles, where N is the number of bits.

Flash ADCs have many high-speed comparators connected in parallel. An 8-bit ADC uses 256 comparators. The digital output is produced by logic gates, which produce a binary coded equivalent of the most significant comparator output. The digital output is latched and is available after a single clock cycle.

All ADCs sample the analogue signal. This is like amplitude modulation and can generate unexpected results unless the analogue signal is band limited to less than half the sampling rate (usually the clock signal). If the signal is not band limited, aliasing can result. Aliasing is where the digitized signals appear to have a lower frequency than the original, and may be frequency inverted (high frequency signals appear at lower frequencies, whilst low frequencies appear at higher frequencies).

The requirements for anti-aliasing filters is reduced if oversampling sigma–delta ADCs are used. These sample the signal at very high rates, up to 256 times the clock frequency. They work by using a single comparator; the logic output from this comparator is integrated and then subtracted from the input voltage. This results in a series of 1s and 0s related to the input voltage. This stream of bits is then used to generate a binary coded output. This can result in an accurate 16-bit ADC.

8.13.1 Integrating ADC

Integrating ADCs are used for high accuracy data conversion when the input is a slowly changing analogue signal. The integrating ADC has low offset errors and can be highly linear. This type of ADC is used in measuring instruments, such as multimeters.

A diagram of the integrating ADC is shown in Figure 8.21. The data conversion process occurs in two stages. In the first stage, the signal to be measured

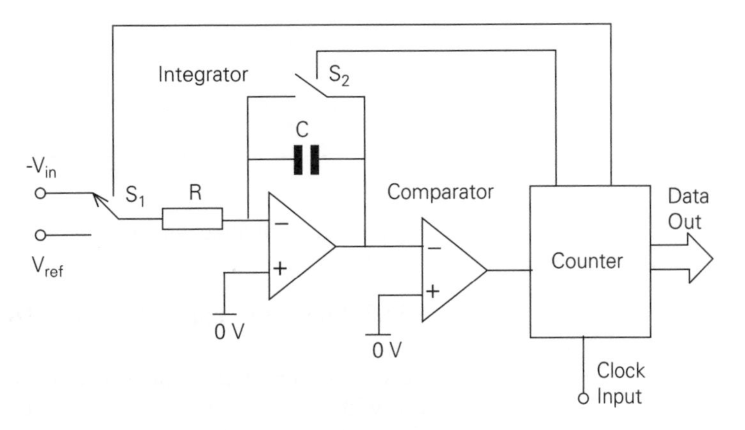

Figure 8.21 *An integrating (dual slope) ADC*

($-V_{in}$) is input for a fixed time interval, T_1. This is achieved by running the clock for $2N$ cycles, where N is the number of bits of the digital output. If the capacitor voltage, V_C, is initially zero, by the end of stage 1, it is:

$$V_C(t) = -\int_0^t \frac{V_{in}}{RC}\,dT = \frac{V_{in}}{RC}t$$

The second stage of conversion takes place over a variable length of time, depending on the value of the input signal. The counter is initially reset to zero and the switch S_1 is connected to the reference voltage, V_{ref}. The control logic will keep the counter running while the comparator output is low. The expression for the input voltage to the comparator is now given as:

$$V_C(t) = -\int_{T_1}^t \frac{V_{ref}}{RC}\,dT + V_C(T_1) = \frac{V_{ref}}{RC}(t - T_1) + \frac{V_{in}T_1}{RC}$$

The comparator output will switch high when its input voltage reaches 0 V (the capacitor is discharged).

$$V_C(T_1 + T_2) = 0 = \frac{V_{ref}}{RC}(T_2) + \frac{V_{in}T_1}{RC}$$

The time period T_2 when the integrator is connected to the reference voltage is given as $T_2 = T_1 (V_{in}/V_{ref})$. The counter is enabled during period T_2 and the output of the counter will be the converted value of the analogue signal. Since T_2 is proportional to T_1, this count is independent of the clock rate. A faster clock rate results in periods T_1 and T_2 both becoming shorter. Figure 8.22 shows this, here $V_{in} < V'_{in}$ and hence $T_2 < T'_2$.

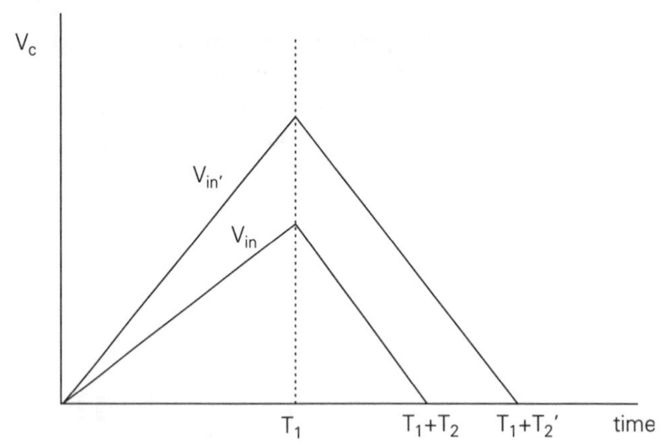

Figure 8.22 *The integrator response with two different values of input signal*

8.13.2 Successive approximation ADC

Successive approximation ADCs are popular because they are moderately fast without being too complex. The basic principle is a 'divide and search' approach in converting analogue voltage to digital form. A binary search approach is used, in which the search space is reduced by half in each clock cycle. Therefore, the successive approximation ADC will require 16 clock cycles to convert an analogue signal into a 16-bit digital equivalent. A diagram of this type of ADC is given in Figure 8.23.

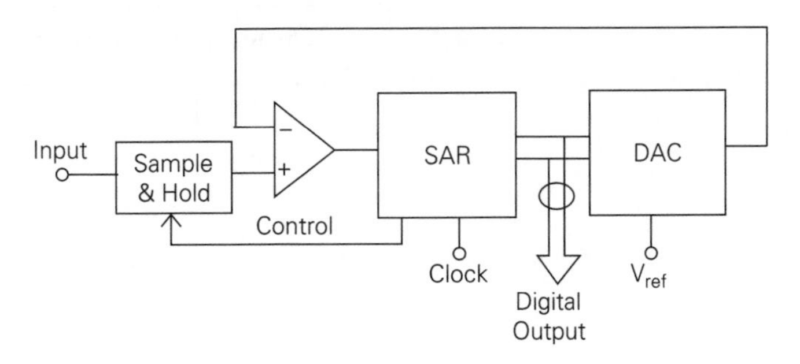

Figure 8.23 *A schematic representation of a successive approximation ADC*

A sample and hold circuit is used to sample the analogue input signal and hold its value during the data conversion process. In the first clock cycle, the most significant bit (MSB) of the successive approximation register (SAR) is set to 1. The SAR is connected to the input of a digital-to-analogue converter (DAC), which converts the binary value of the SAR into an analogue signal.

The analogue signal from the DAC is then applied to the comparator. The comparator compares the output of the DAC with the stored input voltage. If the output from the DAC is smaller than the stored input voltage, the most significant bit of the register remains at logic 1; otherwise it is reset to 0. This bit remains unchanged for the remainder of the data conversion process.

The same process is repeated in the second clock cycle with the next most significant bit (MSB − 1). This bit is set to logic 1, and the output of the SAR is converted to analogue, using the DAC, and then compared with the stored input voltage. If the output from the DAC is smaller than the stored input voltage, the (MSB − 1) bit of the register remains at logic 1; otherwise it is reset to 0. This bit now remains unchanged for the remainder of the data conversion process.

Repeating this process for *N* clock cycles results in converting the analogue signal to its equivalent *N*-bit digital value. At the end of *N* clock cycles, the SAR contains the converted value of the analogue input signal. A flow chart describing the successive approximation ADC process is given in Figure 8.24.

8.14 Digital-to-analogue converter (DAC)

There are several circuit arrangements for digital-to-analogue converters (DACs). The three main classes of DACs are: decoder-based DACs, binary-weighted DACs (including *R*–2*R* converters), and thermometer code DACs.

The output from a DAC should be band limited to prevent switching spikes from appearing at the output. These spikes, or glitches, are usually at the clock frequency or higher.

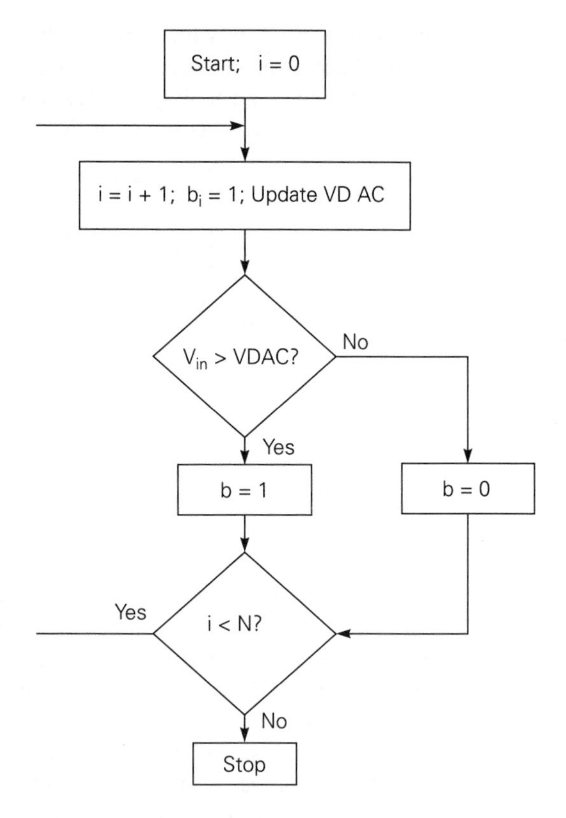

Figure 8.24 *A flow chart describing the successive approximation ADC operation*

8.14.1 Decoder-based DACs

A decoder-based DAC uses a resistor string to produce a number of voltage taps, see Figure 8.25. One of these voltage taps is selected by a decoder, for connection to the output buffer through an analogue gate. Although resistance of the switching network is minimal, the capacitive load is high because one end of all transmission gates is connected to the input of the buffer.

Instead of transmission gates, which use both PMOS and NMOS transistors, we can use just NMOS transistors. This has the disadvantage of limiting the output voltage swing. The speed of operation is improved because of reduced source and drain capacitance, due to the absence of the PMOS transistors.

Using equal values of resistor in the resistor string ensures equal voltage steps in the tapped voltage, and hence for the DAC as a whole. The accuracy of the DAC depends on the matching of these resistors.

8.14.2 Binary-weighted DACs

Since binary numbers represent digital words, individual bits have binary weights depending on their position within the digital word. A simple summing

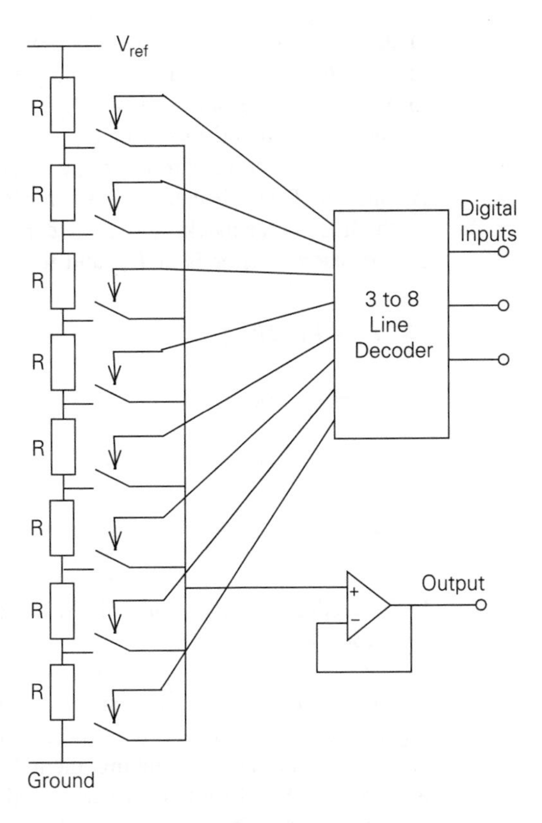

Figure 8.25 *3-bit decoder-based DAC using resistor string*

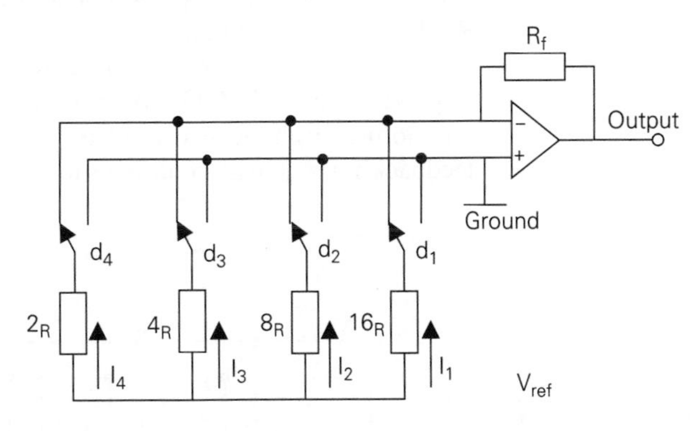

Figure 8.26 *A binary-weighted resistor DAC (4 bit)*

amplifier can be used to convert a digital word to its analogue value, by arranging for each bit to contribute a current equal to its binary weight. This principle is used in binary weighted DACs in two different implementations.

In the first implementation, the resistor values increase by a factor of two as the bit becomes less significant. The circuit of a binary-weighted resistor

DAC is illustrated in Figure 8.26. When the value of the digital bits is at logic 1, the current through the binary-weighted resistance is diverted to flow through the feedback resistor, R_f. When the digital bit is at logic 0, the current flows directly to ground.

The current flowing through the resistors in either position of the switch is constant. This is because the op-amp's inverting input is acting as a 'virtual earth' due to feedback, and is at earth potential. So, the potential on one side of each resistor is at V_{ref} and the other side is at 0 V.

$$I_4 = V_{ref}/2R$$

$$I_3 = V_{ref}/4R$$

$$I_2 = V_{ref}/8R$$

$$I_1 = V_{ref}/16R$$

If the feedback resistor value is R, the buffer's output voltage range is 0 V to $-15/16V_{ref}$.

The accuracy of a binary-weighted resistor DAC is dependent on the matching between different resistors used in the circuit. It is difficult to obtain a good match of resistors in an integrated circuit. When the range of resistor values is very large, matching them becomes the dominant problem. For example, with a 14-bit binary-weighted resistor DAC, the range of resistor values can vary by over four orders of magnitude. Therefore, the use of binary-weighted resistors can result in large mismatch errors.

In the second implementation of binary weighting, only two different values of resistors are used to obtain the binary weighted currents. Using an R–$2R$ network in conjunction with a summing amplifier and CMOS switches, it is possible to implement a binary-weighted DAC as given in Figure 8.27. Depending on the CMOS switch position, the binary-weighted currents either flow to the feedback resistor or to ground. The currents flowing through the feedback resistor will contribute towards the output voltage.

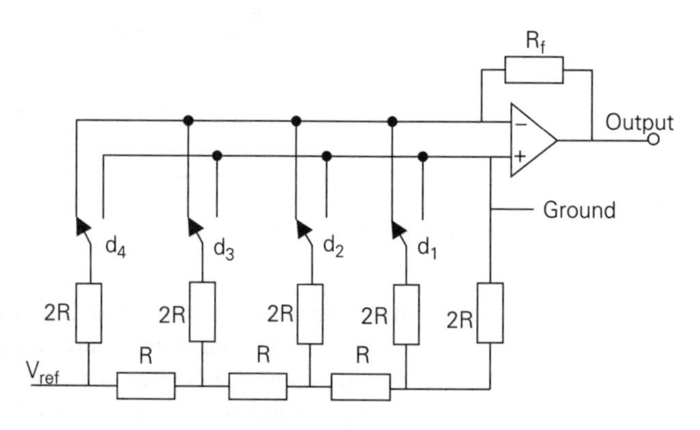

Figure 8.27 *R–2R binary-weighted DAC*

The bit values of the digital word determine the switch position. A logic 1 on the most significant bit causes the corresponding CMOS switch d4 to connect to the buffer's inverting input. Current $V_{ref}/2R$ will flow into the buffer's summing node, thus generating an output voltage. A logic 1 on the next most significant bit operates CMOS switch d3 and causes current $V_{ref}/4R$ to flow into the summing node. The current is halved each time the binary value of the bit is halved, so the output voltage is proportional to the value of the digital word.

The advantage of an R–$2R$ network is that only two resistor values are required. Matching between a number of resistors, of values R and $2R$, is much easier to achieve than by using binary-weighted values.

8.14.3 Thermometer code DACs

Glitches (short pulses) are produced by decoder and binary-weighted converters, and this is a major limitation. When the digital input values changes, it is likely that some of the switches turn ON or OFF faster than others, resulting in glitches in the analogue output. A glitch due to the switch associated with the most significant bit can have an amplitude of almost half V_{ref}.

The production of glitches can be reduced by the use of a thermometer code representation of binary numbers. The disadvantage of this coding scheme is potentially more circuit complexity. The advantages are equal output steps, reduced level of glitches, and linearity.

A thermometer code is an incremental digital output. The number of bits required is equal to the number of voltage steps, thus 7 levels output requires a 7-bit thermometer code, as given in Table 8.1.

Table 8.1 *Thermometer code*

Output	MSB						LSB
0	0	0	0	0	0	0	0
1	0	0	0	0	0	0	1
2	0	0	0	0	0	1	1
3	0	0	0	0	1	1	1
4	0	0	0	1	1	1	1
5	0	0	1	1	1	1	1
6	0	1	1	1	1	1	1
7	1	1	1	1	1	1	1

The number of logic 1s in the thermometer code represents the converted voltage. In a thermometer code-based DAC, the switching network has $(2N - 1)$ switches; each has equal resistance and carries equal current. The current is switched either into the feedback resistor of an op-amp-based summing circuit, or directly to ground. The number of switches contributing to the current flow in the feedback resistor increase in small equal steps, as the binary value of the digital input is increased. This minimizes the amplitude of any glitches, compared to binary-weighted DACs. The schematic of a DAC using thermometer code is given in Figure 8.28.

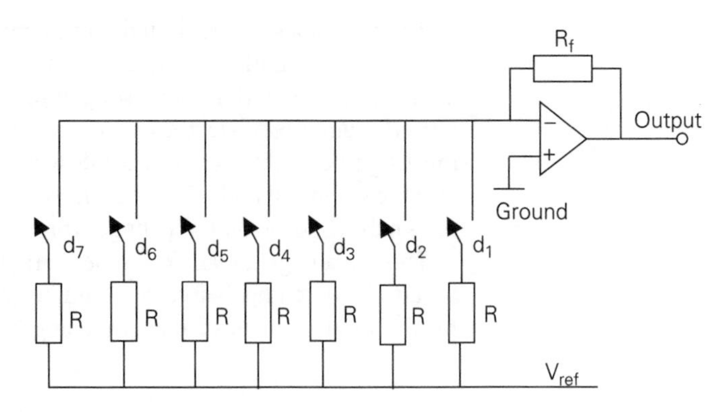

Figure 8.28 *A thermometer code-based DAC*

Exercises

8.1 A charge amplifier (Figure 8.4) has a 100 pF capacitor and a 100 MΩ resistor connected in parallel in the feedback path. The feedback resistor prevents continuous charging of the capacitor. The op-amp used in the circuit is internally frequency compensated and has unity-gain frequency 1 MHz, bias current I_B = 10 pA and input offset voltage V_{io} = 2 mV. The charge amplifier is supplied by a transducer, whose output impedance is capacitive; this capacitance is 900 pF.

(a) Find the upper and lower frequency of the −3 dB bandwidth limits. What is the output offset voltage of the circuit? What would be the effect on circuit performance of connecting the transducer to the amplifier by means of a cable of capacitance 200 pF? Sketch the Bode plots to illustrate your answers.

(b) In order to avoid the use of a very large feedback resistor and yet still maintain the same low frequency bandwidth limit, a 1 MΩ resistor and a resistive T network is used in place of the 100 MΩ resistor (see Figure 4.2). What effect will this have on the output-offset voltage?

8.2 The following component values are used in the circuit of Figure 8.19: R = 100 kΩ, C = 0.1 μF, R_1 = 10 kΩ, R_2 = 22 kΩ, R_3 = 2.2 kΩ. (a) What is the frequency of oscillation when the input voltage is 1 V? (b) What is the amplitude of the triangle wave at the output of the integrator, A_1? Assume ideal op-amp characteristics and a comparator output voltage swing of ±10 V.

8.3 A basic sample hold circuit consisting of a FET switch and a unity-gain buffer stage (Figure 8.17(b)) is supplied by a signal source of output resistance 600 Ω, the FET has an 'on' resistance of 50 Ω and the op-amp has a bias current I_B = 50 pA, C = 10 nF. Find the acquisition time to 1 per cent for a 10 V change in output when switched between the hold and sample mode. (Hint, 4.6 time constants are required to reach 99 per cent of the final value when charging a capacitor through a resistor.)

Assume that the required initial capacitor charging current does not exceed the current output limit of the signal source. What must the current output capability of the source be for this assumption to be valid?

Neglecting switch and capacitor leakage find the output drift rate in the hold mode. (Note: $I = C\, dv/dt$.)

9 Active filters

9.1 Introduction

In the context of this chapter, *filters* are electrical networks that have been designed to pass alternating currents generated at only certain frequencies and to block or attenuate all others. Filters have a wide use in electrical and electronic engineering and are vital elements in many telecommunications and instrumentation systems where the separation of wanted from unwanted signals – including noise – is essential to their success.

There are two generic types of filter: passive and active. The first type comprises simple resistors, capacitors and inductors while the second has the addition of active components, usually in the form of operational amplifiers. Both of these types are sub-divided into the four classes according to their use. These are *low-pass*, *high-pass*, *band-pass* and *band-stop*. This chapter is mainly concerned with active filters employing operational amplifiers (op-amps), but it may serve as a useful introduction for some readers if firstly a brief examination is made of the passive type.

9.2 Passive filters

9.2.1 The low-pass filter

The circuit of a simple *CR* low-pass filter is shown in Figure 9.1. This is essentially a potential divider comprising a resistance in series with a capacitor. The output voltage, e_o, is taken from across the capacitor and is related to the input voltage, e_i, by the equation:

$$e_o = -jX_c\, e_i/(R - jX_c)$$

Algebraic manipulation of this complex number equation shows that the amplitude of e_o is given by the expression:

$$| e_o | = e_I X_c/\sqrt{(R_2 + X_c^2)}$$

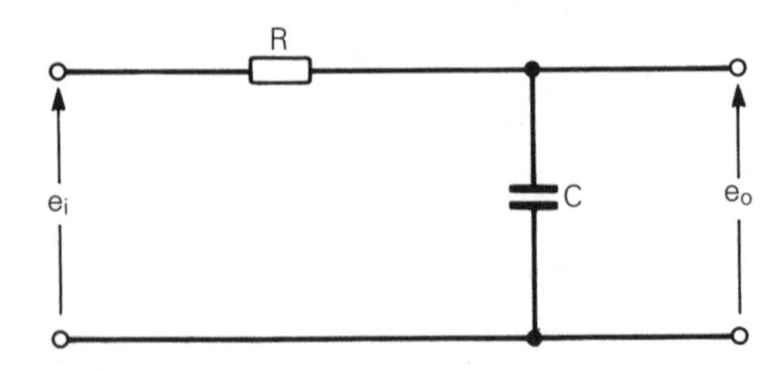

Figure 9.1 *First order low-pass passive CR filter circuit*

Even though e_i may be held constant over a range of input frequencies, the amplitude of e_o decreases as the frequency is increased. This is because the reactance of the capacitor, $X_c = 1/2\pi fC$, varies as the inverse of the frequency, f, and tends from an infinitely high value at zero frequency to zero at an infinitely high frequency. The circuit output is effectively shorted out at very high frequencies. Figure 9.2 shows the response curve for this circuit which is typical of the low-pass filter.

At low frequencies the output volts:input volts ratio remains sensibly level up to a frequency, f_c, at which a marked fall off starts. At about $2f_c$ the fall off (or roll off, as it is usually called) becomes linear at 20 dB per decade (which is the same as 6 dB per octave). The frequency f_c is known as the cut-off frequency and is taken as that frequency at which the reactance of the capacitor has the same magnitude as the resistance in the circuit. Also, f_c is the frequency at which the output voltage has fallen to $1/\sqrt{2}$ times its DC value to give half the DC power output. Simple calculations based on these facts show that the cut-off frequency is given by the equation:

$$f_c = 1/2\pi RC \text{ hertz}$$

For frequencies below f_c the circuit gain (output volts:input volts) is taken as being reasonably constant, while for frequencies higher than f_c the gain is regarded as being so low that the passage of these signals is effectively blocked. The circuit is known as a low-pass filter having a bandwidth extending from DC to f_c.

Because the response of the circuit depends upon frequency to the mathematical first order, the filter is known as a *first order filter*. (Also note that the circuit contains only a single component – the capacitor – the performance of which is frequency conscious.)

The ideal low-pass passive filter frequency response curve or *transfer function* would show no loss of gain for frequencies below f_c and zero output above f_c (see Figure 9.2). Clearly, the first order low-pass filter achieves neither of these ideals. If two *CR* sections are cascaded (see Figure 9.3) to form a second order filter having two frequency dependent capacitors, a steeper roll-off can be obtained, but only at the expense of decreased output. If these two similar sections are used, the roll-off tends to 40 dB per decade but the output is so attenuated as to be of little use.

A better solution for achieving a steep roll-off is still to use two frequency-dependent components but make one a capacitor and the other an inductance. This circuit, shown in Figure 9.4, takes advantage of the ability of the inductance and capacitance to be near their natural resonant frequency at the filter cut-off frequency. This would have the effect of producing an output voltage magnification in the *knee* region of the frequency response curve. By varying the ratio of the values of inductance and capacitance, the shape of the knee can be adjusted. The critical case is where the flat top of the lower frequency response is extended along the frequency scale before failing in a steeper roll-off yet without introducing the undesirable effects of underdamping or overdamping. These include output voltage oscillations before finally settling or having an excessively long response time to transient inputs. The combined high frequency effect of the high inductive reactance coupled with the low

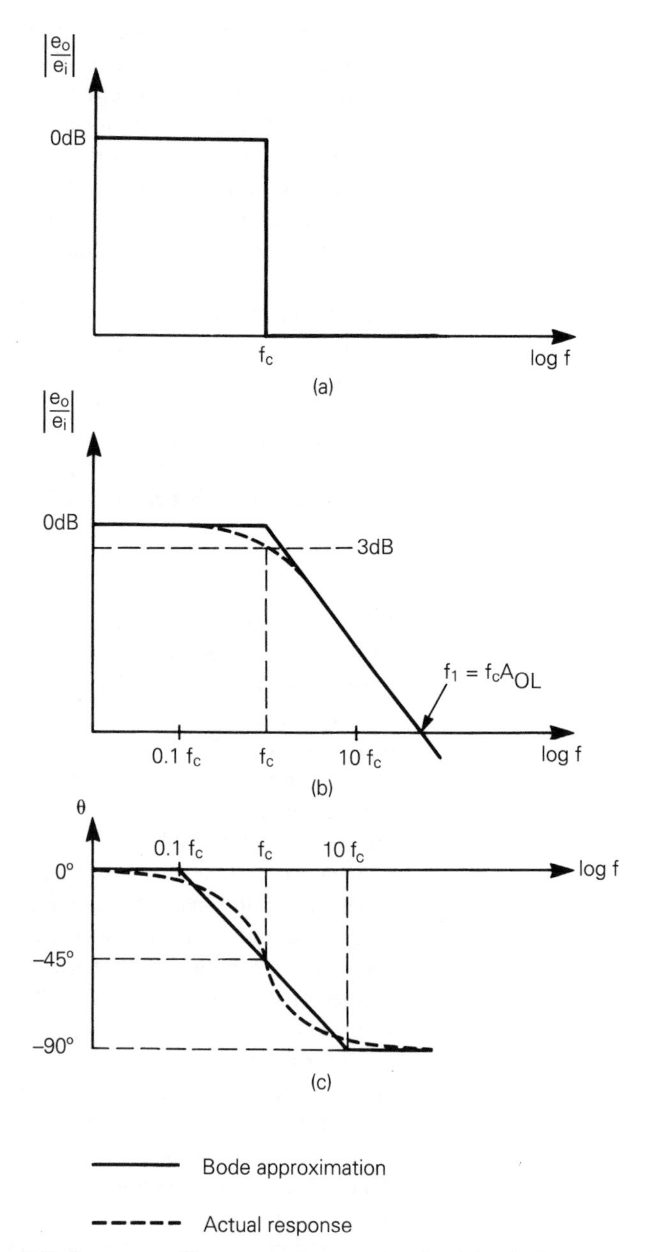

Figure 9.2 *Low-pass filter response curves for (a) ideal magnitude; (b) actual magnitude with Bode approximation; (c) phase shift with Bode approximation*

capacitive reactance eventually produces a second order filter linear roll-off dependent upon the inverse square of the frequency.

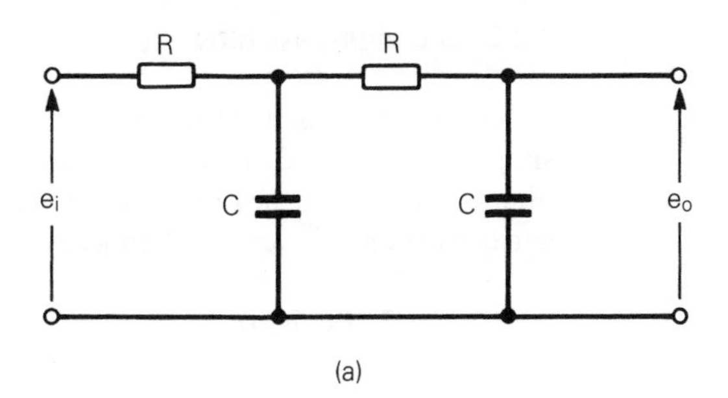

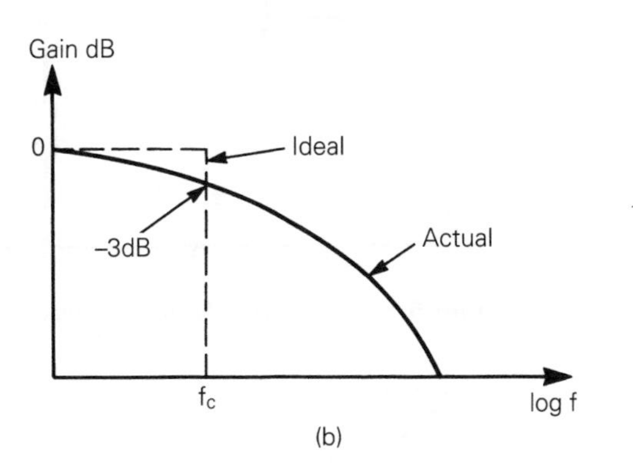

Figure 9.3 *Second order high-pass passive CR filter circuit (a) and response curve (b)*

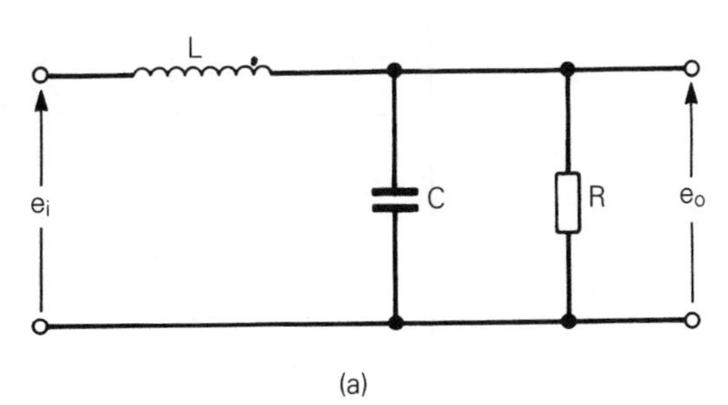

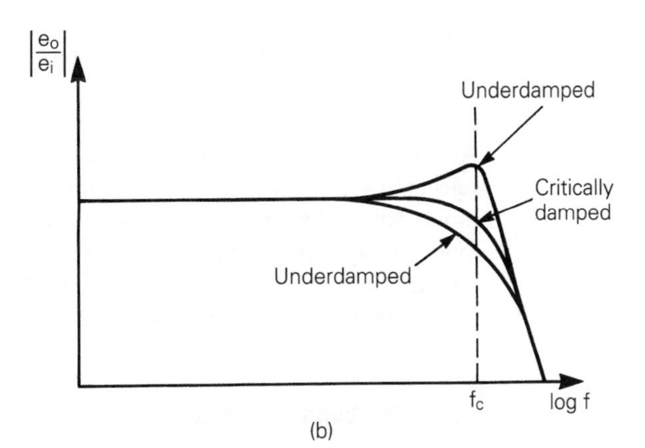

Figure 9.4 *Second order low-pass passive LCR filter circuit (a) and response (b)*

9.2.2 The high-pass filter

To form a high-pass filter, the *CR* components of the low-pass filter are simply interchanged. Figure 9.5 shows the first order high-pass circuit and Figure 9.6 its frequency response curve. The gain roll-off is once again 20 dB per decade and the cut-off frequency is still given by the equation

$$f_c = 1/2\pi RC \text{ hertz}$$

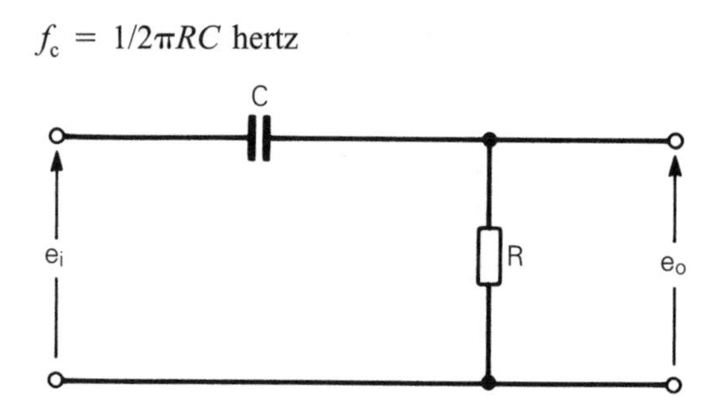

Figure 9.5 *First order high-pass passive CR filter circuit*

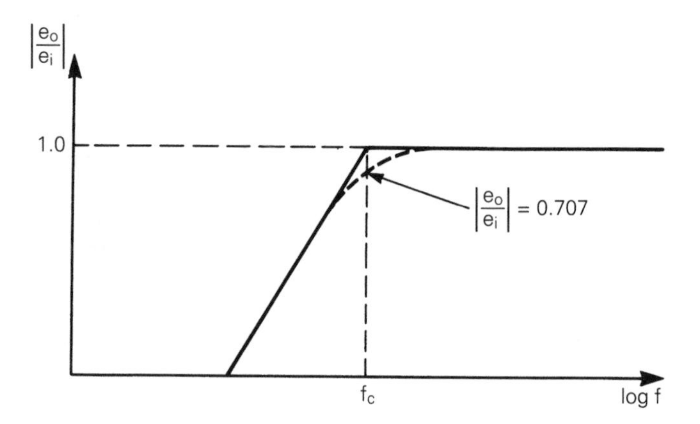

Figure 9.6 *High-pass filter response curve showing the actual (dotted) response and the Bode approximation*

At low input frequencies the capacitor has a high reactance and effectively rejects any input voltage. As the input frequency is increased the capacitor progressively lowers its reactance, allowing an increasing proportion of the input voltage to be developed across the resistor and appear at the circuit output. Frequencies below f_c are regarded as being in a stop-band; those above, as being in the circuit pass-band.

9.2.3 The band-pass filter

A second order band-pass filter can be obtained by using the series LCR circuit arrangement shown in Figure 9.7. At low input frequencies the capacitive

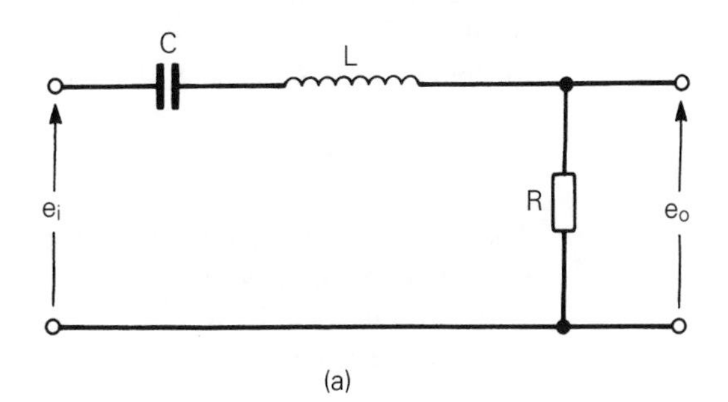

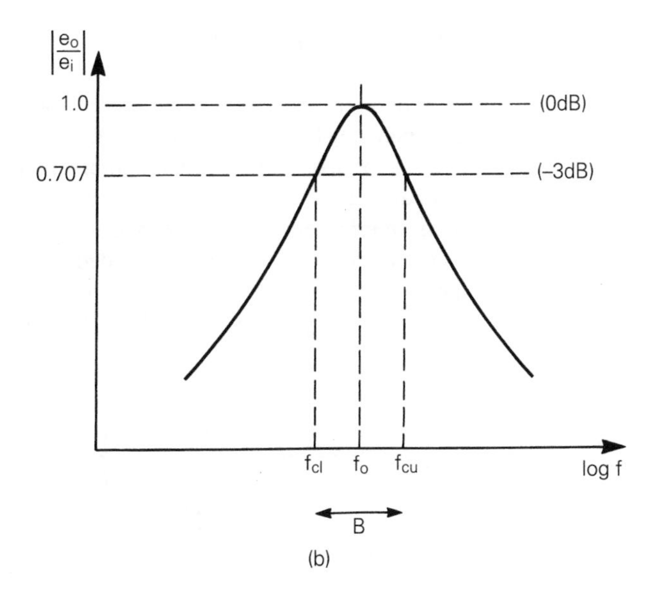

Figure 9.7 *Second order band-pass passive LCR filter circuit (a) and response (b)*

reactance predominates and the circuit behaves as a simple series capacitor with a 6 dB per octave increasing response from DC. As the frequency of the input signal approaches circuit resonance, there is a marked up-turn in the response curve to climax in a peak at:

$$f_o = 1/2\pi\sqrt{(LC)} \text{ hertz}$$

Once the resonant frequency has been exceeded, the inductive reactance becomes increasingly dominant and the response falls away but not as sharply as was the build up from the low frequencies.

Thus, there are two frequencies where the response is 3 dB less than the peak and they are called the upper and lower cut-off frequencies, f_{cu} and f_{cl}. They are equally disposed about the resonant or centre frequency; the centre frequency is always taken as the geometric mean of the two.

$$f_o = \sqrt{(f_{cu} \cdot f_{cl})}$$

The difference between f_{cu} and f_{cl} is taken as the bandwidth or pass-band, B, of the filter and together with the *goodness* or Q of the circuit is related to the centre frequency, f_o, by the following equation:

$$B = f_o/Q \text{ Hz}$$

The higher the Q of the circuit the smaller is its pass-band and the filter is said to be more *selective*.

If several such LCR circuits, each having a slightly different resonant frequency, are connected in series, the resulting circuit is a band-pass filter.

9.2.4 The band-stop filter

A second order band-stop filter can be obtained by using the parallel CL circuit arrangement shown in Figure 9.8. At low input frequencies the circuit is effectively a low-pass arrangement comprising only the L and the R. At the circuit resonant frequency, determined by $f = 1/2\pi\sqrt{(LC)}$, the parallel L and C presents an infinitely high impedance and the circuit output is zero. Once the resonant frequency has been exceeded, the inductive reactance continues to increase while that of the capacitor decreases, making the circuit perform more as if comprising only the C and the R in a simple high-pass filter arrangement.

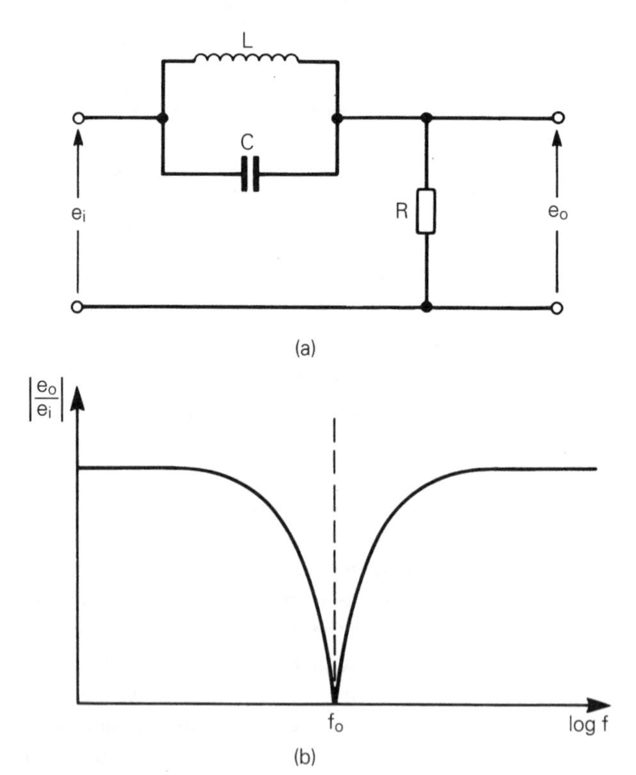

Figure 9.8 *Second order band-stop passive LCR filter circuit (a) and response (b)*

9.2.5 Passive filter summary

The four basic frequency sensitive filter circuits described above can be cascaded using any mix of first and second order variants that is necessary to produce a desired response. The shape of filter response curves has been studied by many eminent people, some of whom have had their names credited to particular circuits which satisfy particular requirements. These names include Butterworth, Bessel, Chebyshev, Cauer and, together with other special filter circuits, they will be discussed later in this chapter.

Passive filter circuits contain various combinations of resistors, capacitors and inductors and in most cases suffer from several shortcomings. Mathematically, they are difficult to design; they are often pulled off frequency by the load current drawn from them; even in their pass-band they usually attenuate signals and are not easily tuned over a wide frequency range without changing their response characteristics. Further problems can be associated with the use of inductors. Not only are they expensive, bulky and heavy; they are prone to magnetic field radiation unless expensive shielding is used to prevent unwanted coupling.

9.3 Active filters

9.3.1 The case for active filters

Op-amps overcome most of the problems associated with the passive filter circuit. Not only will the high input impedance and low output impedance of the op-amp effectively isolate the frequency sensitive filter network from the following load, it can also provide useful current or voltage gain. More significantly, the op-amp can be designed into a *CR* only circuit in such a way as to provide a filter response virtually identical with that of a passive inductive filter network. This means that the use of inductors in filters is now unnecessary. Unlike the inductor, the op-amp does not possess a magnetic field which stores energy, rather it is designed to behave mathematically in the same way as the whole passive circuit it replaces. Any additional circuit energy is obtained from the separate power source used by the op-amp.

9.3.2 Negative impedance conversion

The circuit shown in Figure 9.9 is designed to have an input impedance, Z_i, which appears to be the negative of the impedance Z.

$$i_i = \frac{e_x - e_o}{R} \tag{9.1}$$

Normal OP-AMP action causes

$$e_i = e_x = e_y \text{ and } e_y = \frac{e_o Z}{R + Z}$$

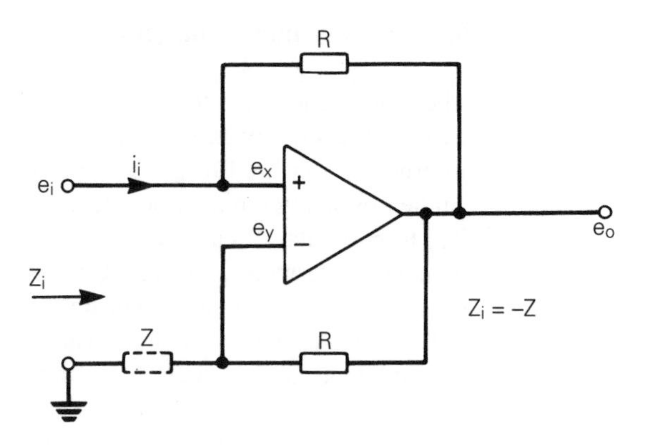

Figure 9.9 *Impedance converter*

Rearranging

$$e_o = \frac{e_i (R + Z)}{Z} \tag{9.2}$$

Substituting (9.2) and $e_x = e_i$ into (9.1)

$$i_i = \frac{e_i - \dfrac{e_i (R + Z)}{Z}}{R} = \frac{e_i \left(1 - \dfrac{R + Z}{Z}\right)}{R}$$

$$\therefore \; i_i = \frac{e_i (Z - R - Z)}{ZR}$$

and rearranging

$$Z_i = \frac{e_i}{i_i} = \frac{ZR}{Z - R - Z}$$

$$\text{or } \underline{\underline{Z_i = -Z}}$$

Suppose Z is a capacitor, C. Then $Z = j/\omega C$ and so it follows that:

$$Z_i = -(-j/\omega C) = +j/\omega C$$

The $+j$ means that the current lags the voltage, that is, has an inductive reactance, jX_L, but where $X_L = 1/\omega C$. However, while the result is to produce an inductive effect, the 'inductive reactance' decreases with increasing frequency rather than increases as would the reactance of a true inductance.

9.3.3 Impedance gyration

A single negative impedance invertor is not capable of simulating the true action of an inductor. However, this effect can be achieved if a pair of negative

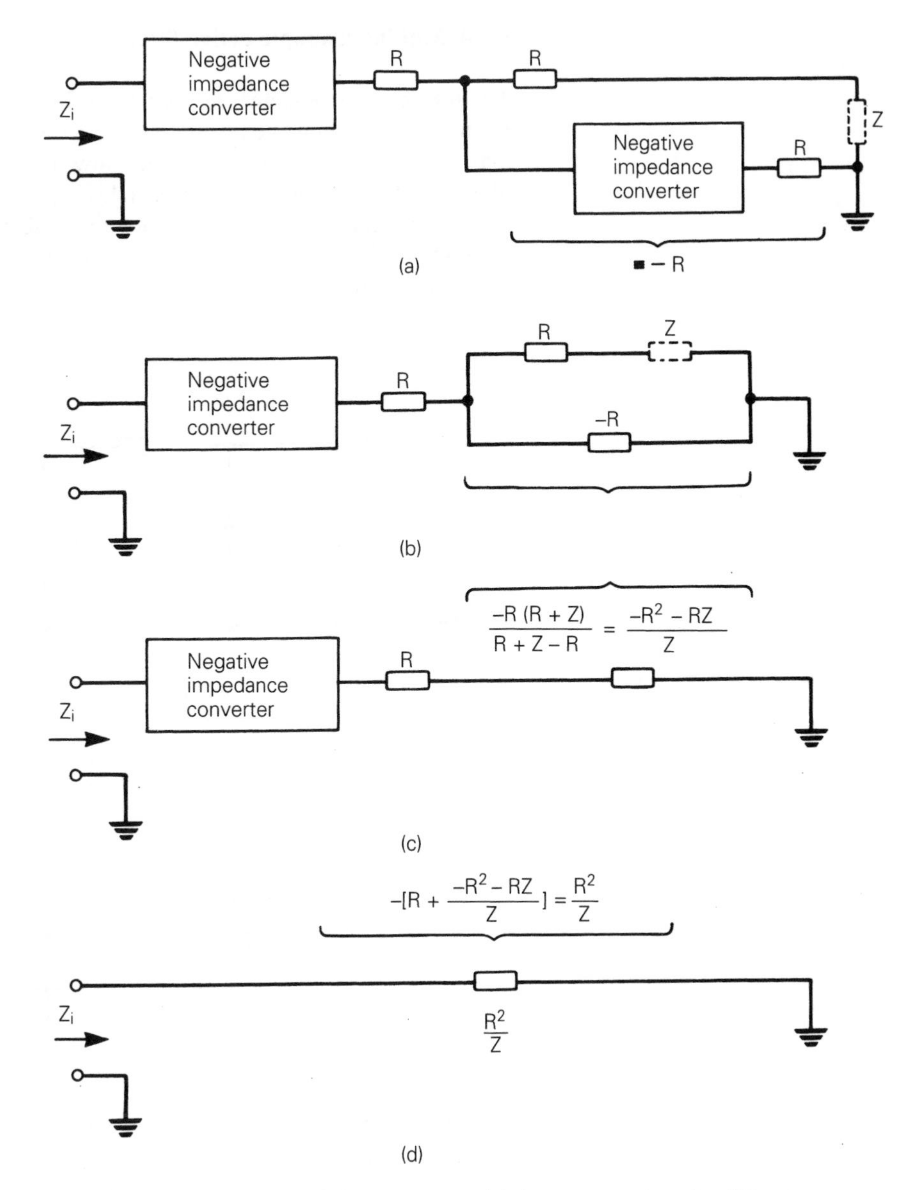

Figure 9.10 *Analysis of the circuit at (a) by progressive simplification through (b) and (c) to (d) shows that the input impedance is proportional to the reciprocal of the load impedance, Z*

impedance convertors are used. Such a circuit is shown in Figure 9.10 where the input impedance is $Z_i = R^2/Z$. Suppose that $Z = -jX_c$. then $Z_i = R^2/-jX_c = +j\omega CR^2$. Now the capacitor C is being made to act as if it were a true inductor of value $L = CR^2$.

Similarly, it can be shown that if Z were an inductive reactance of value jX_L, then the gyrator would make this appear to the preceding circuit as a true capacitance of $C = L/R^2$.

9.3.4 Making a simple active filter

The response curve shown in Figure 9.4 for a passive second order low-pass *LR* filter can be simulated using only resistors and capacitors. A first attempt may include two cascaded first order, low-pass *CR* sections with the addition of an emitter follower. This has a high input impedance but low output impedance and so minimizes any loading effects on the frequency sensitive *CR* sections. This circuit, which produces a highly damped response, is shown in Figure 9.11(a).

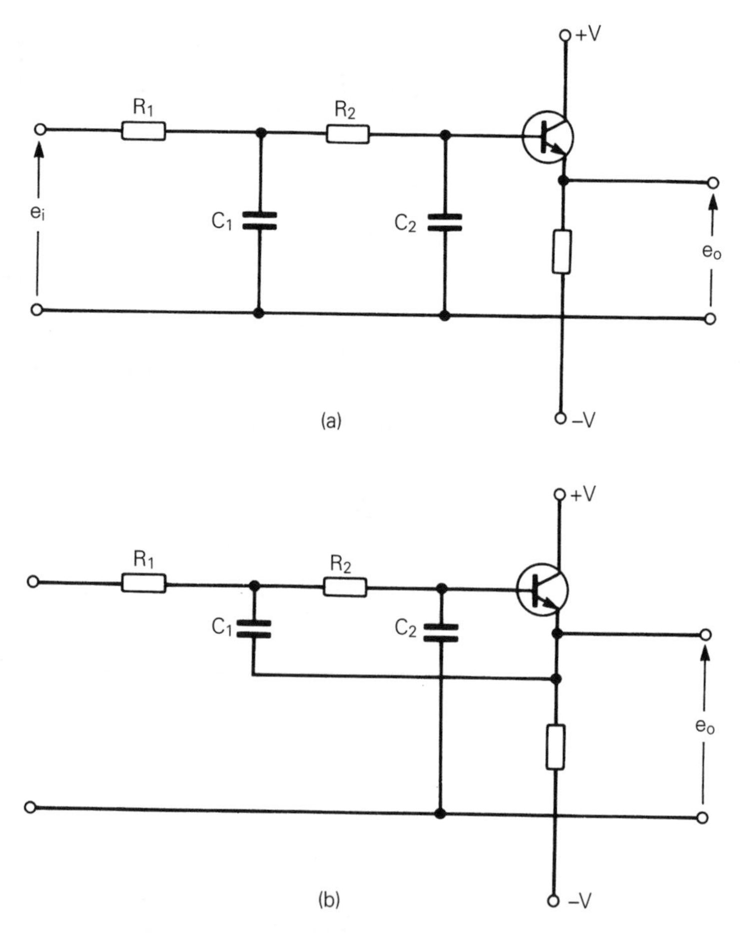

Figure 9.11 *Second order active filter*

Figure 9.11(b) illustrates a major design improvement by the introduction of positive energy feedback to the centre of *CR* section. This 'bootstrapping' has a maximum effect only near the cut-off frequency. At very low frequencies the normal gain enhancement of positive energy feedback is largely negated by the high reactance of C_1 the feedback path. In excess of the cut-off frequency, the low reactance of C_2 allows the signal to leak to earth and attenuate the output accordingly. The values of the filter network capacitors and resistors can be selected to eliminate the damping problem of the previous circuit. While the product of the resistors and capacitors decides the cut-off frequency it is the ratio of the capacitors which affects the circuit response rate. Compared with the values of C_1 and C_2 for critical damping, a large C_1 with a small C_2 will produce an underdamped response while the reverse will cause overdamping.

9.4 Active filters using operational amplifiers (op-amps)

In practical active filters the emitter followers used above are invariably replaced by op-amps in the form of integrated circuits. The frequency sensitive filtering networks are either placed before the op-amp inputs or in the feed-back circuits.

9.4.1 First order high-pass and low-pass filters

Examples of simple first order high- and low-pass active filters are shown in Figure 9.12. As expected, the frequency selective resistor-capacitor circuit elements decide the frequency response. The cut-off frequency is $f_c = 1/2\pi CR$ at which the magnitude of the filter response is 3 dB less than that in the pass-band, and the higher frequency roll-off tends to 20 dB per decade. If a low value of f_c is required, a general purpose Bi-FET operational amplifier should be suitable. This will allow the use of large resistance values without introducing any appreciable bias current off-set error. Resistor values up to 10 MΩ may be used so avoiding the expense of a high value, close tolerance capacitor.

First order low-pass filters are often used to perform a running average of a signal having high frequency fluctuations superimposed upon a relatively slow mean variation; for this purpose it is simply necessary to make the filter time constant CR much greater than the period of the high frequency fluctuations.

A practical point to remember is that all op-amp active high-pass filters show a band-pass characteristic. This is because their response eventually fails at frequencies which exceed the closed loop bandwidth of the op-amp.

9.4.2 Second order low-pass and high-pass filters

Examples of simple second order low-pass and high-pass active filter circuits are shown in Figure 9.13. The second order filter response has a 40 dB per decade roll-off in the stop-band. The sharpness of the response curve knee depends upon the choice of values for the components forming the frequency sensitive element of the filter. In Figure 9.13, the components are proportioned to give a so-called Butterworth response (further details later in this chapter) and the cut-off frequency $f_c = 1/[2\sqrt{2}(\pi CR)]$ hertz.

9.5 Choosing the frequency response of the low-pass filter

Figure 9.2 shows the ideal shape for a low-pass filter. It has a perfectly flat (horizontal) response from zero frequency up to the cut-off frequency where a vertical fall then occurs. In practice this perfectly rectangular shape is

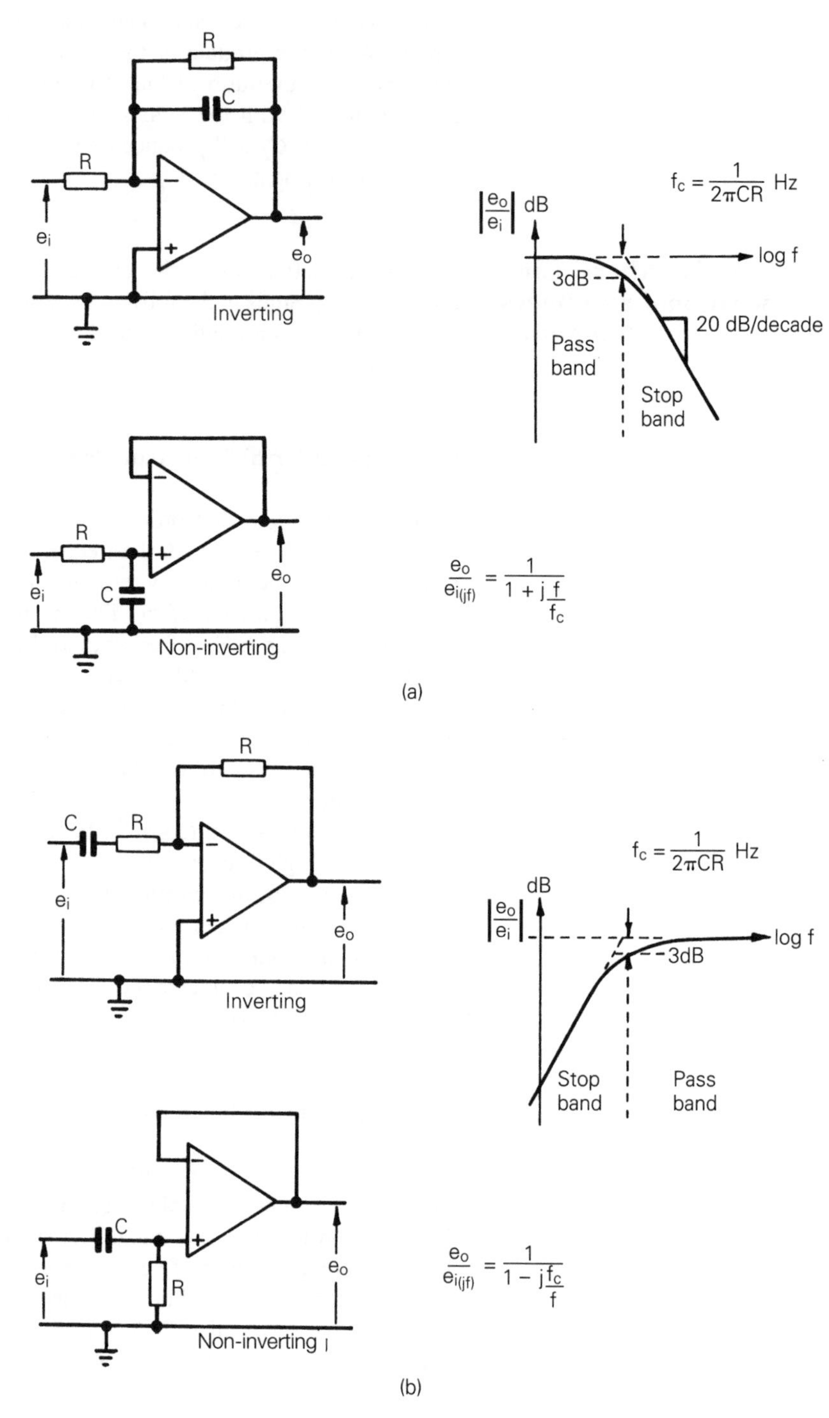

Figure 9.12 *First order low- and high-pass active filters. (a) First order low-pass response. (b) First order high-pass response*

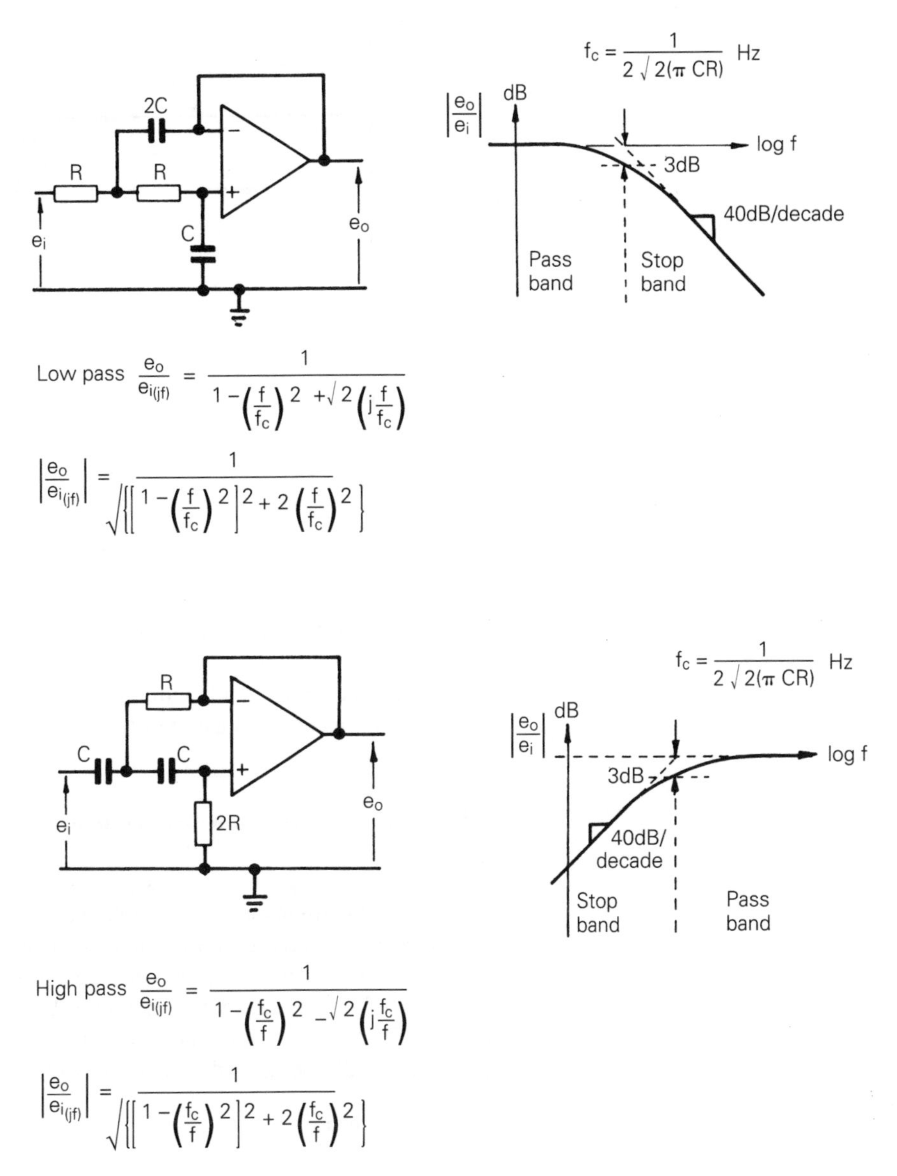

Figure 9.13 *Second order low- and high-pass active filters*

unattainable. Depending upon the intended role of the filter, it can be designed to approximate to the ideal response in varying ways and these are mentioned briefly below.

9.5.1 The Butterworth low-pass response

This response requires that at zero frequency the circuit gain is flat and remains as near flat as possible up to the designed cut-off frequency.

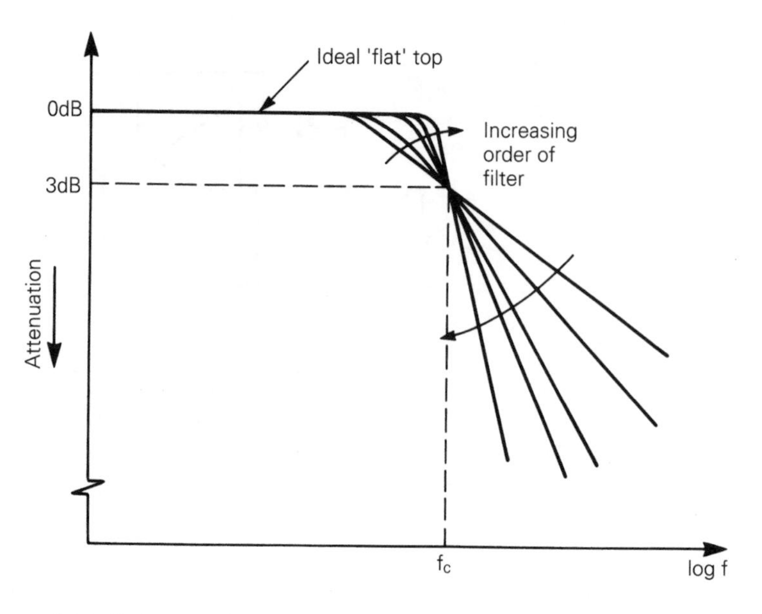

Figure 9.14 *Butterworth response*

The higher the order of filter the more accurately does its response approximate to this ideal, as illustrated by Figure 9.14.

9.5.2 The Chebyshev low-pass response

The Chebyshev approximation is an attempt to overcome the practical failure of the Butterworth response to maintain a truly flat pass-band as the frequency of operation is increased up to the cut-off frequency. The Chebyshev circuit is designed uniformly to spread any deviation of gain over the pass-band in the form of ripples as shown in Figure 9.15. Above the cut-off frequency, like the Butterworth, the Chebyshev roll-off eventually tends to be monotonic at $20n$ dB per decade where n is the order of the filter. Even so, the

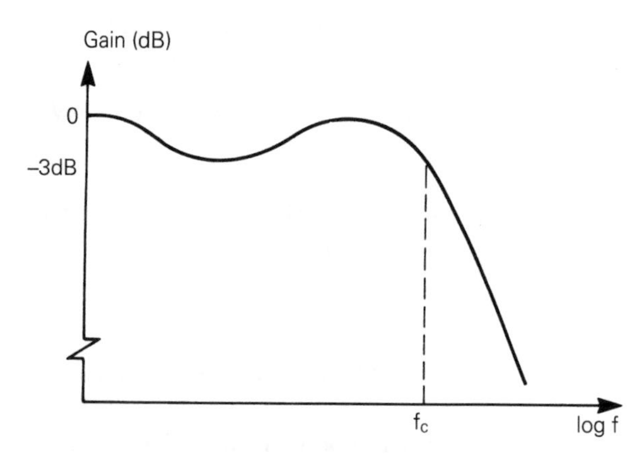

Figure 9.15 *Third order Chebyshev response*

second and third order Chebyshev filters tend to have a less steep initial roll-off than their Butterworth counterparts, whereas comparable fourth order and above filters show the Chebyshev response to have the sharper knee.

9.5.3 The Cauer (or elliptic) low-pass response

Using a Butterworth or Chebyshev filter, a complete signal stop is usually regarded as having been achieved when the filter attenuation has reached a designed level. The frequency at which this degree of attenuation first occurs is taken as the start of the filter stop-band. However, while a continued increase in frequency initially causes further signal attenuation, a practical limit is reached. This is where, owing to unwanted leakage through stray reactances, further increasing the frequency can produce an unwanted output from the filter. The Cauer response is designed to cater for those applications where it is required that an infinite attenuation is achieved at a particular frequency and that for any higher frequencies a designed minimum attenuation is maintained.

Figure 9.16 shows the Cauer filter circuit diagram and the typical response curves it produces. The infinite attenuation is caused at the frequency, f_2, because at this frequency L_2 and C_2 are in resonance and present an infinite impedance to the signal flow.

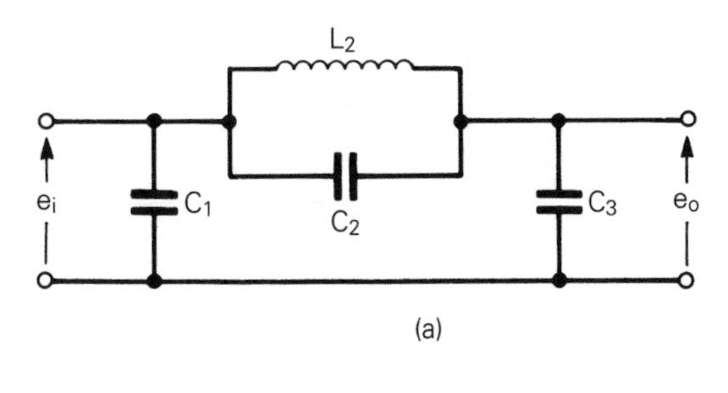

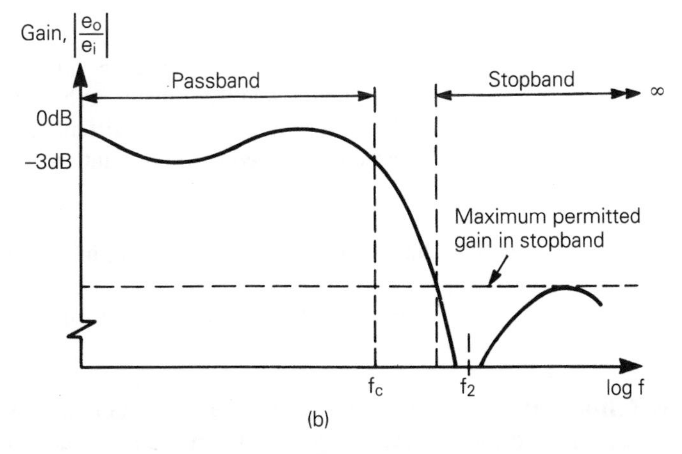

Figure 9.16 *Cauer or elliptic, third order filter circuit (a) and the response it produces (b)*

9.5.4 The Bessel low-pass response

The above studies on the Butterworth, Chebyshev and Cauer filter responses have all emphasized the relative amplitudes of the filter input and output voltages. No mention has been made of the phase shift which occurs as the signal travels through the filter. In applications involving voice or other analogue transmissions, phase shift is not important and optimum amplitude responses are often obtained at the expense of phase shift. However, in the case of digital transmissions it can be important that the pulses are not distorted and linear-phase filters are often used.

Figure 9.17 shows the ideal linear relationship between the signal frequency and the resulting phase shift introduced by the filter. With regard to the signal transit time through the filter, signals of all frequencies should ideally suffer the same time delay and so any signals in phase at the input will still be in phase at the output. But a signal of double the frequency of another will suffer twice its phase shift. This effect is shown at Figure 9.18.

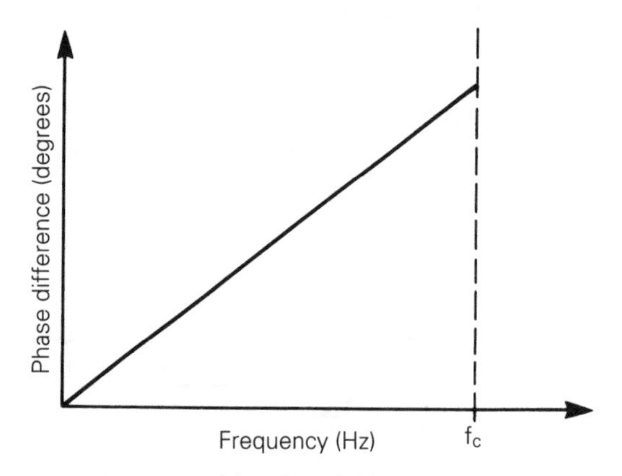

Figure 9.17 *Linear relationship between frequency and phase shift in ideal low-pass filter*

The Bessel approximation is an attempt to produce such a linear-phase filter. The Bessel response circuit has the same appearance as the Butterworth and Chebyshev circuits and differs only in the component values necessary to produce the required constant transit time at all frequencies.

9.5.5 Comparative responses of the different low-pass filters

See Figure 9.19 for a summary of the various low-pass responses.

9.6 Choosing the frequency response of the high-pass filter

Figure 9.6 shows the ideal shape for a high-pass filter response curve. It has a zero output at low frequencies but continued frequency increase eventually causes the response to rise monotonically until just short of the −3 dB cut-off frequency which marks the start of the pass-band. At frequencies higher

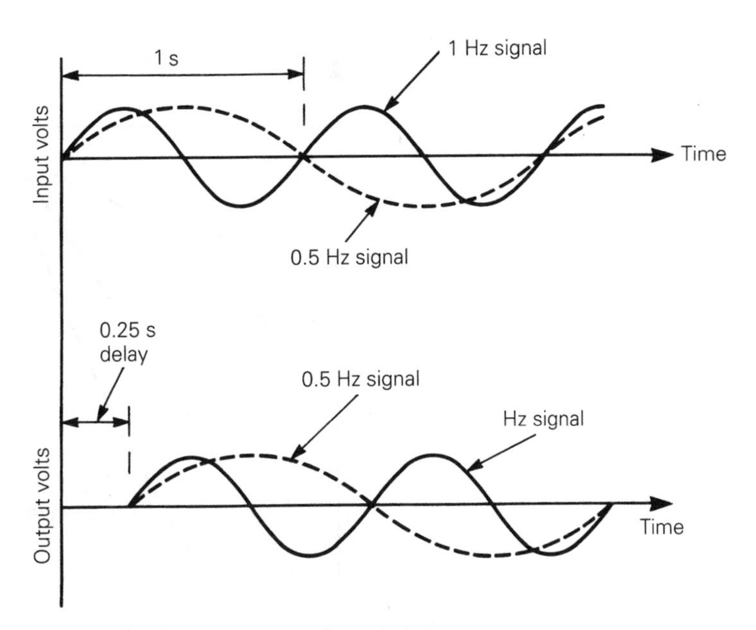

Figure 9.18 *Time-related waveforms showing how a constant time delay of 0.25 s produces a 90° phase shift in a 1 Hz signal but only a 45° shift in a 0.5 Hz signal*

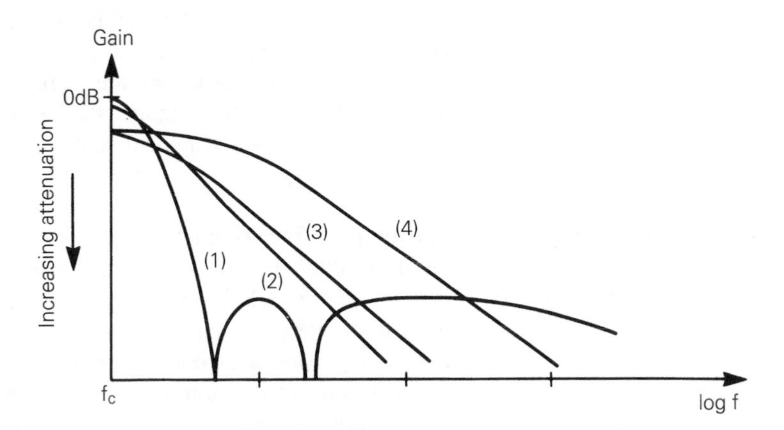

Figure 9.19 *Comparison of the different low-pass filter performance in their stop-bands. (1) Cauer (elliptic); (2) Chebyshev; (3) Butterworth; (4) Bessel*

than this, the response in the pass-band levels at the maximum gain. But, because of practical component inadequacies and stray reactances becoming increasingly significant at the higher frequencies, the flat response of the passive filter circuit element does not extend to infinity and eventually declines. Additionally, in the case of the active filter, the inherent high frequency gain roll-off of the op-amp effectively makes any high-pass filter behave as a form of band-pass filter – but with the upper cut-off frequency being above the highest frequency to be passed.

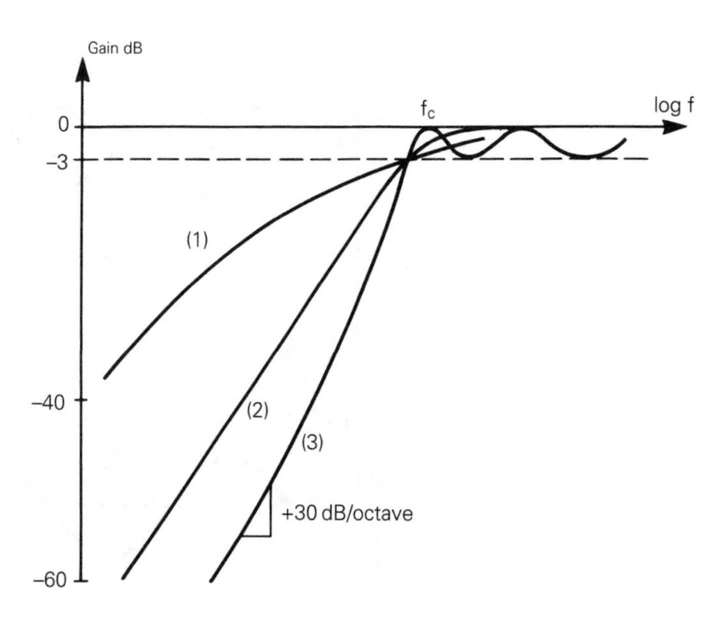

Figure 9.20 *Example of high-pass filter (fifth order) response curves. (1) Overdamped; (2) flattest; (3) Chebyshev*

The studies of the low-pass filtering transfer functions and response curves made in Section 9.5 can be readily modified to suit the high-pass conditions. Basically, the high-pass filter is a mirror image of its low-pass equivalent; the capacitors and resistors are simply interchanged. The mathematical process involved in this change is called *mathematical transformation by 1/f*. Figure 9.20 shows a graphical summary of the high-pass response curves.

9.7 Band-pass filters using the state variable technique

A band-pass filter characterstic can be obtained by cascading a high- and a low-pass filter, but when a highly selective (high Q) band-pass characteristic is required a different approach is necessary. Many examples of active band-pass filters will be found in the literature and in manufacturers' notes, but high Q band-pass filters, based upon a single op-amp, have a Q value which is very sensitive to component variation. The so called state variable filter approach, which is based upon the use of analogue computer techniques (Section 6.10), is less component sensitive although it requires the use of three op-amps.

The circuit for a second order state variable filter is shown in Figure 9.21. It is particularly versatile in that it allows the simultaneous realization of high-pass, low-pass and band-pass characteristics at three separate circuit points. A fourth amplifier can, if required, be used to form a band reject characteristic.

The steady state sinusoidal response equation for the circuit of Figure 9.21 is now derived – op-amps are assumed to behave ideally. It is the action of the feedback loops which forces the desired relationships between inputs and outputs; we derive the band-pass relationship which is exhibited between the input signal and the output of amplifier A_2 (e_{bp}). The relationship between

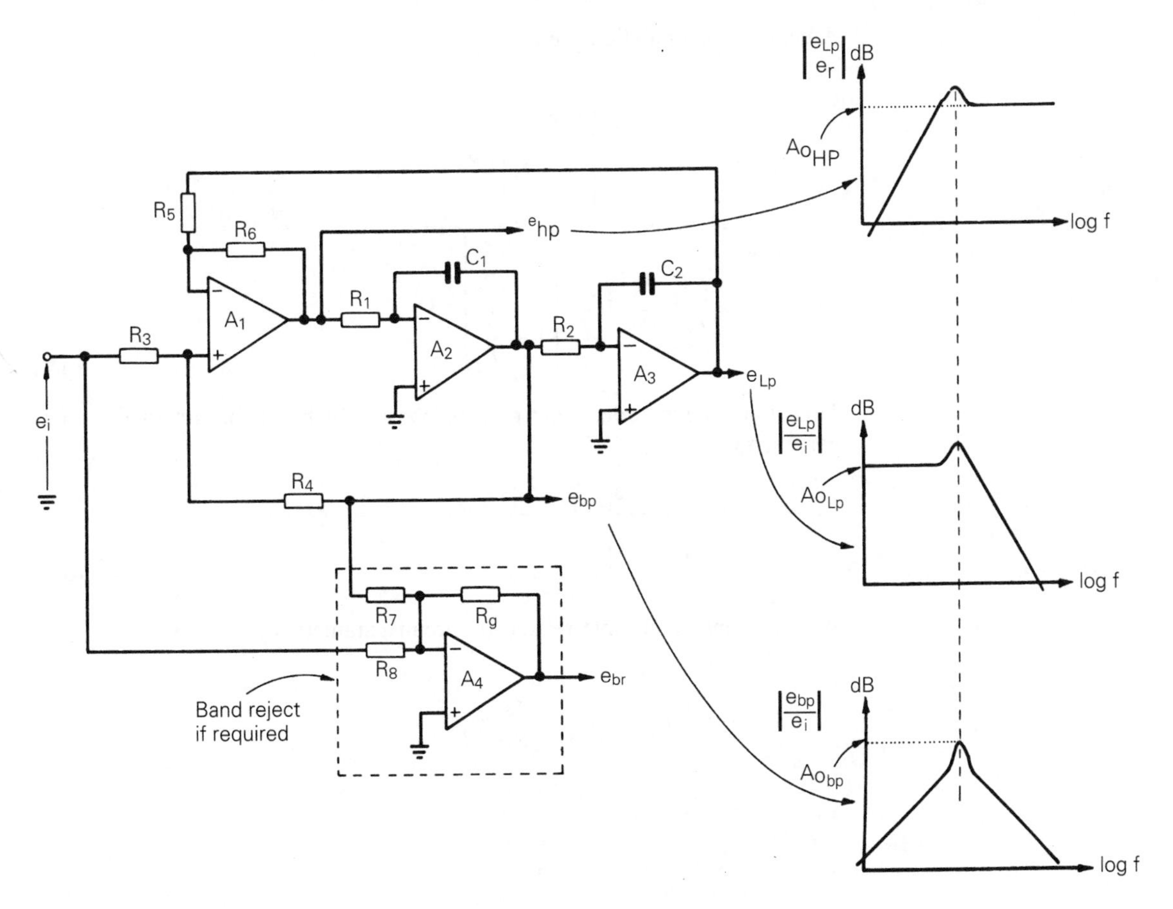

Figure 9.21 *Amplifiers use BiFETS, e.g. LF 356 T1080 singles or TL 084 quad*

e_{bp} and the other output signals is readily found if it is remembered that the action of an integrator is to multiply by

$$-\frac{1}{j\omega T}$$

where $T = CR$ is the integrator time constant.

Amplifier A_1 is connected as an adder–subtractor. It sums the input signal with the output of A_2 and subtracts the output of A_3, thus:

$$e_{o_1} = \frac{e_{bp}}{-\dfrac{1}{j\omega T_1}} = \left[e_i \frac{R_4}{R_3 + R_4} + \frac{e_{bp} R_3}{R_3 + R_4}\right]\left[1 + \frac{R_6}{R_5}\right]$$

$$-\left[-\frac{1}{j\omega T_2} e_{bp} \frac{R_6}{R_5}\right]$$

where $T_1 = C_1 R_1$ and $T_2 = C_2 R_2$.

Algebraic manipulation yields

$$\frac{e_{bp}}{e_{i(j\omega)}} = \frac{-\dfrac{1}{T_1}\dfrac{\left[1+\dfrac{R_6}{R_5}\right]}{\left[1+\dfrac{R_3}{R_4}\right]}j\omega}{\dfrac{1}{T_1T_2}\dfrac{R_6}{R_5}+\dfrac{j\omega}{T_1}\dfrac{\left[1+\dfrac{R_6}{R_5}\right]}{\left[1+\dfrac{R_4}{R_3}\right]}-\omega^2} \tag{9.3}$$

This is a second order band-pass response which can be put in the more general form

$$\frac{e_{bp}}{e_{i(j\omega)}} = -\frac{A_{o(bp)}}{1+jQ\left[\dfrac{\omega}{\omega_o}-\dfrac{\omega_o}{\omega}\right]} \tag{9.4}$$

With response constants related to circuit parameters by

$$A_{obp} = \frac{R_4}{R_3},\ \omega_o = \sqrt{\frac{R_6}{R_5C_1R_1C_2R_2}}$$

$$Q = \sqrt{\frac{C_1R_1R_6}{C_2R_2R_5}}\ \frac{1+\dfrac{R_4}{R_3}}{1+\dfrac{R_6}{R_5}} \tag{9.5}$$

In practice it is convenient to make $R_5 = R_6$, $C_1 = C_2$, $R_1 = R_2$. The centre frequency ω_o can then be tuned without altering the Q by simultaneously changing R_1 and R_2. The Q value can be varied by changing R_4 without altering the centre frequency.

The response at the low-pass and high-pass output is readily derived by the substitution

$$e_{lp} = -\frac{1}{j\omega T_2}e_{bp} \quad \text{and} \quad e_{hp} = -j\omega T_1\, e_{bp}$$

in equation 9.3 yielding, after some algebraic manipulation, equations of the form

Second order low pass

$$\frac{e_{lp}}{e_{i(j\omega)}} = \frac{A_{o(lp)}}{1+2\zeta j\dfrac{\omega}{\omega_o}-\left(\dfrac{\omega}{\omega_o}\right)^2} \tag{9.6}$$

$$\text{With } A_{o(lp)} = \frac{1 + \dfrac{R_5}{R_6}}{1 + \dfrac{R_3}{R_4}}, \ \omega_o = \sqrt{\frac{R_6}{R_5 C_1 R_1 C_2 R_2}}$$

$$\text{and } \zeta = \frac{1}{2} \frac{1 + \dfrac{R_6}{R_5}}{1 + \dfrac{R_4}{R_3}} \sqrt{\frac{C_2 R_2 R_5}{C_1 R_1 R_6}}$$

Second order high pass

$$\frac{e_{hp}}{e_{i(j\omega)}} = \frac{A_{o(hp)}}{1 + 2\zeta j \dfrac{\omega_o}{\omega} - \left(\dfrac{\omega_o}{\omega}\right)^2} \tag{9.7}$$

$$\text{With } A_{o(hp)} = \frac{1 + \dfrac{R_6}{R_5}}{1 + \dfrac{R_3}{R_4}}, \ \omega_o = \sqrt{\frac{R_6}{R_5 C_1 R_1 C_2 R_2}}$$

$$\text{and } \zeta = \frac{1}{2} \frac{1 + \dfrac{R_6}{R_5}}{1 + \dfrac{R_4}{R_3}} \sqrt{\frac{C_2 R_2 R_5}{C_1 R_1 R_6}}$$

9.8 Band reject filter (notch filter)

An active band reject filter based upon a modified twin tee network is given in Figure 9.22. With the components proportioned as shown the performance equation is governed by the relationship

$$\frac{e_o}{e_{i(j\omega)}} = \frac{Q\left[j\dfrac{\omega}{\omega_o} - \dfrac{\omega_o}{\omega}\right]}{1 + jQ\left[\dfrac{\omega}{\omega_o} - \dfrac{\omega_o}{\omega}\right]} \tag{9.8}$$

$$\text{where } \omega_o = \frac{1}{CR} \text{ and } Q = \frac{1}{4(1 - m)}$$

The circuit allows the adjustment of Q by means of a single potentiometer. In practice the depth of the notch obtainable with the filter is very much dependent upon component matching and high Q circuits are very component sensitive.

A band reject filter which is less sensitive to component tolerance than Figure 9.22 can be realized with the state variable filter of the previous section. If a quad op-amp is used for the state variable filter, the fourth amplifier can be used to sum the input signal with the band-pass output giving a band reject response; thus the output of amplifier A_4 in Figure 9.21 is

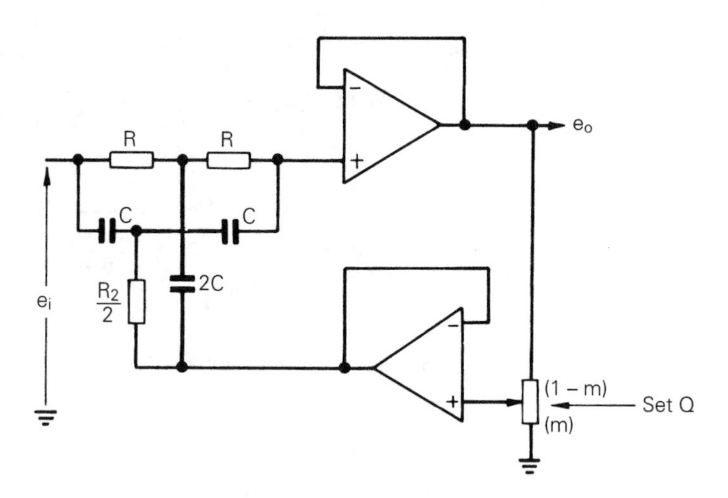

Figure 9.22 *High Q band reject filter (notch filter)*

$$e_{br} = -\left[e_i \frac{R_9}{R_8} + e_{bp} \frac{R_9}{R_7}\right]$$

Substituting for e_{bp} from equation 9.4 gives

$$\frac{e_{br}}{e_{i(j\omega)}} = -\left[\frac{R_9}{R_8} - \frac{A_o \dfrac{R_9}{R_7}}{1 + jQ\left[\dfrac{\omega}{\omega_o} - \dfrac{\omega_o}{\omega}\right]}\right]$$

If resistors are proportional so that $A_o = R_4/R_3 = R_7/R_8$ the response becomes

$$\frac{e_{br}}{e_{i(j\omega)}} = -\frac{R_9}{R_8}\left[\frac{jQ\left[\dfrac{\omega}{\omega_o} - \dfrac{\omega_o}{\omega}\right]}{1 + jQ\left[\dfrac{\omega}{\omega_o} - \dfrac{\omega_o}{\omega}\right]}\right] \qquad (9.9)$$

Note that Q and ω_o have the values for the band-pass function given previously (equation 9.5).

9.9 Phase shifting circuit (all-pass filter)

The circuit shown in Figure 9.23 uses an op-amp to generate an arbitrary phase shift. All frequencies within the closed-loop bandwidth are passed at unity gain, but with a phase shift that varies with frequency. If the resistor R' connected to the non-inverting input terminal is made variable, the circuit provides a convenient means of phase adjustment. A phase variation between 0° and almost 180° is possible.

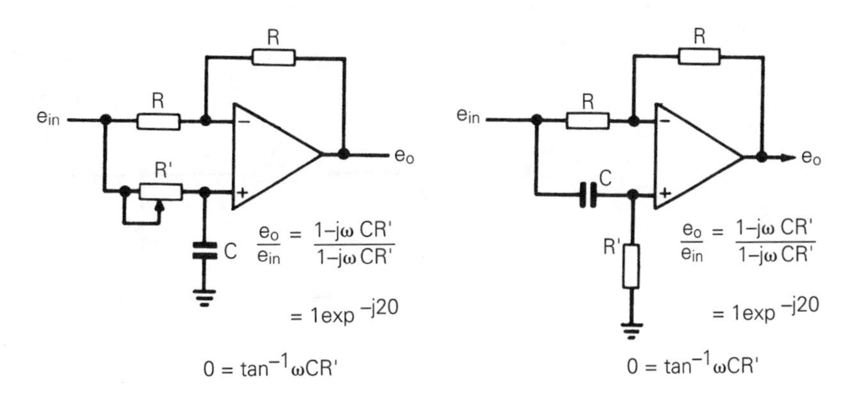

Figure 9.23 *Op-amp phase shifter*

9.10 Filter design

The preceding paragraphs have given an insight into the different shapes of filter response curve which may be obtained by the careful selection of the order of filter required together with the correct component values. The mathematical prediction of a particular response using manual methods becomes lengthy, tedious and error prone as the filter order increases. The recent proliferation of personal computers has made these design calculations a less onerous task, but even more important is the availability of ready-made designs for which tables of 'normalized' frequency against component values are published and which can be used to design a filter having a particular cut-off frequency and input impedance. The tabulated 'normalized' figures are 'scaled' to give practical component values.

9.10.1 Normalization and scaling

Suppose we consider one of the simplest active filter circuits, the single pole low-pass filter, a typical circuit for which is shown in Figure 9.24(a) (the feedback resistor is included for DC off-set purposes). It is shown in Section 9.2.1 that the cut-off frequency for this *CR* circuit is given by:

$$f_c = 1/2\pi CR \text{ Hz}$$

$$\text{or } \omega_c = 1/CR \text{ radians/second}$$

If the circuit were required to have an impedance level of 1 ohm and a cut-off angular frequency of 1 rad/s, then the capacitor would need a value of 1 farad. The circuit would be said to have been 'normalized' to 1 ohm and 1 rad/s and is shown in Figure 9.24(b). With these values for resistance and capacitance the circuit is not of great use but it can be 'scaled' to determine the values of resistance and capacitance which give a particular cut-off frequency and impedance level.

Suppose we wish to raise the impedance level from the normalized 1 ohm to 500 ohms. The rule for this is to raise all the circuit impedances by a

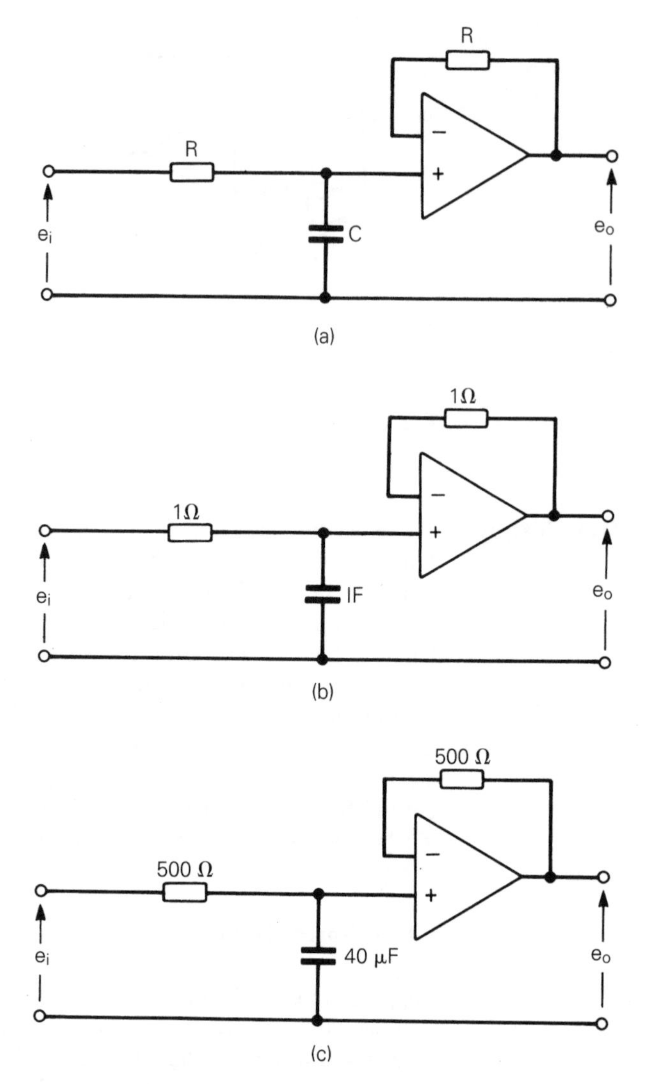

Figure 9.24 *Simple first order low-pass active filter. (a) Typical circuit. (b) Circuit normalized to 1 ohm impedance level and cut-off frequency of 1 rad/s. (c) Circuit scaled to change the impedance level to 500 ohms and cut-off frequency to 50 rad/s (7.96 Hz)*

factor of 500. This means that all the resistances must be multiplied by 500 but that all the capacitances will require their values to be divided by 500, since capacitive reactance is inversely proportional to frequency.

(*Rule No. 1* – To increase the impedance level, multiply the resistors and divide the capacitors by the scaling factor.)

Further, suppose that we now wish to increase the cut-off frequency from 1 rad/s to 50 rad/s without changing the newly adjusted impedance level. The requirement now is for the circuit time constant to be reduced in that same ratio, that is, by 1/50. This means that the product of CR must be reduced by 1/50 without altering the fixed value of R at 500 ohms. Therefore, from the relationship, $\omega = 1/CR$ and knowing that ω must be 50 rad/s and that R is newly fixed at 500 ohms, C becomes $1/(50 \times 500)$ which is 40 F.

(*Rule No. 2* – To increase the cut-off frequency, divide either the resistors or the capacitors, by the scaling factor.)

When the circuits comprise more than a single resistor with a single capacitor, as is the case with the higher order filters, the same basic rules still apply. But remember, for multi-section filters, the ratios of the frequency sensitive capacitor and resistor pairs must remain unchanged if the overall filter frequency response is not to change. Also, if the frequency of one section is altered then all sections must be changed to the same frequency.

It was shown in Figure 9.12(a) that the closed-loop gain for this single pole low-pass filter is:

$$\frac{e_o}{e_i(jf)} = \frac{1}{1 + jf/f_c} = \frac{1}{1 + j\omega/\omega_c}$$

Expressed in polar form, this becomes

$$\frac{e_o}{e_i} = \frac{1}{\sqrt{1 + \left(\frac{\omega}{\omega_c}\right)^2 \angle \tan^{-1} \frac{\omega}{\omega_c}}}$$

$$\text{or } \frac{e_o}{e_i} = \frac{1}{\sqrt{1 + \left(\frac{\omega}{\omega_c}\right)^2}} \angle \tan^{-1} \frac{\omega}{\omega_c}$$

This expression can be used to calculate the circuit gain for varying values of ω, these being made a known fraction of ω_c. The table of data together with the plotted response curve are shown in Figure 9.25.

9.10.2 Sallen–Key second order active filters

There are many circuit configurations which operate successfully as second order filters but perhaps any dissertation on active filters would be incomplete without at least a mention of the circuits jointly attributable to Sallen and Key. There are two basic Sallen–Key designs: the unity-gain filter and the equal-component filter.

While these circuits are relatively simple to construct, in order that they operate as expected, the various component values must have a definite relationship which is a function of the circuit Q-factor. Typical Sallen–Key second order low-pass filter circuits of the two types mentioned are shown in Figure 9.26. It is important to note that both these circuits have constant gains; one being unity, the other 2:1.

An advantage of the unity-gain circuit is that it requires the minimum number of components; even the feedback resistor is not necessary in some circuits and may be omitted. However, the unity-gain circuit does not lend itself to easy conversion to a high-pass or band-pass filter by simply interchanging the circuit positions of some components. On the other hand, the

ω (rad/sec)	$\left\lvert \frac{e_o}{e_i} \right\rvert$	Phase angle (degrees)
$0.0625\ \omega_C$	0.998	− 3.58
$0.100\ \omega_C$	0.995	− 5.71
$0.125\ \omega_C$	0.992	− 7.13
$0.25\ \omega_C$	0.970	−14.0
$0.5\ \omega_C$	0.894	−26.6
ω_C	0.707	−45.0
$2\ \omega_C$	0.447	−63.4
$4\ \omega_C$	0.243	−76.0
$8\ \omega_C$	0.124	−82.9
$10\ \omega_C$	0.100	−84.3
$16\ \omega_C$	0.062	−86.4
$100\ \omega_C$	0.010	−89.4

(a)

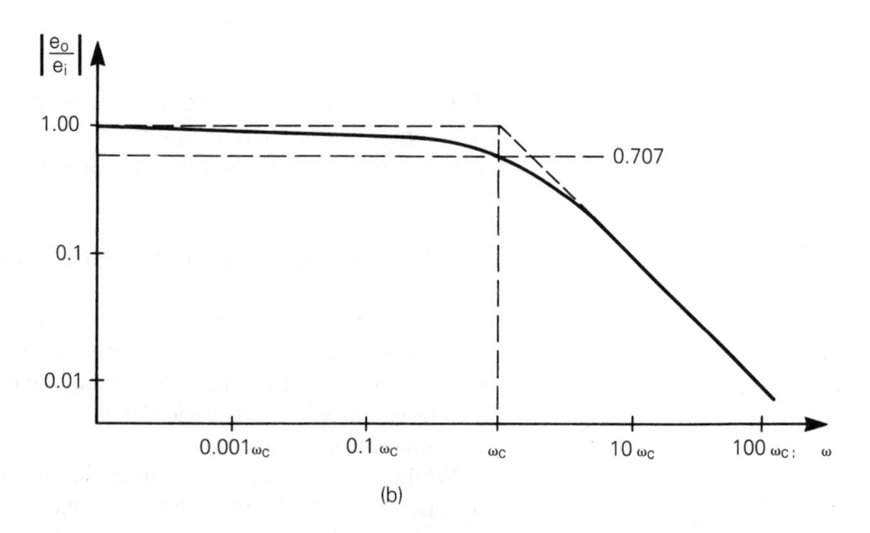

(b)

Figure 9.25 *Response curve and data table for a first order low-pass filter. (a) Table of data for closed-loop magnitude and phase; (b) response curve plotted from data at (a)*

equal-component circuit requires more components but has the advantage of being simple to convert to a high-pass filter by interchanging the frequency determining the resistors and capacitors.

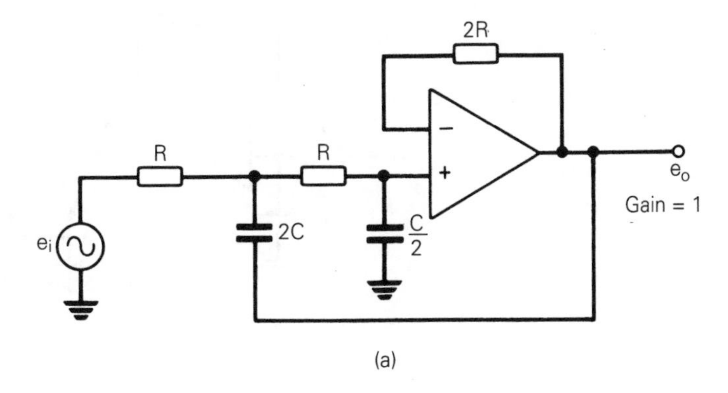

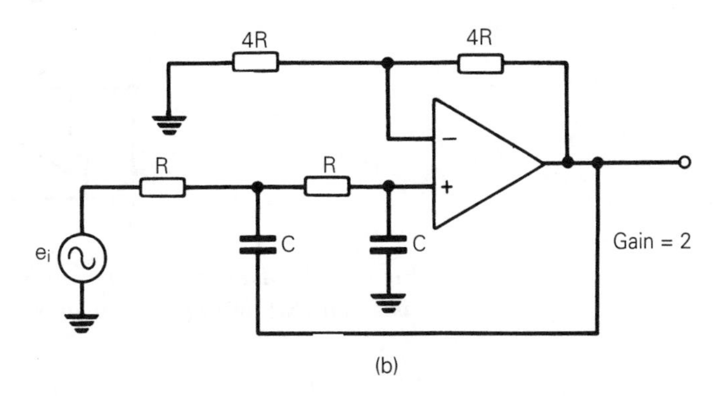

Figure 9.26 *Second order Sallen–Key low-pass active filters. (a) Typical unity-gain circuit; (b) typical equal-component circuit*

9.10.3 Time averaging

Time averaged signals are normally obtained using some form of first order low-pass filter. Consider the arrangement in Figure 9.27(a), in which a pulsed signal is applied to a long time constant *CR* filter. The capacitor charges up during each positive pulse, and discharges when the pulses are absent. If the time constant ($t = CR$) is much greater than the period of the pulses, the capacitor does not have time to discharge fully. Hence the capacitor voltage gradually approaches the mean value of the input signal, albeit with a small fluctuation superimposed.

It is possible to have long time constants without resorting to big *CR* values. This can be realized by making use of the capacitor multiplying principle, as shown in Figure 9.27(b).

A modified circuit is necessary to remove noise from signals that have large variations in mean value (e.g. a noisy square wave). A long time-constant first order low-pass filter is unsuitable, because its output cannot follow the rapid changes of the signal. A modification to the circuit of Figure 9.27(b), which is shown in Figure 9.28, can sometimes be used in such applications.

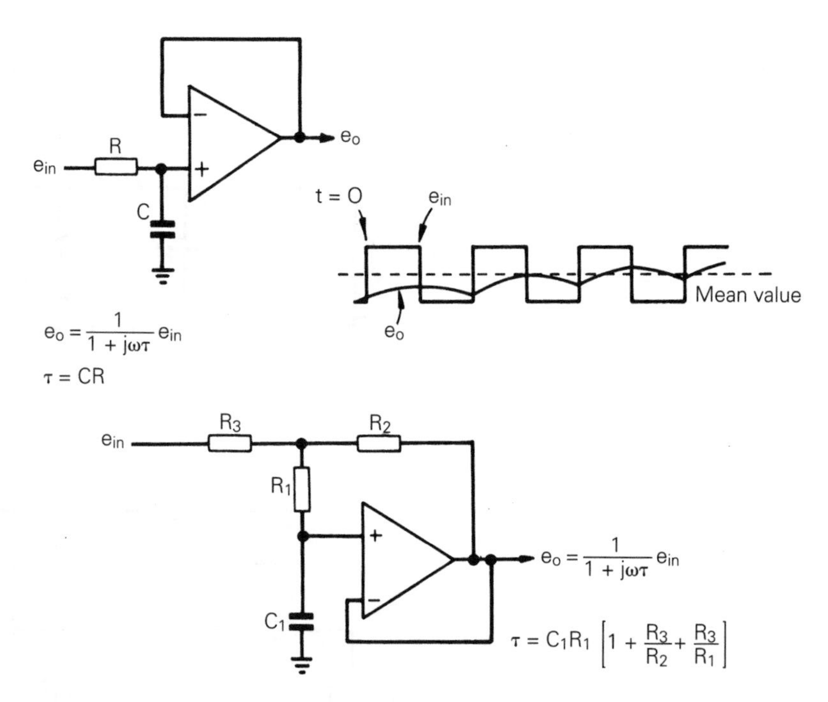

Figure 9.27 *Averaging filters. (a) Simple RC averaging filter. (b) Long time constant averaging filter uses capacitance multiplier*

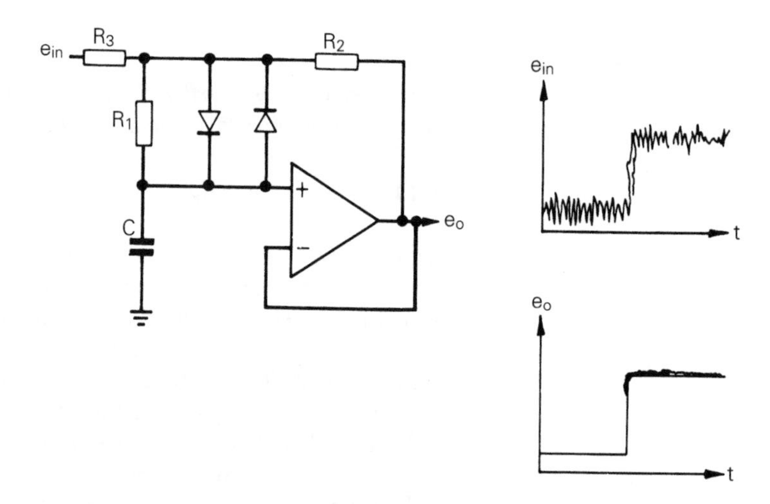

Figure 9.28 *Averaging filter with time constant dependent upon signal amplitude*

The back-to-back diodes bypass resistor R_1 when large changes in the mean level of the input signal occur. This gives the circuit the shorter time constant $t = C \times R_3$. Small noise fluctuations with amplitude below the diode voltage drop (typically 0.6 V) are not passed by the diodes. Hence, the circuit has a much larger time constant for the small noise fluctuations.

9.10.4 FDNR filters

Frequency-dependent negative resistance (FDNR) circuits can be used to make an active filter based on a passive ladder filter. The advantage in doing this is that passive ladder circuits have low sensitivity to component tolerances. However, inductors are bulky and are difficult to obtain; low value inductors for radio applications are reasonably easy to find, but audio frequency applications require much larger values. High-value inductors often have to be specially wound in order to obtain the required inductance. Replacing the inductors and capacitors by resistors and FDNRs gives the same low sensitivity to component tolerances. If there are two signal paths in a system that must be closely matched in terms of amplitude and phase, an FDNR filter is the better choice. FDNR circuits are also a good choice for Cauer (elliptic) filters, rather than using a state variable circuit, because fewer components are required.

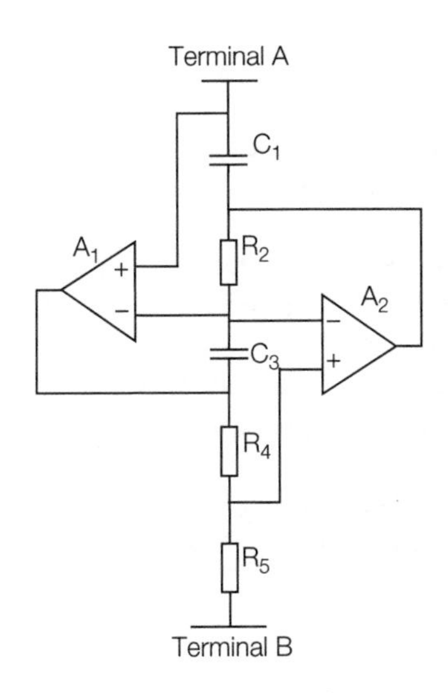

Figure 9.29 *Circuit diagram of an FDNR*

The schematic symbol for an FDNR circuit looks like a capacitor with four plates instead of the usual two and is assigned a letter D. The FDNR is also known as a D element. An FDNR is an active circuit that behaves like an unusual capacitor. In a passive lowpass RC circuit, with a series resistance R and a shunt capacitor C, the voltage drop across the shunt capacitor falls with increasing frequency. Beyond the passband, doubling the frequency halves the voltage across the capacitor. In a low-pass RD circuit, in which the FDNR has replaced the capacitor, the voltage drop across the FDNR falls at double this rate. Thus, above the passband, doubling the frequency quarters the output signal amplitude.

In decibel terms, a signal applied to an RC network has a rate of fall of 6 dB/octave (a first order filter). The same signal applied to an RD network has a rate of fall of 12 dB/octave.of a capacitor. This double rate of fall is the reason for the four plates in the D symbol, rather than the two in a capacitor symbol. The circuit of an FDNR is given in Figure 9.29.

In a simple approach where all resistors are equal to 1 Ω and all capacitors are equal to 1 F, the circuit behaves like a negative resistance of −1 Ω. The equation for the negative resistance is:

$$D = \frac{R_2 \cdot R_4 \cdot C_1 \cdot C_3}{R_5}$$

If $C_1 = C_3 = 1$ F and $R_4 = R_5$, the negative resistance equals R_2.

To use an FDNR, a transformation of the passive components is needed. FDNR elements are used to replace the shunt capacitors in passive lowpass filters. Resistors are used to replace the series inductors. This allows the filter size to be reduced, and a miniature hybrid circuit is possible. The design begins with a conventional double terminated low-pass LC filter design, in the T configuration. This has resistors (for the source and load), series and shunt inductors, and shunt capacitors. Figure 9.30 shows a normalized elliptic low-pass LC filter.

To convert the passive design into an FDNR design, the resistors are replaced by capacitors, the inductors are replaced by resistors and the capacitors are replaced by FDNRs. If the source and load resistor are 1 Ω, these

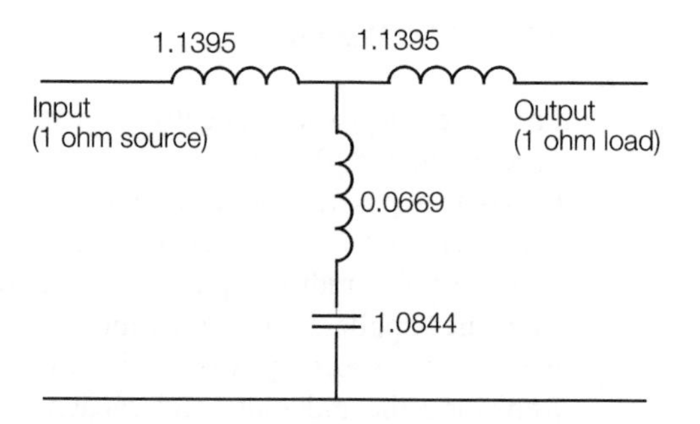

Figure 9.30 *Circuit of normalized low-pass LC filter*

are replaced by capacitors of 1 F. Generally, the capacitor value is $1/R$, so if the load was 0.2 Ω the capacitor would be 5 F.

Inductors are replaced by resistors. A 1 H inductor becomes a 1 Ω resistor. Generally, $R = L$, so a 1.1395 H inductor would be replaced by a 1.1395 Ω resistor.

Capacitors are replaced by FDNRs. In an FDNR, the resistors are normalized to 1 Ω and the capacitors are normalized to 1 F, to replace a 1 F capacitor. If the normalized capacitor is not 1 F, the value of R_2 (in Figure 9.29) is scaled in proportion. Generally, $R_1 = C$. Thus a 1.0844 F capacitor is replaced by an FDNR that has $R_2 = 1.0844$ Ω. The conversion process is illustrated in Figure 9.31.

Applying these simple rules to the normalized low-pass design given in Figure 9.30 gives the FDNR equivalent design, illustrated in Figures 9.32 and 9.33.

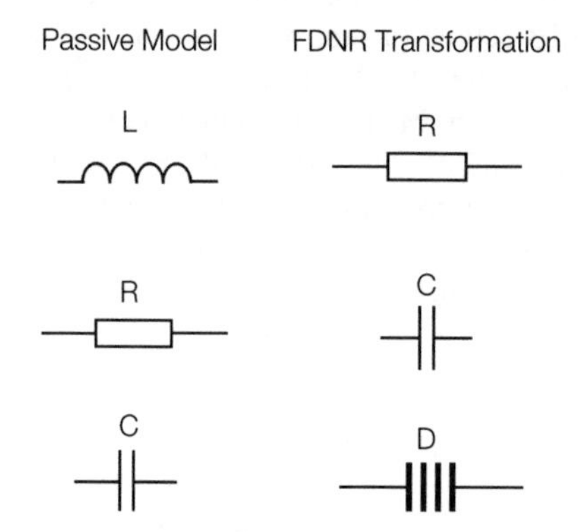

Figure 9.31 *FDNR transformation*

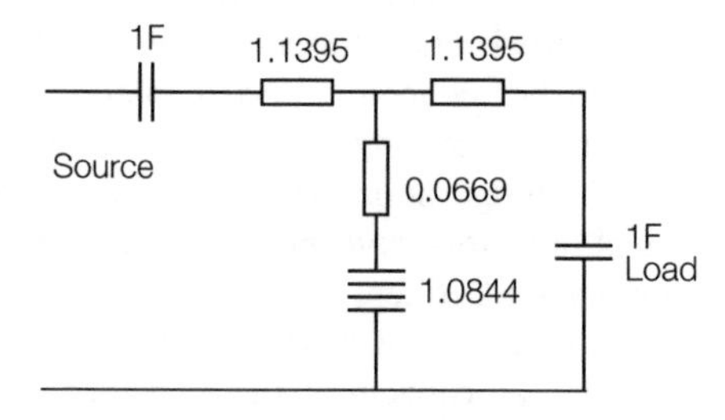

Figure 9.32 *Low-pass filter with D element*

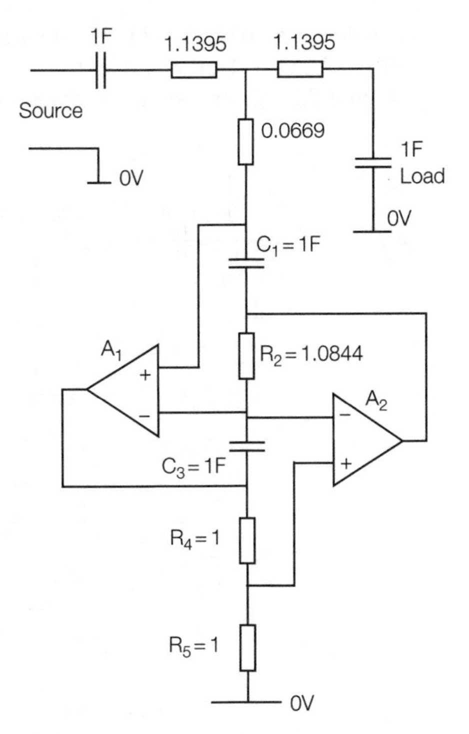

Figure 9.33 *Normalized low-pass FDNR filter*

Frequency scaling is used to obtain practical component values. Consider a third order filter that has a passband of 5 kHz. The normalized design has a passband of 1 rad/s, so the frequency scaling factor is $2 \cdot \mu \cdot F$. The frequency scaling factor is 31415.93 in this case. All capacitor values must now be divided by 31415.93, which makes each one equal to 31.831 μF. This value is a little too large and must be reduced to a more convenient value. Dividing the capacitor value by 3183.1 gives a value of 10 nF for all capacitors in the circuit. Each resistor must now be multiplied by this scaling factor. The values of resistors R_4 and R_5 becomes 3.183 kΩ. The resistors in the series signal path are also scaled. The normalized value of 1.1395 becomes 3.627 kΩ. The shunt resistor becomes 0.0669 times 3183.1 equals 212.95.

Before we redraw the filter, we must define the value of R_2 in the FDNR circuit. If the normalized capacitor is not 1 F, the value of R_2 is given by 3.183 kΩ multiplied by the normalized capacitor value. If, as in this example, the capacitor in the passive filter has a value of 1.0844 F, the value of R_2 in the FDNR will be 3.183 k × 1.0844 = 3.45 kΩ.

Finally, we must allow a d.c. path from the source to the load. This will give us a 6 dB insertion loss, the same as a terminated lossless ladder filter. The output load should be a high value, compared with the other series components; a value of 100 kΩ is often used. The input capacitor must be bypassed by a resistor that has a value less than 100 kΩ. The bypass resistor value should be 100 kΩ minus the sum of other series resistors. Since the other series resistors (replacing series inductors in the passive filter) in our example sum to 7.254 kΩ, the bypass resistor should have a value of (100 − 7.254) kΩ or 92.746 kΩ.

Figure 9.34 gives the circuit diagram of the final FDNR low-pass filter.

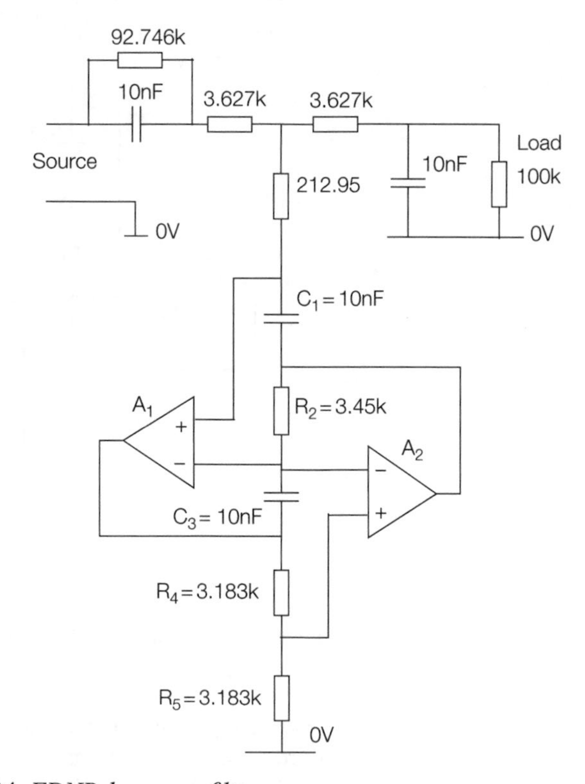

Figure 9.34 *FDNR low-pass filter*

An important point is that the common rail of the filter should be connected to the 0 V rail of the supply. The op-amp should then be powered from positive and negative supply rails.

9.10.5 Gyrator filters

Gyrators are related to the FDNR circuits described in Section 9.10.4 and are used to replace inductors. The gyrator uses two op-amps, four resistors and a capacitor. The gyrator can be smaller than the inductor it replaces, especially if surface mount components are used. Other advantages of using a gyrator instead of an inductor are that temperture effects can be reduced by using suitable components and that the component value can be adjusted easily.

The gyrator has the same structure as the FDNR, that of two op-amps connected to a chain of passive elements. The gyrator only has one capacitor, instead of the two used in the FDNR. All remaining passive components are resistors. The gyrator has a capacitor in place of the fourth element instead of in place of the first and third element.

A circuit diagram for the gyrator is given in Figure 9.35.

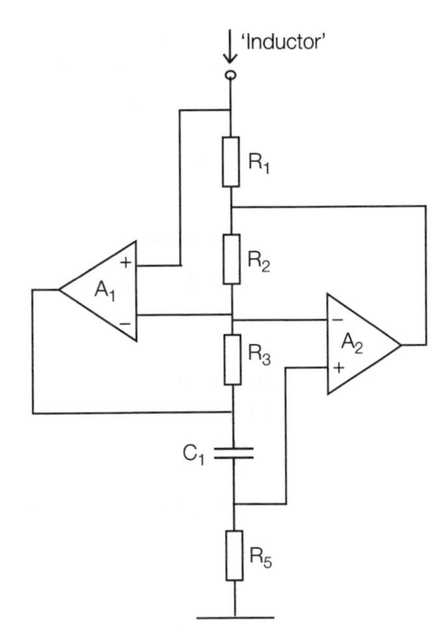

Figure 9.35 *Gyrator circuit*

The gyrator behaves like a shunt inductor, whose value is given by:

$$L = \frac{C \cdot R_1 \cdot R_3 \cdot R_5}{R_2}$$

If C_1, R_1, R_3 and R_2 are all normalized to unity, then $L = R_5$. If all resistors in the gyrator circuit are equal to 'R', then $L = R_2C$.

To design a high-pass filter, first obtain the normalized low-pass passive filter component values and then convert the design into a normalized high-pass circuit by replacing inductors (that have a value L) by capacitors that have a value of $1/L$. Also, replace capacitors (that have a value of C) by inductors that have a value of $1/C$. The gyrator circuit now replaces the inductor, so R_5 in the gyrator circuit has a value of $1/C$. Finally, all component values are normalized. This means that all capacitor values in the final circuit are divided by $Z \cdot 2\pi Fc$ and all resistor values are multiplied by Z.

For example, design a third order high-pass filter using a gyrator. The filter should have a passband cut-off frequency of 10 kHz with an input and output impedance of 600 Ω.

A passive filter must be designed first, and then the gyrator used to replace the inductor. The normalized low-pass model has two inductors in series with a central shunt capacitor. The component values are: $L_1 = 1.4328$; $C_2 = 1.5937$; and $L_3 = 1.4328$. This is shown in Figure 9.36.

The normalized low-pass model is converted into a high-pass equivalent by replacing the series inductors by series capacitors, thus L_1 becomes C_1, etc. The capacitor values in the high-pass model are the inverse of the inductor values in the low-pass model. In this case, $C_1 = 1/1.4328 = 0.697934$.

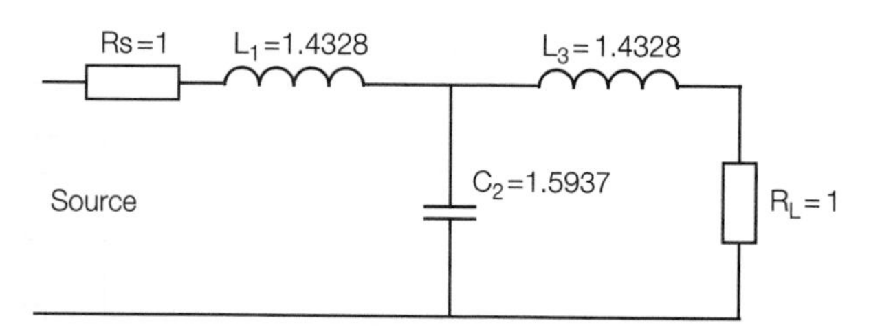

Figure 9.36 *Low-pass model*

Due to symmetry, $C_3 = 0.697934$. The shunt capacitor in the normalized low-pass model becomes a shunt inductor in the high-pass model. The value of the shunt inductor is the inverse of the shunt capacitor in the low-pass model. So, C_2 becomes L_2. The value of $L_2 = 1/1.5937 = 0.627471$. This is illustrated in Figure 9.37.

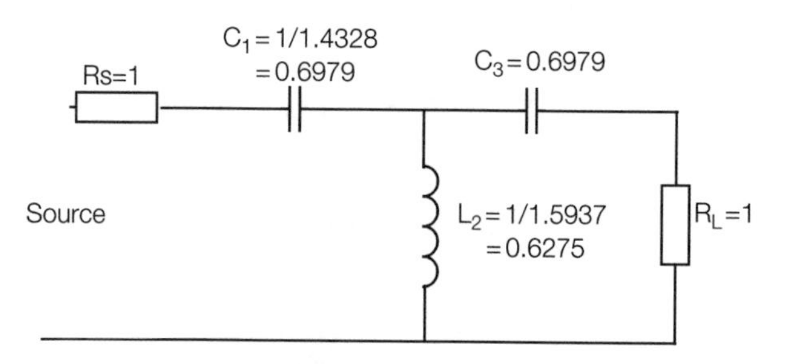

Figure 9.37 *High-pass model*

In order to replace L_2 with a gyrator, as shown in Figure 9.35, the value of R_5 becomes 0.627471, with $R_1 = R_2 = R_3 = 1\ \Omega$ and C_2 of the gyrator circuit equals 1 F.

To denormalize the filter, all resistor values must be multiplied by the load impedance of 600 Ω. Resistors R_1, R_2 and R_3 all become 600 Ω. R_5 becomes 376 Ω. The capacitor values must all be divided by the load impedance and by the cut-off frequency in radians ($2\pi Fc$). Thus, capacitors C_1 and C_3 become 18.5133 nF and C_2 becomes 26.5258 nF. The circuit is given in Figure 9.38.

The gyrator resistors all have a low value, which could be a problem for op-amp drive capability. Although most op-amps do have a reasonable output drive performance, low power devices do not. To overcome this, the resistance values of R_1, R_2, R_3 and R_5 can be increased provided that the combined multiplying factor of R_1, R_3 and R_5 is equal to the multiplying factor of R_2.

Suppose, for example, that R_1, R_3 and R_5 were all multiplied by 2. The value of R_2 would have to be multiplied by 8 to restore the balance of the equation. The modified component values are then: $R_1 = R_3 = 1.2\ \text{k}\Omega$, $R_5 = 752\ \Omega$ and $R_2 = 4.8\ \text{k}\Omega$. The value of C_2 was unchanged for this modification, but it could be reduced so that the value of R_2 would not have to increase by such a large factor.

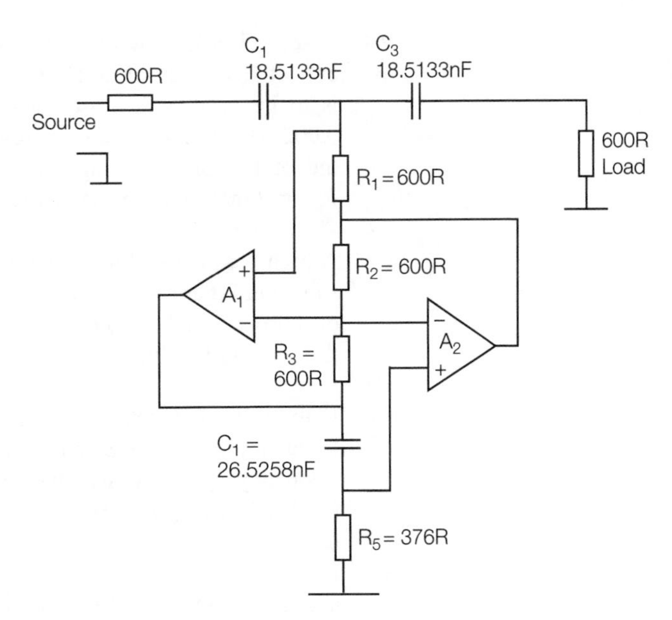

Figure 9.38 *Gyrator high-pass filter*

The secret is to design the filter as initially described, and then modify component values in order to make them practical. Remember to keep the equation for the gyrator inductance (equivalent to the value of L_2) balanced. In practical circuits, the value of C_2 would probably have to be produced by two or more capacitors wired in parallel. Standard capacitor values are usually in the E6 range, which is coursely spaced. It is unlikely that the gyrator capacitor would just happen to fall on one of these E6 values. Fortunately, it is easier to find resistor values that are close tolerance and finely spaced, hence a single precision resistor can be used to 'tune' the gyrator to have the correct inductance value.

Exercises

9.1 Show that for the circuit at Figure 9.1, the cut-off frequency, f_c, is given by the expression:

$$f_c = 1/2\pi RC \text{ Hz}$$

9.2 Also for the circuit at Figure 9.1, show that the magnitude of the voltage output, $| e_o |$, and its phase angle, ϕ, are given respectively by the equations:

$$| e_o | = e_i X_c/(\sqrt{R^2 + X_c^2}) \text{ and } \phi = \tan^{-1}(X_c/R)$$

9.3 Plot a graph of gain, $| e_o/e_i |$, against frequency for a simple first order, passive, low-pass, *CR* filter and use this graph to show that:
(a) at f_c the gain is 3 dB less than its DC value
and
(b) at frequencies higher than $2f_c$ the output declines at 20 dB/decade.

9.4 A second order low-pass filter with a Butterworth response and a 3 dB cut-off frequency of 20 Hz is required. The filter circuit of Figure 9.13 is to be used with close tolerance capacitors of value 0.001 μF and 0.002 μF. What values of resistors are required and what will be the output offset voltage if the op-amp has an input offset voltage V_{io} = 2 mV and the bias current I_B = 50 pA?

9.5 A high-pass active filter is required and the circuit to be used is the one shown in Figure 9.13. If both of the capacitors are to have the same value of 0.001 μF and the 3 dB cut-off frequency is to be 1125 Hz, calculate the necessary values of each of the two resistors.

9.6 If the high-pass active filter circuit of Figure 9.13 has a 3 dB cut-off frequency of 1125 Hz and its input is connected to a 100 mV variable frequency signal, estimate the voltage output from the filter at (a) 100 Hz, (b) 1 kHz and (c) 2 kHz.

9.7 It is required to produce a unity-gain low-pass Sallen–Key active filter based upon the circuit shown in Figure 9.25(a). If the filter input resistance is to be 20 kΩ and the 3 dB cut-off frequency 1 kHz, suggest suitable values for the capacitors and resistors.

9.8 Component values $R_3 = R_5 = R_6$ = 10 kΩ, $C_1 = C_2$ = 0.001 μF, $R_1 = R_2 = R_4$ = 1 MΩ are used in the state variable filter of Figure 9.14. Calculate the constants of the equation (equation 9.9) relating the output of the amplifier A_2 to the input signal. Sketch the response on a dB against log f plot. In order to obtain a band-pass response, a fourth op-amp is used to sum the input signal with the output of A_2. Suggest suitable values for the components to be connected to this fourth amplifier. If the output voltage limits of the op-amp are ±10 V, calculate the maximum allowable input signal amplitude at the rejection frequency. Explain this limitation. Suggest modifications to component values in order to give the band-pass response a centre frequency of 200 Hz and a Q-factor of 100.

10 Practical considerations

This chapter describes how to set about choosing an op-amp and other components. It also discusses some of the more important practical points that should be considered when designing and using op-amp circuits.

The main applications for op-amps have been outlined in the preceding chapters and more applications will be found in Appendix A1. In some circuits, specific op-amps and component values have been given but this is simply for the convenience of the reader. In general, applications are not confined to particular op-amp types and specific component values. The op-amp type and component values suited to a particular application are dictated by the performance requirements of that application.

10.1 Op-amp selection and design specification

The performance of general-purpose op-amps is normally adequate for the majority of circuits. With experience, the designer will be in a better position to start designing his own op-amp circuits and these may require the use of more specialized op-amp types. Chapter 3 has already discussed the various technologies used to make integrated circuit op-amps, i.e. BiFET, bipolar, linear CMOS. The advantages and disadvantages of current and voltage feedback op-amps were also given. This chapter describes how a particular op-amp and associated components are selected.

The choice of op-amp type and other design decisions are made easier by a systematic design approach. To choose an op-amp for a given circuit, the designer must consider the function that it is required to perform. It is important to consider all the op-amp parameters that may influence the performance. Essential to the proper formulation of any op-amp design is knowledge of the following.

What is the nature of the signal source? Is it a voltage or current source? What is the source impedance? What is the expected amplitude range of the input signal? What are the expected time/frequency characteristics of the input signal?

What is the nature of the load? What is the load impedance? What output voltage and current are required? Remember that it is always possible to increase the output current capability of an op-amp by the addition of a suitable booster op-amp (see later in this section).

What is the required accuracy? Accuracy must of course be defined with respect to bandwidth, DC offset and other parameters. It is important to make a realistic estimate of accuracy requirements. The op-amp circuit may represent a subsystem of a complete measurement or instrumentation system, and accuracy must be related to the required overall system accuracy. A mistake often made by beginners is either to start a design with little more than the haziest idea of performance errors, or to specify a far greater accuracy than is really required (over-engineered solution).

What are the environmental conditions? What is the maximum range of temperature and supply voltage over which the circuit must operate with accuracy and without readjustment? In assessing the relevant factors it is necessary to consider the total environment in which the circuit is required to operate. The physical environment includes temperature, humidity, mechanical vibration, the presence of near-by sources of interference noise, etc.

Having carefully considered the details of the circuit's specification, the designer must then decide how best to meet it. The decision is usually made from a cost/performance standpoint; 'best' is regarded as that which achieves a desired performance specification at minimum cost.

In making a realistic cost estimate, the designer should look carefully for hidden costs. Many factors, other than the price of an op-amp, components, or complete circuit modules, can contribute to the overall cost of implementing an application. Cost relates to the external requirements that may be necessary to make a device compatible with other elements in the system. This includes the time factor involved in any setting-up procedure.

For example, there may be a specific circuit function that can be performed by a low cost general-purpose op-amp, but which requires the use of trim potentiometers. This circuit then requires an adjustment procedure that must be performed by a skilled operator. Under such circumstances there could well be an overall cost advantage in using a higher performance, more expensive op-amp if the adjustment requirement were thereby eliminated. Savings would be in the price of the potentiometer (perhaps a precision one), space, and skilled operator's time. In addition, there could well be an added bonus in the superior performance provided by the more expensive op-amp.

10.2 Selection processes

Op-amp performance parameters will influence its behaviour in a specific application. Performance errors determine the ability of a particular op-amp to meet a desired accuracy requirement. In most applications, not all op-amp parameters have equal importance. The ability to recognize the important performance limiting specifications does simplify op-amp selection.

A general-purpose op-amp, if it can meet desired performance requirements, is likely to be the best choice calculated on a cost/performance basis. Where the use of a general-purpose op-amp is not possible, it is generally because of limitations encountered in two areas: (1) DC offset and drift performance, and/or (2) bandwidth and slew rate requirements.

10.2.1 DC and low frequency applications

Some op-amp applications are concerned with slowly varying signals in which knowledge of the DC level of the signal is important. In such applications, it is the op-amp's offset voltage, bias current and drift parameters that largely influence the final selection of an op-amp.

The non-inverting feedback configuration is usually best for processing the signal from voltage sources (where performance requirements allow its use). High input impedance minimizes loading errors, and can be obtained without the use of high value resistors. It is the op-amp's input offset voltage and drift that are of prime importance in determining DC errors. Common mode errors and limits are important considerations in non-inverting applications.

Voltage summing applications, with isolation between signal sources, require the use of the inverting feedback configuration. The resistor values required to minimize loading errors normally make op-amp bias current and offset voltage important in determining DC errors. Inverting configurations do not require a consideration of common mode errors and limitations.

Applications in which the input signals are essentially currents applied by current sources invariably use the inverting op-amp configuration. In such applications it is usually op-amp bias current and drift which dominate the DC errors.

In examining offset and drift specifications the designer must consider how much offset and drift error can be tolerated. This is related to the input signal level and the required accuracy. For example, to amplify or otherwise manipulate a DC input signal of 1 V with an accuracy of 0.1 per cent, the offset error must be 1 mV or less. Note that the offset error is a combination of the effects of op-amp input offset voltage and bias current (see Chapter 2); values are made up of initial values plus drift. Initial offsets can be balanced out by a suitable trimming arrangement (see Section 10.6); errors are then due to drift. This of course assumes that other sources of error such as input loading, noise and gain errors have already been assessed.

10.2.2 Wide-band applications

Applications in which DC levels are not of interest can use a capacitor to block out DC offset. Op-amp offset and drift specifications can then largely be ignored. An exception to this is where a large DC output offset might restrict the dynamic swing of alternating output signals.

Some significant points relating to the selection of an op-amp for amplifying or manipulating continuous sinusoidal, complex or random waveforms are as follows:

1. What closed-loop bandwidth is required?
   Closed-loop bandwidth is determined by the intersection of the open-loop and $1/\beta$ frequency response plots (see Chapter 2).
2. What loop gain βA_{OL} is required?
   The available loop gain at a particular frequency or over a range of frequencies is often more important than closed-loop bandwidth in an application. Closed-loop gain stability, output impedance and non-linearity all depend upon loop gain.

   Closed-loop gain stability

$$\frac{\Delta A_{CL}}{A_{CL}} \cong \frac{\Delta A_{OL}}{A_{OL}} \frac{1}{\beta A_{OL}} \tag{10.1}$$

   $\Delta A_{OL}/A_{OL}$ is the open-loop gain stability, which is dependent upon temperature and power supply voltages.

Closed-loop output impedance (see Section 2.2.2)

$$Z_{oCL} \cong \frac{Z_{oCL}}{\beta A_{OL}} \tag{10.2}$$

Closed-loop distortion (non-linearity)

$$D_{CL} \cong \frac{D_{OL}}{A_{OL}} \tag{10.3}$$

The open-loop input/output transfer curve for an op-amp may exhibit non-linearity, but the effects of this on the closed-loop behaviour are reduced to an extent dependent upon the magnitude of the loop gain.

A loop gain of 100 (40 dB) is normally adequate for most applications but remember that loop gain decreases with increase in frequency (see Section 2.5). This makes it difficult to obtain large loop gain at high frequencies. For this reason it may be necessary to use an op-amp with a 10 MHz unity-gain bandwidth in order to achieve adequate loop gain over a 10 kHz bandwidth. In high gain wide-band applications it may be necessary, and more economical, to use two op-amps in cascade each at lower gain.

3. What full-power bandwidth and/or slew rate is required? An op-amp should be selected whose slew rate exceeds the maximum expected rate of change of the output signal. An op-amp should be chosen whose power bandwidth is not exceeded at the highest operating frequency.

   In applications in which large amplitude sinusoidal signals are processed, a wide bandwidth is not much help if the available output signal is only a fraction of a volt.

   In applications such as sample and hold circuits used in A/D and D/A converters, the transient response of the wide-band op-amp is generally more important than gain bandwidth characteristics. Slew rate, overload recovery and settling time are the important op-amp specifications to consider.

10.3 Attention to external circuit details

Having selected an op-amp to reduce the errors in an application, it is important not to degrade performance by improper attention to external circuit details.

The experienced electronics designer will already be aware of the importance of the physical layout of a circuit and of the quirks and idiosyncrasies of practical electronic components. Op-amp circuits, in common with most other electronic circuits, give best results if some care is taken over the physical arrangement of components. A neat circuit layout, which minimizes the effects of stray capacitance, should be sought.

The following are some points worth observing:

- Low resistance and low inductance earth and power supply leads should be used.
- Power supplies should be AC coupled to earth.
- Bypass capacitors should be connected at or as near as possible to the device socket; values of ceramic bypass capacitors in the range 10 nF to 100 nF in parallel with 10 μF tantalum capacitors are normally satisfactory.

- Proper attention should be paid to frequency compensation of the op-amp.
- Input and output leads should be kept as short as possible and shielded if required.
- It is advisable to use one common tie point for all earth connections and this should be near to the op-amp.

10.4 Avoiding unwanted signals

A practical op-amp circuit is prone to disturbances that are not obvious from its circuit diagram. If the sources of such disturbances are identified and understood, their ill effects can be considerably reduced by the use of appropriate circuit techniques. The smaller the expected input signal, the greater attention to detail must be paid. Particular care is required in the measurement of very small currents from high impedance sources.

10.4.1 Earth loops

In an electronic system, providing separate earth connections to individual subsystems can create large area circuit loops. Stray magnetic fields at power line frequencies can induce currents in such loops; thus injecting unwanted signals into the system.

In op-amp applications where both signal source and op-amp are separately earthed, as shown in Figure 10.1, an earth loop is created. The obvious remedy is to earth the system at one point only, usually at the op-amp input. If this is not possible a differential input circuit configuration should be used in order to reject the unwanted pick-up signal (see Section 4.4).

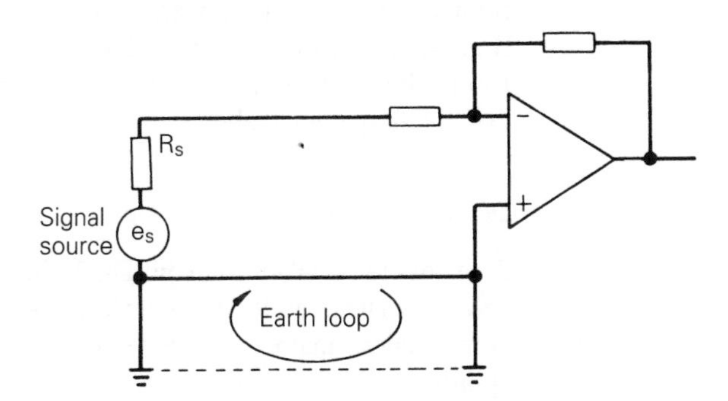

Figure 10.1 *Separate earth connections to the source and op-amp creates earth loop*

Another earth loop error, which is noticeable when measuring small signals, occurs when the power supply current or load current is allowed to flow through the input signal return connection. If this happens, an error voltage is applied to the input of the op-amp via the signal source; this results in an error at the output. The proper connection to avoid this effect is shown in Figure 10.2. The signal return and load return should be connected to power supply common as close to the op-amp pins as possible.

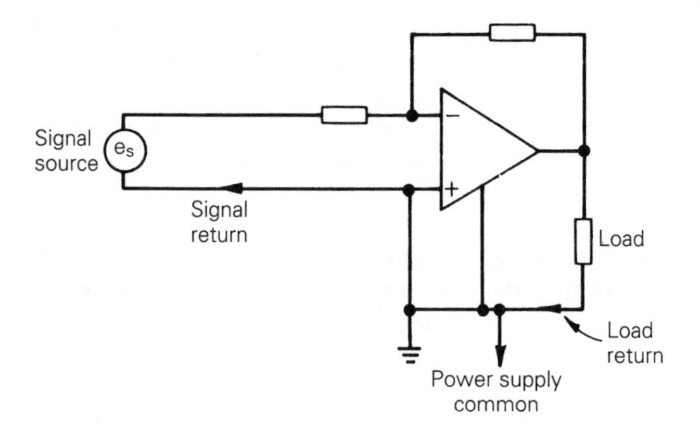

Figure 10.2 *Proper connections to avoid earth loops*

10.4.2 Interference noise/shielding and guarding

Unwanted alternating signals can be introduced into the input of an op-amp through capacitive or inductive coupling. Take, for example, the inverting op-amp shown in Figure 10.3. If e_s = 100 mV and R_1 = 1 MΩ it only needs a leakage coupling of 10^{11} Ω from a 200 V AC supply to introduce an unwanted noise signal equivalent to 2 per cent of the input signal. A capacitance as small as 0.03 pF will provide this at 50 Hz.

Capacitive coupling of signals can be minimized by surrounding the input circuitry of the op-amp, or preferably the whole of the amplifier, with an earthed electrostatic shield. The shield should be ferromagnetic as well as conductive when the unwanted signal is introduced by inductive coupling. High permeability magnetic shielding is best for shielding away signals under 100 Hz but high conductivity shielding (aluminium or copper) is more effective for frequencies above 100 Hz.

DC leakage paths

Care should be taken to avoid DC leakage paths when using low bias current op-amps. The overall circuit performance is often limited by leakage in capacitors, diodes, analogue switches or printed circuit boards rather than by the op-amp itself.

Printed circuit boards must be clean and solder fluxes should be removed. Solder fluxes may be good insulators in low impedance circuits, but can cause gross errors in low current high impedance circuits and erratic behaviour as the temperature is changed. Even the leakage of properly cleaned boards can be troublesome at elevated temperatures. At 125°C the leakage resistance between adjacent tracks on a clean, high quality FR4 epoxy glass board (1 mm separation, running parallel for 25 mm) may be no more than 10^{11} Ω.

The leakage becomes worse if the board becomes contaminated. To prevent this the boards should be coated with silicone rubber or some form of lacquer, after being cleaned. Many printed circuit board manufacturers apply a solder

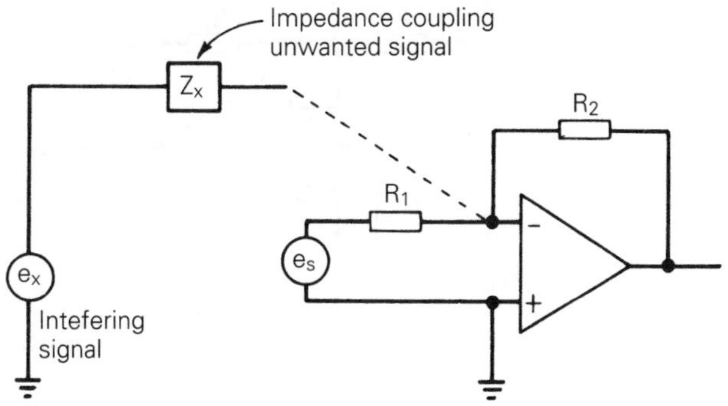

Figure 10.3 *Unwanted signal can be picked up at the op-amp input because of stray coupling impedance*

resist coating that prevents solder flow between component pads; this coating also reduces leakage between tracks.

Guarding

Guards can be used to interrupt the leakage paths to the input terminals of an op-amp. The standard pin configuration used with most integrated circuit op-amps has the non-inverting input pin adjacent to one of the power supply pins. A guard is a conducting ring surrounding the input pins. This ring can be formed using a circuit board track.

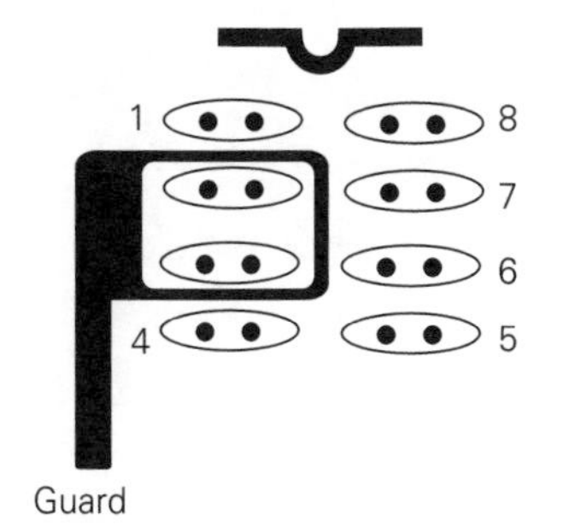

Figure 10.4 *Printed circuit board input guard for standard op-amp*

A suitable board layout for input guarding of an 8-pin dual-in-line device is shown in Figure 10.4. Note that if the integrated circuit leads or the leads of other components connected to the input of the op-amp go through the board it may be necessary to guard both sides of the board.

In order to be effective, the potential of the guard must be equal to that of the input terminals. Proper guard connections for the common feedback configurations are shown in Figure 10.5.

In circuit configurations in which the input terminals are close to earth potential, e.g. inverters and integrators, a guard is simply connected to earth. In voltage follower configurations, in which the input terminals are above earth potential, the guard is maintained at the same potential as the input terminals by driving it by a signal derived from the op-amp output. A low impedance drive should be used calling for relatively low input, and feedback resistors as shown by R_1 and R_2 in Figure 10.5(c).

Another consideration in high impedance circuits is the capacitance of cable used for input shielding. Capacitance should be minimized by using as short a length of cable as possible. In high input impedance follower applications, cable capacitance lowers the effective input impedance. Cable capacitance also forms a long time constant *CR* low-pass filter with the high resistance signal source, which restricts bandwidth and increases rise time.

If the cable shield is driven (rather than connecting it to earth) by connecting it to the guard drive as in Figure 10.5(c), cable capacitance is uncharged and its effect on the input is reduced. It is possible to decrease input capacitance

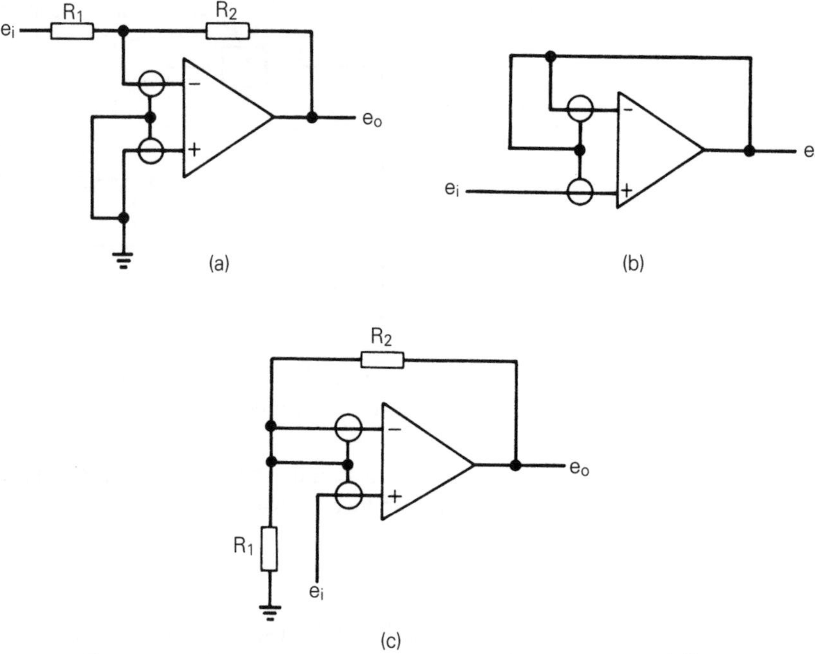

Figure 10.5 *Guard connections. (a) Inverter. (b) Unity-gain follower. (c) Follower with gain*

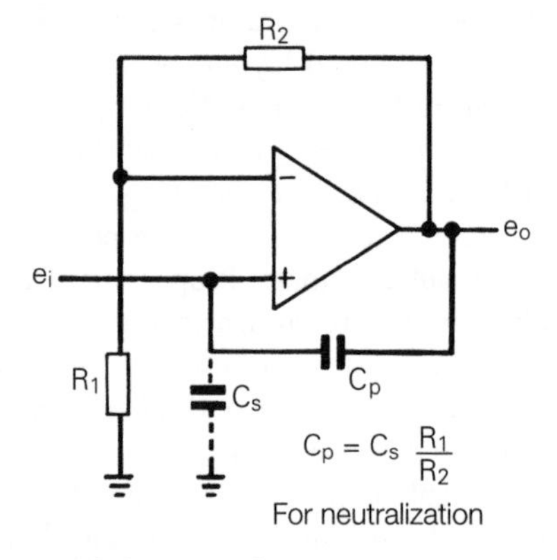

Figure 10.6 *Neutralizing input capacitance in follower with gain*

in the follower with gain circuit by connecting a capacitor C_p as shown in Figure 10.6 as a positive feedback path from output to input.

A further point about input cable and input leads in high input impedance circuits is that they should be kept as mechanically rigid as possible. Movement causes capacitor changes, which in turn cause changes in charge and a spurious signal flow.

10.5 Ensure closed-loop stability

Precise performance of an op-amp circuit is obtained as a result of negative feedback. An op-amp circuit is essentially a feedback system and, like other feedback systems, it can become unstable and oscillate as a result of excess phase shift in the feedback loop. The factors and techniques governing closed-loop stability were discussed in detail in Chapter 2.

Op-amps are available as internally compensated or externally compensated. When using externally compensated op-amps, it is best to follow the manufacturer's recommendations for providing compensation. If speed and bandwidth are not a design limitation, greater stability can generally be obtained by increasing the size of compensating capacitors.

External circuit influences can decrease stability phase margin. Some external factors which can adversely affect closed-loop stability are: load capacitance (stray wiring capacitance or an actual capacitance at the op-amp output); capacitance at the inverting input of the op-amp; a large resistance at the non-inverting input terminal; and supply voltage not adequately bypassed.

10.5.1 Effect of load capacitance

Capacitance between the output terminal of an op-amp and earth forms a first order lag network with the op-amp output resistance. This introduces an extra break in the op-amp open-loop frequency response. If this break occurs at a frequency before the loop gain is reduced to unity (or at frequencies near to it), the extra phase lag decreases the phase margin and can destabilize the circuit.

The output resistance of an op-amp is generally low and therefore most op-amps will tolerate quite a few picofarads at the output, but when load capacitance reaches 100 pF or more it may be necessary to take appropriate steps to ensure closed-loop stability. For example, directly driving a coaxial cable can cause instability.

Compensation for a capacitance load can be achieved using the circuit arrangements shown in Figure 10.7. Resistor R_3 (value about 100 Ω) is used to isolate the capacitive load from the output, and a feedback capacitor C_f is connected directly from the op-amp's output to its inverting input. The value of C_f should be chosen so that its capacitive reactance at the unity-gain crossover frequency is no more than 1/10 of the resistance of R_2. At high frequencies, feedback is predominantly through C_f making the high frequency value of $1/\beta$ approach the limiting value of unity. This means that when using this capacitive load isolation scheme, the op-amp must be frequency compensated for unity-gain operation regardless of the closed-loop signal gain.

10.5.2 Effect of input capacitance

Any capacitance between the inverting input of an op-amp and earth decreases the amount of feedback at high frequencies. It also introduces a phase lag into the feedback loop, thus reducing phase margin and leading to possible instability.

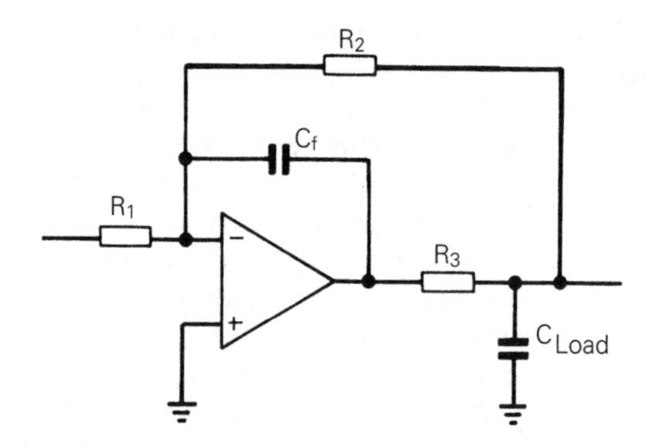

Figure 10.7 *Preventing instability due to capacitive loading*

The stability problem is most conveniently examined in terms of Bode plots as shown in Figure 10.8. The break in $1/\beta$ occurs at the frequency

$$f = \frac{1}{2\pi C_i\left(\dfrac{R_1 R_2}{R_1 + R_2}\right)}$$

If this is less than the frequency at which the loop gain becomes unity (the $1/\beta$ and A_{OL} intersection) the phase margin is reduced. Clearly the effects of stray capacitance at the input are likely to be significant in circuits that use large value input and feedback resistors.

A simple method of compensating for the effect of input capacitance is to connect a phase lead capacitor C_f in parallel with the feedback resistor R_2. This causes $1/\beta$ to break back at the frequency $1/(2\pi C_f R_2)$, the leading phase shift thus introduced cancelling the lag due to C_S. The value of the capacitor C_f should be chosen so that the frequency $1/(2\pi C_f R_2)$ is at least an octave before the $1/\beta$ and A_{OL} intersection frequency. The closed-loop signal bandwidth is set at $1/(2\pi C_f R_2)$ by the value used for C_f.

10.5.3 Supply bypassing

Normally power supplies have very low impedance. However, the inductance of the supply leads or circuit board tracks can present appreciable impedance at the high frequencies. A signal on the op-amp's output will cause additional current drain from the supply, due to the load current. If the supply impedance is significant, the supply voltage will be modulated by the signal. Since the supply is also coupled to the op-amp's input, positive feedback and hence oscillation is possible. A simple cure is to bypass the positive and negative supply terminals of the op-amp to earth with capacitors of value at least 10 nF. Power supply decoupling is generally good practice.

Power supply bypassing and decoupling are particularly important when current drive circuits are used in conjunction with op-amps to boost the output current capabilities of the op-amp. Current boosters can feed a high amplitude signal back into the supply lines and bypass capacitor values should be increased accordingly. Suitable bypass capacitors are tantalum (10 μF to 100 μF) in parallel with a small (10 nF) ceramic. The high value tantalum de-couples low

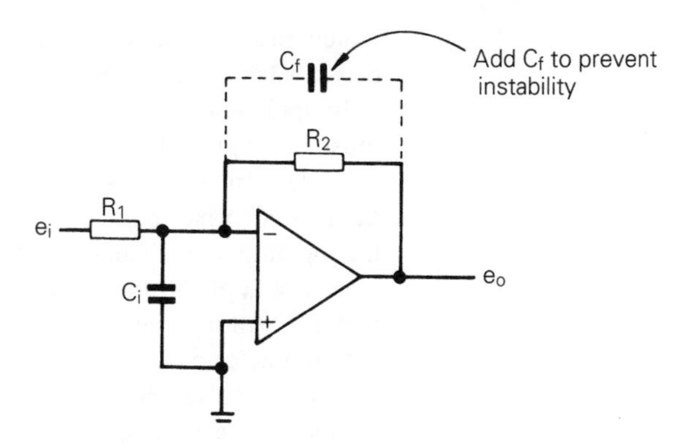

$$\text{Without } C_f \quad \frac{1}{\beta_{(j\omega)}} = \left[1 + \frac{R_2}{R_1}\right]\left[1 + j\omega\, C_i\, R_1 // R_2\right]$$

$$\text{With } C_f \quad \frac{1}{\beta_{(j\omega)}} = \left[1 + R_2\right]\left[\frac{1 + j\omega(C_1 + C_f)\, R_1 // R_2}{1 + j\omega\, C_f\, R_2}\right]$$

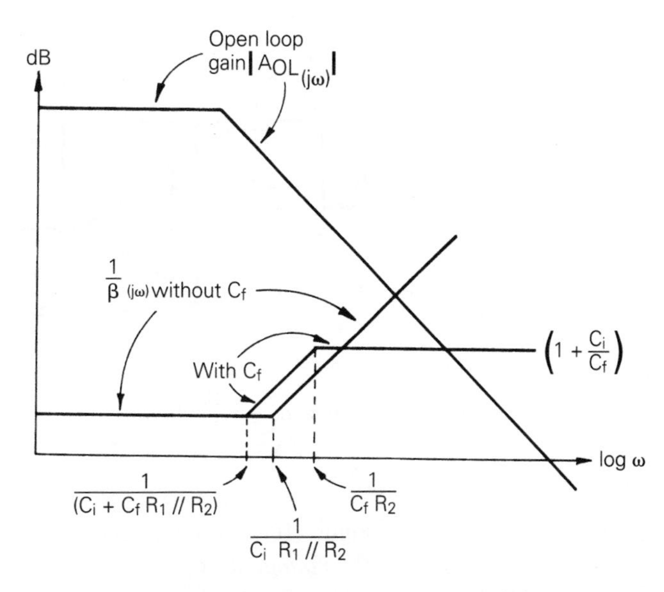

Figure 10.8 *Capacitance at the input can cause instability*

frequency signals and the small ceramic de-couples high frequency signals. Tantalum capacitors have high inductance at frequencies above 1 MHz.

10.6 Offset nulling techniques

Most DC applications require zero output voltage with zero input. Unfortunately DC offsets arise at the output of an op-amp as a result of the effects of op-amp input offset voltage and bias current.

There are a variety of circuit techniques that may be used to balance out the effects of initial offset. The most suitable technique depends upon the nature of the circuit. In some applications a small initial offset may be tolerable, in which case choose an op-amp type that will achieve this without using an offset balance circuit. Savings will be in material costs and adjustment time.

In applications requiring zero initial offset, some form of offset balancing must be employed. Most op-amps have provision for adjustment of V_{io} with a single trim potentiometer connected directly to the op-amp. This is known as internal offset balancing because offsets within the internal circuitry of the op-amp are changed. Note that published drift specifications do not normally apply to the op-amp when it is connected to this offset potentiometer. Nulling the input offset voltage with the recommended trim potentiometer can induce additional voltage drift with temperature.

External offset-balancing techniques apply an adjustable DC signal directly to one of the op-amp's inputs. Because this does not affect the op-amp's internal circuit conditions, it cannot introduce any extra drift. There are several external offset-balancing methods, and in many cases there is little to choose between them. Some of the more commonly used offset-balancing techniques are now given.

Input offset error voltage in the inverting op-amp configuration can be balanced by supplying a small additional current to the op-amp summing point using the arrangement shown in Figure 10.9(a).

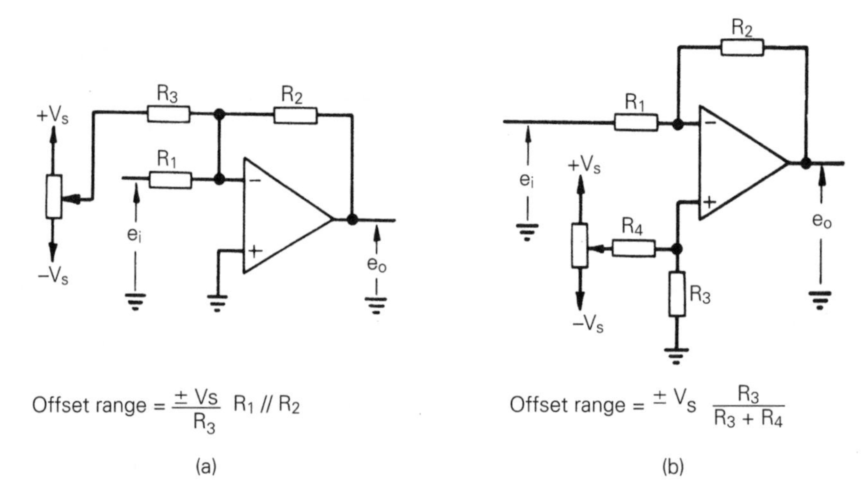

Figure 10.9 *Offset balancing – inverting configuration. (a) Current bias. (b) Voltage bias*

If a fine offset balance is to be possible the total range should not be significantly greater than the expected maximum value of E_{OS}. This normally dictates that R_3 should be made several thousand times greater than the parallel combination of R_1 and R_2. In applications using large values of R_1 and R_2, the resistance value required for R_3 becomes excessive. In this case, offset balancing is more readily achieved by adding a small adjustable voltage to the non-inverting input terminal of the op-amp, using the balancing circuit of Figure 10.9(b).

In the non-inverting op-amp configuration, as shown in the circuit of Figure 10.10(a), offset balancing is (effectively) a small adjustable voltage in series with the input. In this circuit, the correction voltage is developed across the

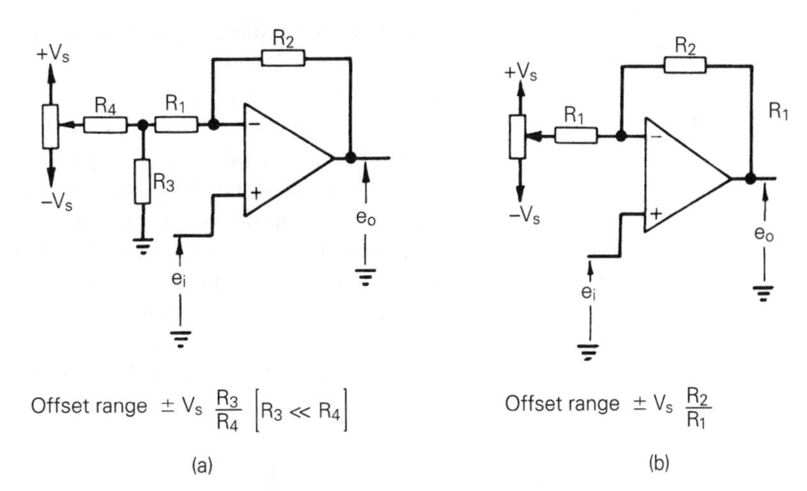

Figure 10.10 *Offset balancing – follower configuration. (a) Follower. (b) Unity-gain follower*

resistor R_3. The balancing adjustment alters the gain but, provided $R_3 << R_1$, the effect is not significant.

A balancing arrangement suitable for a unity-gain follower is shown in Figure 10.10(b). The value of resistor R_1 should be made very much greater than R_2, but this still leaves the circuit with a gain slightly greater than unity. For example, with typical values, say $R_1 = 2.2\ \text{M}\Omega$, $R_2 = 2.2\ \text{k}\Omega$, the gain error is 0.1 per cent.

The arrangement illustrated in Figure 10.11 provides a method of offset balancing for a differential amplifier configuration. The small bias voltage developed across the resistor R_4 applies the offset correction. A disadvantage of this method is that unless $R_5 >> R_4$ the offset nulling procedure will degrade the CMRR of the circuit because of the resistive imbalance introduced.

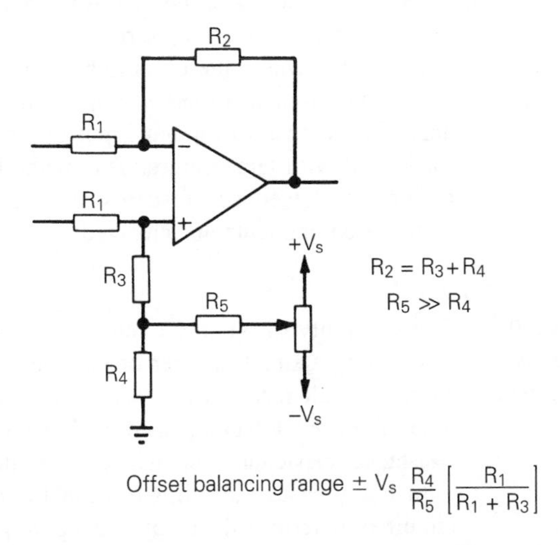

Figure 10.11 *Offset balancing a differential amplifier*

10.6.1 Offset balancing with drift compensation

A variety of circuit techniques have been devised to balance op-amp bias currents and offset voltage, and at the same time to provide compensation for the temperature drift of these parameters. Most of these techniques require extensive external circuitry and require considerable care and time spent in adjustment procedures. The whole-life cost of buying an expensive close-tolerance device will be less than buying a low cost device that needs additional circuitry and adjustment.

The arrangement shown in Figure 10.12 is a simple offset-balancing technique which provides a measure of compensation against the bias current temperature drift encountered in bipolar input op-amps.

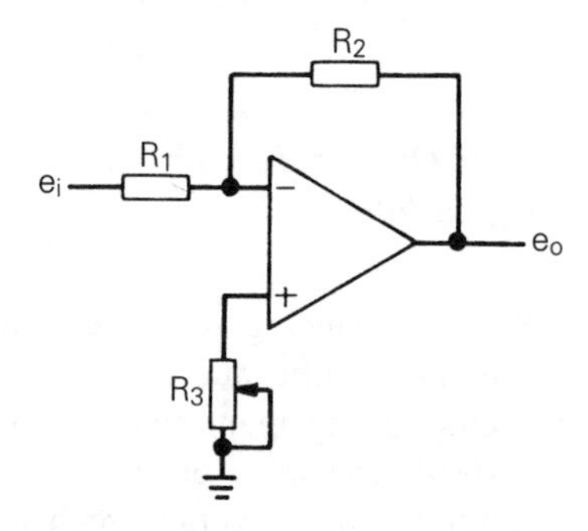

Figure 10.12 *Offset balancing with bias current drift compensation*

The technique is easy to apply. The total equivalent input offset error voltage for the circuit is

$$E_{os} = \pm V_{io} + [I_b^- \, R_1//R_2] - I_b^+ R_3$$

R_3 is adjusted to make the output voltage zero when the input signal is zero; the adjustment makes E_{os} zero.

In circuits where the op-amp bias current is a major source of offset, this technique provides a marked improvement in the temperature drift performance. In the case of bipolar input op-amps the currents I_b^- and I_b^+ tend to track well with temperature. The method is only applicable for fixed values of input and feedback resistors, and readjustment is required if the values of these components are changed.

10.7 Importance of external passive components

The versatility of op-amp circuits is the ability to set performance characteristics using a small number of passive components. The designer should not forget that all resistors and capacitors have deficiencies. Component magnitudes all have a tolerance factor and they exhibit temperature dependence. Pure resistance, capacitance or inductance is not found in a practical component.

In many cases the ultimate limit to accuracy and stability in an op-amp circuit is determined not by the op-amp and its power supply, but by the external components. The tolerance on the component values used in a circuit sets a limit on closed-loop gain accuracy. The designer should also look very closely at other component characteristics.

10.7.1 Fixed resistors

Resistors are the most frequently used passive components in op-amp circuits. There are many different kinds of resistors available, differing widely in performance characteristics and cost. In the case of fixed value resistors, initial tolerance is the greatest source of error, followed by the temperature coefficient of resistance (TCR). Other sources of error are leakage current (particularly in high value resistors), humidity effects, and drift with time and voltage. Resistors have some series inductance (particularly wire-wound types) and stray capacitance across their terminals. They also exhibit several kinds of noise generation effects, dependent on their construction. Surface mount resistors have low series inductance and low shunt capacitance because of their leadless and planar construction.

Widely used types of fixed value discrete resistors are, in ascending order of temperature stability (decreasing TCR), carbon composition, carbon film, metal film (including thick film surface mount devices) and wire-wound. Although wire-wound resistors have the greatest temperature stability, they are not often used because they have large values of inherent inductance and shunt capacitance. The noise produced by wire-wound resistors approaches the theoretical minimum, but they are expensive.

Metal film resistors are superior to carbon film types, although they are marginally more expensive. In critical gain setting and filter applications metal film resistors are used. Metal film resistors are available with TCR in the range 100 ppm/°C to 25 ppm/°C.

Carbon film resistors are an excellent choice for experimental and not so critical applications. Carbon film resistors can be usefully employed when the greater TCR causes only a small percentage change in the overall value.

10.7.2 Potentiometers

Potentiometers are another important component used in some op-amp circuits. Potentiometers are essentially resistive dividers and as such must be considered to suffer from the same deficiencies found in fixed value resistors. In addition, other effects must be considered such as linearity error, end resistance and resolution. There are also stray capacitance and inductance effects, which vary with potentiometer setting. Over a period of time these devices are prone to become noisy.

10.7.3 Capacitors

Capacitors are important components in op-amp applications such as integrators, track and hold circuits and active filters. They are also important in frequency compensating circuits.

The performance of a capacitor varies widely, and depends on the dielectric used in its construction. A number of factors therefore have to be considered in assessing the most suitable capacitor dielectric for a particular application. The choice of capacitor will normally be dependent upon a required combination of the following parameters: capacity/size, voltage and/or current rating, tolerance/stability/temperature coefficient, frequency, power factor/insulation resistance/Q, environmental conditions, shape, finish and cost. In addition to all these factors, the method of construction may also be important; rolled foil capacitors have high inductance that is undesirable in high frequency applications.

Power supply decoupling is often achieved using tantalum capacitors. These are small, but have a high capacitance per unit volume. Unfortunately, tantalum capacitors have a self-resonant frequency of about 1 MHz. This means that their impedance rises at frequencies above 1 MHz, which results in a corresponding reduction in their effectiveness as a decoupling component. To overcome this, it is normal procedure to connect a 10 nF ceramic capacitor in parallel with a 10 μF tantalum capacitor for decoupling.

Dielectric absorption has the greatest effect after a capacitor has been charged for a prolonged period (10 minutes or more). This gives enough time for the individual charges to have distributed themselves evenly within the dielectric material. If the capacitor is then quickly discharged, some of the charges do not have enough time to move through the dielectirc. Once the discharge path is removed, these 'sluggish' charges cause the capacitor to regain some of its terminal voltage (maybe ~30 mV after 1 minute for polyester or Mylar). The shorter the discharge time, the greater the 'memory' of the capacitor.

The worst culprits for high dielectric absorption are electrolytic or tantalum capacitors. The lowest dielectric absorption occurs in polypropylene, polycarbonate and polystyrene capacitors. Not all 'poly' capacitors have low dielectric absorption: polyester (or Mylar) is mediocre.

In a sample-and-hold or integrator circuit, where the capacitor must store a charge, dielectric absorption can be a problem. The 'memory' of the dielectric can cause drift in the capacitor voltage. For example, after discharging the capacitor in an integrator circuit, the capacitor voltage will rise and give an incorrect output. It will have the same effect as charging the capacitor with an external current.

10.8 Avoiding fault conditions

Most modern op-amps incorporate internal circuit protection against accidental overloads, but it is important to examine those overloads that cannot be tolerated. Find out what are the maximum differential input voltage, the maximum common mode voltage, and the output voltage limits. Remember that these limits only apply for the op-amp connected to rated values of power supplies. Damage can be caused if signals are applied to the op-amp's inputs before the power supplies are switched on.

It might seem, at first sight, that nothing could go wrong if an op-amp's inputs are internally protected and if its output is current limited. However, under certain external circuit conditions, faults can arise which might lead to device destruction. Look out for voltages retained by capacitors that can apply input signals when the power supplies are switched off, or for conditions under which capacitor voltages can cause input voltage limits to be exceeded. Also, guard against conditions which might allow the op-amp's output to be presented with a voltage higher than the supply voltage because, say, of back EMF from an inductive load.

An example of a capacitor discharge that can cause an op-amp input limit to be exceeded is shown in Figure 10.13. The op-amp is connected as an integrator. The integrating capacitor, assumed charged up to the positive output limit of the op-amp, is to be discharged to the negative limit of the op-amp by connecting the op-amp output to the negative supply.

Closing the discharge switch in Figure 10.13 will almost certainly damage the op-amp. In effect, it puts a transient −30 V on the op-amp's inverting input. The inverting input is made negative with respect to the negative supply voltage and the capacitor's discharge current is supplied by the low impedance negative supply line. Fortunately, it is not difficult to prevent the condition from destroying the op-amp. It is merely necessary to limit the capacitor discharge current to a safe value of a few milliamps, and a resistor (say 10 kΩ) connected directly to the op-amp's input terminal will normally provide the necessary protection.

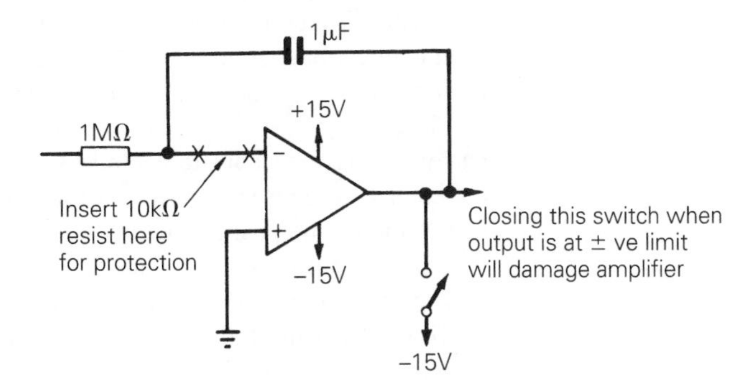

Figure 10.13 *An op-amp input limit can be exceeded by transient capacitor discharge*

Diode clamps can be used to protect against input overloads (see Section 2.1) and diodes in series with power supply leads can be used to guard against inadvertent power supply reversal. Rather more elaborate circuits are required to guard against transient supply overvoltages.

The output circuitry of most op-amps is internally protected by some means of output current limiting. This prevents excessive current being drawn because of an accidental short to either earth or power supply lines. In such op-amps, damage to the output stage can still be caused because of voltage breakdown if the op-amp's output is taken to a potential higher than that of the supply line (say because of an inductive load). The zener diode output clamp shown in Figure 10.14 can be used to provide protection against such output over voltage.

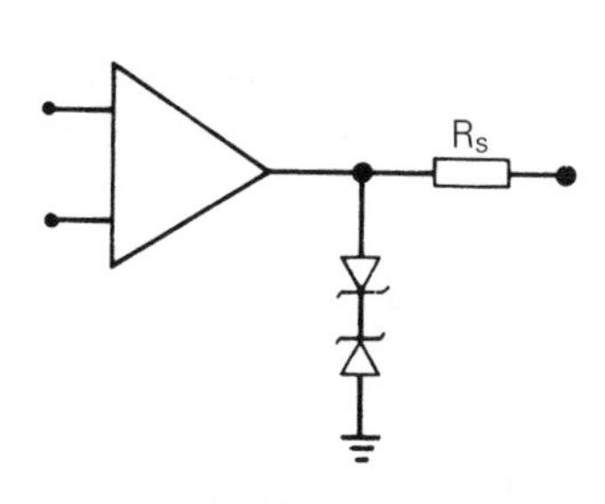

Figure 10.14 *Zener diodes protect against output overvoltage*

The output signal from the op-amp is taken from the side of R_s remote from the op-amp. R_s is included within any feedback loop; its value must be sufficient to prevent excessive current being passed through the zener diodes under fault conditions.

10.9 Modifying an op-amp's output capability

Op-amps have a high voltage gain and are designed for low power consumption. They are not intended to deliver an appreciable amount of output power. Traditionally, op-amps have worked off ±15 V power supplies to give a rated output voltage of ±10 V with output currents limited to something of the order of ±5 mA. Op-amps working off lower power supply voltages are also common, but they have reduced output capabilities.

Greater than normal output voltage and/or current swing may be required in some applications. The options are to use an op-amp with greater than normal output capabilities, or add external voltage or current boost circuits.

10.9.1 Output current boosting

Connecting a current amplifier, with unity voltage gain, directly to the op-amp's output can increase capability. The current amplifier is normally included within any feedback loop in which the op-amp is used. Feedback then reduces any non-linearity in the current amplifier in proportion to the loop gain of the circuit. Provided that the bandwidth of the current amplifier extends beyond the unity loop-gain frequency, the inclusion of the current amplifier in the feedback loop does not alter the closed-loop stability phase margin.

In applications requiring only a single polarity output current and moderate current gain, a single transistor emitter follower can be used as a current amplifier. In order to protect the transistor against excessive short-circuit current, it is advisable to include a resistor in series with the transistor's collector as shown in Figure 10.15(a). This resistor inevitably imposes some restriction on the positive output voltage swing. For negative output currents, reverse the supply voltages and use a pnp transistor.

The current booster circuit shown in Figure 10.16 eliminates crossover distortion by using transistor T_3 to establish class AB biasing of the complementary output pair. The action of T_3 is to hold the voltage across the 1 kΩ potentiometer at some multiple of the base-emitter voltage of T_3, determined by the potentiometer setting.

Larger output currents are obtained in the circuit of Figure 10.17 by adding further current gain in the form of Darlington pairs. Transistors T_6 and T_7 are used to set an output current limit. If the bias regulator transistor T_3 is mounted on the same heat sink as the output transistors T_1 and T_2, thermal feedback ensures bias stability. The circuit gives ±10 V across a 10 Ω load.

Note that offset drift in current boosters is of little significance. This is because offset drift is divided by the open-loop gain of the op-amp, when it is referred to the input of the composite amplifier.

10.9.2 Output voltage boosting

High voltage op-amps are commercially available. However, it is also possible to add a discrete amplifier stage, employing high voltage transistors, at the output of a general-purpose op-amp. Either circuit can be used for higher than normal output voltage swings. The simple circuit shown in Figure 10.18 can be used if only single polarity output voltages are required.

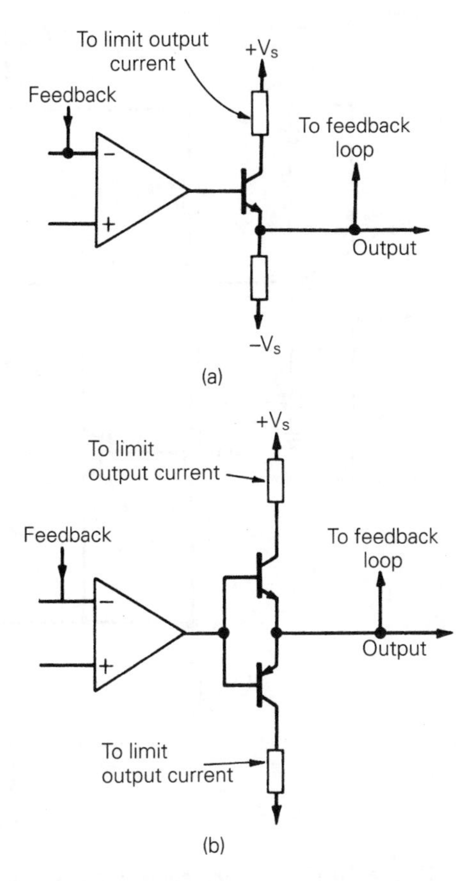

Figure 10.15 *Output current boosters: (a) simple emitter follower, (b) simple booster for dual polarity output current*

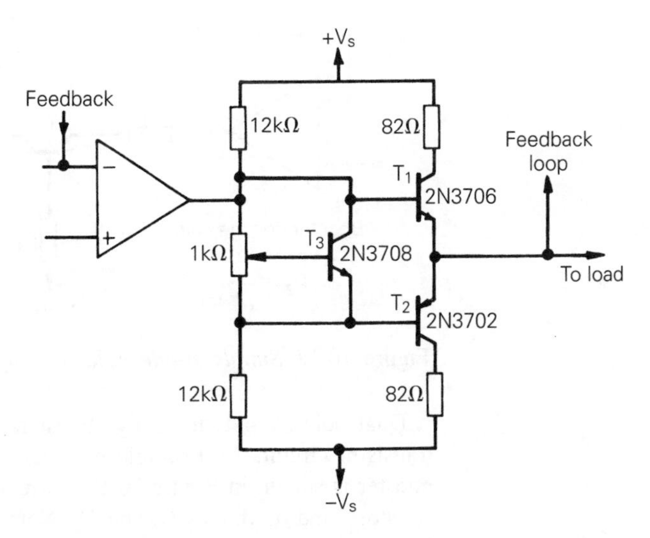

Figure 10.16 *Class AB biasing of current op-amp eliminates crossover distortion*

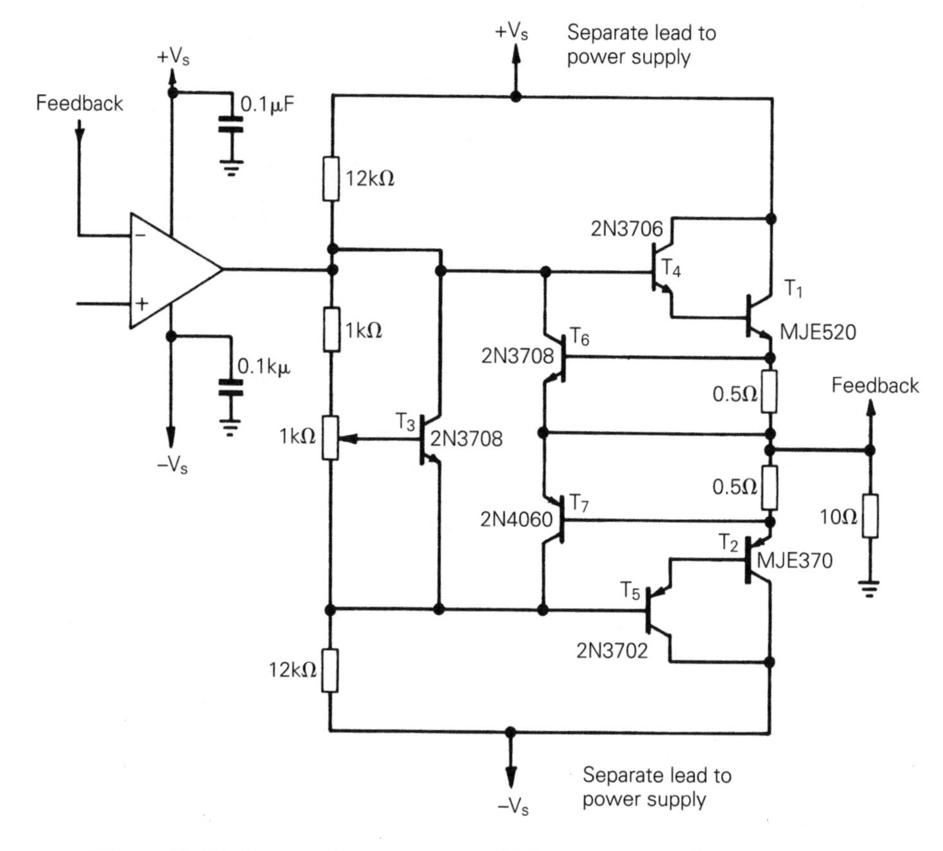

Figure 10.17 *Current booster gives ±10 V across a 10 Ω load*

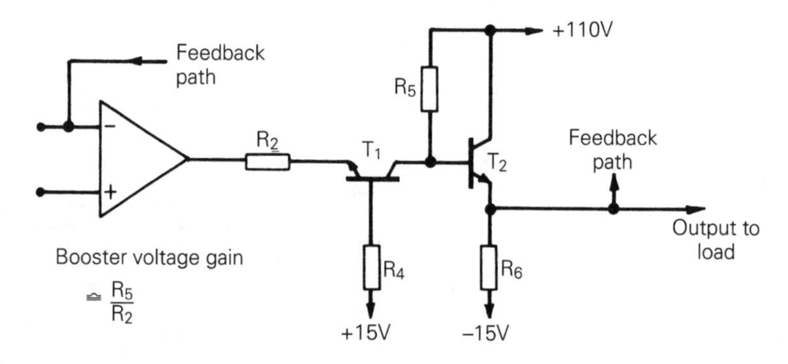

Figure 10.18 *Simple single polarity output voltage: 0–100 V*

Dual polarity output voltage boosters normally employ a complementary transistor circuit. An example of a complementary transistor output voltage booster is shown in Figure 10.19, current limiting is provided by the emitter resistors, and by diodes D_3 and D_4. Note that this voltage gain stage is phase inverting, so that the overall feedback circuit around the composite op-amp must be returned to the non-inverting input of the op-amp.

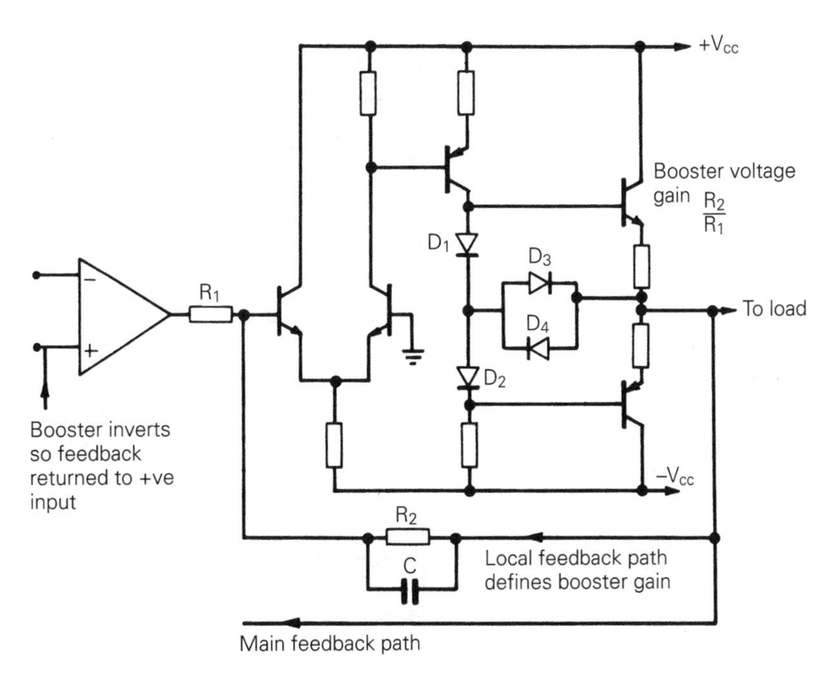

Figure 10.19 *Voltage booster with class AB biasing and output current limit*

Voltage boosters normally have their gain defined by a local feedback path connected around them. They are then connected to the output of the op-amp and, like current boosters, are included within any feedback path around the op-amp.

The closed-loop stability of an op-amp voltage booster circuit requires particular attention. Unity-gain frequency compensation of the op-amp does not ensure closed-loop stability of the cascaded arrangement. In many cases the op-amp requires greater than unity gain frequency compensation. The effect can be examined in terms of the appropriate Bode plots.

Unity-gain frequency compensation of an op-amp requires a frequency compensating capacitor that reduces the open-loop gain down to unity before the second break in the op-amp's open-loop response. When combined with a voltage boosting amplifier, any additional gain has to be considered. It is reasonable to assume that the bandwidth of the voltage booster is greater than this second break frequency. The frequency compensating capacitor value must be chosen so that the gain of the *composite amplifier* is reduced to unity before the second break frequency in the op-amp response. Thus, if the voltage booster has a gain of 10 (20 dB), the composite amplifier requires a value of frequency compensating capacitor which is ten times greater than that required for the op-amp alone. See Figure 10.20.

10.9.3 Output level biasing

It is sometimes necessary to develop the output signal of an op-amp about a reference level other than zero. A reference level that is considerably greater

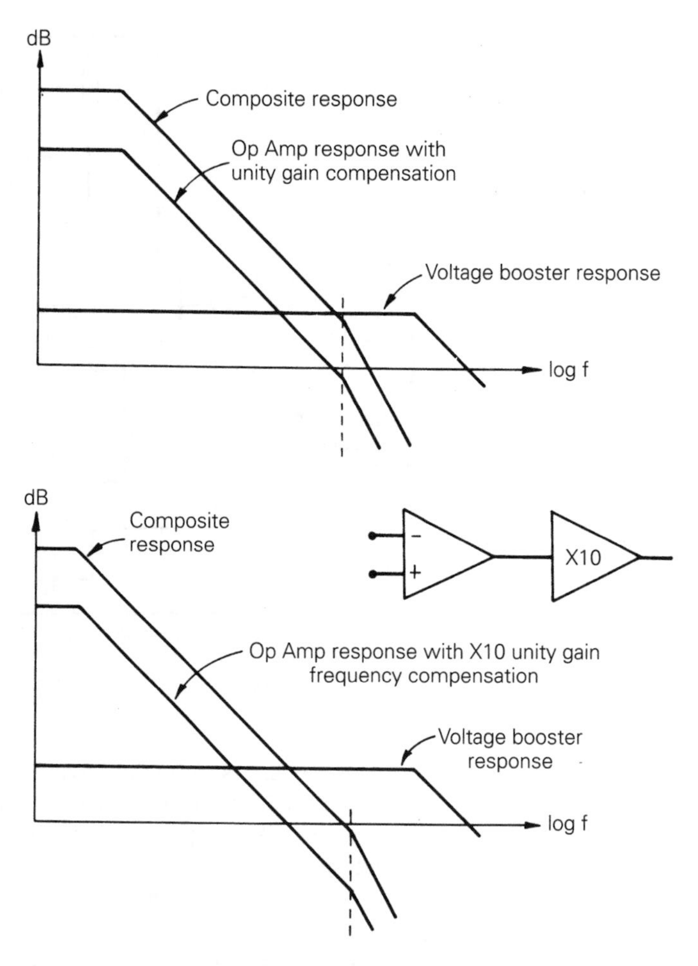

Figure 10.20 *Op-amp must be used with greater than unity gain frequency compensation when followed by a voltage booster*

than an op-amp's output rating can be produced without the use of an output voltage booster. This can be obtained by the appropriate addition of a fixed DC bias to the output of the op-amp. In the circuit shown in Figure 10.21 the zener diode serves to hold the op-amp output voltage within its rated limits, and the zener diode's breakdown voltage should be selected accordingly. The output voltage that is applied to the load appears about the fixed positive level $E_{ref}R_2/R_3$.

If the output to the load is to be developed about a fixed negative reference level, the zener diode should be reversed and the opposite polarity used for E_{ref} and the zener bias supply. The output current rating of the op-amp must be sufficient to supply the load, the feedback and the zener bias. Note that, since the zener is connected within the feedback loop, the effects of its internal impedance and any drift in zener voltage are divided by the loop gain.

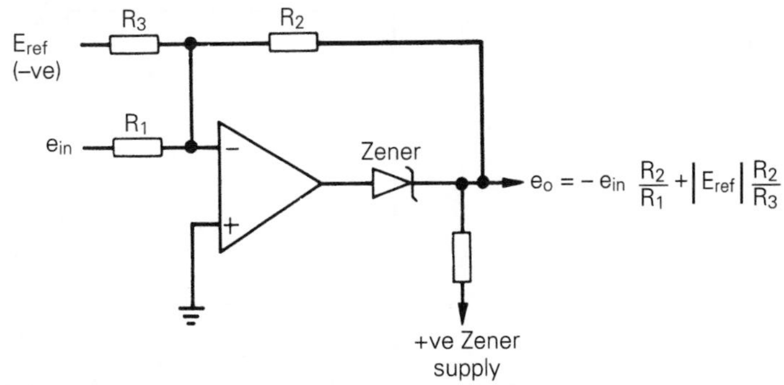

Figure 10.21 *Output level biasing with zener included within feedback loop*

10.10 Speeding up a low drift op-amp

Op-amps designed for low drift and DC stability are generally somewhat limited in their bandwidth and slew rate. Conversely, op-amps which have a high slew rate and wide bandwidth tend to exhibit considerable offset voltage drift with both time and temperature. It is possible to combine a low drift op-amp with a high speed op-amp in such a way that the composite op-amp has the DC stability of the low drift op-amp, combined with the fast response of the high-speed op-amp.

A composite op-amp connection is shown in Figure 10.22. The low drift op-amp A_1 is connected as an integrator, so that its closed-loop signal gain rolls off at 20 dB/decade. Unity gain (0 dB) is reached at the frequency $1/(2\pi CR)$. The output of op-amp A_1 is connected to the non-inverting input of the wide-band op-amp A_2. Diodes D_1 and D_2 are included to prevent a possible latch-up condition, which can exist in the composite op-amp if the output limits of the low drift op-amp exceed the common mode range of the wide band op-amp.

The open-loop gain of the composite op-amp is

$$A_{OL} = A_2(jf)\left[1 + A_1 \frac{1}{j\dfrac{fA_1}{f_o}}\right]$$

where: $A_2(jf)$ is the frequency dependent open-loop gain of the wide-band op-amp, A_1 is the zero frequency open-loop gain of the low drift op-amp and $f_o = 1/(2\pi CR)$ is the frequency at which the signal gain of the integrator-connected low drift op-amp is unity.

A choice of component values CR, which makes $f_o = f_1/A_2$, gives the composite op-amp a single 20 dB/decade roll-off down to unity gain. The wide-band op-amp has an open-loop DC gain of A_2, and a unity-gain frequency f_1. It is assumed that the wide-band op-amp has a unity gain compensated open-loop response.

At frequencies higher than f_o there is negligible signal transmission through the low drift op-amp, which connected as an integrator. At these frequencies, the gain of the composite amplifier is provided by the wide-band op-amp. There is a direct path to the inverting input of the wide-band op-amp (see Figure 10.22).

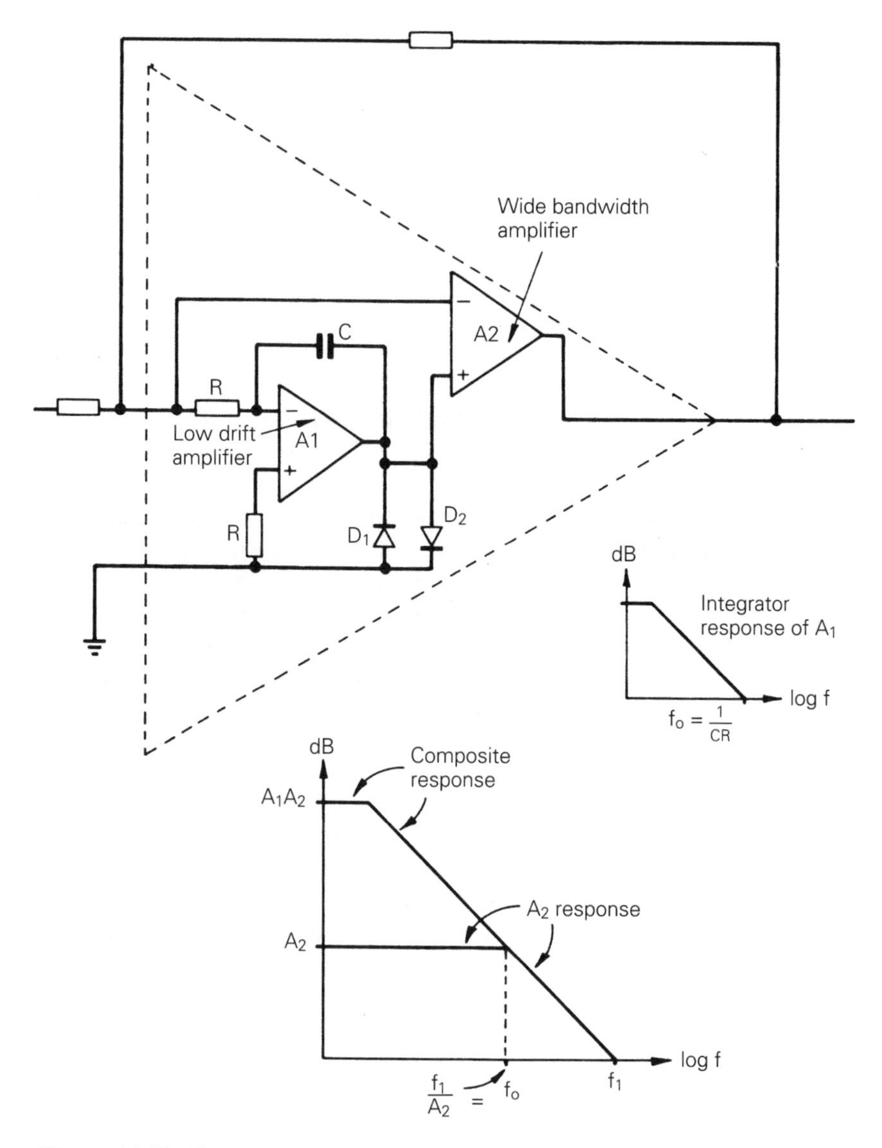

Figure 10.22 *Composite connections used to speed up a low drift op-amp in inverting applications*

DC offset and drift in the wide-band op-amp, when it is referred to the input of the composite amplifier, is effectively divided by the open-loop gain of the low drift op-amp. The DC offset and drift characteristics of the composite amplifier are essentially those of the low drift op-amp. The integrator connected low drift op-amp senses and integrates any offset voltage from earth present at the inverting input terminal of the composite amplifier. It then applies an offset compensating voltage to the non-inverting input of the fast op-amp.

Note that if A_1 is an externally frequency compensated type, an alternative to integrator connection is to use A_1 with a much greater than normal frequency compensating capacitor, so that its open-loop unity-gain frequency is set at the value f_o.

The interconnection technique of Figure 10.22 is essentially feedforward frequency compensation (see Section 2.7.2). It is also the same principle that underlies the low drift characteristics of chopper stabilized op-amps.

10.11 Single power supply operation for op-amps

Op-amps are generally designed to operate from symmetrical positive and negative power supplies. Most of the circuits given in this book do not show power supply connections, but dual supplies have been assumed. The use of dual supplies permits the op-amp's output voltage swing to be both positive and negative, with respect to the potential of the power supply common terminal (earth). Applications not requiring a response down to zero frequency, or in which only single polarity output signals are of interest, can be implemented with an op-amp powered by a single voltage supply.

The problem of single power supply operation is simply that of maintaining the DC voltage levels in the circuit at their proper values. Most op-amps have three reference levels when operated from dual supplies, these are: $+V$, earth and $-V$ (V is the value of the supply voltage). For single supply operation these reference levels can be maintained by using $2V$, V and earth, where V is obtained from a resistive divider network or split zener diode biasing system. Decoupling capacitors across the zener diodes are necessary to reduce noise. The negative supply terminal of the op-amp is connected to earth, the positive supply terminal is connected to $2V$, and the differential input terminals are biased up to V.

Connections for single power supply operation of an AC inverter and an AC follower (see Section 4.7) are shown in Figure 10.23. The DC blocking capacitors at input and output determine the low frequency pass-band edge. Their presence means that op-amp offsets and their temperature drift are of no great importance, except that output voltage offset may reduce the output voltage swing capability of the op-amp. Offset due to bias current can be minimized if $R_3 = R_2 // R_1$. The input impedance of the AC follower can be increased using the bootstrapping technique previously mentioned in Section 4.7.

In applications that require a response down to DC, blocking capacitors cannot be used. The design problem with single supply op-amp circuits is to make sure that the potential at the op-amp's inputs (measured with respect to the voltage midway between the supply pins) remains within the allowable common mode range of the op-amp. Single power supply operation of course allows only single polarity output signals.

Some op-amps have a common mode range that extends down to the negative supply rail. These op-amps are specifically designed with single power supply operation in mind. Figure 10.24 shows one circuit arrangement that is made possible by single supply op-amps. This circuit cannot be implemented directly with general-purpose op-amps, although the circuit does not work very well when the output potential is near earth. Op-amps designed specifically for single supply operation can also be used with dual or referenced supplies, but they do not normally perform as well as most general-purpose types.

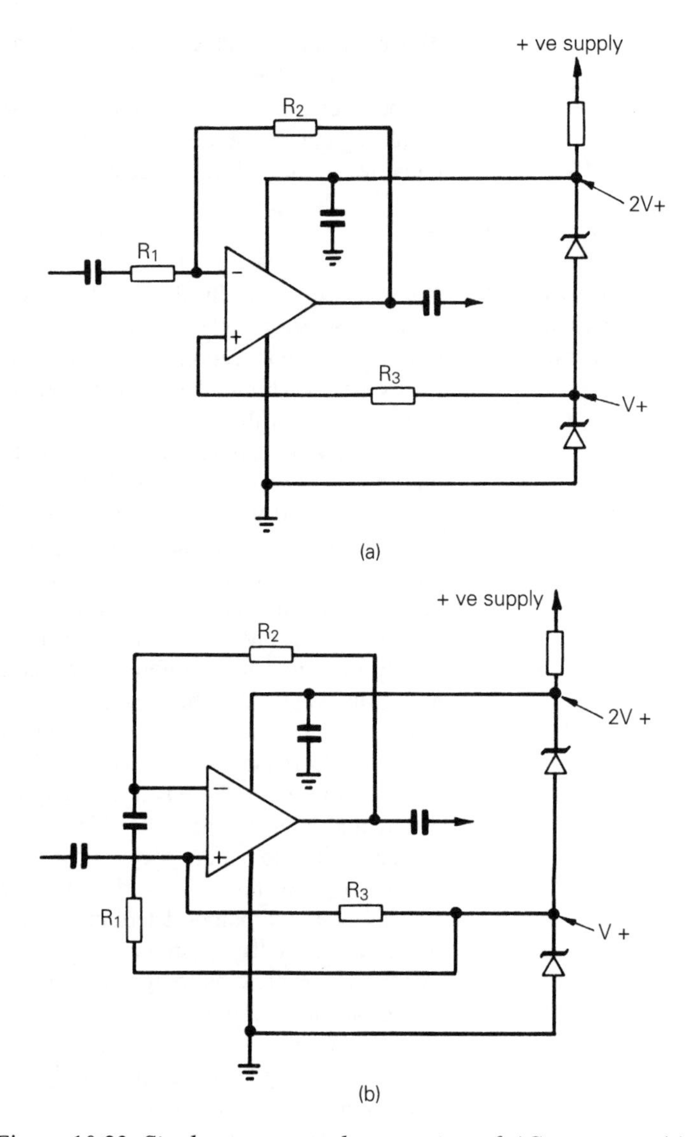

Figure 10.23 *Single power supply operation of AC op-amps. (a) Inverting AC op-amp with single supply operation. (b) AC follower with gain – single supply operation*

A method of referencing a differential amplifier for single supply operation is shown in Figure 10.25.

The common mode range at the differential amplifier's input terminals is:

$$V_{ref} \pm (\text{Common mode range of op-amp}) \times (R_1 + R_2)/R_2$$

For low gains, this can include earth or even negative common mode signals. Either input of the differential amplifier may be earthed. A voltage that is

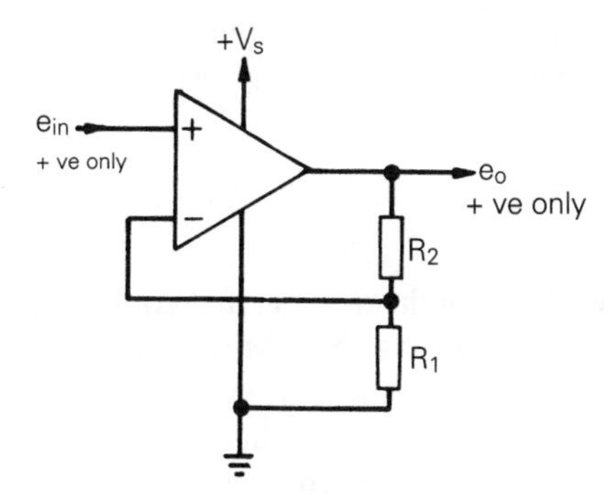

Figure 10.24 *Connections possible with op-amp designed for single power supply operation*

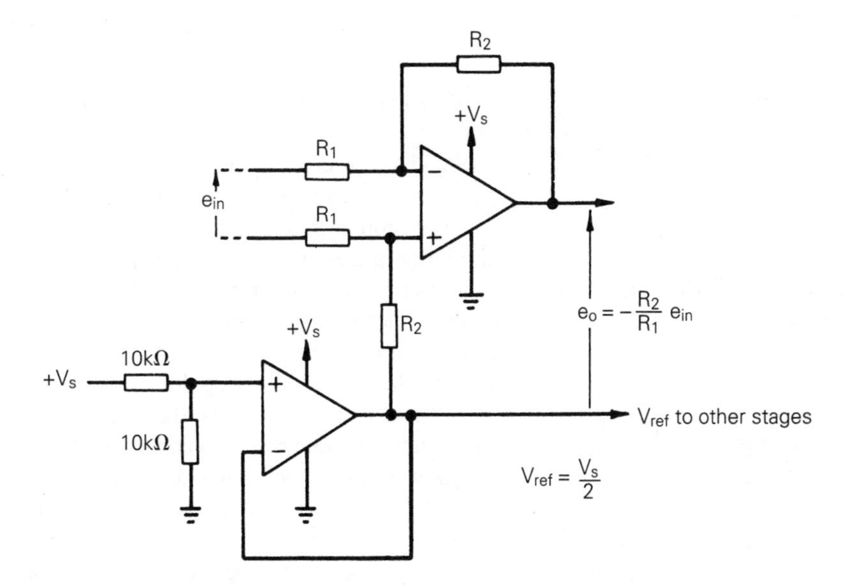

Figure 10.25 *Single power supply operation of general-purpose op-amp with a second op-amp used to supply a reference*

positive or negative with respect to earth may be applied to the other input. The output can swing both positive and negative with respect to the reference level, and the reference voltage V_{ref} can be easily connected to other amplifier stages. Texas Instruments produce a 'virtual ground generator'. This device is capable of overcoming many of the problems encountered by the existing biasing methods discussed above and is described in Section 10.12.1.

10.12 Voltage regulator circuits

Most op-amp circuits are powered from voltage regulators. This reduces the problem of limitations in performance due to power supply variations. A simple voltage regulator can be built using a voltage reference, with an op-amp driving a power transistor; Figure 10.26 shows this.

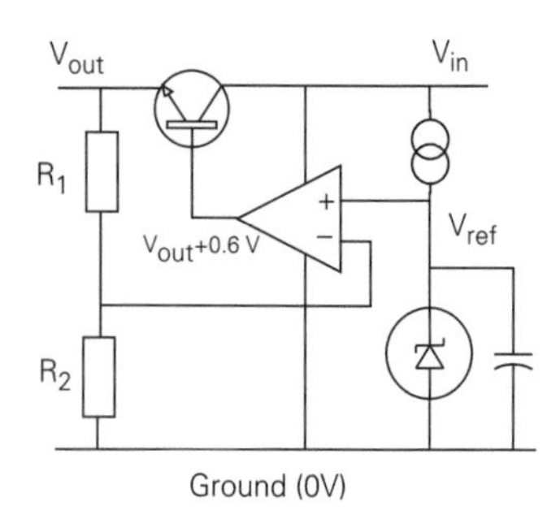

Figure 10.26 *Simple voltage regulator circuit*

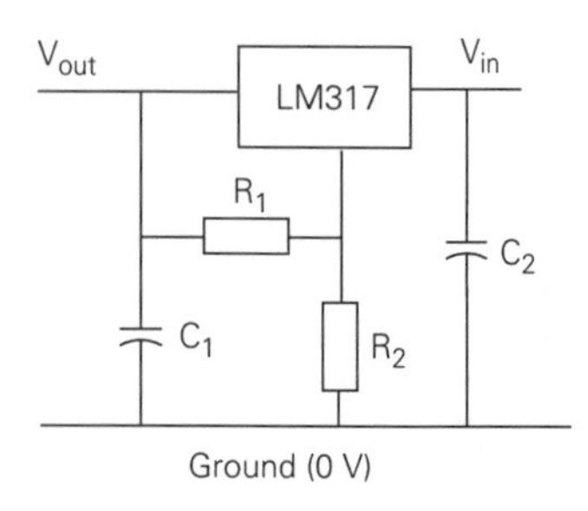

Figure 10.27 *An integrated circuit voltage regulator circuit*

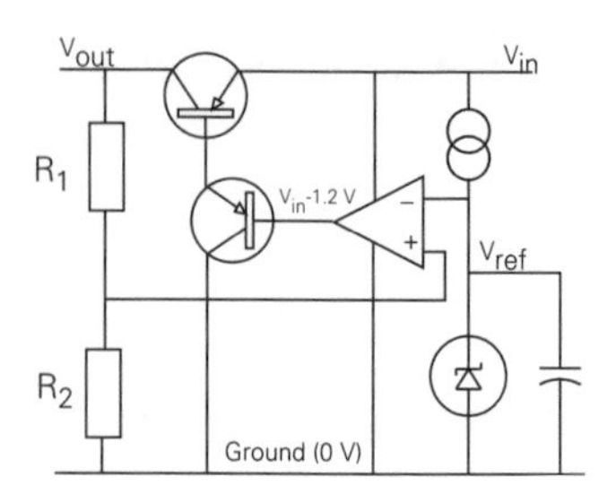

Figure 10.28 *Low dropout voltage regulator*

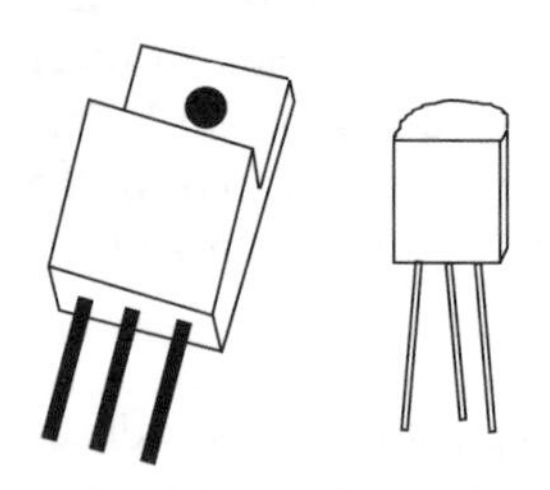

Figure 10.29 *Voltage regulator packages*

A fraction of the output voltage is compared with V_{ref}. Any difference is amplified by the op-amp, which maintains the condition:

$$V_{out} = V_{ref}(1 + R_1/R_2)$$

Voltage regulator integrated circuits have simplified power supply design. A circuit, such as the LM317, can be set to produce any voltage within its rated limits, simply by adding two external resistors. These resistors perform a similar function to those in Figure 10.26, by proving feedback from the output. An integrated circuit voltage regulator circuit is given in Figure 10.27. However, the equation for determining the output voltage is modified because the reference voltage is relative to V_{out}, rather than earth:

$$V_{out} = V_{ref}(1 + R_2/R_1), \text{ where } R_1 \text{ is usually a value} \sim 180\ \Omega.$$

Note that in Figure 10.27, capacitors are connected from V_{in} to earth, and from V_{out} to earth. These reduce noise and prevent instability.

Standard voltage regulators require about 2 V drop between V_{in} and V_{out}, to ensure correct biasing of internal circuits. Regulation may suffer if the voltage drop is reduced below this figure. As shown in Figure 10.26, the limitation arises because an npn power transistor is used. The output from the op-amp must be greater than 0.6 V above V_{out} in order to overcome the base-emitter voltage drop. There are additional voltage drops within the op-amp's output stage.

Low dropout voltage regulators overcome this limitation by using a pnp power transistor, instead of the npn type used in standard regulators. The minimum voltage drop between V_{in} and V_{out} is the saturation voltage across the power transistor, which is about 0.2 V. The output of the op-amp is $V_{in} - 0.6$ V in this configuration. The circuit for a low dropout regulator is shown in Figure 10.28.

Voltage regulator integrated circuits are available in TO-220, TO-92, and SO-8 surface mount packages; see Figure 10.29.

10.12.1 The 'virtual ground generator'

In Section 10.11 the problems associated with the correct biasing of an op-amp powered from a single supply rail were discussed. For example, an op-amp in an inverting amplifier circuit, powered from a single supply, requires the non-inverting input to be biased at half the supply voltage. This will then allow the amplified signal at the output to swing evenly about half the supply voltage. Figure 10.30 illustrates a suitable circuit.

Some of the traditional methods of providing bias (see Figure 10.25 for one method) have the disadvantages of a high standing power dissipation in the bias resistors. They also have poor input regulation, because of any supply voltage variations being automatically passed on to biasing network. The use

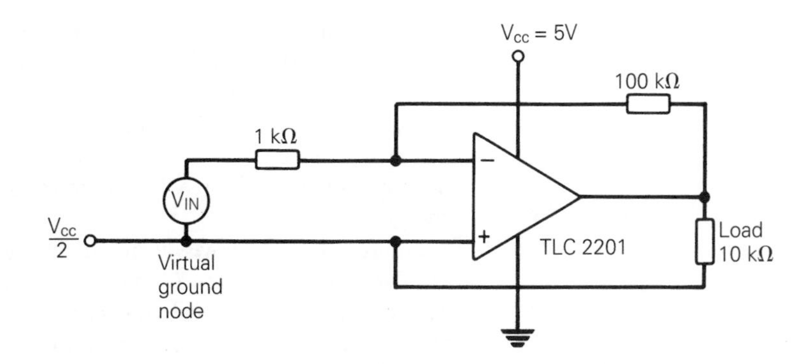

Figure 10.30 *Op-amp properly biased for single supply applications*

of additional buffer op-amps and zener diode voltage references helps to reduce but does not eliminate these disadvantages.

The introduction, by Texas Instruments, of 'virtual ground generator' integrated circuits has made life easier for the designer (Figure 10.31). One device, the TLE2425, is designed to operate with +5 V systems and produces a virtual ground voltage of +2.5 V. A variant of this device is the TLE2426, which is a mid-rail voltage generator that can operate at supply voltages of up to 40 V.

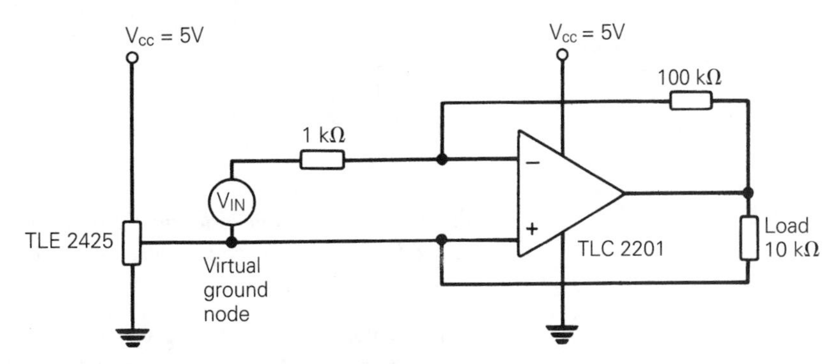

Figure 10.31 *Use of virtual ground generator for the biasing of single supply op-amps*

10.12.2 Voltage reference devices

Precision voltage references ICs, with very low temperature coefficients, have been developed and produced by several companies, such as National Semiconductor and Texas Instruments. These devices are known as band-gap voltage references.

The temperature coefficient (TC) of the voltage across a silicon junction is about −2 mV/°C, and this is inversely proportional to the current density through it. It is necessary to produce a circuit to cancel this TC. By manipulating the current sharing through two junctions, and using the difference in forward bias voltages, it is possible to create a circuit that produces an output voltage with a positive TC of 2 mV/°C. The output voltage from this circuit can then be added to the forward bias voltage of a third junction, which will have a TC of –2 mV/°C. The positive TC of the first two junctions cancels the negative TC of the third junction, to produce a circuit with zero TC.

Figure 10.32(a) shows a circuit that generates a voltage with a positive TC. Note that transistor T_1 may be considered as a diode. Resistor R_1 connected to T_2's emitter causes less current flow though the base-emitter junction of T_2 than the base-emitter junction of T_1. Therefore the base-emitter junction TC of T_2 is more negative than that of T_1.

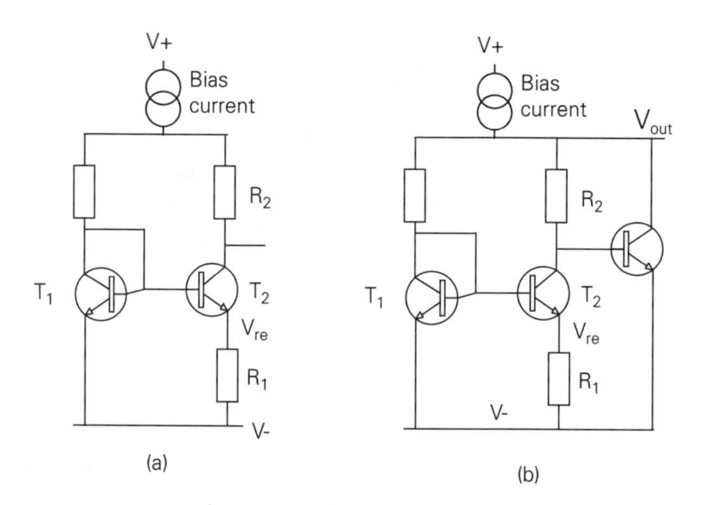

Figure 10.32 *Voltage reference circuits with: (a) positive temperature coefficient; (b) zero temperature coefficient*

The voltage across R_1 is given by $V_{R_1} = V_{BE_1} - V_{BE_2}$, so that the TC of V_{R_1} is positive. As an example, let the TC for transistor T_1 be exactly -2 mV/°C, and the TC of transistor T_2 be -2.2 mV/°C. Then, if the temperature rises 10°C, V_{BE_1} will drop by 20 mV and V_{BE_2} will drop by 22 mV. The voltage across R_1 will rise by 2 mV; thus a positive temperature coefficient of 0.2 mV/°C. Since the collector and emitter currents of T_2 are virtually equal, the voltage across R_2 will rise by the ratio R_2/R_1.

By adjusting the R_2/R_1 ratio, voltages with controlled positive TC can be generated. The same positive TC with a smaller R_2/R_1 ratio can be obtained by decreasing the current density in the base-emitter junction of transistor T_2. This is achieved by increasing its base-emitter area during manufacture.

The next step is to add the voltage across R_2 (with positive TC) to the base-emitter junction voltage (with negative TC) of another transistor, in order to produce a voltage with zero TC. This is shown in Figure 10.32(b). The output voltage is $V_{R_2} + V_{BE_3}$.

When the current densities in the transistors and R_2/R_1 are adjusted so that the TC of V_{out} is 0, then V_{out} = 1.2 V. This is close to the band-gap of silicon, hence the name of this technique.

The circuit just described is helpful to explain the basic idea. In practice producing an output voltage other than 1.2 V, with zero TC, is very difficult unless large R_2/R_1 ratios are used. Large resistor ratios are difficult to manufacture, so transistors with different junction areas are used to create a TC difference between the two base-emitter voltages.

The band-gap reference building block at the heart of many voltage references is shown in Figure 10.33. The two transistors, T_1 and T_2, are made with different base-emitter junction areas. Although the feedback loop forces identical collector (and emitter) currents through both transistors, the differing areas of the base-emitter junctions mean that the current densities and hence the TCs of the junctions are different. The difference voltage has a positive TC. In Figure 10.33, the sum voltage is amplified with the ratio R_4/R_5.

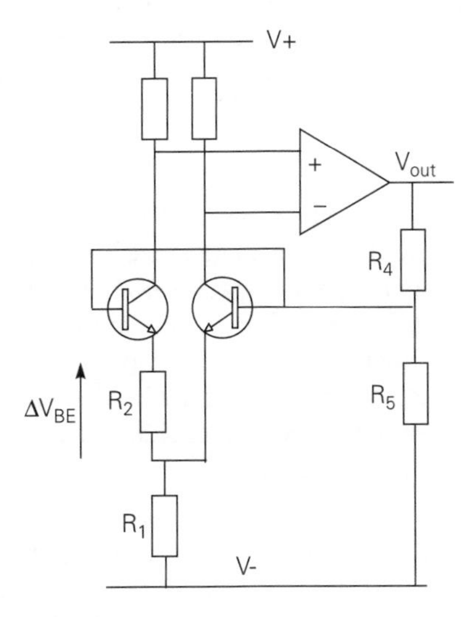

Figure 10.33 *2.5 V band-gap reference*

10.12.3 Constant current diodes

Constant current diodes are actually junction field effect transistors (JFETs). A JFET allows current to flow between the drain and source when there is no gate-source voltage applied. However, the drain-source channel is resistive and, when current flows through this channel, a voltage is developed across it. If the gate is connected to the source, there is a limit to the current flow between drain and source (the drain-source saturation current, I_{DSS}).

Consider the operation of an n-channel JFET. An n-channel JFET conducts when the gate-source voltage is zero, but stops conducting when the gate is made negative relative to the source by an amount V_P. The voltage V_P is the pinch-off voltage. Drain-source saturation occurs because, in effect, the potential of the channel is more positive than the gate (due to the voltage drop across the channel). Relative to the channel, the gate will have a negative potential.

As the current through the drain-source channel increases, so does the voltage drop across it. Hence the gate-channel voltage becomes more negative with increasing drain-source current. Regulation is achieved by negative feedback: as the voltage drop across the channel approaches V_P, the JFET tries to reduce the drain-source current; as the drain-source current reduces, the channel voltage drop reduces. The current limit depends on the characteristics of the JFET used, but 5 mA is typical.

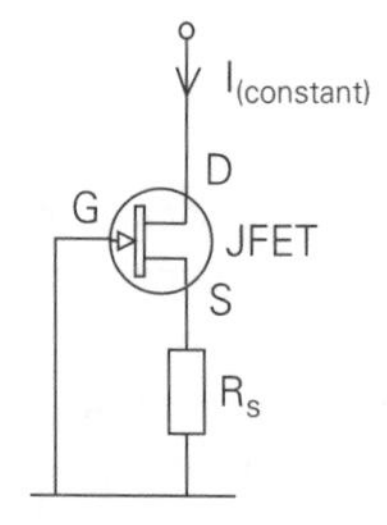

Figure 10.34 *Constant current 'diode'*

The pinch-off voltages of JFETs have a wide tolerance, typically ±50 per cent, and a precise current regulator using just a JFET may be impractical. Adding a resistor R_s in series with the source, as shown in Figure 10.34, allows the current limit to be adjusted for each JFET. Resistor R_s can also be used to set much lower current limits than the 5 mA quoted above. Current limits of 100 μA to 10 mA are commonly available. The CR series of devices (Siliconix) are temperature compensated.

The term constant current diode is used because the gate-drain junction behaves like a diode if the voltage polarity of the limiting circuit is reversed.

Exercises

10.1 A photodiode is used to detect variations in the intensity of the light emitted by a modulated light source, which produces a photo current with a peak-to-peak variation of 0.01 μA. This current is converted into a voltage using an op-amp current-to-voltage converter (Figure 4.10). What stray coupling capacitance between the inverting input terminal of the op-amp and a 230 V, 50 Hz supply line is sufficient to introduce an unwanted output signal with an amplitude 10 per cent of that of the desired signal?

10.2 An op-amp is used as a follower with a gain of 11, and it has an input capacity of 47 pF between its non-inverting input and earth. How would you reduce the capacitive loading imposed on the signal source? (See Figure 10.6.)

10.3 Component values R_1 = 5.1 kΩ, R_2 = 22 kΩ are used in the circuit of Figure 10.9(a). The op-amp has a maximum input offset voltage V_{io} = 10 mV and a maximum bias current I_B = 0.5 μA; ±15 V supplies are used. Find a suitable value for the bias current supply resistor R_3.

10.4 The following components are used in the circuit of Figure 10.11: R_1 = 10 kΩ, R_2 = 100 kΩ, R_4 = 1 kΩ. Find suitable values for resistors R_3 and R_5 assuming the op-amp has a worst case input offset voltage V_{io} = 2 mV and input different current I_{io} = 5 nA, and that ±15 V supplies are used.

10.5 Component values $R_1 = 10\ \mathrm{k\Omega}$, $R_2 = 22\ \mathrm{k\Omega}$, and a single $+30$ V supply, are used in the circuit of Figure 10.25. The rated common mode range of the op-amp when worked off ± 15 V supplies is ± 12 V. What is the allowable common mode range at the input terminals of the circuit? If the gain of the circuit is increased (by increasing R_2) what is the maximum gain for which connecting the non-inverting input of the circuit to earth will not exceed the common mode range of the op-amp?

Answers to exercises

Chapter 1

1.1 (a) Circuit of Figure 1.2(a) with $R_1 = 100\ k\Omega$, $R_2 = 500\ k\Omega$
(b) Circuit of Figure 1.2(a) with $R_1 = 2\ k\Omega$, $R_2 = 40\ k\Omega$
(c) Circuit of Figure 1.2(b) with $R_2/R_1 = 99$
(d) Circuit of Figure 1.8 with $R = 100\ k\Omega$, $C = 0.1\ \mu F$
(e) Circuit of Figure 1.5 with $R_1 = 400\ \Omega$

1.2 (a) −6 V, (b) +6 V, (c) −2 V, (d) −2 V, (e) +6 V, (f) +10 V, (g) −1 V

Chapter 2

2.1 (a) $R_{in} = 10\ k\Omega$, $R_f = 1\ M\Omega$
(b) −1% ($A_{CL} = 99.99$), (c) 0.05%

2.2 $A_{CL} = 4.9$, $R = 1000\ M\Omega$

2.3 $\beta = R_S/(R_S + R_f)$, $1/\beta = 1 + R_f/R_S$, where R_S is the resistance of the current source. If $R_S \to \infty$, $\beta \to 1$, $1/\beta \to 1$.
$\beta = R_p/(R_p + R_f)$, $1/\beta = 1 + R_f/R_p$, where $R_p = R_1//R_2//R_3$
$\beta = 1$, $1/\beta = 1$
$\beta = R_1/(R_1 + R_2)$, $1/\beta = 1 + R_2/R_1$
$\beta = R/(R + 1/[j\omega C])$, $1/\beta = 1 + 1/j\omega CR$

2.4 (a) 0 dB, (b) 6 dB, (c) 10 dB, (d) 20 dB, (e) 40 dB, (f) 60 dB, (g) 120 dB

2.5 (a) 16 dB, (b) 24 dB, (c) 10 dB, (d) 50 dB, (e) 20 dB, (f) −40 dB, (g) −26 dB, (h) 3 dB, (i) −3 dB

2.6 (a) (i) 500 kHz, (ii) 100 kHz, (iii) 20 kHz
(b) (i) $e = -(4.444e_1 + 3.076e_2 + 2.143e_3)$, (ii) $1/\beta = 10.66$,
(iii) 93.77 kHz, (iv) 2%

2.7 (a) 7.96 kHz, (b) 15.9 kHz

2.8 $\zeta = 0.294$, overshoot 38%
$C_{f(min)} = 58$ pF, overshoot 4.3%

2.9 5.73 MHz (equation A2.6), 35° phase margin (equation A2.8), 4.44 dB, 5.66 MHz, 58 μs
(a) 5, 2.83 × 106 Hz
(b) 10, 1.29 MHz

2.10 (i) (a) 0.7 V, (b) 0.02 V, (c) 0.02 V, (d) 9.9 kΩ
(ii) (a) 0.25 V, (b) 0.011 V, (c) 0.011 V

2.11 (a) 0.57 V
(b) 0.11 V, 2.19 V, 0.28 V, 0.14 V, 0.07 V, 0.1 V, 0.05 V, 0.025 V

2.12 CMRR = 10^4, 0.01%

2.13 (a) 2.4 μV, (b) 2.4 μV, (c) 4.1 μV

2.14 (a) (i) 310 nV, (ii) 363 nV, (iii) 734 nV, (iv) 2.05 μV
(b) (i) 20.4 pA, (ii) 20.6 pA, (iii) 26.7 pA, (iv) 41.6 pA
(c) (i) 0.31 μV, 0.37 μV, 2.05 μV; (ii) 0.36 μV, 0.43 μV, 2.1 μV; (iii) 0.74 μV, 0.87 μV, 3.1 μV; (iv) 2 μV, 2.4 μV, 6.1 μV

Chapter 3

3.1 LG = 1333, A_{CL} = 8.4936

3.2 LG = 995, A_{CL} = 8.4915, Gain error = −0.025%

3.3 Input referred noise = 2.21 nV/$\sqrt{\text{Hz}}$

Chapter 4

4.1 $e_o = -(e_1 + 2.128e_2 + 10e_3)$, 0.14%

4.2 (a) 101 (40 dB), (b) 202 (46 dB), (c) 19.8 kHz, (d) 0.41 V

4.3 (a) 66 dB, (b) 10 kHz, (c) 0.25 V

4.4 62 dB, 40 kHz

4.5 10°, 90°, 16 kHz

4.6 (i) (a) 1 μA, (b) 53 nA
(ii) (a) 1.1 μA, (b) 0.4 μA

Chapter 5

5.1 Break points, 0 V, 0.94 V, 2.64 V. Slopes, −1, −0.5, −0.2423

5.2 180 mV, 167 mV

5.4 150 mV

5.5 R_2 = 15 MΩ, R_4 = 54.6 kΩ

5.6 70 mV

5.7 (a) 2.5 nA, (b) 0.1 μA

5.8 $R_2 = 10\ \text{k}\Omega$, $R_4 = 54.7\ \text{k}\Omega$

5.9 $e = 1/(10^4 I_o)e_{in}^2$, $e_o = e_{in}^2/10$

Chapter 6

6.1 70 mV/s

6.2 Circuit of Figure 6.8, with $R_1 = 1\ \text{M}\Omega$, $R_2 = 500\ \text{k}\Omega$, $R_3 = 100\ \text{k}\Omega$; 2.1 mV/s

6.3 0.018 Hz, 0.38%, 10°, 1.5%

6.4 5000

6.5 16 Hz

6.6 4 kHz, 160 Hz, 14 mV

Chapter 7

7.1 See Figure 7.3, make $R_2 = 19 \times R_1$, $E_{ref} = 2.89$ V and clamp output levels to +5 V and −1 V

7.2 See Figure 7.2; input signals applied through resistors *R*, *R*/2, *R*/3 and a reference −5 V applied through a resistor *R*/6

7.3 $t_1 = 1.302$ ms, $t_2 = 2.32$ ms

7.4 $R_3 = 56\ \text{k}\Omega$, $R_4 = 224.9\ \text{k}\Omega$

7.5 −4.2 V to −0.95 V

7.6 2667 Hz, 59%

Chapter 8

8.1 (a) 100 kHz, 16 Hz, 3 mV, (b) noise gain $1/\beta$ increases, upper frequency bandwidth limit becomes 83.3 kHz
(b) 200 mV

8.2 11 Hz

8.3 29.9 μs, $I > 15.4$ mA, 5 mV/s

Chapter 9

9.4 5.6 mΩ, 2.582 mV

9.5 100 kΩ, 200 kΩ

9.6 (a) 79 μV, (b) 62 mV, (c) 95.34 mV

9.7 10 kΩ, 10 kΩ, 0.32 μF, 7.95 nF

9.8 A_o = 100, f_o = 159 Hz, Q = 50.5, R_g = 1 MΩ, R_7 = 1 MΩ, R_8 = 10 kΩ, e_{max} = 0.1 V, R_1 = R_2 = 796 kΩ, R_4 = 1.99 MΩ

Chapter 10

10.1 0.013 pF

10.2 Refer to Figure 10.6, let C_p = 4.7 pF

10.3 R_3 = 5.1 MΩ (a slightly lower value than this would be used in practice, say 4.7 MΩ, in order to give a margin of adjustment)

10.4 R_3 = 99 kΩ, R_5 = 690 kΩ

10.5 −2.45 V to +32.45 V, 4

Appendix A1
Operational amplifier applications and circuit ideas

The circuits given in this appendix represent extensions or modifications to the circuits given in the main body of the text. The reader conversant with the factors controlling accuracy and performance limitations (Chapters 2 and 9) should be able to use them as a basis for practical designs. Most circuits will function with a general purpose operational amplifier (use a BI-FET) say) but the amplifier type used will inevitably govern performance limits. In all cases care should be taken to ensure that applied signals do not exceed allowable amplifier limits.

Scaling circuits

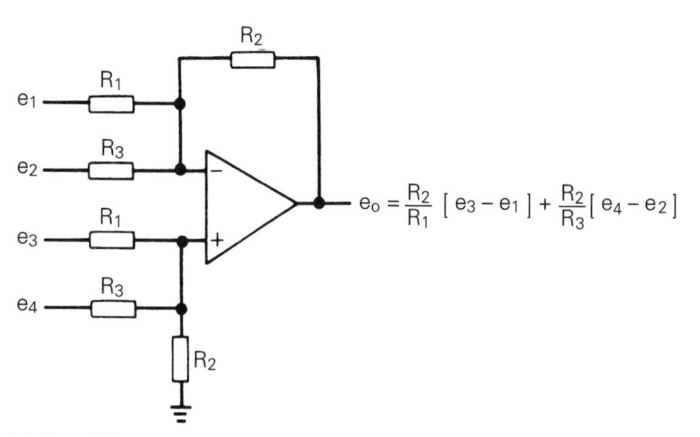

Figure A1.1 *Adder–subtractor*

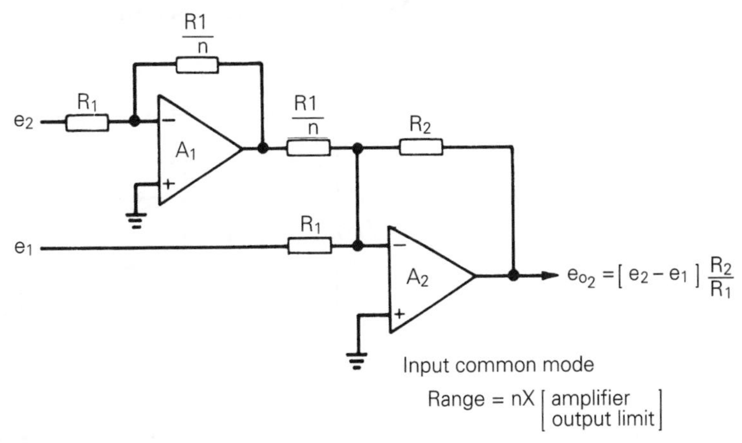

Figure A1.2 *Differential input amplifier configuration with large common mode range*

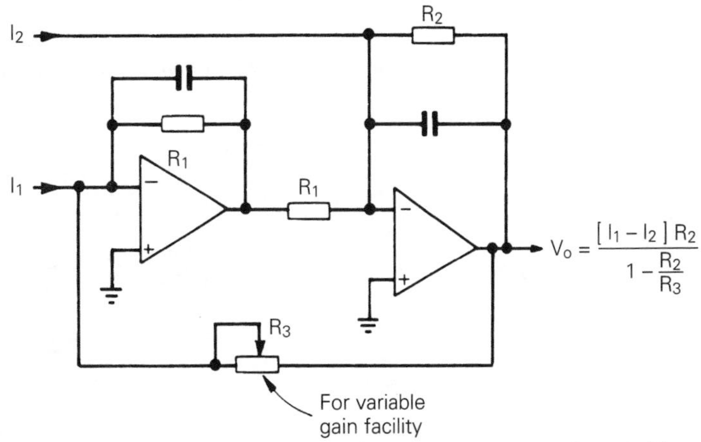

Figure A1.3 *Current difference-to-voltage conversion with variable scaling factor*

Signal sources

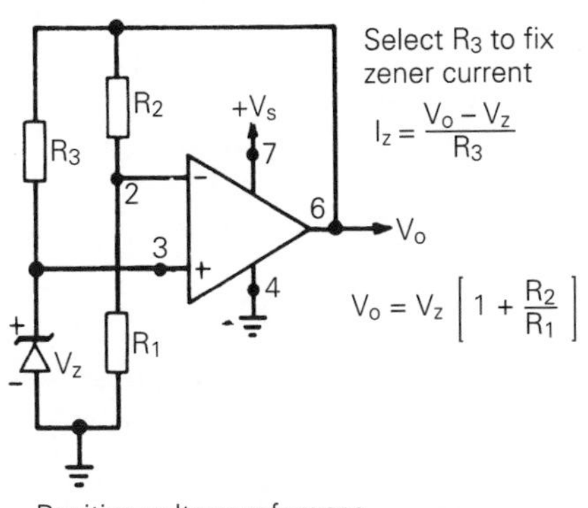

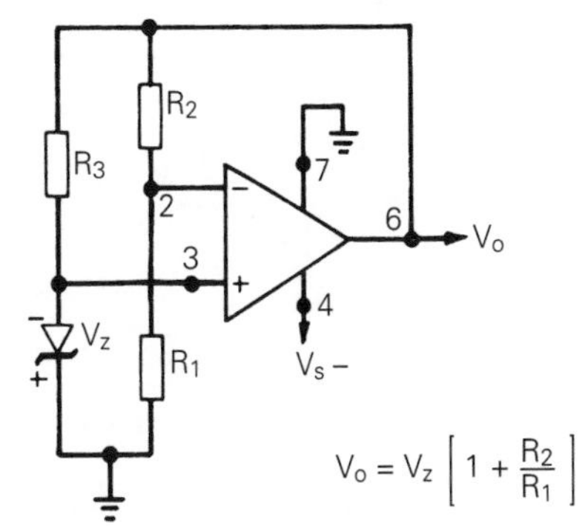

Figure A1.4 *Voltage references*

Measurement and processing

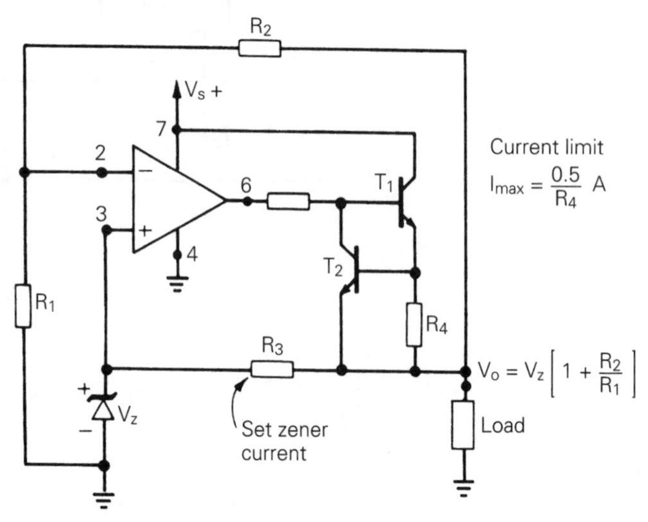

Figure A1.5 *Regulated voltage supply*

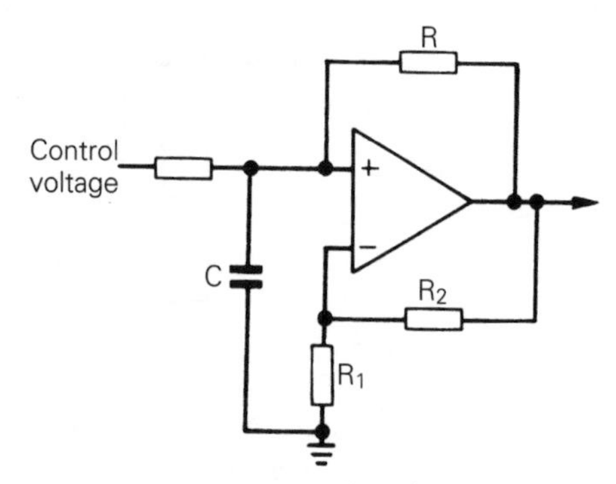

Figure A1.6 *Square wave generator with voltage control of pulse width (see Section 7.2.1)*

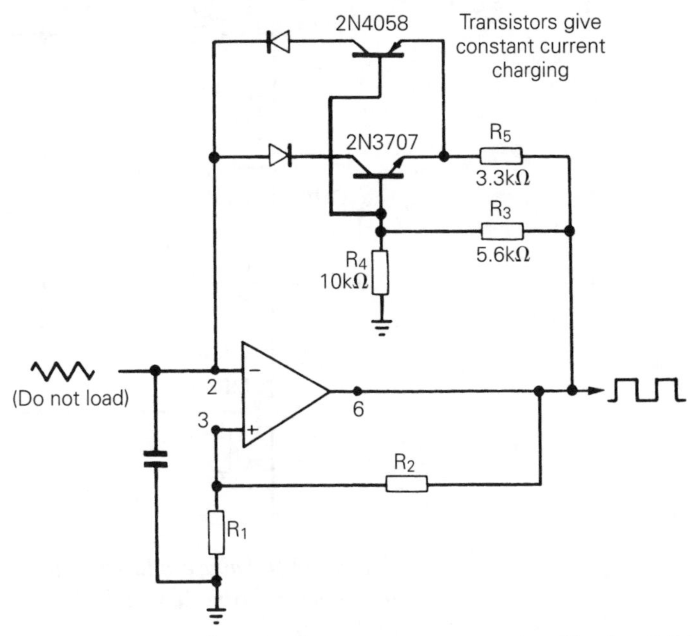

Figure A1.7 *Square and triangular wave generator (see Section 7.2.1)*

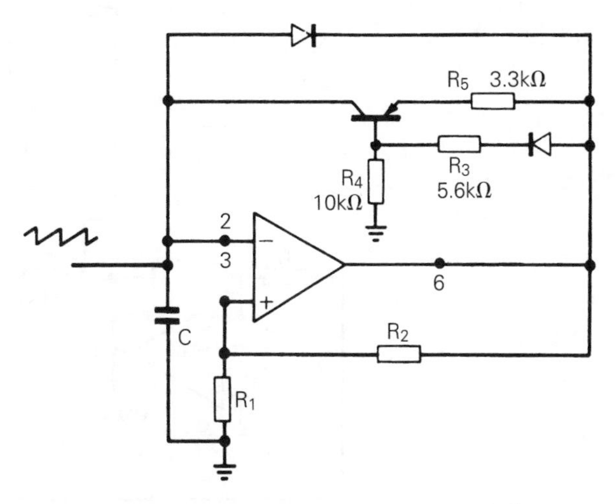

Figure A1.8 *Positive ramp generator (see Section 7.2.1)*

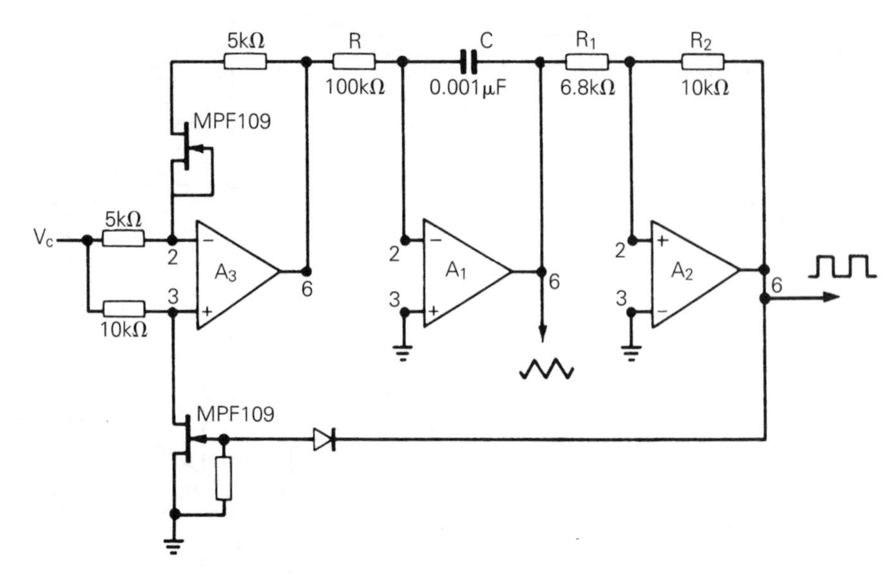

Figure A1.9 *Square and triangular wave generator with voltage control of frequency using a switched gain polarity amplifier (see Sections 7.4.3 and 8.12)*

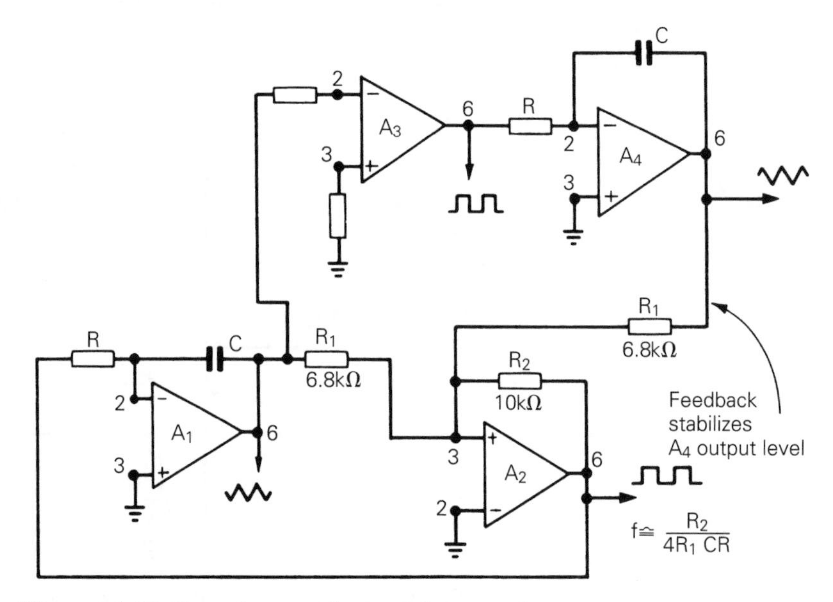

Figure A1.10 *Two-phase and triangular waveform generator with waveforms in quadrature (see Section 7.4.1)*

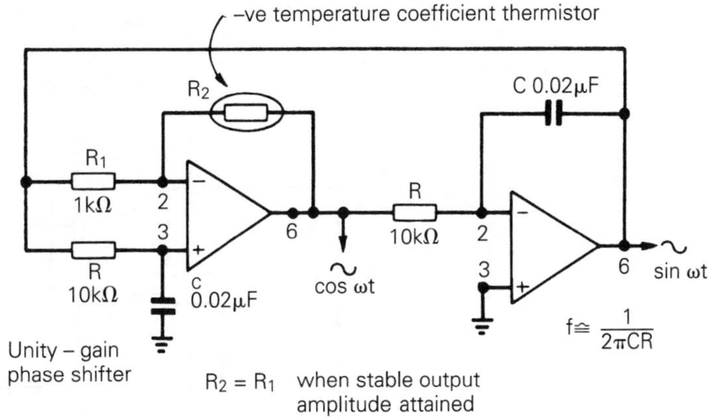

Figure A1.11 *Sine, cosine and wave quadrature oscillator using phase shifter*

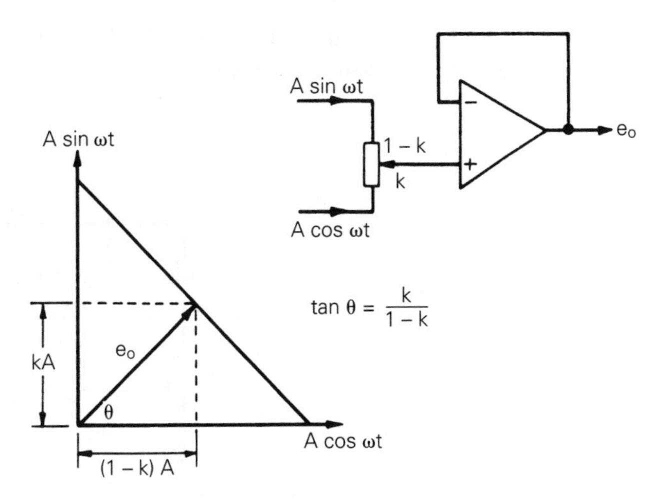

Figure A1.12 *Adjustable phase circuit for use with quadrature oscillator*

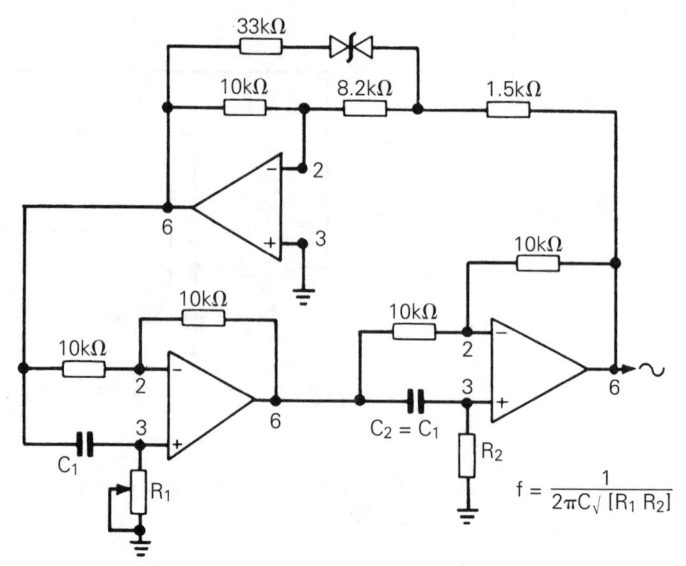

Figure A1.13 *Phase shift oscillator with single resistor frequency control and zener amplitude stabilization*

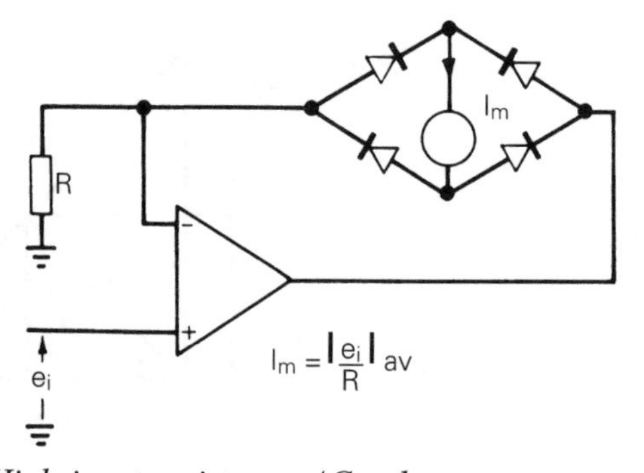

Figure A1.14 *High input resistance AC voltmeter*

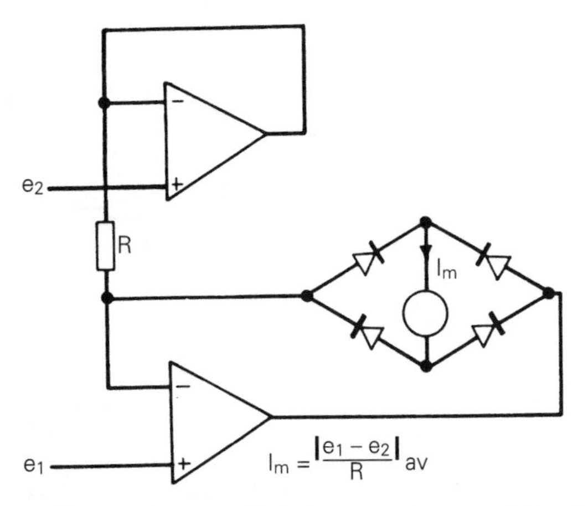

Figure A1.15 *Differential input, high input resistance AC voltmeter*

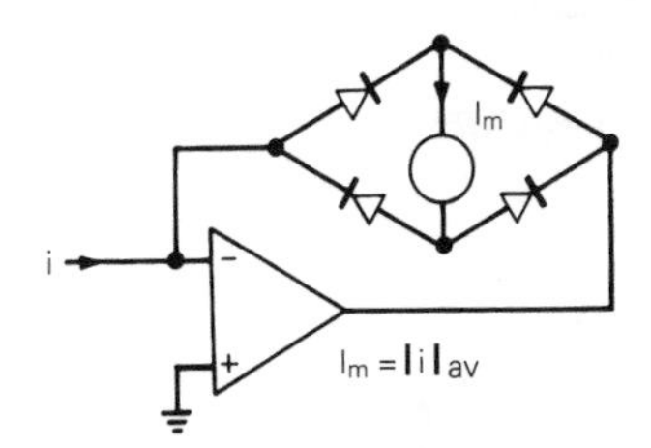

Figure A1.16 *Average reading AC current meter*

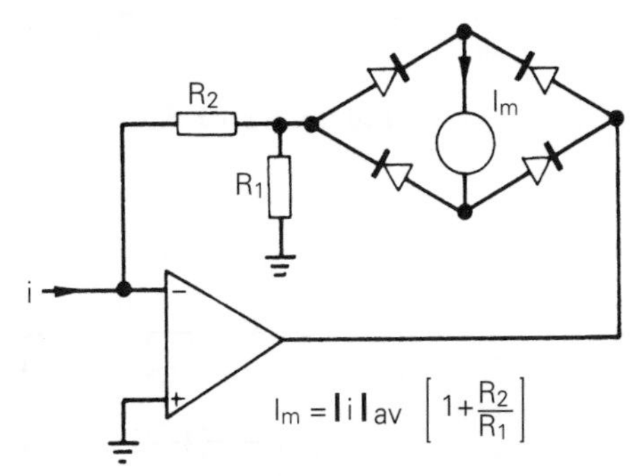

Figure A1.17 *Average reading AC current meter with current amplification*

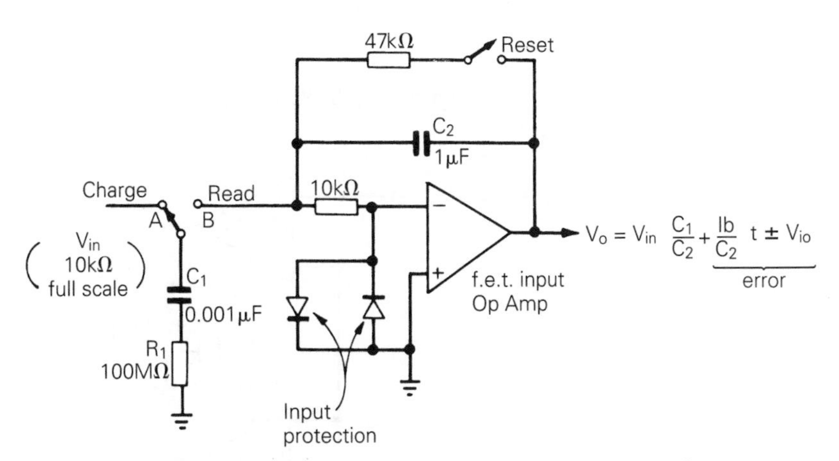

Figure A1.18 *Measurement of high DC voltage with low reading voltmeter*

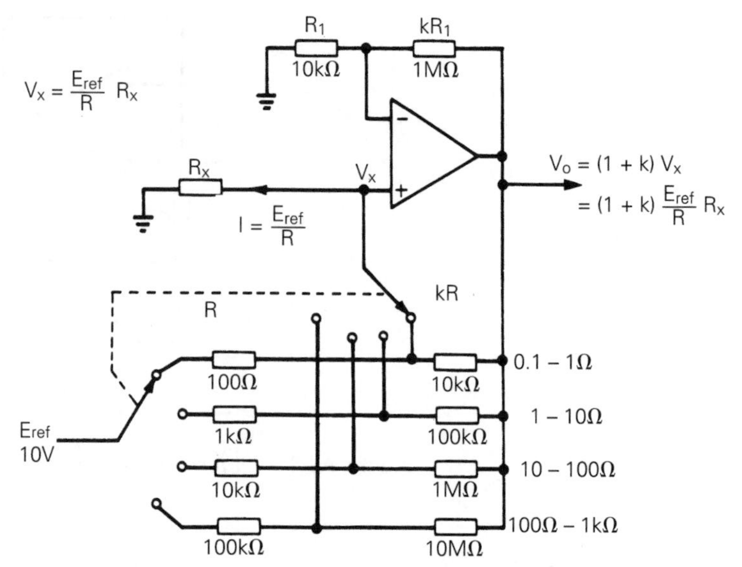

Figure A1.19 *Resistance measurement, earthed resistor*

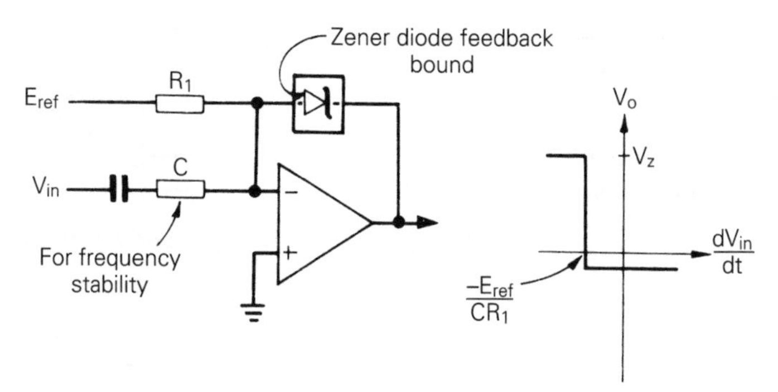

Figure A1.20 *Rate comparator*

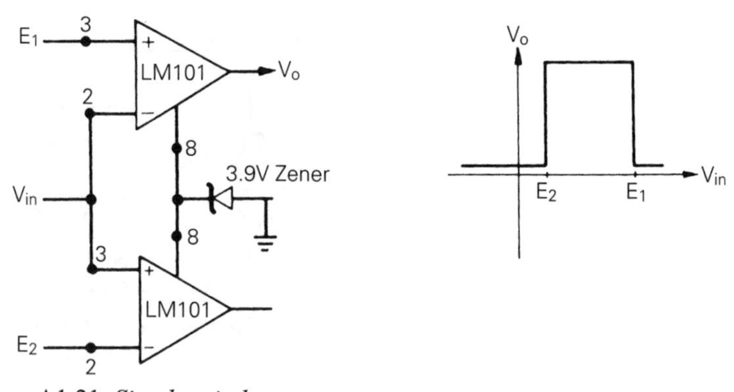

Figure A1.21 *Simple window comparator*

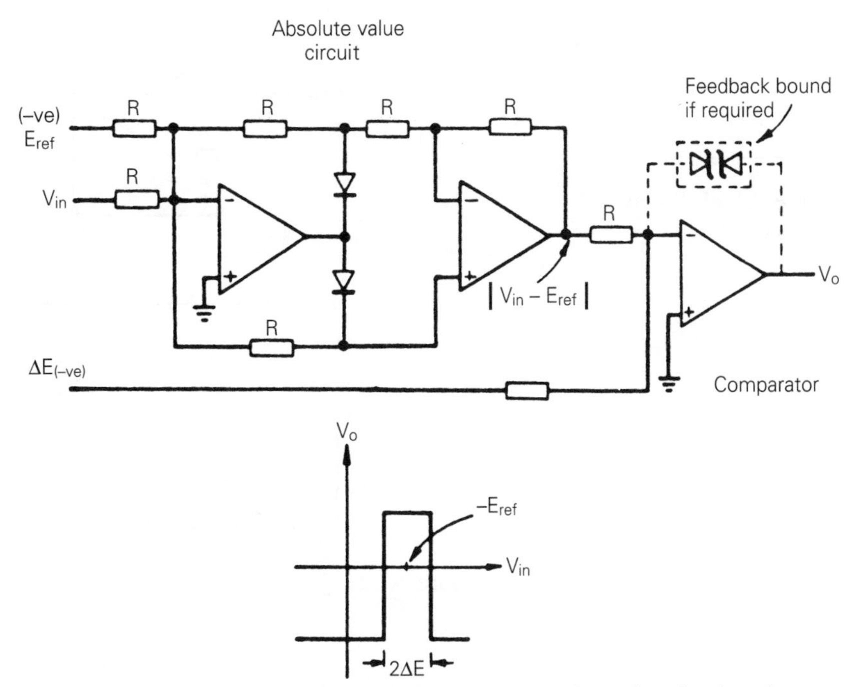

Figure A1.22 *Window comparator with control of window level and window width*

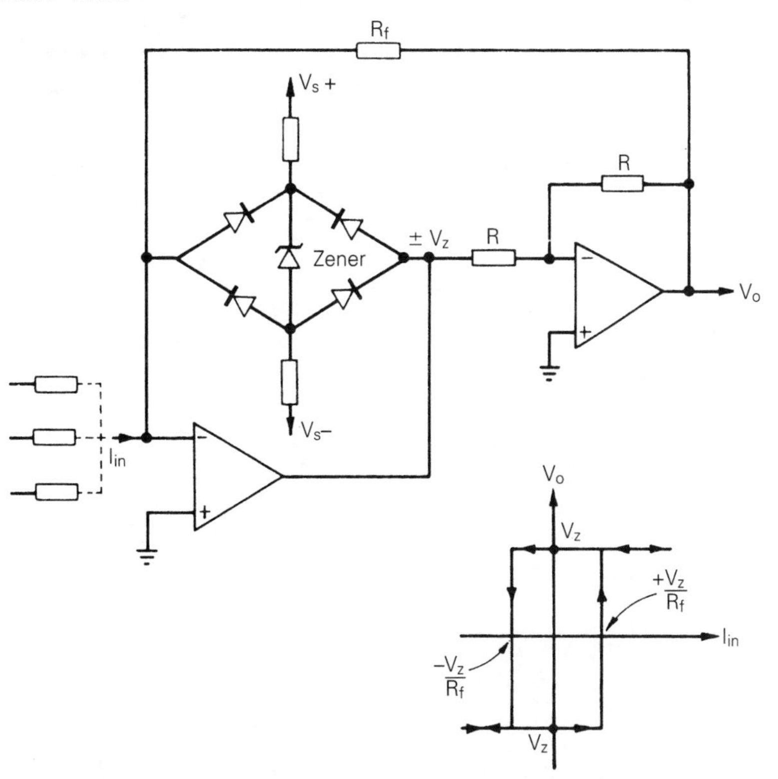

Figure A1.23 *Two-amplifier regenerative comparator with feedback bound and summing capability*

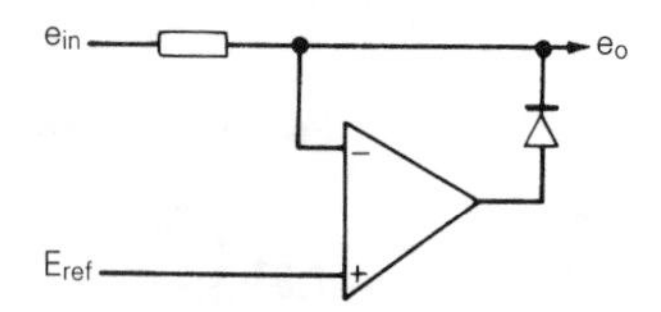

Figure A1.24 *Precise clipping circuit*

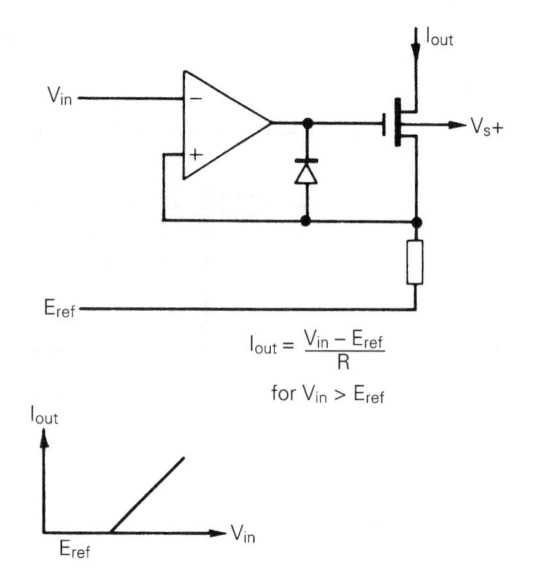

Figure A1.25 *Ideal diode with current output*

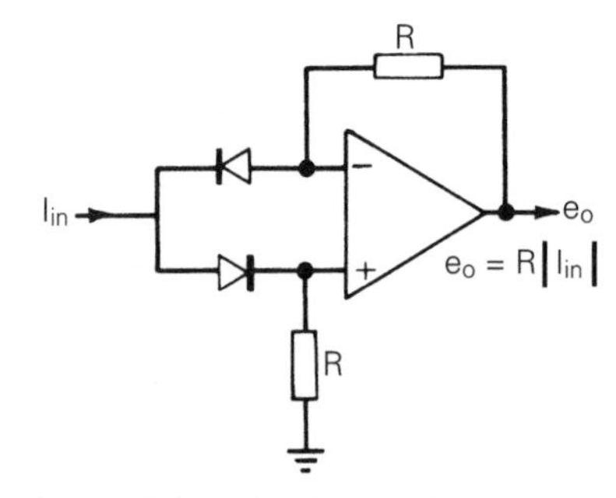

Figure A1.26 *Single amplifier absolute value circuit for current input*

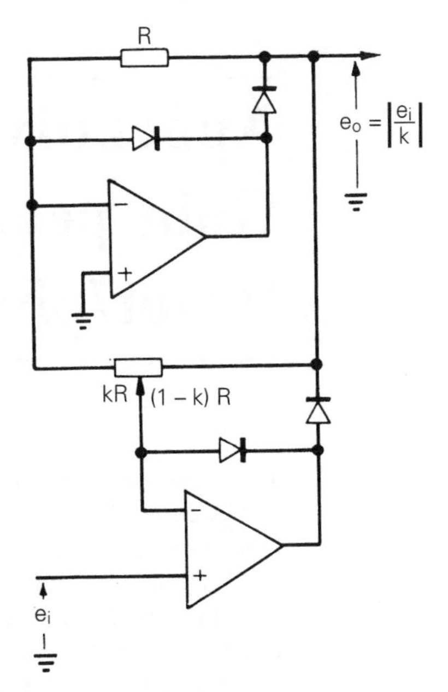

Figure A1.27 *High input impedance absolute value circuit with variable gain*

Appendix A2 Gain peaking/damping factor/phase margin

The closed-loop gain-peaking and lightly damping transient response exhibited by closed-loop configurations having an inadequate stability phase margin is in many cases due to the phase shift in the loop gain introduced by two break frequencies, one of which is remote (more than a decade away) from the other. Bode plots for commonly encountered situations are shown in Figure A2.1 and in both the cases considered the frequency dependence of the loop gain can be expressed by the relationship

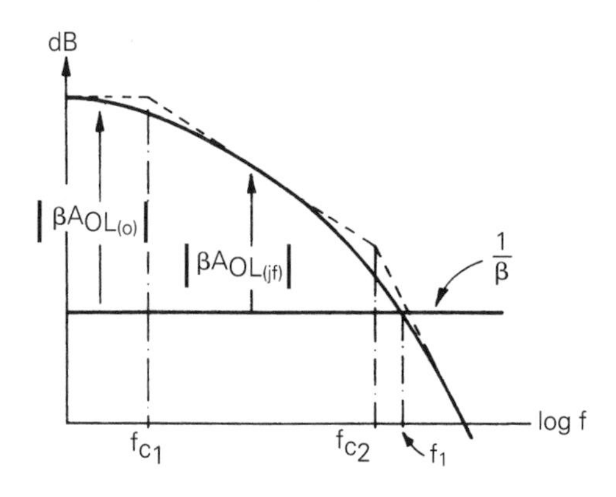

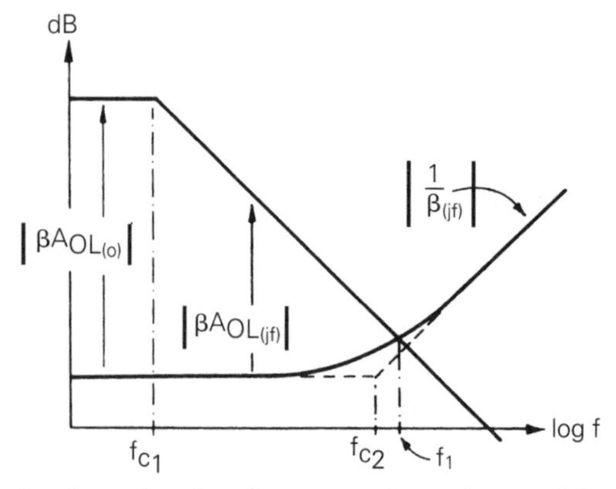

Figure A2.1 *Bode plots showing frequency dependence of loop gain*

$$\beta A_{OL(jf)} \cong -j\,|\beta A_{OL(o)}|\frac{f_{c1}}{f}\frac{1}{1 + j\dfrac{f}{f_{c_2}}} \qquad \text{(A2.1)}$$

$(f_{c_2} >> 10f_{c_1})$

The closed-loop signal gain of an op-amp feedback circuit can be expressed in the form (see Section 2.3.1)

$$A_{CL(f)} = A_{CL(o)}\left[\frac{1}{1 + \dfrac{1}{\beta A_{OL(jf)}}}\right]$$

$A_{CL(o)}$ is the ideal frequency independent closed-loop signal gain. Substitution for $\beta A_{OL(jf)}$ and rearrangement give

$$A_{CL(jf)} = \frac{A_{CL(o)}}{1 + j\dfrac{f}{|\beta A_{OL(o)}|f_{c_1}} - \dfrac{f^2}{|\beta A_{OL(o)}|f_{c_1}f_{c_2}}} \qquad \text{(A2.2)}$$

Equation A2.2 represents the closed-loop sinusoidal response.

The more general closed-loop transfer function is obtained in terms of the amplex variable s by the substitution, $s = jf$, $s^2 = -f^2$ giving

$$A_{CL(s)} = \frac{A_{CL(o)}}{1 + j\dfrac{s}{|\beta A_{OL(o)}|f_{c_1}} - \dfrac{s^2}{|\beta A_{OL(o)}|f_{c_1}f_{c_2}}} \qquad \text{(A2.3)}$$

Equation A2.3 represents a second order transfer function. Comparison with the general second order function

$$T_{(s)} = \frac{1}{1 + 2\dfrac{\zeta}{\omega_o}s + \dfrac{s^2}{\omega_o^2}}$$

gives the relationships between the damping factor ζ, natural frequency f_o and amplifier parameters as

$$\zeta = \frac{\sqrt{f_{c_2}}}{2\sqrt{(|\beta A_{OL(o)}|f_{c_1})}} \qquad \text{(A2.4)}$$

$$\text{and } f_o = \sqrt{(|\beta A_{OL(o)}|f_{c_1}f_{c_2})} \qquad \text{(A2.5)}$$

A2.1 Damping factor and phase margin

At the frequency f_1 at which the $1/\beta$ and the open-loop gain frequency plots intersect the magnitude of the loop gain is unity. Equation A2.1 gives the magnitude as

$$\begin{matrix}|\beta A_{(jf)}| \\ \text{at } f = f_1\end{matrix} = 1 = |\beta A_{OL(o)}| \frac{f_{c_1}}{f_1} \frac{1}{\sqrt{\left[1 + \left(\frac{f_1}{f_{c_2}}\right)^2\right]}} \tag{A2.6}$$

Combining equations A2.4 and A2.6 gives

$$\zeta = \frac{1}{2\sqrt{\left\{\frac{f_1}{f_{c_2}}\left[1 + \left(\frac{f_1}{f_{c_2}}\right)^2\right]^{1/2}\right\}}} \tag{A2.7}$$

Phase margin is related to f_1 and the break frequency f_{c2} by the relationships

$$\frac{f_1}{f_{c_2}} = \frac{1}{\tan\theta_m}; \left[1 + \left(\frac{f_1}{f_{c_2}}\right)^2\right]^{1/2} = \frac{1}{\sin\theta_m} \tag{A2.8}$$

$$(f_{c_2} >> f_{c_1})$$

Substitution in equation A2.7 gives the relationship between damping factor and phase margin as

$$\zeta = \frac{1}{2\sqrt{\left(\frac{\cos\theta_m}{\sin^2\theta_m}\right)}} \tag{A2.9}$$

Gain peaking

The magnitude of the closed-loop signal gain is, from equation A2.2

$$|A_{CL(jf)}| = \frac{A_{CL_{(o)}}}{\sqrt{\left\{\left[1 - \left(\frac{f}{f_o}\right)^2\right]^2 + \left(2\zeta\frac{f}{f_o}\right)^2\right\}}} \tag{A2.10}$$

where ζ and f_o are determined by equations A2.4 and A2.5.

The magnitude peaks for $\zeta < 1/\sqrt{2}$ and the frequency at which the gain peak occurs can be found by differentiating equation A2.10 with respect to f and equating to zero. This gives the frequency at which the gain peak occurs as

$$f_p = f_o\sqrt{(1 - 2\zeta^2)} \text{ (For } \zeta < 1/\sqrt{2}) \tag{A2.11}$$

Substituting this value of f_p in equation A2.10 gives

$$\begin{matrix}|A_{CL(jf)}| \\ \text{at peak}\end{matrix} = \frac{A_{CL(o)}}{2\zeta\sqrt{(1 - \zeta^2)}} \tag{A2.12}$$

The extent of the magnitude peaking may be expressed as

$$P_{(\text{dB of peaking})} = 20\log_{10}\frac{1}{2\zeta\sqrt{(1 - \zeta^2)}} \tag{A2.13}$$

The relationship between gain peaking and phase margin may be obtained by substituting the value of ζ from equation A2.9, thus

$$P_{(\text{dB of peaking})} = 20 \log_{10} \frac{\dfrac{2 \cos \theta m}{\sin^2 \theta m}}{\sqrt{\left(4 \dfrac{\cos \theta m}{\sin^2 \theta m} - 1\right)}} \tag{A2.14}$$

Appendix A3
Effect of resistor tolerance on CMRR of one amplifier differential circuit

In Figure A3.1 the amplifier is assumed ideal, resistors have tolerance $100 \times$ per cent per cent and worst case CMRR is considered. An input common mode signal e_{cm} gives rise to an output signal

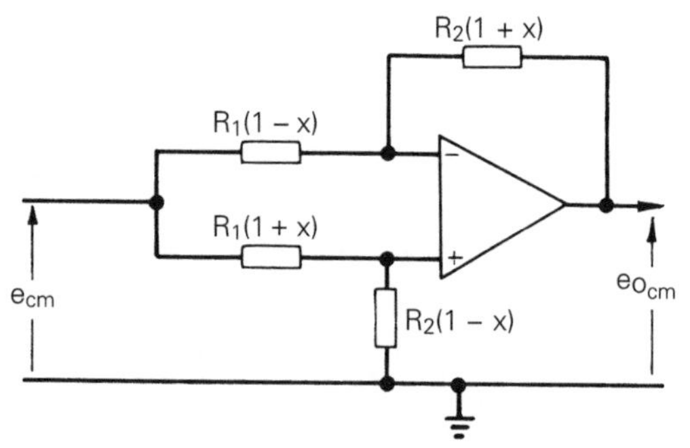

Figure A3.1 *CMRR due to resistor tolerance with worst case distribution*

$$e_{o_{cm}} = e_{cm}\left[\frac{R_2(1-x)}{R_2(1-x) + R_1(1+x)}\,\frac{R_1(1-x) + R_2(1+x)}{R_1(1-x)} - \frac{R_2(1+x)}{R_1(1-x)}\right]$$

$$= e_{cm}\frac{R_2}{R_1}\left[\frac{R_1(1-x) + R_2(1+x)}{R_2(1-x) + R_1(1+x)} - \frac{1+x}{1-x}\right]$$

$$= e_{cm}\frac{R_2}{R_1}\,\frac{R_1\,4x}{R_2(1-x)^2 + R_1(1+x^2)}$$

$$\cong e_{cm}\frac{R_2}{R_1}\,\frac{4x\,R_1}{R_2 + R_1}$$

Thus common mode gain

$$\frac{e_{o_{cm}}}{e_{cm}} = \frac{R_2}{R_1} \frac{4x\,R_1}{R_2 + R_1}$$

$$\text{and} \quad \text{CMRR} = \frac{\text{differential gain}}{\text{common mode gain}} \cong \frac{\dfrac{R_2}{R_1}}{\dfrac{R_2}{R_1}\dfrac{4x\,R_1}{R_2 + R_1}}$$

$$\text{CMRR} \cong \frac{1 + \dfrac{R_2}{R_1}}{4x} \tag{A3.1}$$

A3.1 CMRR of one amplifier differential circuit due to non-infinite CMRR of operational amplifier

Common mode signal applied to op-amp

$$= e_{cm} \frac{R_2}{R_1 + R_2}$$ (see Figure A3.2)

Non-infinite CMRR of an op-amp is represented by an equivalent input error signal $e_{\epsilon_{cm}}$ applied directly to the input terminal of the op-amp

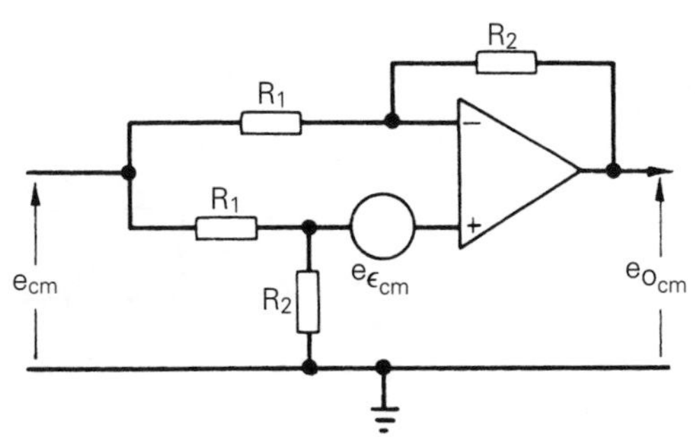

Figure A3.2 *CMRR of circuit due to non-infinite CMRR of amplifier*

$$e_{\epsilon_{cm}} = \frac{e_{cm}\dfrac{R_2}{R_1 + R_2}}{\text{CMRR}_{(A)}}$$

$e_{\epsilon cm}$ gives an output signal

$$e_{o_{cm}} = e_{\epsilon_{cm}}\left(1 + \frac{R_2}{R_1}\right)$$

Thus, common mode gain of circuit

$$= \frac{e_{o_{cm}}}{e_{cm}} = \frac{\dfrac{R_2}{R_1}}{\text{CMRR}_{(A)}}$$

and CMRR of the circuit = differential gain/common mode gain = $\text{CMRR}_{(A)}$.

A3.2 Overall CMRR due to resistor mismatch and non-infinite CMRR of operational amplifier

Effects of resistor tolerance and $\text{CMRR}_{(A)}$ are represented by separate input error generators (Figure A3.1). Output signal $e_{o_{cm}}$ due to input signal e_{cm} is

$$e_{o_{cm}} = e_{cm}\left[\frac{1}{\text{CMRR}_{(R)}} \pm \frac{1}{\text{CMRR}_{(A)}}\right]A_{\text{diff}}$$

$$\text{Overall CMRR} = \frac{\text{differential gain}}{\text{common mode gain}} = \frac{A_{\text{diff}}}{\left[\dfrac{1}{\text{CMRR}_{(R)}} \pm \dfrac{1}{\text{CMRR}_{(A)}}\right]A_{\text{diff}}}$$

$$= \frac{\text{CMRR}_{(R)} \times \text{CMRR}_{(A)}}{\text{CMRR}_{(A)} \pm \text{CMRR}_{(R)}} \qquad \text{(A3.2)}$$

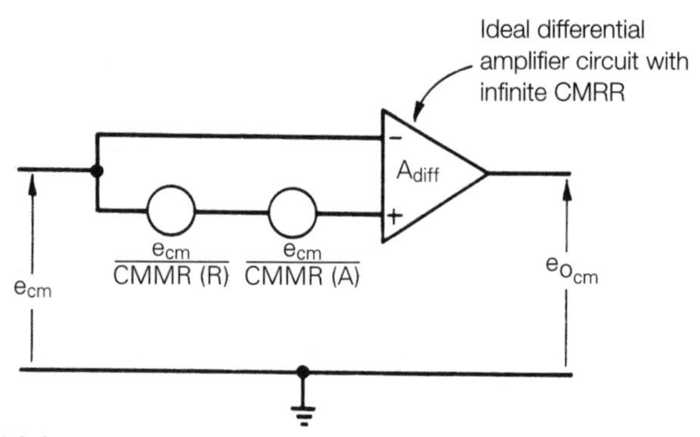

Figure A3.3

Appendix A4
Instrumentation transducers

A4.1 Introduction

Because the 'front end' of an instrumentation system frequently consists of one or more sensors, the purpose of this appendix is to give the reader a brief introduction to these devices which often provide the input signal to op-amps.

Most instrumentation systems comprise three basic sections: one for sensing the measurand, the next for conditioning the sensed signal and finally one for displaying or recording the conditioned signal. Figure A4.1 shows the block diagram of this arrangement.

The sensing element, known as a sensor or transducer, simply converts one form of energy into another. In this appendix all the transducers considered produce an electrical output when stimulated. However, the transduced electrical output may be of insufficient power and require amplification or other modification before it can be displayed or recorded. The necessary amplification, shaping, mixing or other such processing is undertaken in the signal conditioning section using the techniques variously described elsewhere in this book. Finally, the conditioned signal is recorded or displayed.

A simple example of the above basic system is that of a tank containing a hot liquid, the temperature of which needs to be monitored and recorded continually. The temperature sensing element could be a thermocouple (described later), the electrical output of which is conditioned by an op-amp to raise it to the necessary power level required to drive a chart recorder.

The remainder of this appendix will describe some of the more popular transducers used with instrumentation systems and will not consider signal conditioning or display and recording techniques.

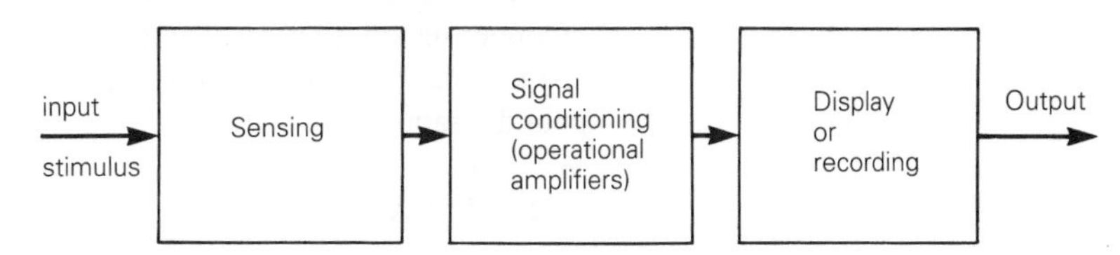

Figure A4.1 *Block diagram of basic instrumentation system*

A4.2 Resistance strain gauges

These are devices which when subjected to mechanical strain are deformed, within their elastic limit, and change their ohmic resistance. The strain gauge therefore is most suitable for detecting and measuring small mechanical displacements. The strain gauge is firmly secured to the test piece which when strained under load causes the attached strain gauge also to distort.

The strain gauge is usually part of an initially balanced resistive bridge network (see Chapter 8) and the accompanying change of gauge resistance causes an imbalance and an output signal from the bridge indicative of the amount of strain in the workpiece.

Figure A4.2 shows typical foil type strain gauges used for general engineering strain analysis.

The strain gauge measuring grid is manufactured from a copper nickel alloy which has a low and controllable temperature coefficient. The actual form of the metal grid, which changes its resistance under strain, is accurately produced by photo-etching techniques. A thermoplastic film is used to encapsulate the grid which helps to protect the gauge from mechanical and environmental damage. It also acts as a medium to transmit the strain from the test piece to the gauge material.

The principle of operation of this device is based on the fact that the resistance of an electrical conductor changes with a ratio of $\delta R/R$ if a stress is applied such that its length changes by a factor $\delta L/L$. This is where δR is the change in resistance from the unstressed value R and δL is the corresponding change in the unstressed length L.

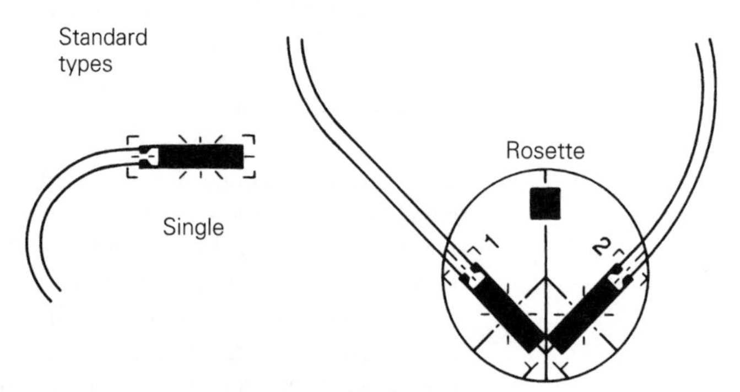

Figure A4.2 *Typical foil strain gauges*

The change in resistance is brought about mainly by the change in physical size of the conductor and, because of changes in its physical structure, an alteration in the conductivity in the material.

Copper nickel alloy is commonly used in the construction of strain gauges because the resistance change of the foil is virtually proportional to the applied strain, i.e.

$$\delta R/R = K \,.\, E$$

where K is a constant known as the gauge factor and E is the applied strain.

Therefore

$$\delta R/R = \text{gauge factor} \times \delta L/L$$

The change in resistance of the strain gauge can thus be utilized accurately to measure strain when connected to an appropriate measuring and indicating circuit.

Strain gauges are available commercially in a wide variety of preferred sizes and specifications. However, a typical specification for a small foil type strain gauge is as follows:

Measurable strain	2 to 4% maximum
Thermal output at 20–160°C	±2 μstrain/°C*
at 160–180°C	±5 μstrain/°C*
Gauge factor change with temperature	±0.015%/°C max.
Gauge resistance	120 Ω
Gauge resistance tolerance	±0.5%
Fatigue life	⩾ 10^5 reversals at 1000 μstrain*
Foil material	copper nickel alloy
Temperature range	−30°C to +80°C
Gauge length	8 mm
Gauge width	2 mm
Gauge factor	2.1
Base length (single types)	13.0 mm
Base width (single types)	4.0 mm
Base diameter (rosettes)	21.0 mm

* 1 μstrain is equivalent to an extension of 0.0001%.

While the strain gauge is basically a displacement type transducer, because strain is caused by force the gauge can readily be adapted to measure force, torque, weight, acceleration and many other quantities.

It should be noted that because the resistance of strain gauges is affected by changes in the temperature, they are often used in pairs; one in each of the balancing limbs of the measuring bridge circuits. Only one gauge is fixed to the test piece to act as the sensor, the other, being connected into the balancing limb of the bridge, is alongside but not fixed to the workpiece and is purely for temperature compensation purposes. Figure A4.3 illustrates this connection.

Also, because the strain gauge resistance is quite low, typically 120 Ω, the remote connection of a sensor gauge away from the instrumentation bridge circuit can cause problems. This is because the resistance of the long connecting leads may have significant resistance compared with that of the sensor gauge itself. The problem can be overcome using three connecting leads to the remote sensor gauge as shown in Figure A4.4. The extra lead is connected so as effectively to place two equal resistance connecting leads in series with each of the bridge limbs AC and DC without disturbing the electrical balance.

A4.3 Platinum resistance temperature detectors

Like strain gauges, these are passive transducers which change their resistance when stimulated. They are also known as resistance thermometers because they suffer a change of resistance with change of temperature. The change of resistance can be detected using similar bridge circuits and op-amp conditioning circuits as are used with strain gauges. The voltage output from these circuits is calibrated to indicate temperature.

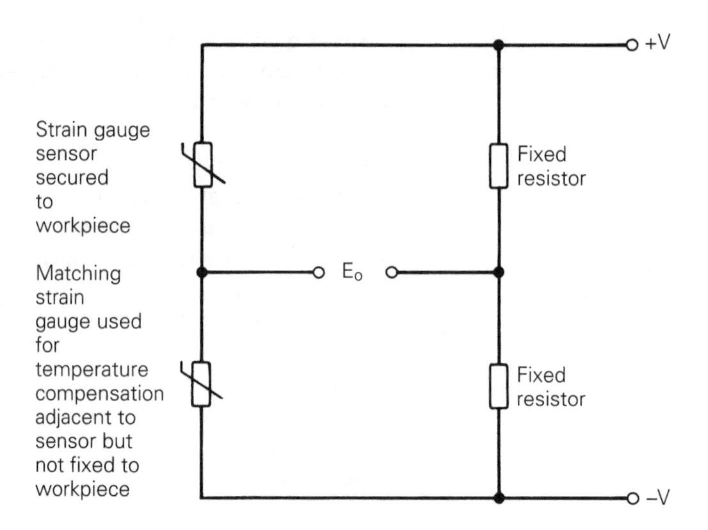

Figure A4.3 *Strain gauge bridge circuit with temperature compensation included*

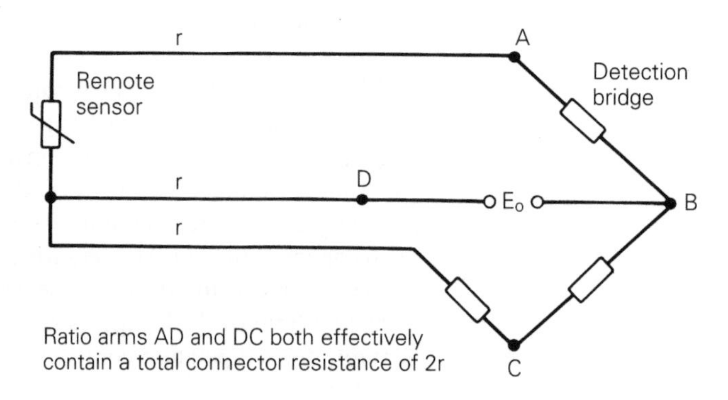

Figure A4.4 *Three-wire compensating lead connection*

The resistive element is made from platinum because not only does this metal exhibit a near linear variation of resistance with temperature change, it also shows a large (38.5 per cent) change of resistance for a 100°C temperature change. The platinum temperature sensing element is usually sheathed for protection and may be mounted in a variety of probes, some being hand-held.

The platinum thermometers can be very accurate and typically are manufactured to conform to BS 1904 Grade 2 and DIN 43 760. A typical specification is as follows:

Resistance at 0°C	100 ± 0.1 Ω
Temperature coefficient	0.385 Ω/°C
Maximum temperature	500°C
Minimum temperature	−50°C
Resistance tolerance at 0°C	±0.2 Ω (±0.3°C)
at 500°C	±0.8 Ω (±2.4°C)

Platinum resistance thermometers can be manufactured to give very accurate and long-term stable readings. They are often employed as laboratory temperature standards and where accurate temperature control is required. However, they can be fragile and have a slow response time of up to a second.

A4.4 Thermistors

These are devices which may have positive or negative temperature coefficients (ptc or ntc). They are manufactured from semiconductor materials and are often packaged as small discs, some 5–12 mm in diameter, or at the tip of a probe.

The principle of operation of the thermistor is that its resistance changes with temperature. However, unlike the platinum resistance thermometer, the resistance: temperature relationship of the thermistor is very non-linear and its upper working temperature is usually much lower. The thermistor is quite robust, especially in disc form, and its small size and sensitivity make it very suitable for the temperature control of ovens, deep freezers, rooms, process control, temperature compensation, high temperature protection, high current protection and the like. But, because of its non-linearity and fairly wide tolerances, care should be taken to check the calibration of circuits after a thermistor change has been made. A typical specification is as follows:

Resistance at 25°	10 kΩ
at +125°C	260 Ω
Maximum temperature range	−30°C to +125°C
Maximum dissipation	900 mW
Thermal time constant	30 s

A4.5 Pressure transducers

There are several designs available and the purpose of each is to convert a fluid pressure into an analogous electrical signal. Some designs have a diaphragm which is moved by the measurand pressure and this movement is translated into a change of resistance, inductance or capacitance. Other pressure transducers use the piezo-resistive effect. Advanced manufacturing techniques include laser trimmed bridge resistors for close tolerance on null and sensitivity. The sensing element is a 0.1 inch square silicon chip with integral sensing diaphragm and four piezo resistors. When pressure is applied to the diaphragm it is caused to flex, changing the resistance, which results in an output voltage proportional to pressure when a suitable excitation voltage is applied to the device. The sensing resistors are connected as a four-active element bridge for best linearity and sensitivity. A typical technical specification for one of these transducers are as follows:

Pressure range	0–30 psi
Full scale output	79 mV
Sensitivity/psi	2.63 mV
Excitation	10 V DC
Overpressure	60 psi max.

A4.6 Thermocouples

These are active transducers which convert the difference between two temperatures into a proportional voltage. The thermocouple principle is based upon the Seebeck effect which is simply the generation of a voltage by a heated junction of dissimilar metals. Figure A4.5 shows how a couple of two such metal junctions, at different temperatures, can be connected to cause a meter to indicate the voltage potential between the two junctions. One junction, the 'cold' or 'reference' junction, is usually held at 0°C (although room temperature suffices for some applications) while the other junction is used as a 'temperature sensor'.

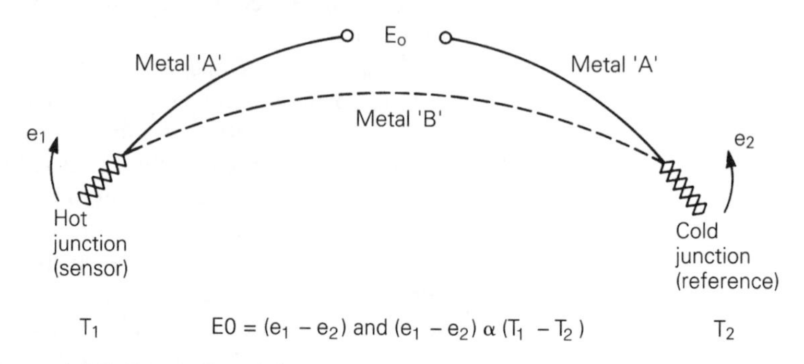

Figure A4.5 *Principle of thermocouple action*

The metals used include nickel, chromium, iron, platinum, rhodium, aluminium, constantan, manganese and silicon. Most thermocouple manufacturers use different pairs of these metals, or their alloys, to produce a selection of small, robust devices capable of measuring temperatures ranging from −230°C to +1300°C. To indicate their designed temperature ranges and other characteristics, thermocouples are usually classified as being Type J, K, N or T. However, because the thermocouple sensor junction is placed in physical contact with the measurand, for temperatures above +1300°C, which is higher than the freezing point of most metals, optical pyrometer temperature measurement becomes more appropriate.

The device usually comprises two metres of thermocouple wire insulated with varnish-impregnated glass fibre sleeving having an overall diameter of 1.5 mm. The hot junction tip is welded in an argon atmosphere to eliminate any oxidization of the junction. It has an operating range of −50°C to +400°C. Its small size and flexibility make it suitable for temperature measurement in confined places such as electronic assemblies.

A high quality hand-held probe can be used not only for general purpose temperature measurements in the range −100°C to +600°C by, say, immersion in liquids, but also because of its sharpened stainless steel probe tip and robust construction it can penetrate solids for internal temperature measurements. This makes it ideal for use in the food industry for checking the temperature of frozen foods, or for general measurement of below-surface temperatures of soil, grain, and powders.

A4.7 Linear variable differential transformers (LVDT)

A miniature AC energized LVDT is one of a range of the most common forms of displacement transducer. LVDTs typically comprise three coils wound in-line on an insulating hollow former inside which is a movable nickel iron core. The centre of the three coils is energized by an alternating current and, with the movable core in the centre position, induces equal emfs across the other two coils which effectively form the secondary windings of a transformer. Movement of the core disturbs the balance of the magnetic coupling between the three coils. By comparison of the now unbalanced secondary coil output voltages, the magnitude and direction of the core movement can be determined.

The displacement to be measured is applied to the movable core and because it has no direct sliding contact is virtually friction free. This gives the LVDT an advantage over the resistive potentiometric transducer which is sometimes used in similar applications. The LVDT can be used to detect movements in the range 0.5–25 mm.

A4.8 Capacitive transducers

These are passive displacement transducers which also require an AC excitation. A basic capacitor comprises a pair of parallel metal places between which is either a space or a solid dielectric material in which energy is stored when a voltage is applied between the plates. The capacitance, in farads, of a device is a measure of its ability to store this energy and depends upon the area of the plates, their spacing and the nature of the dielectric. The mechanical variation of any one of these three parameters will cause a sympathetic variation in the capacitance of the device. Since the spacing between the plates of a capacitor is usually less than 1.0 mm, a very detectable 10 per cent variation in capacitance requires a change in the plate spacing of less than 100 microns. This sensitivity makes the capacitive transducer one of the most suitable sensors for the measurement of small displacements.

A4.9 Tachometers

These are used for the measurement of shaft angular velocity and are available in two basic types:

Pulse tachometers. These have a toothed ferromagnetic disc which must be coupled to and rotated by the shaft, the speed of which is to be measured. The ferromagnetic disc may be manufactured with only a single tooth-like protrusion which is arranged to fall close to a pick-up head once in each revolution of the disc. The pick-up head comprises a permanent magnet around which a coil is wound. The passage of the tooth through the magnetic field causes a distorting movement of the field and this flux movement induces an emf across the pick-up coil. The number of pulses counted or the average DC produced in a given time from a train of these pulses is indicative of the shaft speed.

Tachogenerators. These are really no more than either DC machines which produce a direct voltage proportional to their angular velocity or AC alternators which have a direct relationship between their speed of rotation and their output frequency.

A4.10 Electromagnetic flowmeters

Figure A4.6 shows the typical layout of an electromagnetic flowmeter as used for measuring the rate of flow of a wide range of liquids. The basic principle involved is that of a current being induced into a conductor which is moving through a magnetic flux. In this case the moving conductor is the fluid itself and the magnetic flux is produced by external excitation. The fluid must have a resistance per cm^3 of less than 10 MΩ to permit the generation of a satisfactory signal (domestic tap water has a resistance per cm^3 of about 50 kΩ). The moving fluid is contained within a smooth-bore plastic pipe into which two pick-up electrodes are inserted and which collect the current generated, it being proportional to the rate of fluid flow.

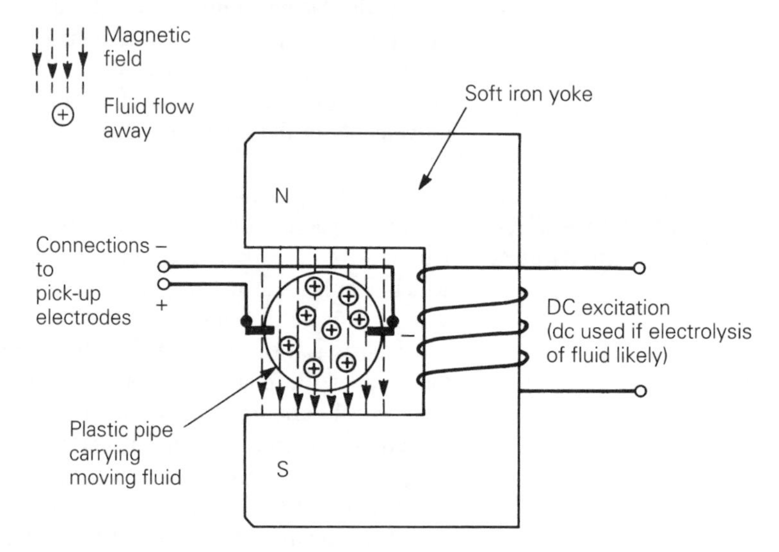

Figure A4.6 *Principle of electromagnetic flowmeter*

Electromagnetic flowmeters are available commercially with diameters ranging from 3 mm to 2000 mm. They have been used successfully to measure the flow of tap water, sea water, mercury, blood, chemicals and, because of the use of smooth-bored pipes, slurries and liquids containing solids.

A4.11 Hall effect transducers

Figure A4.7 shows how a current-carrying conductor situated in a perpendicular magnetic field experiences a transverse voltage which is proportional to the product of the current and the magnetic field flux density. The voltage so established is present in all conductors but is of particular significance in semiconductor materials. It can be shown that the current, I the flux density, B, and the voltage generated, E, are related by the expression:

$$E = -R_H(I \times B)$$

R_H is known as the Hall coefficient and is given by $1/ne$ where n is the number of charge carriers per unit volume which constitute the current and e is the charge on the carriers.

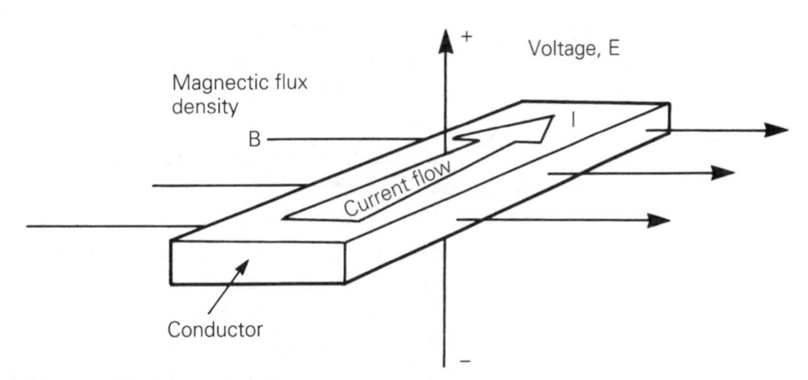

Figure A4.7 *The Hall effect*

The Hall effect voltage in semiconductors in the presence of a magnetic field is used to produce 'bounce-free' switching. A switched transistor is turned on by a Hall effect device under the influence of a magnetic field exceeding a designed 'operate' strength. However, the Hall effect device does not 'release' until the magnetic field strength is reduced to a level below the operate level. It is this hysteresis effect that produces the 'bounce-free' switching.

The Hall effect vane switch transducer makes use of bounce-free switching in its operation of transducing the presence of a ferrous metal into an electrical signal. A Hall effect sensor and magnet are housed in a pcb mounting package which will detect the presence of a ferrous metal vane passing through the gap between the sensor and the magnet. The device, which operates from 5 V direct (7 mA quiescent current), features two independent TTL compatible outputs capable of sinking up to 4 mA each or 8 mA combined. The switching time is less than 3 μs and the operating frequency can be up to 100 kHz. The device is useful in many position or counting operations, particularly in dusty or high ambient light environments, where an optical switch would be unsuitable.

The miniature linear Hall effect IC is a magnetic field sensor in a moulded 4-pin dil plastic package less than 8 mm square. This device features a differential output stage. One output increases linearly in voltage, while the other decreases, for a linear increase in magnetic flux density over a ±40 mT range. Typically, the output voltage varies linearly between 1.0 V and 3.0 V with a sensitivity of 1.0 mV/Gauss. Typical applications for this device include the investigation of magnetic fields in the vicinity of transformers and cables and as current sensors with high isolation and in linear feedback elements in analogue control systems. The sensor is immune from damage by high values of flux density.

A4.12 Opto transducers

Opto transducers are devices which change one or more of their electrical characteristics when struck by light. The light is not necessarily visible to the human eye; it may be infrared. Outlined below are brief details of a small selection of the many opto devices available commercially.

The *light dependent resistor* (LDR) uses a small strip of cadmium sulphide which may be illuminated by light passing through a clear window in the casing of the device. The resistance of this particular device can vary from about 500 Ω in bright sunlight to 1.0 Ω in darkness. It can be used at mains voltages (320 V DC or AC peak), it is cheap and sensitive but can have a slow response time of 120 ms. Typical applications for the LDR are the automatic control of public lighting, in intruder sensing devices and reflective smoke alarms.

All silicon diode junctions are affected by incident light and the *photodiode* is little more than a conventional silicon diode placed in a casing which is fitted with a window to allow the diode junction to be illuminated. The leakage current of the diode is very small but this increases when the junction is struck by light. The photodiode is operated in a reversed bias mode and in series with a load resistor through which the light dependent leakage current flows. The voltage drop across the load resistor is analogous to the intensity of the light striking the photodiode. Compared with the LDR, the photodiode is similarly packaged, will not operate with such high supply voltages, is not as sensitive to light stimulation but its response is much faster, being in the order of a few microseconds. Typical applications for photodiodes are in fast response AC circuits, in infrared beam switching and with photographic flash circuits.

The *phototransistor* operates in much the same way as the photodiode, the base-collector junction being effectively reverse biased and stimulated by light. However, the amplifying effect of the transistor makes the sensitivity of this device more than ten times that of the photodiode. But it cannot operate at such high frequencies; typically up to 200 kHz rather than the 500 MHz of the photodiode.

Optical shaft encoders are now available for sensing shaft position or angular velocity. Typically, these devices are 50 mm long and 50 mm wide and 50 mm in diameter and contain a light source beamed through a perforated rotating disc and detected by a light sensor. Rotation of the input shaft causes the energized encoder to produce an output comprising a number of TTL compatible pulses for each complete revolution. The device typically requires a DC excitation of 5–30 V and, depending upon the specification chosen, will provide resolutions of 100, 1250, 2000 or 2500 pulses per revolution. Typical applications are in machine tool control, robotics and position sensors for feedback on mechanical valve openings.

Appendix A5
Integrated circuit datasheets

19-3731; Rev 0; 10/92

Low-Noise Precision Operational Amplifiers

OP27/OP37

General Description

The OP27/OP37 precision operational amplifiers provide lower noise and higher speed with the same input offset and drift specifications as the OP07. Both parts have a 10µV offset, 0.2µV/°C drift, and 1.8 million gain. Coupled with a low-voltage noise of 3.5nV/√Hz at 10Hz and a low 1/f noise corner frequency of 2.7Hz, the OP27/OP37 are optimized for accurate amplification of low-level signals. The OP27 features an 8MHz gain-bandwidth product and a 2.8V/µs slew rate. For applications demanding higher speed, the OP37 has a 63MHz gain-bandwidth product, 17V/µs slew rate, and is stable at gains of five or more.

An output swing of ±10V into 600Ω together with low distortion make the OP27/OP37 ideal for professional audio applications.

For applications requiring greater precision or lower noise than the OP27 or OP37, see the MAX427/MAX437 and the MAX410/MAX412/MAX414 data sheets.

Applications

- Low-Noise DC Amplifiers
- Microphone Amplifiers
- Precision Amplifiers
- Tape-Head Preamplifiers
- Thermocouple Amplifiers
- Low-Level Signal Processing
- Medical Instrumentation
- Strain-Gauge Amplifiers
- High-Accuracy Data Acquisition

Features

- ♦ **10µV Input Offset Voltage**
- ♦ **0.2µV/°C Drift**
- ♦ **3nV/√Hz Input Noise Voltage (1kHz)**
- ♦ **80nV$_{p-p}$ Noise (0.1Hz to 10Hz)**
- ♦ **2.8V/µs Slew Rate (OP27)**
- ♦ **17V/µs Slew Rate (OP37)**
- ♦ **8MHz Gain-Bandwidth Product (OP27)**
- ♦ **63MHz Gain-Bandwidth Product (OP37)**

Ordering Information

PART	TEMP. RANGE	PIN-PACKAGE
OP27EP	0°C to +70°C	8 Plastic DIP
OP27FP	0°C to +70°C	8 Plastic DIP
OP27GP	-40°C to +85°C	8 Plastic DIP
OP27GS	-40°C to +85°C	8 SO
OP27EZ	-40°C to +85°C	8 CERDIP
OP27FZ	-40°C to +85°C	8 CERDIP
OP27GZ	-40°C to +85°C	8 CERDIP
OP27EJ	-40°C to +85°C	8 TO-99
OP27FJ	-40°C to +85°C	8 TO-99

Ordering Information continued on last page.

Typical Application Circuit

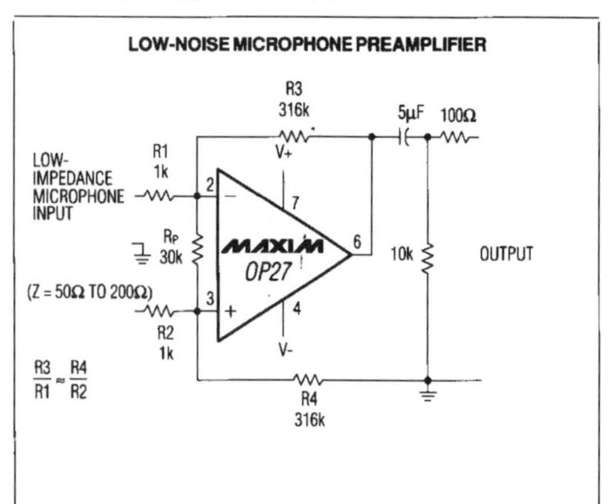

Pin Configurations

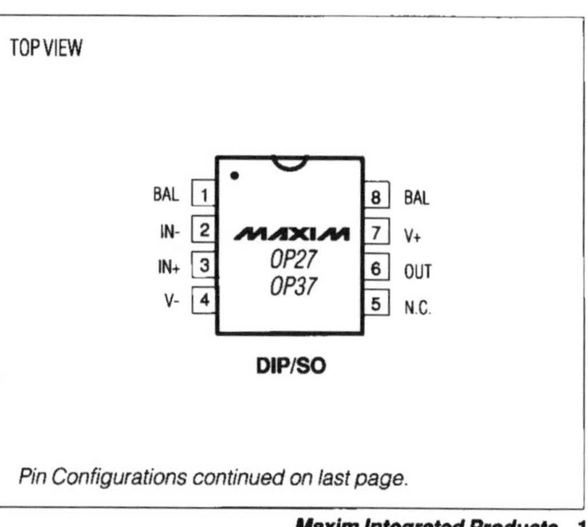

Pin Configurations continued on last page.

Low-Noise Precision Operational Amplifiers

OP27/OP37

ABSOLUTE MAXIMUM RATINGS

Supply Voltage .... ±22V
Input Voltage (Note 1) .... ±22V
Output Short-Circuit Duration .... Continuous
Differential Input Voltage (Note 2) .... ±0.7V
Differential Input Current (Note 2) .... ±25mA
Continuous Power Dissipation (T_A = +70°C)
  Plastic DIP (derate 9.09mW/°C above +70°C) 727mW
  SO (derate 5.88mW/°C above +70°C) 471mW
  CERDIP (derate 8.00mW/°C above +70°C) 640mW
  TO-99 (derate 6.67mW/°C above +70°C) 533mW

Operating Temperature Ranges:
  OP27/OP37EP/FP .... 0°C to +70°C
  OP27/OP37G_/EZ/EJ/FZ/FJ .... -40°C to +85°C
  OP27/OP37A_/B_/C_ .... -55°C to +125°C
Junction Temperature Range .... -65°C to +150°C
Storage Temperature Range .... -65°C to +150°C
Lead Temperature (soldering, 10 sec) .... +300°C

Note 1: For supply voltages less than ±22V, the absolute maximum input voltage is equal to the supply voltage.
Note 2: OP27/OP37 inputs are protected by back-to-back diodes. Current-limiting resistors are not used in order to achieve low noise. If differential input voltage exceeds ±0.7V, the input current should be limited to 25mA.

Stresses beyond those listed under "Absolute Maximum Ratings" may cause permanent damage to the device. These are stress ratings only, and functional operation of the device at these or any other conditions beyond those indicated in the operational sections of the specifications is not implied. Exposure to absolute maximum rating conditions for extended periods may affect device reliability.

ELECTRICAL CHARACTERISTICS

(V_S = ±15V, T_A = +25°C, unless otherwise noted.)

PARAMETER	SYMBOL	CONDITIONS	OP27A/E OP37A/E			OP27B/F OP37B/F			OP27C/G OP37C/G			UNITS
			MIN	TYP	MAX	MIN	TYP	MAX	MIN	TYP	MAX	
Input Offset Voltage (Note 3)	V_{OS}			10	25		20	60		30	100	µV
Long-Term V_{OS} Stability (Notes 4, 5)	V_{OS}/TIME			0.2	1.0		0.3	1.5		0.4	2.0	µV/Mo
Input Bias Current	I_B			±10	±40		±12	±55		±15	±80	nA
Input Offset Current	I_{OS}			7	35		9	50		12	75	nA
Input Voltage Range	I_{VR}		±11.0	±12.3		±11.0	±12.3		±11.0	±12.3		V
Input Resistance - Differential Mode (Note 6)	R_{IN}		1.3	6		0.94	5		0.7	4		MΩ
Input Resistance - Common Mode	R_{INCM}			3			2.5			2		GΩ
Input Noise Voltage (Notes 5, 7)	e_{np-p}	0.1Hz to 10Hz		0.08	0.18		0.08	0.18		0.09	0.25	$µV_{P-P}$
Input Noise-Voltage Density (Note 5)	e_n	f_O = 10Hz		3.5	5.5		3.5	5.5		3.8	8.0	nV/√Hz
		f_O = 30Hz		3.1	4.5		3.1	4.5		3.3	5.6	
		f_O = 1kHz		3.0	3.8		3.0	3.8		3.2	4.5	
Input Noise-Current Density (Notes 5, 8)	i_n	f_O = 10Hz		1.7	4.0		1.7	4.0		1.7		pA/√Hz
		f_O = 30Hz		1.0	2.3		1.0	2.3		1.0		
		f_O = 1kHz		0.4	0.6		0.4	0.6		0.4	0.6	
Large-Signal Voltage Gain	A_{VO}	R_L ≥ 2kΩ, V_O = ±10V	1000	1800		1000	1800		700	1500		V/mV
		R_L ≥ 1kΩ, V_O = ±10V	800	1500		800	1500		400	1500		
		R_L ≥ 600Ω, V_O = ±1V, V_S = ±4V (Note 5)	250	700		250	700		200	500		
Output Voltage Swing	V_O	R_L ≥ 2kΩ	±12.0	±13.8		±12.0	±13.8		±11.5	±13.5		V
		R_L ≥ 600Ω	±10.0	±11.5		±10.0	±11.5		±10.0	±11.5		

Low-Noise Precision Operational Amplifiers

ELECTRICAL CHARACTERISTICS (continued)

(V_S = ±15V, T_A = +25°C, unless otherwise noted.)

PARAMETER	SYMBOL	CONDITIONS	OP27A/E OP37A/E			OP27B/F OP37B/F			OP27C/G OP37C/G			UNITS
			MIN	TYP	MAX	MIN	TYP	MAX	MIN	TYP	MAX	
Open-Loop Output Resistance	R_O	V_O = 0, I_O = 0		70			70			70		Ω
Common-Mode Rejection Ratio	CMRR	V_{CM} = ±11V	114	126		106	123		100	120		dB
Power-Supply Rejection Ratio	PSRR	V_S = ±4V to ±18V		1	10		1	10		2	20	µV/V
Gain-Bandwidth Product (Note 5)	GBP	f_O = 100kHz, OP27	5.0	8.0		5.0	8.0		5.0	8.0		MHz
		f_O = 10kHz, A_{VCL} ≥ 5, OP37	45	63		45	63		45	63		
		f_O = 1MHz, A_{VCL} ≥ 5, OP37		40			40			40		
Slew Rate (Note 5)	SR	R_L ≥ 2kΩ, OP27	1.7	2.8		1.7	2.8		1.7	2.8		V/µs
		R_L ≥ 2kΩ, A_{VCL} ≥ 5, OP37	11	17		11	17		11	17		
Power Dissipation	PD	V_O = 0		90	140		90	140		100	170	mW
Offset Adjustment Range		R_P = 10kΩ		±4.0			±4.0			±4.0		mV

ELECTRICAL CHARACTERISTICS

(V_S = ±15V, T_A = T_{MIN} to T_{MAX}, unless otherwise noted.)

PARAMETER	SYMBOL	CONDITIONS	OP27A OP37A			OP27B OP37B			OP27C OP37C			UNITS
			MIN	TYP	MAX	MIN	TYP	MAX	MIN	TYP	MAX	
Input Offset Voltage (Note 3)	V_{OS}			30	60		50	200		70	300	µV
Average-Offset Voltage Drift (Note 9)	TCV_{OS}			0.2	0.6		0.3	1.3		0.4	1.8	µV/°C
Input Bias Current	I_B			±20	±60		±28	±95		±35	±150	nA
Input Offset Current	I_{OS}			10	50		14	85		20	135	nA
Input Voltage Range	I_{VR}		±10.3	±11.5		±10.3	±11.5		±10.2	±11.5		V
Large-Signal Voltage Gain	A_{VO}	R_L ≥ 2kΩ, V_O = ±10V	600	1200		500	1000		300	800		V/mV
Maximum Output-Voltage Swing	V_O	R_L ≥ 2kΩ	±11.5	±13.5		±11.0	±13.2		±10.5	±13.0		V
Common-Mode Rejection Ratio	CMRR	V_{CM} = ±10V	108	122		100	119		94	116		dB
Power-Supply Rejection Ratio	PSRR	V_S = ±4.5V to ±18V		2	16		2	20		4	51	µV/V

Low-Noise Precision Operational Amplifiers

OP27/OP37

ELECTRICAL CHARACTERISTICS (continued)

(V_S = ±15V, T_A = T_{MIN} to T_{MAX}, unless otherwise noted.)

PARAMETER	SYMBOL	CONDITIONS	OP27E OP37E MIN	TYP	MAX	OP27F OP37F MIN	TYP	MAX	OP27G OP37G MIN	TYP	MAX	UNITS
Input Offset Voltage (Note 3)	V_{OS}			20	50		40	140		55	220	µV
Average Offset-Voltage Drift (Note 9)	TCV_{OS}			0.2	0.6		0.3	1.3		0.4	1.8	µV/°C
Input Bias Current	I_B			±14	±60		±18	±95		±25	±150	nA
Input Offset Current	I_{OS}			10	50		14	85		20	135	nA
Input Voltage Range	I_{VR}		±10.5	±11.8		±10.5	±11.8		±10.5	±11.8		V
Large-Signal Voltage Gain	A_{VO}	R_L ≥ 2kΩ, V_O = ±10V	750	1500		700	1300		450	1000		V/mV
Output Voltage Swing	V_O	R_L ≥ 2kΩ	±11.7	±13.6		±11.4	±13.5		±11.0	±13.3		V
Common-Mode Rejection Ratio	CMRR	V_{CM} = ±10V	110	124		102	121		96	118		dB
Power-Supply Rejection Ratio	PSRR	V_S = ±4.5V to ±18V		2	15		2	16		2	32	µV/V

Note 3: V_{OS} is measured approximately 0.5 seconds after application of power.
Note 4: Long-term input offset voltage stability refers to the average trend line of V_{OS} vs. Time over extended periods after the first 30 days of operation.
Note 5: Guaranteed by design.
Note 6: Guaranteed by input bias current.
Note 7: See test circuit and frequency response curve for 0.1Hz to 10Hz tester (Figures 1, 6).
Note 8: See test circuit for current-noise measurement (Figure 2).
Note 9: The TCV_{OS} performance is within the specifications unnulled or when nulled with R_P = 8kΩ to 20kΩ. TCV_{OS} is sample tested to 0.1% AQL for A/E grades. B/C/F/G are guaranteed by design.

Typical Operating Characteristics

(T_A = +25°C, unless otherwise noted)

VOLTAGE-NOISE DENSITY vs. FREQUENCY

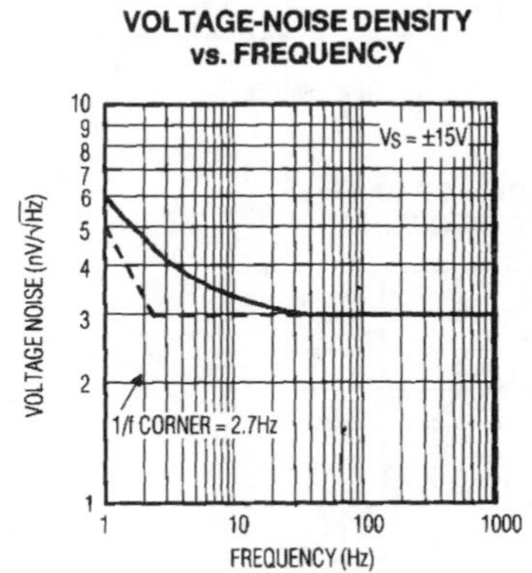

INPUT WIDEBAND VOLTAGE NOISE vs. BANDWIDTH (0.1Hz TO FREQUENCY INDICATED)

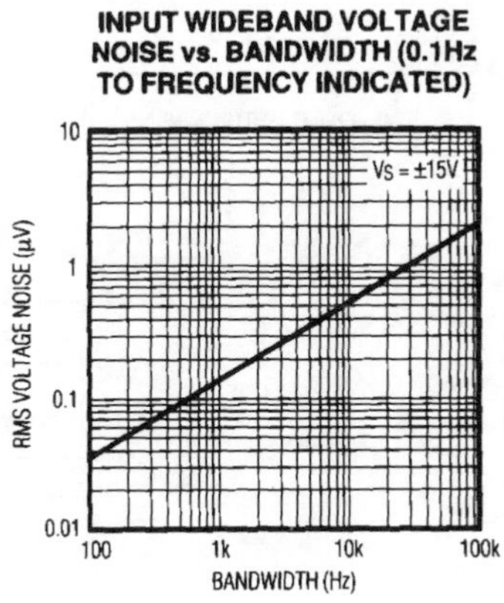

TOTAL NOISE vs. SOURCE RESISTANCE

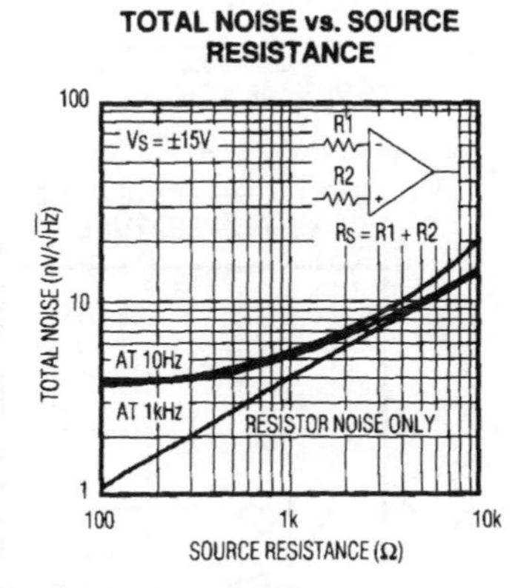

Low-Noise Precision Operational Amplifiers

Typical Operating Characteristics (continued)

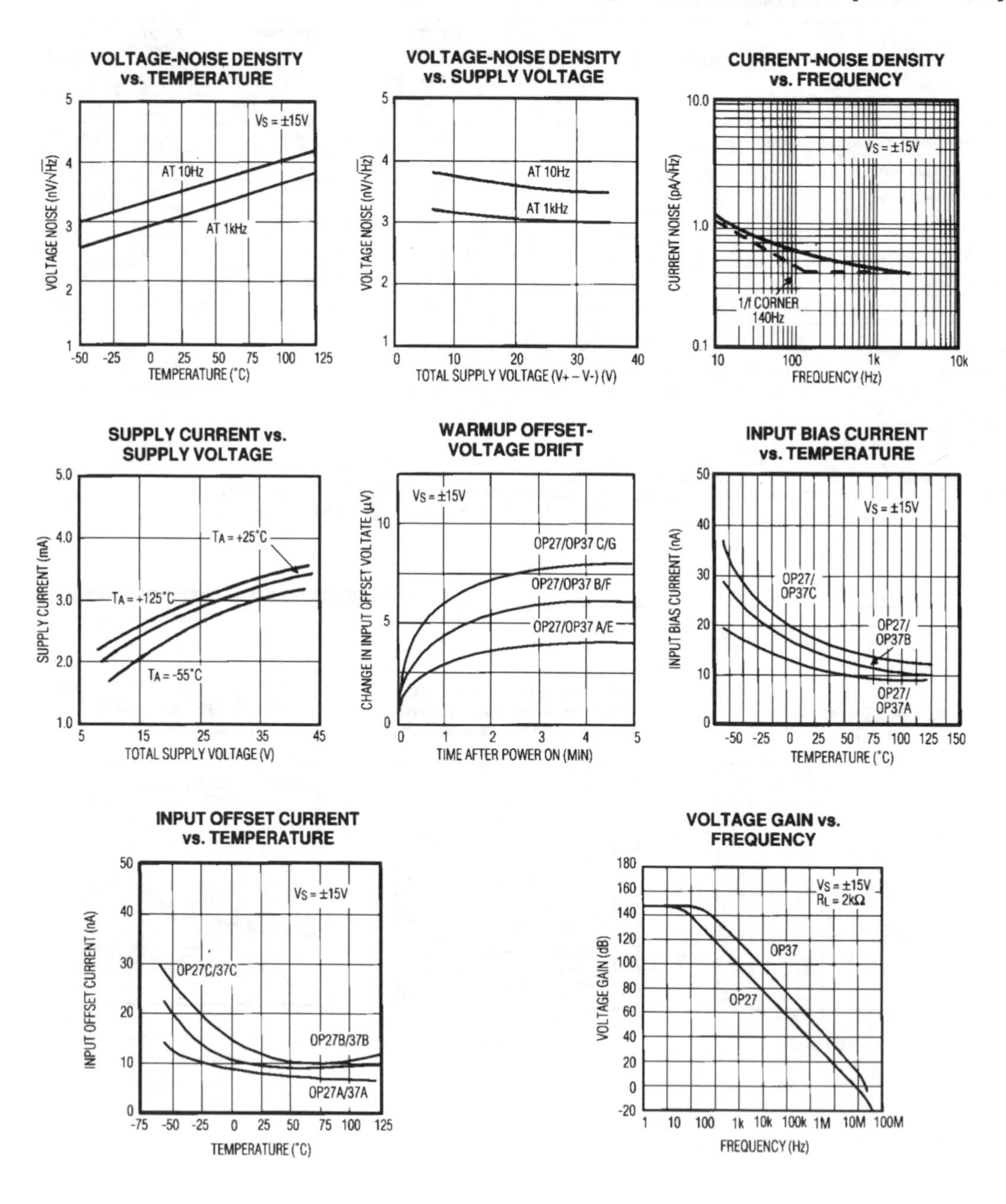

Low-Noise Precision Operational Amplifiers

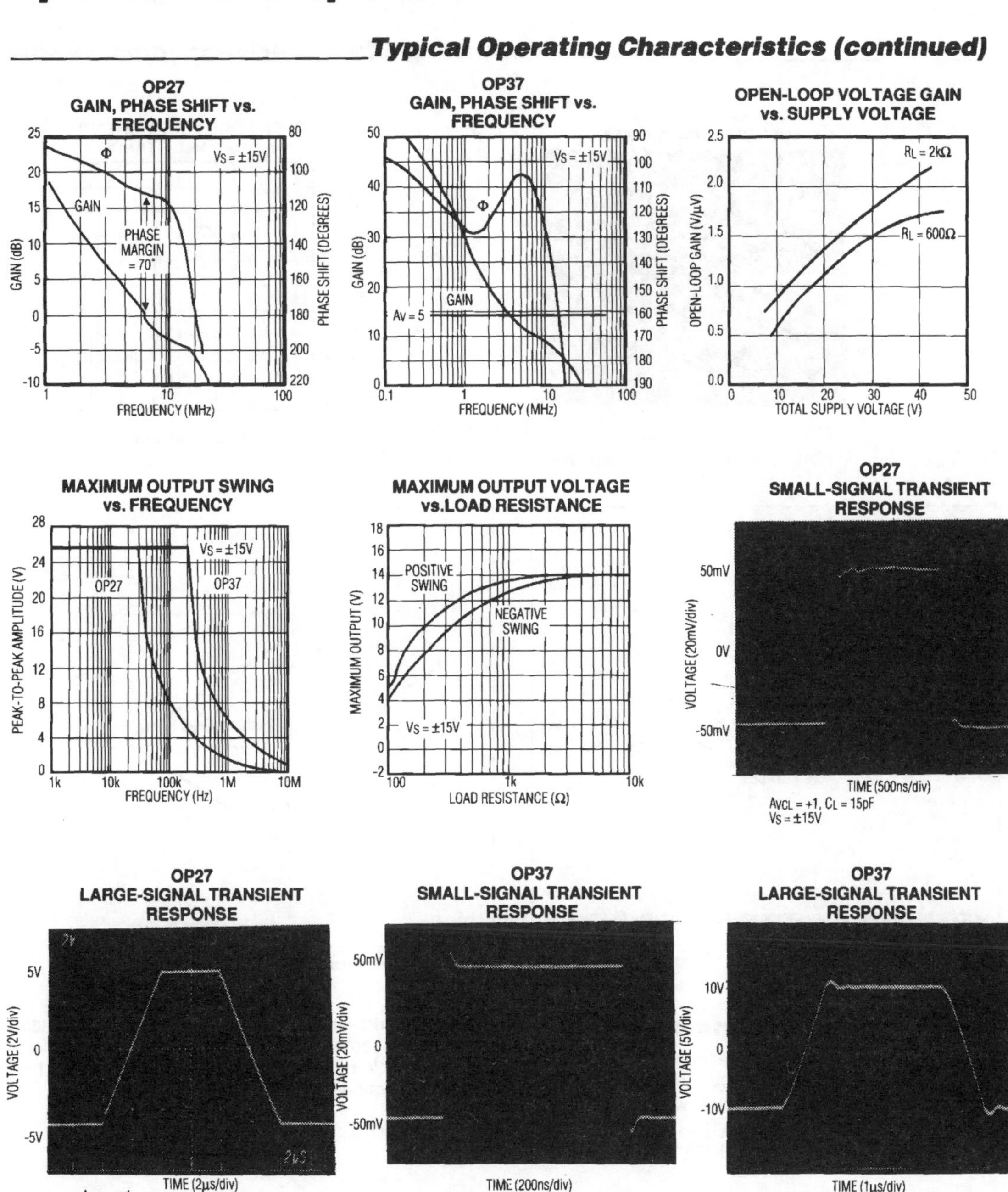

Low-Noise Precision Operational Amplifiers

Typical Operating Characteristics (continued)

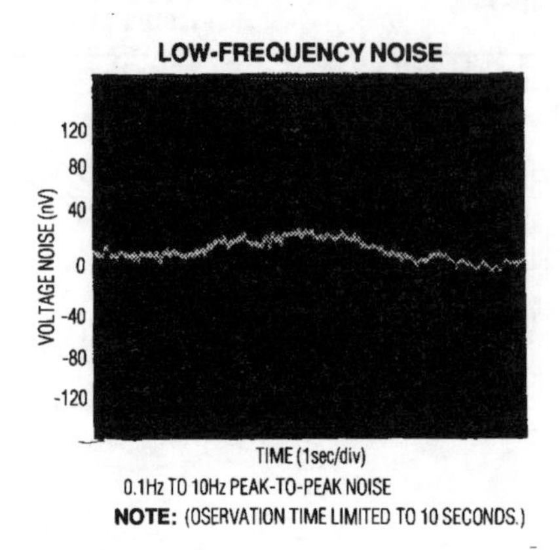

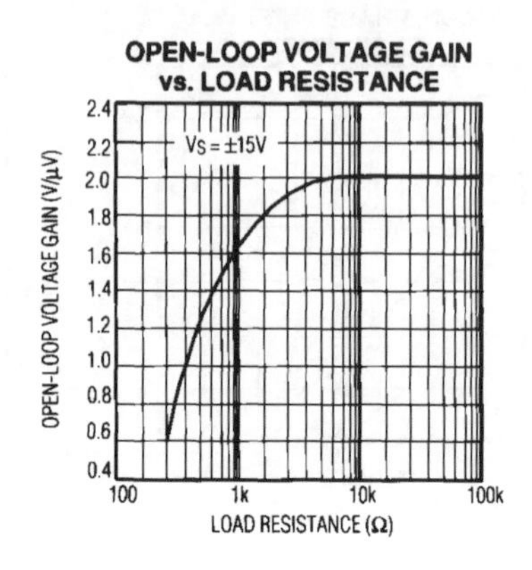

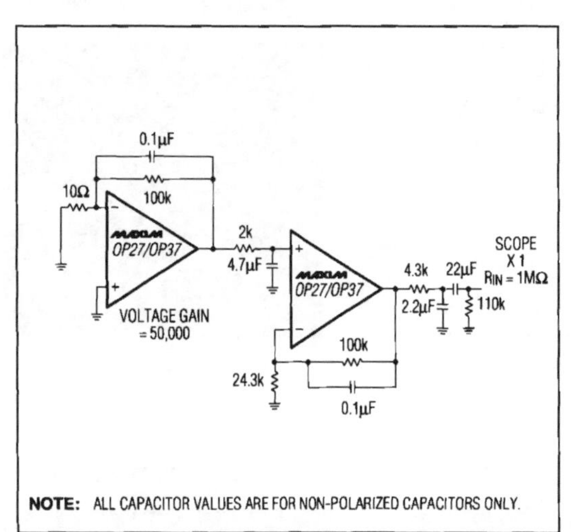

Figure 1. Voltage-Noise Test Circuit (0.1Hz to 10Hz)

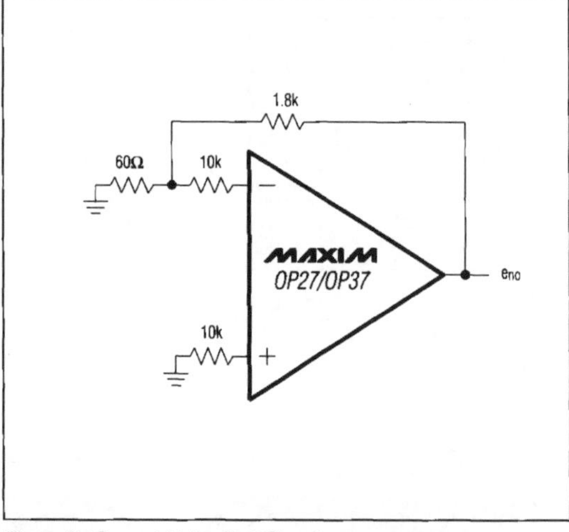

Figure 2. Current-Noise Test Circuit

Applications Information

The OP27/OP37 provide stable operation with load capacitances of up to 2nF and ±10V output swings; larger capacitances should be decoupled with a 50Ω series resistor inside the feedback loop. The OP27 is unity-gain stable and the OP37 is stable at gains of five or greater.

Thermoelectric voltages generated by dissimilar metals at the input terminals degrade the drift performance. Connections to both inputs should be maintained at the same temperature for best operation.

Low-Noise Precision Operational Amplifiers

OP27/OP37

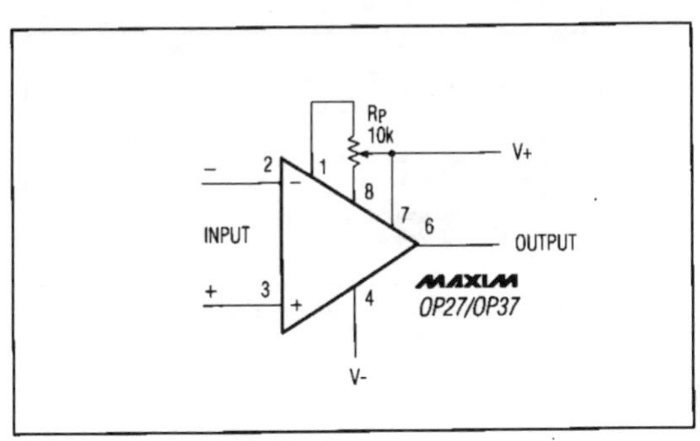

Figure 3. Offset Nulling Circuit

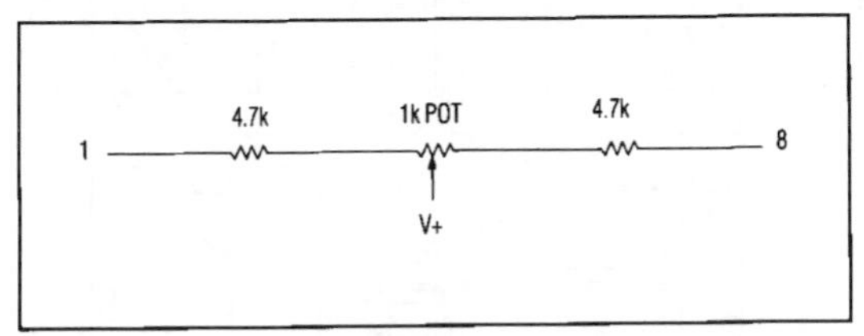

Figure 4. Alternate Offset-Voltage Adjustment

Offset-Voltage Adjustment

Input offset voltage (VOS) is trimmed at the wafer level. If VOS adjustment is necessary, a 10kΩ trim potentiometer (pot) may be used and will not degrade TCVOS (Figure 3). Other trim pot values from 1kΩ to 1MΩ can be used with a slight degradation (0.1µV/°C to 0.2µV/°C) of TCVOS. Adjusting, but not zeroing, VOS creates a drift of approximately (VOS/300)µV/°C. For example, the change in TCVOS is 0.33µV/°C if VOS is adjusted to 100µV. The adjustment range with a 10kΩ trim pot is ±4mV. For a smaller range, reduce nulling sensitivity by connecting a smaller pot in series with fixed resistors; for example, Figure 4 has a ±280µV adjustment range.

Noise Measurements

To measure the 80nVp-p noise specification of the OP27/OP37 in the 0.1Hz to 10Hz range, observe the following precautions:

1. The device must warm up for at least five minutes. Figure 5 shows how VOS typically increases 4µV with increases in chip temperature after power-up. In the 10sec measurement interval, temperature-induced effects can exceed 10nV.

2. For similar reasons, the device must be well-shielded from air currents, including those caused by motion. This minimizes thermocouple effects.

3. As shown in Figure 6, the 0.1Hz corner is defined by only one zero. A maximum test time of 10sec acts as an additional zero to eliminate noise contributions from the frequency band below 0.1Hz.

4. A noise-voltage-density test is recommended when measuring noise on a large number of units. A 10Hz noise-voltage-density measurement correlates well with a 0.1Hz to 10Hz peak-to-peak noise reading, since both results are determined by the white noise and the location of the 1/f corner frequency.

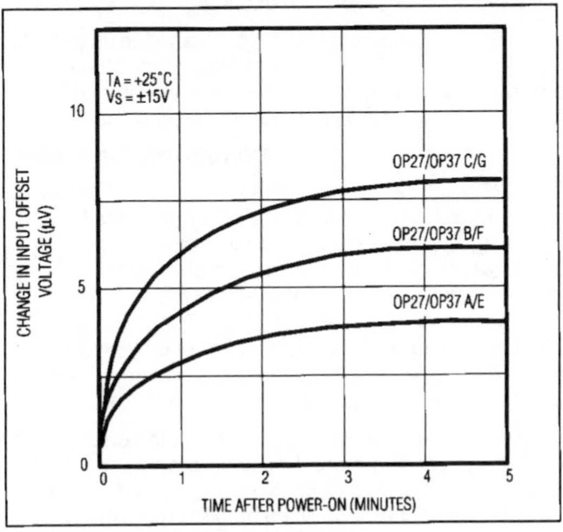

Figure 5. Warm-Up Offset Voltage Drift

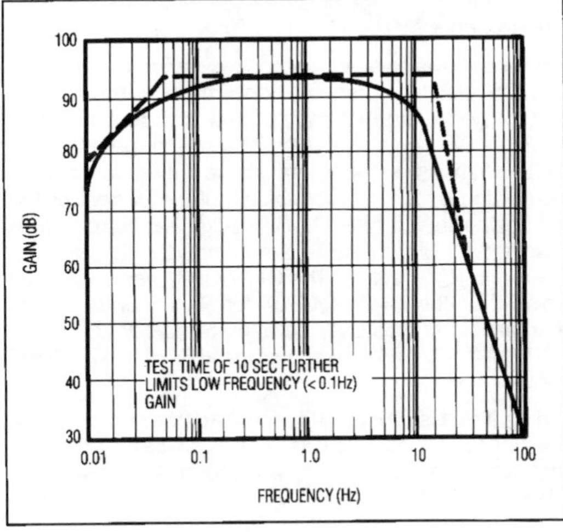

Figure 6. 0.1Hz to 10Hz V_{p-p} Noise Tester Frequency Response

Low-Noise Precision Operational Amplifiers

Unity-Gain Buffer Applications (OP27 Only)

Figure 7 shows the circuit and output waveform with R1 ≤ 100Ω, and the input driven with a fast, large signal pulse (>1V).

During the fast rise portion of the output, the input protection diodes short the output to the input, and a current, limited only by the output short-circuit protection, is drawn by the signal generator. With $R_f \geq 500\Omega$, the output is capable of handling the current required ($I_L \leq 20mA$ at 10V) and a smooth transition occurs.

When $R_f \geq 2k\Omega$, a pole created with R_f and the amplifier's input capacitance (8pF) causes additional phase shift and reduces phase margin. A small capacitor (20pF to 50pF) in parallel with R_f eliminates this problem.

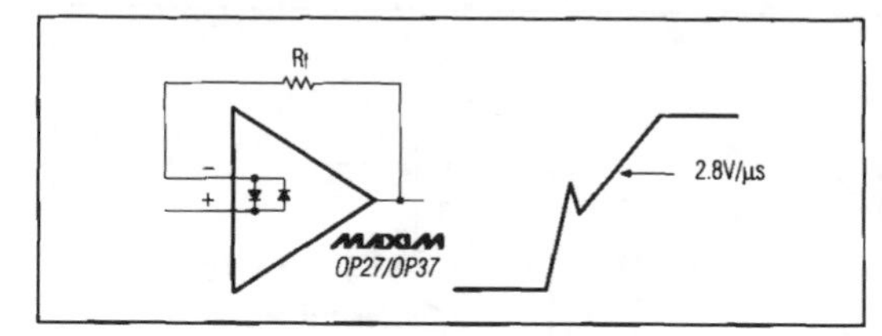

Figure 7. Pulsed Operation of Unity-Gain Buffer

Comments on Noise

The OP27/OP37 are very low-noise amplifiers. They have outstanding input voltage noise characteristics by operating the input stage at a high quiescent current. Input bias and offset currents, which would normally increase with the quiescent current, are minimized by bias-current cancellation circuitry. The OP27/OP37A and E grade devices have I_B and I_{OS} of only ±40nA and 35nA respectively at +25°C. This is particularly important with high source-resistances.

Voltage noise is inversely proportional to the square-root of bias current, but current noise is proportional to the square-root of bias current. The OP27/OP37 low-noise advantages are reduced when high source resistors are used.

Total noise = $[(\text{voltage noise})^2 + (\text{current noise} \times R_S)^2 + (\text{resistor noise})^2]^{1/2}$

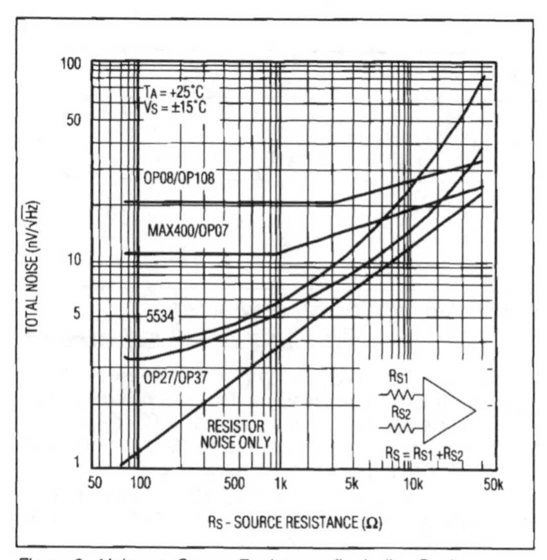

Figure 8. Noise vs. Source Resistance (Including Resistor Noise) at 1kHz

Figure 8 shows noise vs. source resistance at 1kHz. To use this plot for wideband noise, multiply the vertical scale by the square-root of the bandwidth. The OP27/OP37 maintains low input noise voltage with $R_S < 1k\Omega$. With $R_S > 1k\Omega$, total noise increases and is dominated by the resistor noise, not the current or the voltage noise. It is only with $R_S \geq 20k\Omega$ that current noise dominates. Current noise is not important for applications with $R_S < 20k\Omega$. The OP27/OP37 has lower total noise than the MAX400/OP07 for $R_S < 10k\Omega$. As R_S increases, the crossover between the OP27/OP37 and the MAX400/OP07 noise occurs in the R_S = 15kΩ to 40kΩ region.

Figure 9 shows 0.1Hz to 10Hz peak-to-peak noise. Here, resistor noise is negligible and current noise (i_n) becomes important, because $i_n \propto 1/\sqrt{f}$. The crossover with the MAX400/OP07 occurs in the R_S = 3kΩ to 5kΩ range, depending on whether balanced or unbalanced source resistors are used (at 3kΩ the I_B and I_{OS} error can be three times the V_{OS} specification). For low-frequency applications, the MAX400/OP07 is better than the OP27/OP37 when R_S > 3kΩ, except when gain error is important. Figure 10 illustrates the 10Hz noise. As expected, the results fall between those of the previous two figures.

For reference, typical source resistances of some signal sources are listed in Table 1.

OP27/OP37

Low-Noise Precision Operational Amplifiers

Table 1. Signal Source vs. Source Impedance

DEVICE	SOURCE IMPEDANCE	COMMENTS
Strain Gauge	<500Ω	Typically used in low-frequency applications.
Magnetic Tapehead	< 1500Ω	Low I_B is very important to reduce self-magnetization problems when direct coupling is used. OP27 I_B can be neglected.
Linear Variable Differential Transformer	< 1500Ω	Used in rugged servo-feedback applications. Bandwidth of interest is 400Hz to 5kHz.

Table 2. Open-Loop Gain vs. Frequency

OPEN-LOOP GAIN			
FREQUENCY AT:	OP07	OP27	OP37
3Hz	100dB	124dB	125dB
10Hz	100dB	120dB	125dB
30Hz	90dB	110dB	124dB

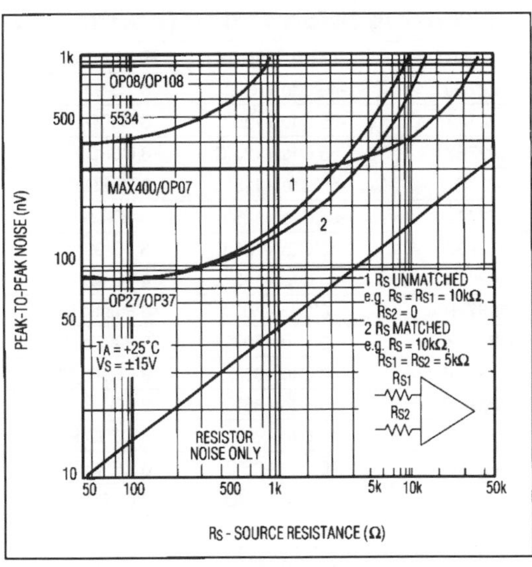

Figure 9. Peak-to-Peak Noise (0.1 to 10Hz) vs. Source Resistance (Includes Resistor Noise)

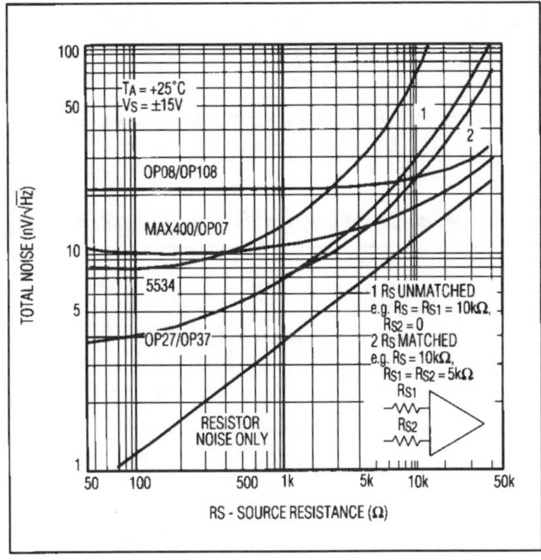

Figure 10. 10Hz Noise vs. Source Resistance (Includes Resistor Noise)

Low-Noise Precision Operational Amplifiers

OP27/OP37

Ordering Information (continued)

PART	TEMP. RANGE	PIN-PACKAGE
OP27GJ	-40°C to +85°C	8 TO-99
OP27AZ	-55°C to +125°C	8 CERDIP*
OP27BZ	-55°C to +125°C	8 CERDIP*
OP27CZ	-55°C to +125°C	8 CERDIP*
OP27AJ	-55°C to +125°C	8 TO-99*
OP27BJ	-55°C to +125°C	8 TO-99*
OP27CJ	-55°C to +125°C	8 TO-99*
OP37EP	0°C to +70°C	8 Plastic DIP
OP37FP	0°C to +70°C	8 Plastic DIP
OP37GP	-40°C to +85°C	8 Plastic DIP
OP37GS	-40°C to +85°C	8 SO
OP37EZ	-40°C to +85°C	8 CERDIP
OP37FZ	-40°C to +85°C	8 CERDIP
OP37GZ	-40°C to +85°C	8 CERDIP
OP37EJ	-40°C to +85°C	8 TO-99
OP37FJ	-40°C to +85°C	8 TO-99
OP37GJ	-40°C to +85°C	8 TO-99
OP37AZ	-55°C to +125°C	8 CERDIP*
OP37BZ	-55°C to +125°C	8 CERDIP*
OP37CZ	-55°C to +125°C	8 CERDIP*
OP37AJ	-55°C to +125°C	8 TO-99*
OP37BJ	-55°C to +125°C	8 TO-99*
OP37CJ	-55°C to +125°C	8 TO-99*

**Contact factory for availability and processing to MIL-STD-883.*

Pin Configurations (continued)

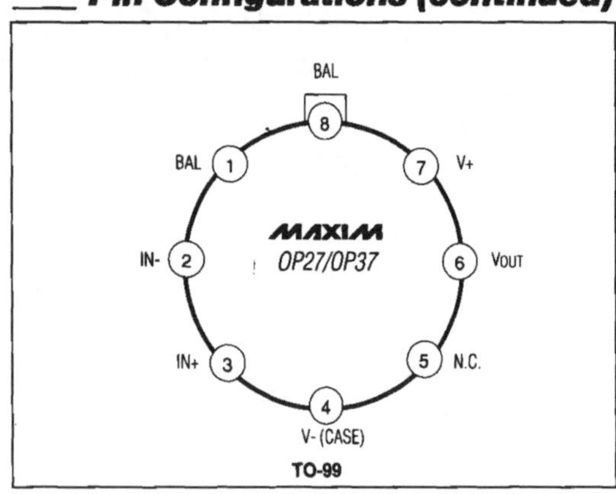

TO-99

Chip Topography

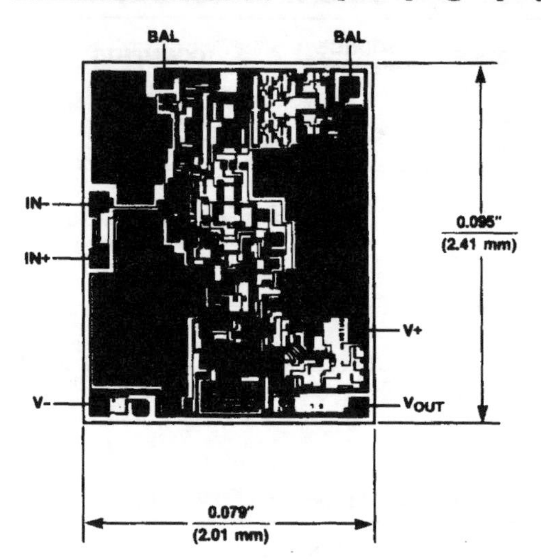

OP27/OP37

SUBSTRATE CONNECTED TO V-

Low-Noise Precision Operational Amplifiers

OP27/OP37

Package Information

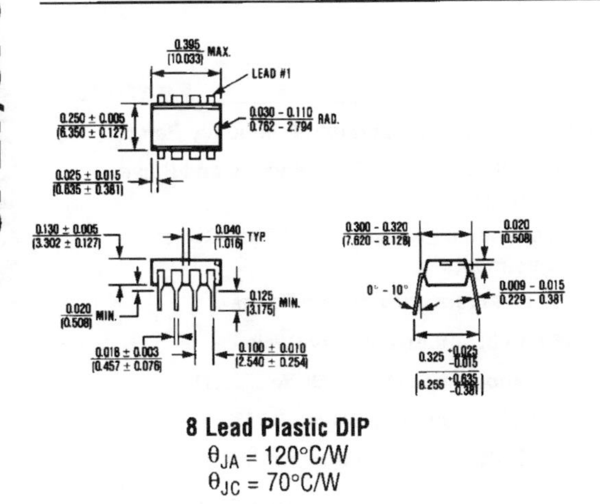

8 Lead Plastic DIP
θ_{JA} = 120°C/W
θ_{JC} = 70°C/W

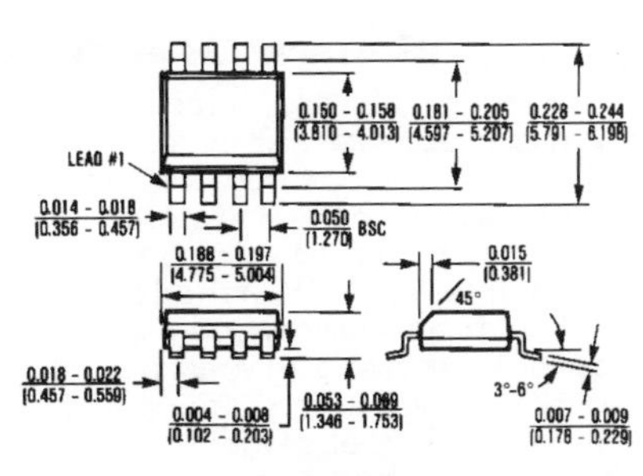

8 Lead Small Outline
θ_{JA} = 170°C/W
θ_{JC} = 80°C/W

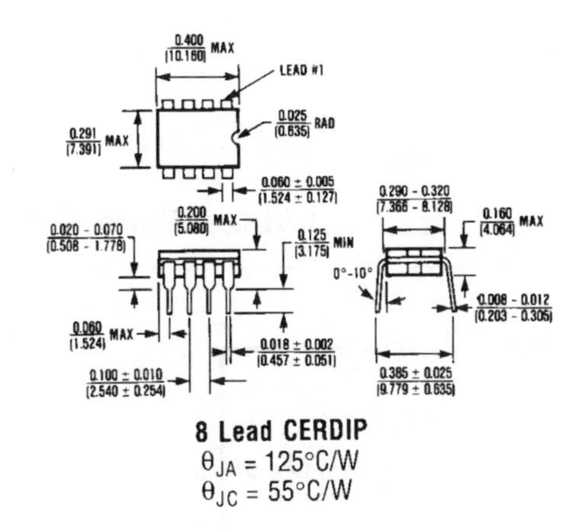

8 Lead CERDIP
θ_{JA} = 125°C/W
θ_{JC} = 55°C/W

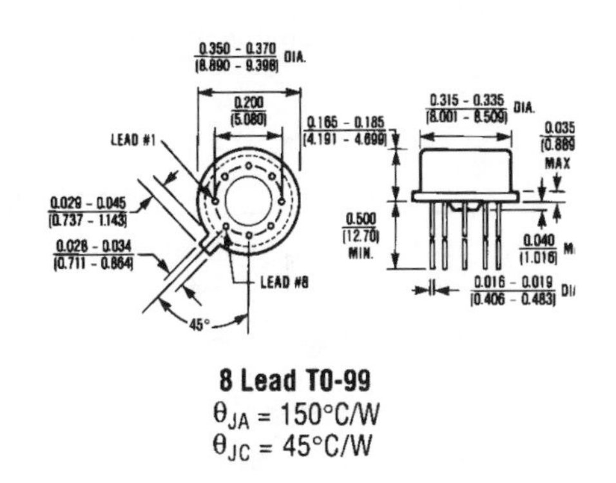

8 Lead TO-99
θ_{JA} = 150°C/W
θ_{JC} = 45°C/W

19-0266; Rev 2a; 9/96

High-Frequency Waveform Generator

General Description

The MAX038 is a high-frequency, precision function generator producing accurate, high-frequency triangle, sawtooth, sine, square, and pulse waveforms with a minimum of external components. The output frequency can be controlled over a frequency range of 0.1Hz to 20MHz by an internal 2.5V bandgap voltage reference and an external resistor and capacitor. The duty cycle can be varied over a wide range by applying a ±2.3V control signal, facilitating pulse-width modulation and the generation of sawtooth waveforms. Frequency modulation and frequency sweeping are achieved in the same way. The duty cycle and frequency controls are independent.

Sine, square, or triangle waveforms can be selected at the output by setting the appropriate code at two TTL-compatible select pins. The output signal for all waveforms is a $2V_{P-P}$ signal that is symmetrical around ground. The low-impedance output can drive up to ±20mA.

The TTL-compatible SYNC output from the internal oscillator maintains a 50% duty cycle—regardless of the duty cycle of the other waveforms—to synchronize other devices in the system. The internal oscillator can be synchronized to an external TTL clock connected to PDI.

Features

- ♦ 0.1Hz to 20MHz Operating Frequency Range
- ♦ Triangle, Sawtooth, Sine, Square, and Pulse Waveforms
- ♦ Independent Frequency and Duty-Cycle Adjustments
- ♦ 350 to 1 Frequency Sweep Range
- ♦ 15% to 85% Variable Duty Cycle
- ♦ Low-Impedance Output Buffer: 0.1Ω
- ♦ Low-Distortion Sine Wave: 0.75%
- ♦ Low 200ppm/°C Temperature Drift

Ordering Information

PART	TEMP. RANGE	PIN-PACKAGE
MAX038CPP	0°C to +70°C	20 Plastic DIP
MAX038CWP	0°C to +70°C	20 SO
MAX038C/D	0°C to +70°C	Dice*
MAX038EPP	-40°C to +85°C	20 Plastic DIP
MAX038EWP	-40°C to +85°C	20 SO

**Contact factory for dice specifications.*

Applications

Precision Function Generators
Voltage-Controlled Oscillators
Frequency Modulators
Pulse-Width Modulators
Phase-Locked Loops
Frequency Synthesizer
FSK Generator—Sine and Square Waves

Pin Configuration

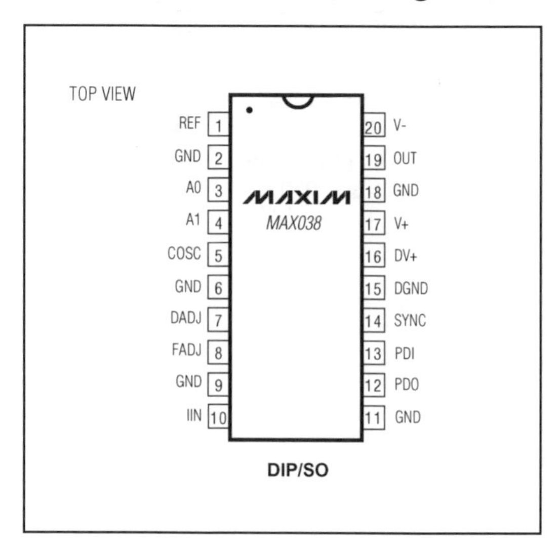

High-Frequency Waveform Generator

MAX038

ABSOLUTE MAXIMUM RATINGS

$V+$ to GND ....-0.3V to +6V
$DV+$ to DGND ....-0.3V to +6V
$V-$ to GND ....+0.3V to -6V
Pin Voltages
IIN, FADJ, DADJ, PDO ....$(V- - 0.3V)$ to $(V+ + 0.3V)$
COSC ....+0.3V to V-
A0, A1, PDI, SYNC, REF ....-0.3V to V+
GND to DGND ....±0.3V
Maximum Current into Any Pin ....±50mA
OUT, REF Short-Circuit Duration to GND, V+, V- 30sec

Continuous Power Dissipation (T_A = +70°C)
Plastic DIP (derate 11.11mW/°C above +70°C) 889mW
SO (derate 10.00mW/°C above +70°C) 800mW
CERDIP (derate 11.11mW/°C above +70°C) 889mW
Operating Temperature Ranges
MAX038C_ _ ....0°C to +70°C
MAX038E_ _ ....-40°C to +85°C
Maximum Junction Temperature ....+150°C
Storage Temperature Range ....-65°C to +150°C
Lead Temperature (soldering, 10sec) ....+300°C

Stresses beyond those listed under "Absolute Maximum Ratings" may cause permanent damage to the device. These are stress ratings only, and functional operation of the device at these or any other conditions beyond those indicated in the operational sections of the specifications is not implied. Exposure to absolute maximum rating conditions for extended periods may affect device reliability.

ELECTRICAL CHARACTERISTICS

(Circuit of Figure 1, GND = DGND = 0V, V+ = DV+ = 5V, V- = -5V, $V_{DADJ} = V_{FADJ} = V_{PDI} = V_{PDO}$ = 0V, C_F = 100pF, R_{IN} = 25kΩ, R_L = 1kΩ, C_L = 20pF, $T_A = T_{MIN}$ to T_{MAX}, unless otherwise noted. Typical values are at T_A = +25°C.)

PARAMETER	SYMBOL	CONDITIONS	MIN	TYP	MAX	UNITS
FREQUENCY CHARACTERISTICS						
Maximum Operating Frequency	F_O	15pCF ≤ 15pF, I_{IN} = 500µA	20.0	40.0		MHz
Frequency Programming Current	I_{IN}	V_{FADJ} = 0V	2.50		750	µA
		V_{FADJ} = -3V	1.25		375	
IIN Offset Voltage	V_{IN}			±1.0	±2.0	mV
Frequency Temperature Coefficient	$\Delta F_O/°C$	V_{FADJ} = 0V		600		ppm/°C
	$F_O/°C$	V_{FADJ} = -3V		200		
Frequency Power-Supply Rejection	$\frac{(\Delta F_O/F_O)}{\Delta V+}$	V- = -5V, V+ = 4.75V to 5.25V		±0.4	±2.00	%/V
	$\frac{(\Delta F_O/F_O)}{\Delta V-}$	V+ = 5V, V- = -4.75V to -5.25V		±0.2	±1.00	
OUTPUT AMPLIFIER (applies to all waveforms)						
Output Peak-to-Peak Symmetry	V_{OUT}			±4		mV
Output Resistance	R_{OUT}			0.1	0.2	Ω
Output Short-Circuit Current	I_{OUT}	Short circuit to GND		40		mA
SQUARE-WAVE OUTPUT (R_L = 100Ω)						
Amplitude	V_{OUT}		1.9	2.0	2.1	V_{P-P}
Rise Time	t_R	10% to 90%		12		ns
Fall Time	t_F	90% to 10%		12		ns
Duty Cycle	dc	V_{DADJ} = 0V, dc = t_{ON}/t x 100%	47	50	53	%
TRIANGLE-WAVE OUTPUT (R_L = 100Ω)						
Amplitude	V_{OUT}		1.9	2.0	2.1	V_{P-P}
Nonlinearity		F_O = 100kHz, 5% to 95%		0.5		%
Duty Cycle	dc	V_{DADJ} = 0V (Note 1)	47	50	53	%
SINE-WAVE OUTPUT (R_L = 100Ω)						
Amplitude	V_{OUT}		1.9	2.0	2.1	V_{P-P}
Total Harmonic Distortion	THD	Duty cycle adjusted to 50%		0.75		%
		Duty cycle unadjusted		1.50		

High-Frequency Waveform Generator

ELECTRICAL CHARACTERISTICS (continued)

(Circuit of Figure 1, GND = DGND = 0V, V+ = DV+ = 5V, V- = -5V, V_{DADJ} = V_{FADJ} = V_{PDI} = V_{PDO} = 0V, C_F = 100pF, R_{IN} = 25kΩ, R_L = 1kΩ, C_L = 20pF, T_A = T_{MIN} to T_{MAX}, unless otherwise noted. Typical values are at T_A = +25°C.)

PARAMETER	SYMBOL	CONDITIONS	MIN	TYP	MAX	UNITS
SYNC OUTPUT						
Output Low Voltage	V_{OL}	I_{SINK} = 3.2mA		0.3	0.4	V
Output High Voltage	V_{OH}	I_{SOURCE} = 400µA	2.8	3.5		V
Rise Time	t_R	10% to 90%, R_L = 3kΩ, C_L = 15pF		10		ns
Fall Time	t_F	90% to 10%, R_L = 3kΩ, C_L = 15pF		10		ns
Duty Cycle	dc_{SYNC}			50		%
DUTY-CYCLE ADJUSTMENT (DADJ)						
DADJ Input Current	I_{DADJ}		190	250	320	µA
DADJ Voltage Range	V_{DADJ}			±2.3		V
Duty-Cycle Adjustment Range	dc	-2.3V ≤ V_{DADJ} ≤ 2.3V	15		85	%
DADJ Nonlinearity	dc/V_{FADJ}	-2V ≤ V_{DADJ} ≤ 2V		2	4	%
Change in Output Frequency with DADJ	F_O/V_{DADJ}	-2V ≤ V_{DADJ} ≤ 2V		±2.5	±8	%
Maximum DADJ Modulating Frequency	F_{DC}			2		MHz
FREQUENCY ADJUSTMENT (FADJ)						
FADJ Input Current	I_{FADJ}		190	250	320	µA
FADJ Voltage Range	V_{FADJ}			±2.4		V
Frequency Sweep Range	F_O	-2.4V ≤ V_{FADJ} ≤ 2.4V		±70		%
FM Nonlinearity with FADJ	F_O/V_{FADJ}	-2V ≤ V_{FADJ} ≤ 2V		±0.2		%
Change in Duty Cycle with FADJ	dc/V_{FADJ}	-2V ≤ V_{FADJ} ≤ 2V		±2		%
Maximum FADJ Modulating Frequency	F_F			2		MHz
VOLTAGE REFERENCE						
Output Voltage	V_{REF}	I_{REF} = 0	2.48	2.50	2.52	V
Temperature Coefficient	V_{REF}/°C			20		ppm/°C
Load Regulation	V_{REF}/I_{REF}	0mA ≤ I_{REF} ≤ 4mA (source)		1	2	mV/mA
		-100µA ≤ I_{REF} ≤ 0µA (sink)		1	4	
Line Regulation	V_{REF}/V+	4.75V ≤ V+ ≤ 5.25V (Note 2)		1	2	mV/V
LOGIC INPUTS (A0, A1, PDI)						
Input Low Voltage	V_{IL}				0.8	V
Input High Voltage	V_{IH}		2.4			V
Input Current (A0, A1)	I_{IL}, I_{IH}	V_{A0}, V_{A1} = V_{IL}, V_{IH}			±5	µA
Input Current (PDI)	I_{IL}, I_{IH}	V_{PDI} = V_{IL}, V_{IH}			±25	µA
POWER SUPPLY						
Positive Supply Voltage	V+		4.75		5.25	V
SYNC Supply Voltage	DV+		4.75		5.25	V
Negative Supply Voltage	V-		-4.75		-5.25	V
Positive Supply Current	I+			35	45	mA
SYNC Supply Current	I_{DV+}			1	2	mA
Negative Supply Current	I-			45	55	mA

Note 1: Guaranteed by duty-cycle test on square wave.
Note 2: V_{REF} is independent of V-.

High-Frequency Waveform Generator

Typical Operating Characteristics

(Circuit of Figure 1, V+ = DV+ = 5V, V- = -5V, V_{DADJ} = V_{FADJ} = V_{PDI} = V_{PDO} = 0V, R_L = 1kΩ, C_L = 20pF, T_A = +25°C, unless otherwise noted.)

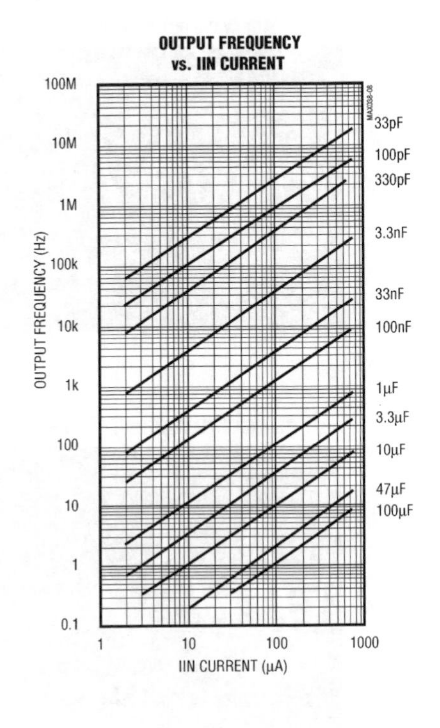

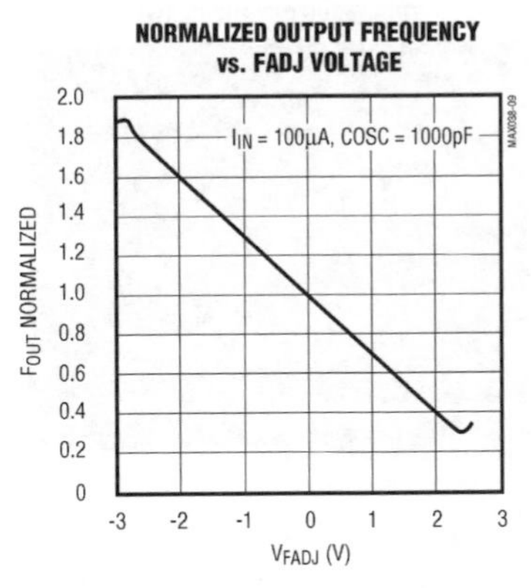

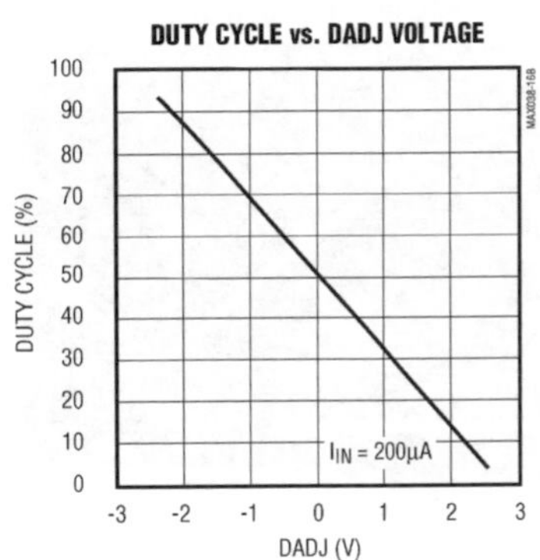

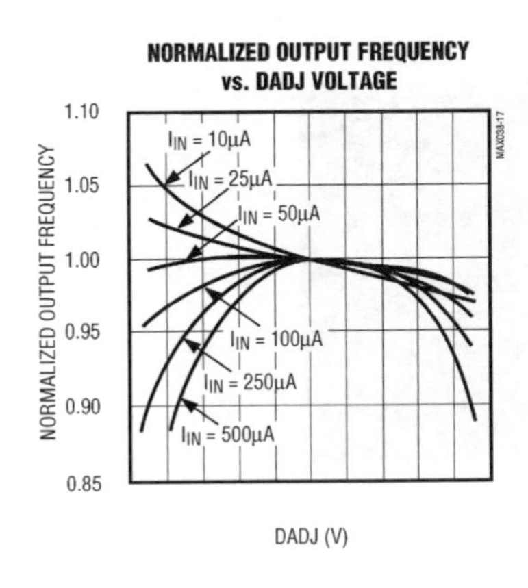

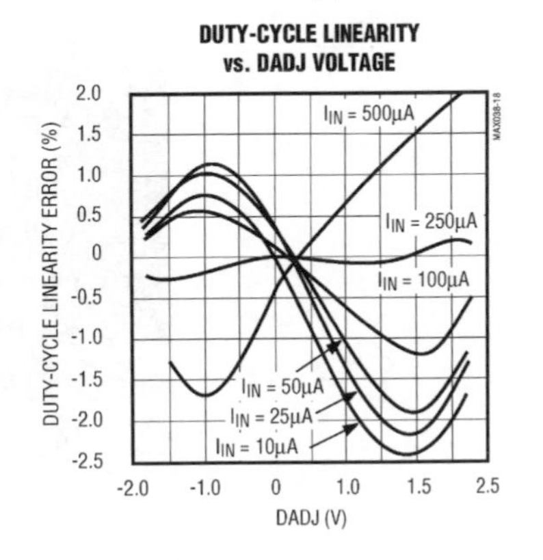

High-Frequency Waveform Generator

Typical Operating Characteristics (continued)

MAX038

(Circuit of Figure 1, V+ = DV+ = 5V, V- = -5V, $V_{DADJ} = V_{FADJ} = V_{PDI} = V_{PDO}$ = 0V, R_L = 1kΩ, C_L = 20pF, T_A = +25°C, unless otherwise noted.)

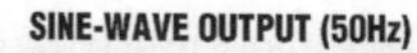
SINE-WAVE OUTPUT (50Hz)

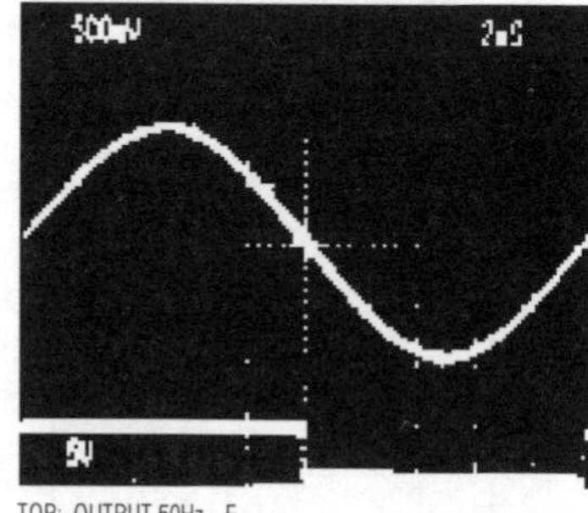

TOP: OUTPUT 50Hz = F_0
BOTTOM: SYNC
I_{IN} = 50μA
C_F = 1μF

SINE-WAVE OUTPUT (20MHz)

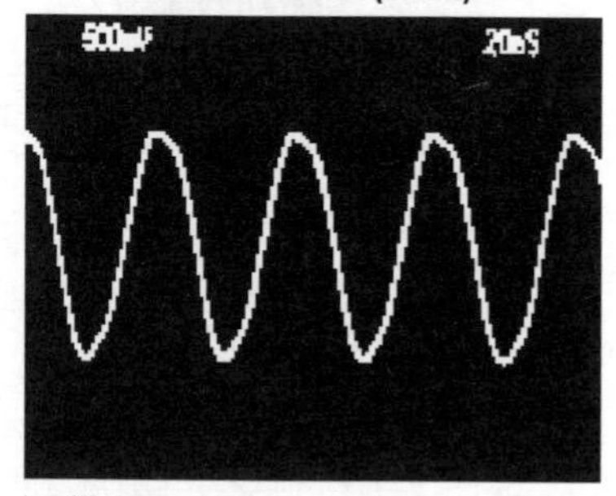

I_{IN} = 400μA
C_F = 20pF

TRIANGLE-WAVE OUTPUT (50Hz)

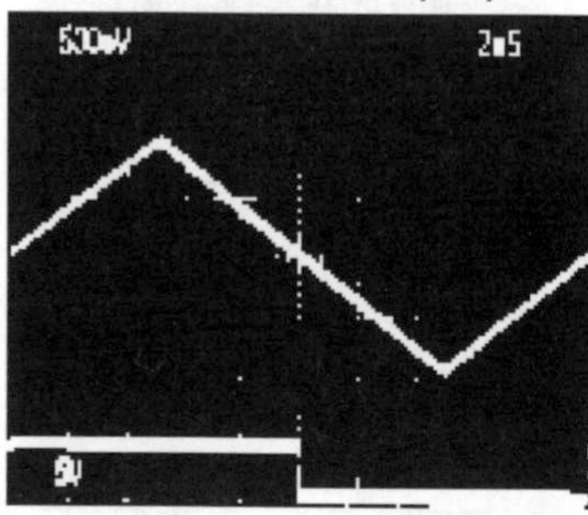

TOP: OUTPUT 50Hz = F_0
BOTTOM: SYNC
I_{IN} = 50μA
C_F = 1μF

TRIANGLE-WAVE OUTPUT (20MHz)

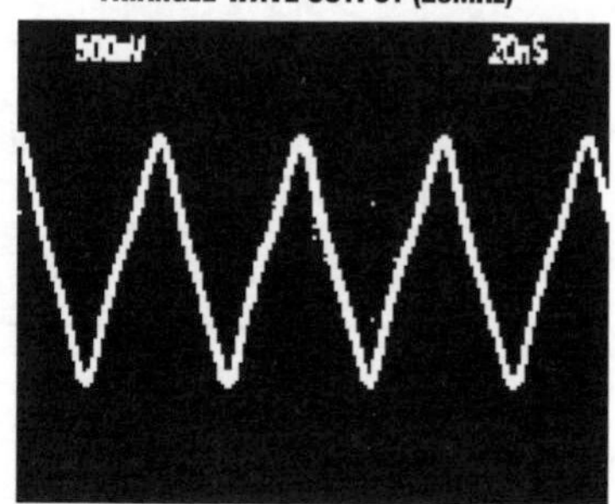

I_{IN} = 400μA
C_F = 20pF

SQUARE-WAVE OUTPUT (50Hz)

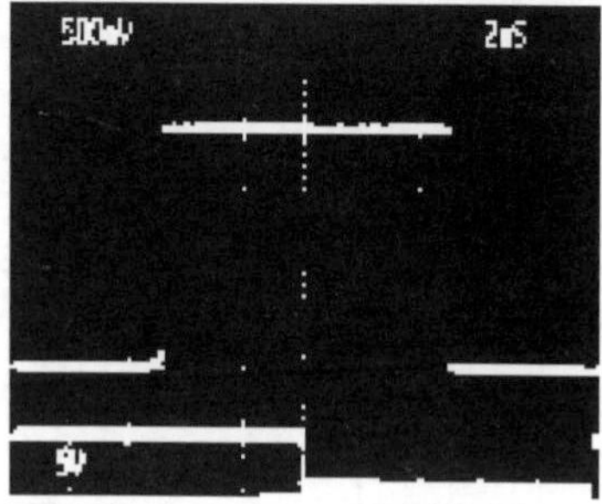

TOP: OUTPUT 50Hz = F_0
BOTTOM: SYNC
I_{IN} = 50μA
C_F = 1μF

High-Frequency Waveform Generator

MAX038

Typical Operating Characteristics (continued)

(Circuit of Figure 1, V+ = DV+ = 5V, V- = -5V, $V_{DADJ} = V_{FADJ} = V_{PDI} = V_{PDO}$ = 0V, R_L = 1kΩ, C_L = 20pF, T_A = +25°C, unless otherwise noted.)

SQUARE-WAVE OUTPUT (20MHz)

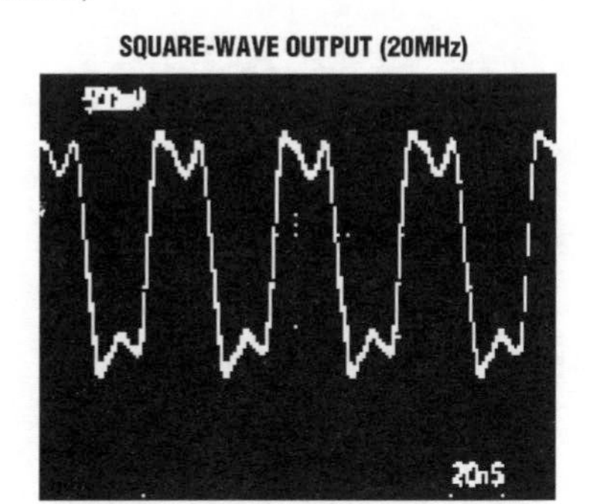

I_{IN} = 400μA
C_F = 20pF

FREQUENCY MODULATION USING FADJ

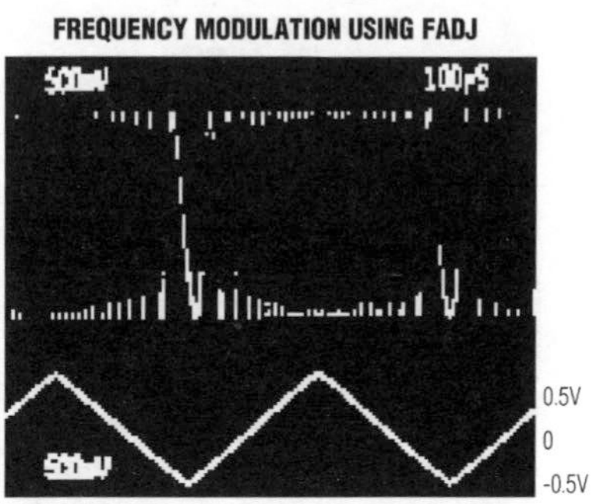

TOP: OUTPUT
BOTTOM: FADJ

FREQUENCY MODULATION USING I_{IN}

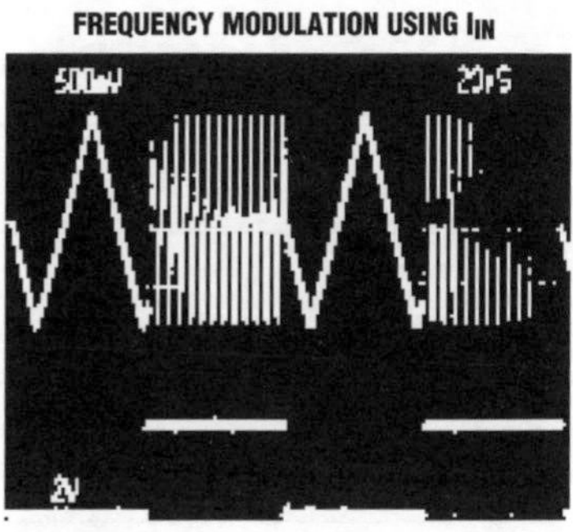

TOP: OUTPUT
BOTTOM: I_{IN}

FREQUENCY MODULATION USING I_{IN}

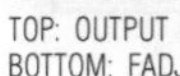

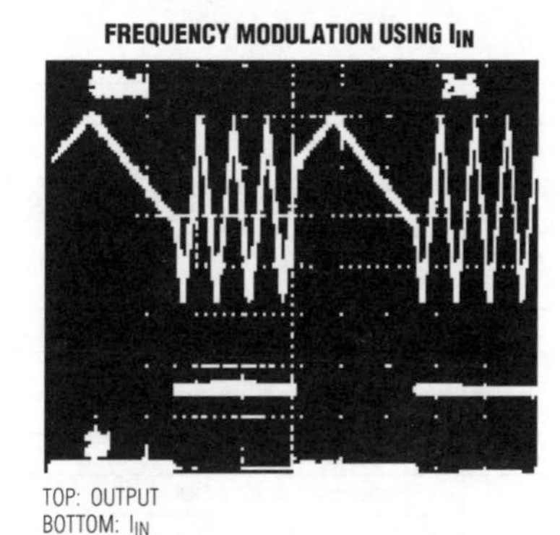

TOP: OUTPUT
BOTTOM: I_{IN}

PULSE-WIDTH MODULATION USING DADJ

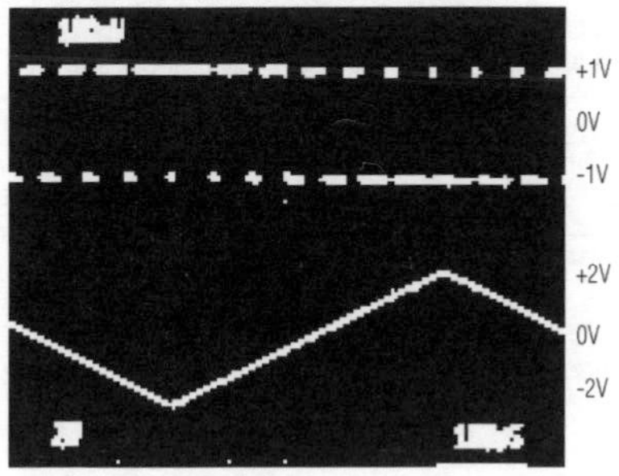

TOP: SQUARE-WAVE OUT, $2V_{P-P}$
BOTTOM: V_{DADJ}, -2V to +2.3V

High-Frequency Waveform Generator

Typical Operating Characteristics (continued)

(Circuit of Figure 1, V+ = DV+ = 5V, V- = -5V, V_{DADJ} = V_{FADJ} = V_{PDI} = V_{PDO} = 0V, R_L = 1kΩ, C_L = 20pF, T_A = +25°C, unless otherwise noted.)

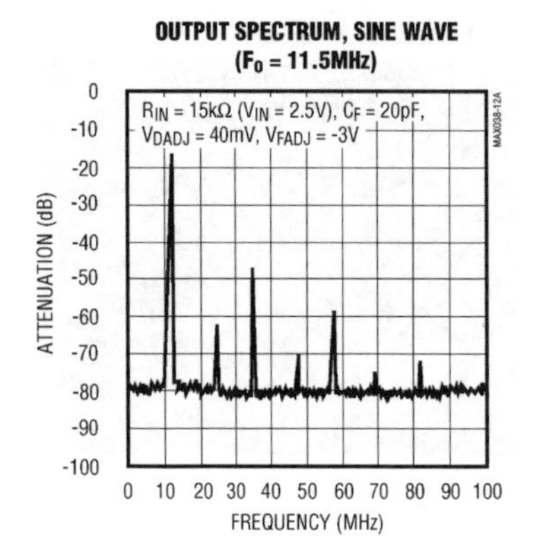

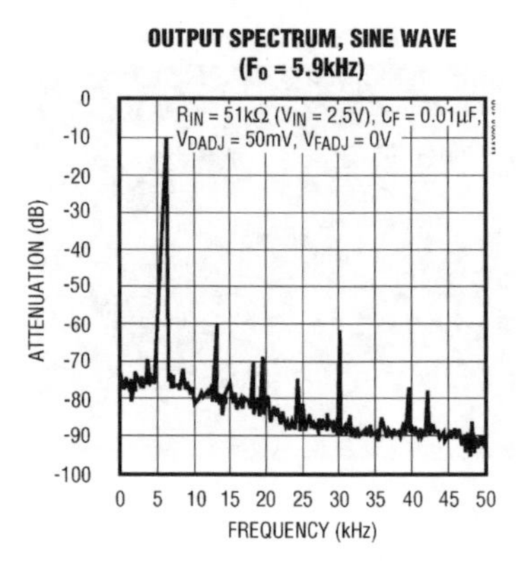

Pin Description

PIN	NAME	FUNCTION
1	REF	2.50V bandgap voltage reference output
2, 6, 9, 11, 18	GND	Ground*
3	A0	Waveform selection input; TTL/CMOS compatible
4	A1	Waveform selection input; TTL/CMOS compatible
5	COSC	External capacitor connection
7	DADJ	Duty-cycle adjust input
8	FADJ	Frequency adjust input
10	IIN	Current input for frequency control
12	PDO	Phase detector output. Connect to GND if phase detector is not used.
13	PDI	Phase detector reference clock input. Connect to GND if phase detector is not used.
14	SYNC	TTL/CMOS-compatible output, referenced between DGND and DV+. Permits the internal oscillator to be synchronized with an external signal. Leave open if unused.
15	DGND	Digital ground
16	DV+	Digital +5V supply input. Can be left open if SYNC is not used.
17	V+	+5V supply input
19	OUT	Sine, square, or triangle output
20	V-	-5V supply input

**The five GND pins are not internally connected. Connect all five GND pins to a quiet ground close to the device. A ground plane is recommended (see Layout Considerations).*

MAX038

High-Frequency Waveform Generator

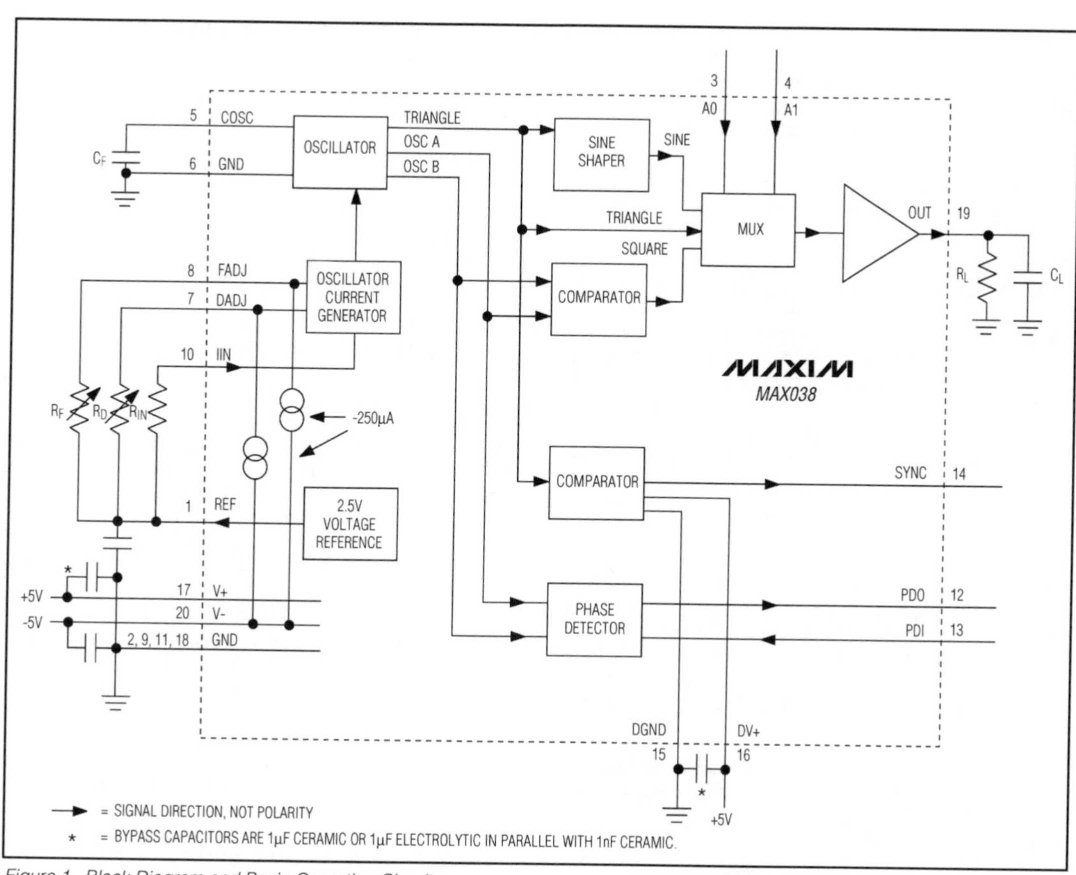

Figure 1. Block Diagram and Basic Operating Circuit

Detailed Description

The MAX038 is a high-frequency function generator that produces low-distortion sine, triangle, sawtooth, or square (pulse) waveforms at frequencies from less than 1Hz to 20MHz or more, using a minimum of external components. Frequency and duty cycle can be independently controlled by programming the current, voltage, or resistance. The desired output waveform is selected under logic control by setting the appropriate code at the A0 and A1 inputs. A SYNC output and phase detector are included to simplify designs requiring tracking to an external signal source.

The MAX038 operates with ±5V ±5% power supplies. The basic oscillator is a relaxation type that operates by alternately charging and discharging a capacitor, C_F, with constant currents, simultaneously producing a triangle wave and a square wave (Figure 1). The charging and discharging currents are controlled by the current flowing into IIN, and are modulated by the voltages applied to FADJ and DADJ. The current into IIN can be varied from 2μA to 750μA, producing more than two decades of frequency for any value of C_F. Applying ±2.4V to FADJ changes the nominal frequency (with V_{FADJ} = 0V) by ±70%; this procedure can be used for fine control.

Duty cycle (the percentage of time that the output waveform is positive) can be controlled from 10% to 90% by applying ±2.3V to DADJ. This voltage changes the C_F charging and discharging current ratio while maintaining nearly constant frequency.

High-Frequency Waveform Generator

A stable 2.5V reference voltage, REF, allows simple determination of IIN, FADJ, or DADJ with fixed resistors, and permits adjustable operation when potentiometers are connected from each of these inputs to REF. FADJ and/or DADJ can be grounded, producing the nominal frequency with a 50% duty cycle.

The output frequency is inversely proportional to capacitor C_F. C_F values can be selected to produce frequencies above 20MHz.

A sine-shaping circuit converts the oscillator triangle wave into a low-distortion sine wave with constant amplitude. The triangle, square, and sine waves are input to a multiplexer. Two address lines, A0 and A1, control which of the three waveforms is selected. The output amplifier produces a constant $2V_{P-P}$ amplitude (±1V), regardless of wave shape or frequency.

The triangle wave is also sent to a comparator that produces a high-speed square-wave SYNC waveform that can be used to synchronize other oscillators. The SYNC circuit has separate power-supply leads and can be disabled.

Two other phase-quadrature square waves are generated in the basic oscillator and sent to one side of an "exclusive-OR" phase detector. The other side of the phase-detector input (PDI) can be connected to an external oscillator. The phase-detector output (PDO) is a current source that can be connected directly to FADJ to synchronize the MAX038 with the external oscillator.

Waveform Selection

The MAX038 can produce either sine, square, or triangle waveforms. The TTL/CMOS-logic address pins (A0 and A1) set the waveform, as shown below:

A0	A1	WAVEFORM
X	1	Sine wave
0	0	Square wave
1	0	Triangle wave

X = Don't care

Waveform switching can be done at any time, without regard to the phase of the output. Switching occurs within 0.3µs, but there may be a small transient in the output waveform that lasts 0.5µs.

Waveform Timing

Output Frequency

The output frequency is determined by the current injected into the IIN pin, the COSC capacitance (to ground), and the voltage on the FADJ pin. When V_{FADJ} = 0V, the fundamental output frequency (F_O) is given by the formula:

$$F_O\,(\text{MHz}) = I_{IN}\,(\mu A) \div C_F\,(\text{pF}) \quad [1]$$

The period (t_O) is:

$$t_O\,(\mu s) = C_F\,(\text{pF}) \div I_{IN}\,(\mu A) \quad [2]$$

where:

I_{IN} = current injected into IIN (between 2µA and 750µA)

C_F = capacitance connected to COSC and GND (20pF to >100µF).

For example:

0.5MHz = 100µA ÷ 200pF

and

2µs = 200pF ÷ 100µA

Optimum performance is achieved with I_{IN} between 10µA and 400µA, although linearity is good with I_{IN} between 2µA and 750µA. Current levels outside of this range are not recommended. For fixed-frequency operation, set I_{IN} to approximately 100µA and select a suitable capacitor value. This current produces the lowest temperature coefficient, and produces the lowest frequency shift when varying the duty cycle.

The capacitance can range from 20pF to more than 100µF, but stray circuit capacitance must be minimized by using short traces. Surround the COSC pin and the trace leading to it with a ground plane to minimize coupling of extraneous signals to this node. Oscillation above 20MHz is possible, but waveform distortion increases under these conditions. The low frequency limit is set by the leakage of the COSC capacitor and by the required accuracy of the output frequency. Lowest frequency operation with good accuracy is usually achieved with 10µF or greater non-polarized capacitors.

An internal closed-loop amplifier forces IIN to virtual ground, with an input offset voltage less than ±2mV. IIN may be driven with either a current source (I_{IN}), or a voltage (V_{IN}) in series with a resistor (R_{IN}). (A resistor between REF and IIN provides a convenient method of generating I_{IN}: $I_{IN} = V_{REF}/R_{IN}$.) When using a voltage in series with a resistor, the formula for the oscillator frequency is:

$$F_O\,(\text{MHz}) = V_{IN} \div [R_{IN} \times C_F\,(\text{pF})] \quad [3]$$

and:

$$t_O\,(\mu s) = C_F\,(\text{pF}) \times R_{IN} \div V_{IN} \quad [4]$$

High-Frequency Waveform Generator

When the MAX038's frequency is controlled by a voltage source (V_{IN}) in series with a fixed resistor (R_{IN}), the output frequency is a direct function of V_{IN} as shown in the above equations. Varying V_{IN} modulates the oscillator frequency. For example, using a 10kΩ resistor for R_{IN} and sweeping V_{IN} from 20mV to 7.5V produces large frequency deviations (up to 375:1). Select R_{IN} so that I_{IN} stays within the 2µA to 750µA range. The bandwidth of the IIN control amplifier, which limits the modulating signal's highest frequency, is typically 2MHz.

IIN can be used as a summing point to add or subtract currents from several sources. This allows the output frequency to be a function of the sum of several variables. As V_{IN} approaches 0V, the I_{IN} error increases due to the offset voltage of IIN.

Output frequency will be offset 1% from its final value for 10 seconds after power-up.

FADJ Input

The output frequency can be modulated by FADJ, which is intended principally for fine frequency control, usually inside phase-locked loops. Once the fundamental, or center frequency (F_O) is set by I_{IN}, it may be changed further by setting FADJ to a voltage other than 0V. This voltage can vary from -2.4V to +2.4V, causing the output frequency to vary from 1.7 to 0.30 times the value when FADJ is 0V (F_O ±70%). Voltages beyond ±2.4V can cause instability or cause the frequency change to reverse slope.

The voltage on FADJ required to cause the output to deviate from F_O by D_X (expressed in %) is given by the formula:

$$VFADJ = -0.0343 \times D_X \qquad [5]$$

where V_{FADJ}, the voltage on FADJ, is between -2.4V and +2.4V.

Note: While I_{IN} is directly proportional to the fundamental, or center frequency (F_O), V_{FADJ} is linearly related to % deviation from F_O. V_{FADJ} goes to either side of 0V, corresponding to plus and minus deviation.

The voltage on FADJ for any frequency is given by the formula:

$$V_{FADJ} = (F_O - F_X) \div (0.2915 \times F_O) \qquad [6]$$

where:

F_X = output frequency

F_O = frequency when V_{FADJ} = 0V.

Likewise, for period calculations:

$$V_{FADJ} = 3.43 \times (t_X - t_O) \div t_X \qquad [7]$$

where:

t_X = output period

t_O = period when V_{FADJ} = 0V.

Conversely, if V_{FADJ} is known, the frequency is given by:

$$F_X = F_O \times (1 - [0.2915 \times V_{FADJ}]) \qquad [8]$$

and the period (t_X) is:

$$t_X = t_O \div (1 - [0.2915 \times V_{FADJ}]) \qquad [9]$$

Programming FADJ

FADJ has a 250µA constant current sink to V- that must be furnished by the voltage source. The source is usually an op-amp output, and the temperature coefficient of the current sink becomes unimportant. For manual adjustment of the deviation, a variable resistor can be used to set V_{FADJ}, but then the 250µA current sink's temperature coefficient becomes significant. Since external resistors cannot match the internal temperature-coefficient curve, using external resistors to program V_{FADJ} is intended only for manual operation, when the operator can correct for any errors. This restriction does not apply when V_{FADJ} is a true voltage source.

A variable resistor, R_F, connected between REF (+2.5V) and FADJ provides a convenient means of manually setting the frequency deviation. The resistance value (R_F) is:

$$R_F = (V_{REF} - V_{FADJ}) \div 250\mu A \qquad [10]$$

V_{REF} and V_{FADJ} are signed numbers, so use correct algebraic convention. For example, if V_{FADJ} is -2.0V (+58.3% deviation), the formula becomes:

$$R_F = (+2.5V - (-2.0V)) \div 250\mu A$$
$$= (4.5V) \div 250\mu A$$
$$= 18k\Omega$$

Disabling FADJ

The FADJ circuit adds a small temperature coefficient to the output frequency. For critical open-loop applications, it can be turned off by connecting FADJ to GND (not REF) through a 12kΩ resistor (R1 in Figure 2). The -250µA current sink at FADJ causes -3V to be developed across this resistor, producing two results. First, the FADJ circuit remains in its linear region, but disconnects itself from the main oscillator, improving temperature stability. Second, the oscillator frequency doubles. If FADJ is turned off in this manner, be sure to correct equations 1-4 and 6-9 above, and 12 and 14 below by doubling F_O or halving t_O. Although this method doubles the normal output frequency, it does not double the upper frequency limit. Do not operate FADJ open circuit or with voltages more negative than -3.5V. Doing so may cause transistor saturation inside the IC, leading to unwanted changes in frequency and duty cycle.

High-Frequency Waveform Generator

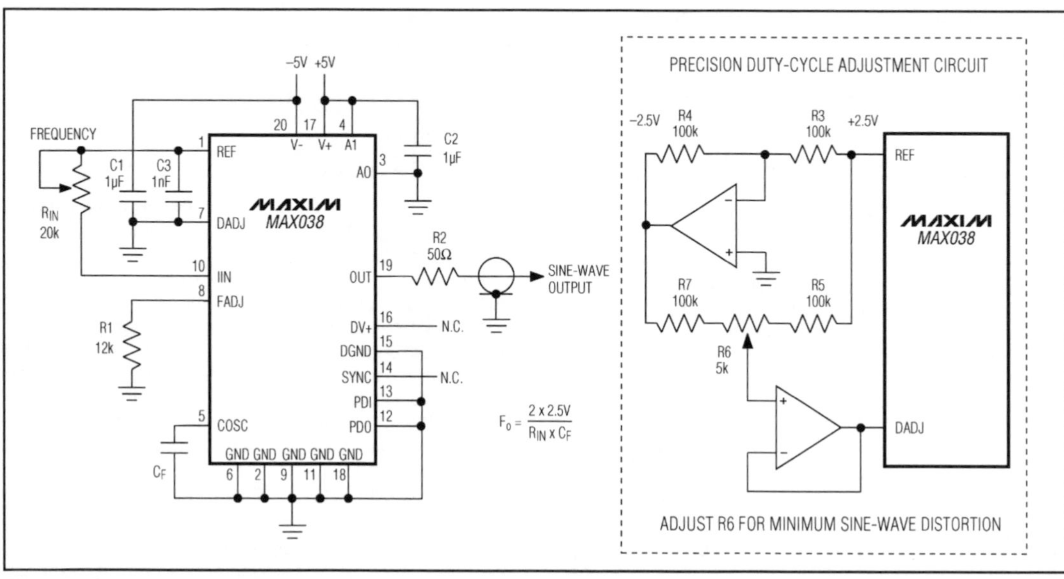

Figure 2. Operating Circuit with Sine-Wave Output and 50% Duty Cycle; SYNC and FADJ Disabled

With FADJ disabled, the output frequency can still be changed by modulating I_{IN}.

Swept Frequency Operation

The output frequency can be swept by applying a varying signal to IIN or FADJ. IIN has a wider range, slightly slower response, lower temperature coefficient, and requires a single polarity current source. FADJ may be used when the swept range is less than ±70% of the center frequency, and it is suitable for phase-locked loops and other low-deviation, high-accuracy closed-loop controls. It uses a sweeping voltage symmetrical about ground.

Connecting a resistive network between REF, the voltage source, and FADJ or IIN is a convenient means of offsetting the sweep voltage.

Duty Cycle

The voltage on DADJ controls the waveform duty cycle (defined as the percentage of time that the output waveform is positive). Normally, V_{DADJ} = 0V, and the duty cycle is 50% (Figure 2). Varying this voltage from +2.3V to -2.3V causes the output duty cycle to vary from 15% to 85%, about -15% per volt. Voltages beyond ±2.3V can shift the output frequency and/or cause instability.

DADJ can be used to reduce the sine-wave distortion. The unadjusted duty cycle (V_{DADJ} = 0V) is 50% ±2%; any deviation from exactly 50% causes even order harmonics to be generated. By applying a small adjustable voltage (typically less than ±100mV) to V_{DADJ}, exact symmetry can be attained and the distortion can be minimized (see Figure 2).

The voltage on DADJ needed to produce a specific duty cycle is given by the formula:

$$V_{DADJ} = (50\% - dc) \times 0.0575 \quad [11]$$

or:

$$V_{DADJ} = (0.5 - [t_{ON} \div t_O]) \times 5.75 \quad [12]$$

where:

V_{DADJ} = DADJ voltage (observe the polarity)

dc = duty cycle (in %)

t_{ON} = ON (positive) time

t_O = waveform period.

Conversely, if V_{DADJ} is known, the duty cycle and ON time are given by:

$$dc = 50\% - (V_{DADJ} \times 17.4) \quad [13]$$

$$t_{ON} = t_O \times (0.5 - [V_{DADJ} \times 0.174]) \quad [14]$$

High-Frequency Waveform Generator

Programming DADJ

DADJ is similar to FADJ; it has a 250µA constant current sink to V- that must be furnished by the voltage source. The source is usually an op-amp output, and the temperature coefficient of the current sink becomes unimportant. For manual adjustment of the duty cycle, a variable resistor can be used to set V_{DADJ}, but then the 250µA current sink's temperature coefficient becomes significant. Since external resistors cannot match the internal temperature-coefficient curve, using external resistors to program V_{DADJ} is intended only for manual operation, when the operator can correct for any errors. This restriction does not apply when V_{DADJ} is a true voltage source.

A variable resistor, R_D, connected between REF (+2.5V) and DADJ provides a convenient means of manually setting the duty cycle. The resistance value (R_D) is:

$$R_D = (V_{REF} - V_{DADJ}) \div 250\mu A \qquad [15]$$

Note that both V_{REF} and V_{DADJ} are signed values, so observe correct algebraic convention. For example, if V_{DADJ} is -1.5V (23% duty cycle), the formula becomes:

$$R_D = (+2.5V - (-1.5V)) \div 250\mu A$$
$$= (4.0V) \div 250\mu A = 16k\Omega$$

Varying the duty cycle in the range 15% to 85% has minimal effect on the output frequency—typically less than 2% when $25\mu A < I_{IN} < 250\mu A$. The DADJ circuit is wideband, and can be modulated at up to 2MHz (see photos, *Typical Operating Characteristics*).

Output

The output amplitude is fixed at $2V_{P-P}$, symmetrical around ground, for all output waveforms. OUT has an output resistance of under 0.1Ω, and can drive ±20mA with up to a 50pF load. Isolate higher output capacitance from OUT with a resistor (typically 50Ω) or buffer amplifier.

Reference Voltage

REF is a stable 2.50V bandgap voltage reference capable of sourcing 4mA or sinking 100µA. It is principally used to furnish a stable current to IIN or to bias DADJ and FADJ. It can also be used for other applications external to the MAX038. Bypass REF with 100nF to minimize noise.

Selecting Resistors and Capacitors

The MAX038 produces a stable output frequency over time and temperature, but the capacitor and resistors that determine frequency can degrade performance if they are not carefully chosen. Resistors should be metal film, 1% or better. Capacitors should be chosen for low temperature coefficient over the whole temperature range. NPO ceramics are usually satisfactory.

The voltage on COSC is a triangle wave that varies between 0V and -1V. Polarized capacitors are generally not recommended (because of their outrageous temperature dependence and leakage currents), but if they are used, the negative terminal should be connected to COSC and the positive terminal to GND. Large-value capacitors, necessary for very low frequencies, should be chosen with care, since potentially large leakage currents and high dielectric absorption can interfere with the orderly charge and discharge of C_F. If possible, for a given frequency, use lower IIN currents to reduce the size of the capacitor.

SYNC Output

SYNC is a TTL/CMOS-compatible output that can be used to synchronize external circuits. The SYNC output is a square wave whose rising edge coincides with the output rising sine or triangle wave as it crosses through 0V. When the square wave is selected, the rising edge of SYNC occurs in the middle of the positive half of the output square wave, effectively 90° ahead of the output. The SYNC duty cycle is fixed at 50% and is independent of the DADJ control.

Because SYNC is a very-high-speed TTL output, the high-speed transient currents in DGND and DV+ can radiate energy into the output circuit, causing a narrow spike in the output waveform. (This spike is difficult to see with oscilloscopes having less than 100MHz bandwidth). The inductance and capacitance of IC sockets tend to amplify this effect, so sockets are not recommended when SYNC is on. SYNC is powered from separate ground and supply pins (DGND and DV+), and it can be turned off by making DV+ open circuit. If synchronization of external circuits is not used, turning off SYNC by DV+ opening eliminates the spike.

Phase Detectors

Internal Phase Detector

The MAX038 contains a TTL/CMOS phase detector that can be used in a phase-locked loop (PLL) to synchronize its output to an external signal (Figure 3). The external source is connected to the phase-detector input (PDI) and the phase-detector output is taken from PDO. PDO is the output of an exclusive-OR gate, and produces a rectangular current waveform at the MAX038 output frequency, even with PDI grounded. PDO is normally connected to FADJ and a resistor, R_{PD}, and a capacitor C_{PD}, to GND. R_{PD} sets the gain of the phase detector, while the capacitor attenuates high-frequency components and forms a pole in the phase-locked loop filter.

High-Frequency Waveform Generator

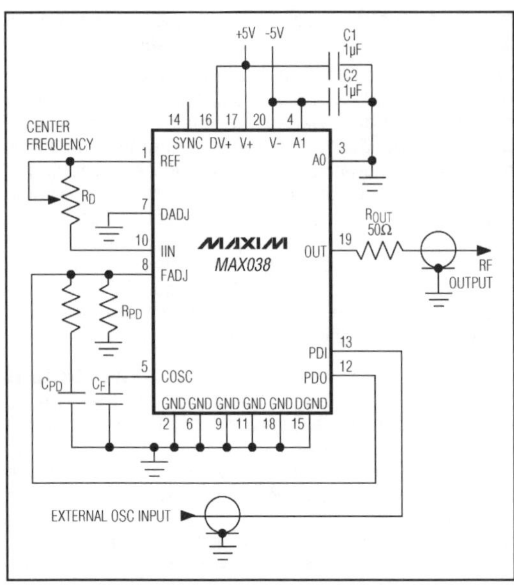

Figure 3. Phase-Locked Loop Using Internal Phase Detector

PDO is a rectangular current-pulse train, alternating between 0µA and 500µA. It has a 50% duty cycle when the MAX038 output and PDI are in phase-quadrature (90° out of phase). The duty cycle approaches 100% as the phase difference approaches 180° and conversely, approaches 0% as the phase difference approaches 0°. The gain of the phase detector (K_D) can be expressed as:

$$K_D = 0.318 \times R_{PD} \text{ (volts/radian)} \quad [16]$$

where R_{PD} = phase-detector gain-setting resistor.

When the loop is in lock, the input signals to the phase detector are in approximate phase quadrature, the duty cycle is 50%, and the average current at PDO is 250µA (the current sink of FADJ). This current is divided between FADJ and R_{PD}; 250µA always goes into FADJ and any difference current is developed across R_{PD}, creating V_{FADJ} (both polarities). For example, as the phase difference increases, PDO duty cycle increases, the average current increases, and the voltage on R_{PD} (and V_{FADJ}) becomes more positive. This in turn decreases the oscillator frequency, reducing the phase difference, thus maintaining phase lock. The higher R_{PD} is, the greater V_{FADJ} is for a given phase difference; in other words, the greater the loop gain, the less the capture range. The current from PDO must also charge C_{PD}, so the rate at which V_{FADJ} changes (the loop bandwidth) is inversely proportional to C_{PD}.

The phase error (deviation from phase quadrature) depends on the open-loop gain of the PLL and the initial frequency deviation of the oscillator from the external signal source. The oscillator conversion gain (K_O) is:

$$K_O = \Delta\omega_O \div \Delta V_{FADJ} \quad [17]$$

which, from equation [6] is:

$$K_O = 3.43 \times \omega_O \text{ (radians/sec)} \quad [18]$$

The loop gain of the PLL system (K_V) is:

$$K_V = K_D \times K_O \quad [19]$$

where:

K_D = detector gain

K_O = oscillator gain.

With a loop filter having a response F(s), the open-loop transfer function, T(s), is:

$$T(s) = K_D \times K_O \times F(s) \div s \quad [20]$$

Using linear feedback analysis techniques, the closed-loop transfer characteristic, H(s), can be related to the open-loop transfer function as follows:

$$H(s) = T(s) \div [1+ T(s)] \quad [21]$$

The transient performance and the frequency response of the PLL depends on the choice of the filter characteristic, F(s).

When the MAX038 internal phase detector is not used, PDI and PDO should be connected to GND.

External Phase Detectors

External phase detectors may be used instead of the internal phase detector. The external phase detector shown in Figure 4 duplicates the action of the MAX038's internal phase detector, but the optional ÷N circuit can be placed between the SYNC output and the phase detector in applications requiring synchronizing to an exact multiple of the external oscillator. The resistor network consisting of R4, R5, and R6 sets the sync range, while capacitor C4 sets the capture range. Note that this type of phase detector (with or without the ÷N circuit) locks onto harmonics of the external oscillator as well as the fundamental. With no external oscillator input, this circuit can be unpredictable, depending on the state of the external input DC level.

Figure 4 shows a frequency phase detector that locks onto only the fundamental of the external oscillator. With no external oscillator input, the output of the frequency phase detector is a positive DC voltage, and the oscillations are at the lowest frequency as set by R4, R5, and R6.

High-Frequency Waveform Generator

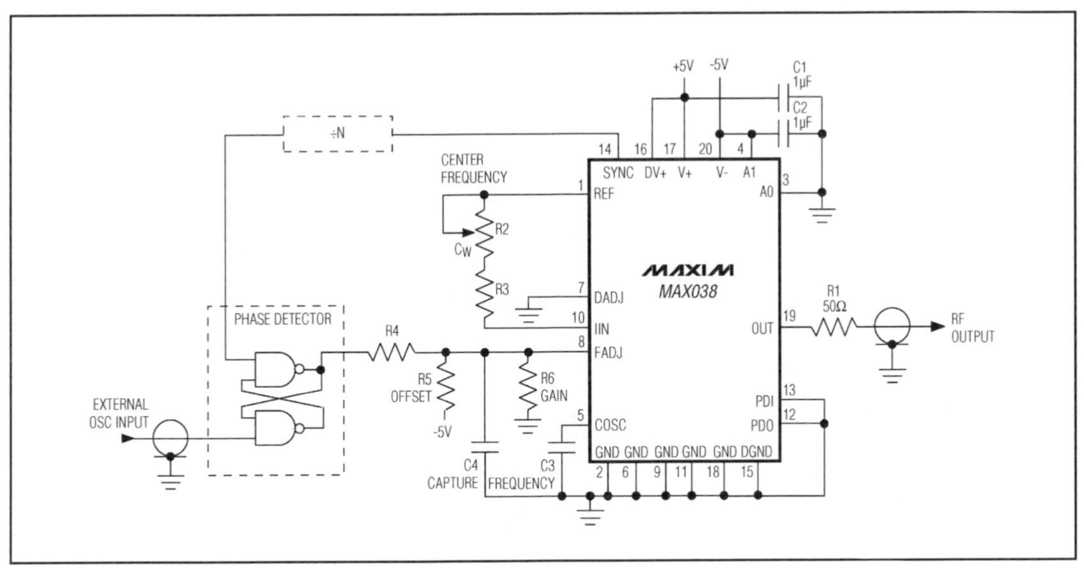

Figure 4. Phase-Locked Loop Using External Phase Detector

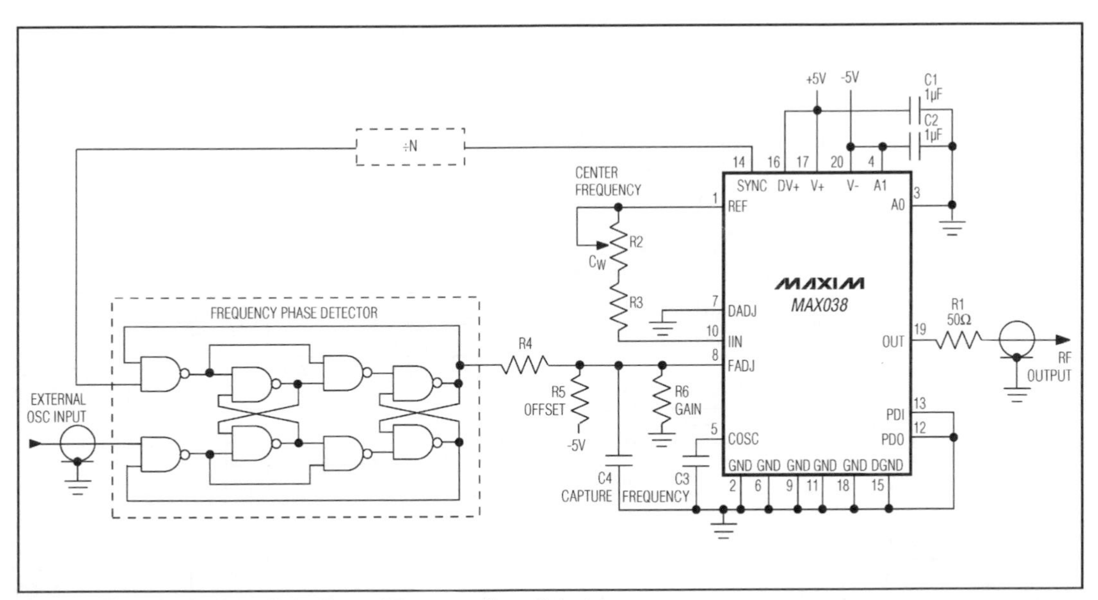

Figure 5. Phase-Locked Loop Using External Frequency Phase Detector

High-Frequency Waveform Generator

MAX038

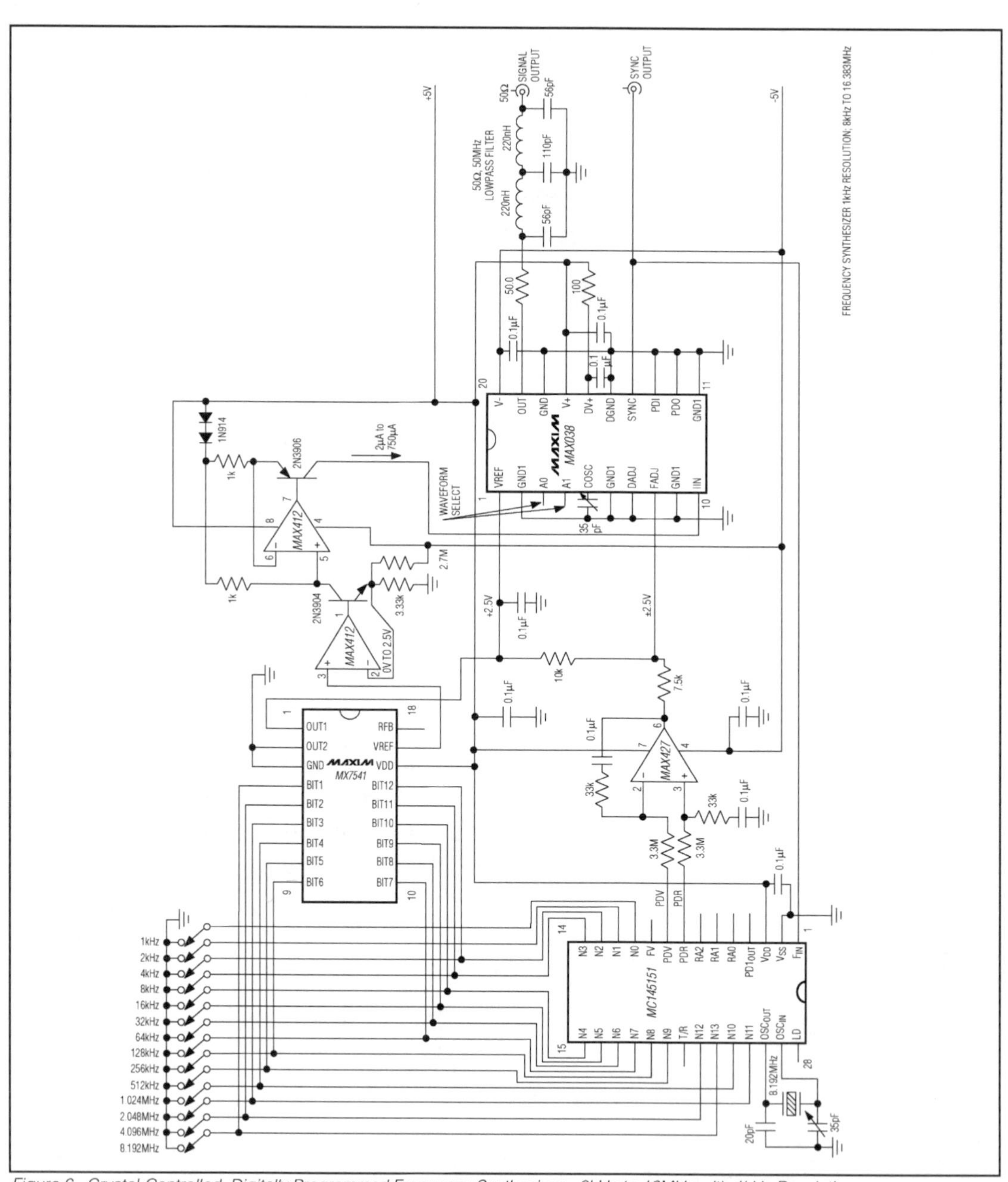

Figure 6. Crystal-Controlled, Digitally Programmed Frequency Synthesizer—8kHz to 16MHz with 1kHz Resolution

High-Frequency Waveform Generator

Layout Considerations

Realizing the full performance of the MAX038 requires careful attention to power-supply bypassing and board layout. Use a low-impedance ground plane, and connect all five GND pins directly to it. Bypass V+ and V- directly to the ground plane with 1µF ceramic capacitors or 1µF tantalum capacitors in parallel with 1nF ceramics. Keep capacitor leads short (especially with the 1nF ceramics) to minimize series inductance.

If SYNC is used, DV+ must be connected to V+, DGND must be connected to the ground plane, and a second 1nF ceramic should be connected as close as possible between DV+ and DGND (pins 16 and 15). It is not necessary to use a separate supply or run separate traces to DV+. If SYNC is disabled, leave DV+ open. Do not open DGND.

Minimize the trace area around COSC (and the ground plane area under COSC) to reduce parasitic capacitance, and surround this trace with ground to prevent coupling with other signals. Take similar precautions with DADJ, FADJ, and IIN. Place C_F so its connection to the ground plane is close to pin 6 (GND).

Applications Information

Frequency Synthesizer

Figure 6 shows a frequency synthesizer that produces accurate and stable sine, square, or triangle waves with a frequency range of 8kHz to 16.383MHz in 1kHz increments. A Motorola MC145151 provides the crystal-controlled oscillator, the ÷N circuit, and a high-speed phase detector. The manual switches set the output frequency; opening any switch increases the output frequency. Each switch controls both the ÷N output and an MX7541 12-bit DAC, whose output is converted to a current by using both halves of the MAX412 op amp. This current goes to the MAX038 IIN pin, setting its coarse frequency over a very wide range.

Fine frequency control (and phase lock) is achieved from the MC145151 phase detector through the differential amplifier and lowpass filter, U5. The phase detector compares the ÷N output with the MAX038 SYNC output and sends differential phase information to U5. U5's single-ended output is summed with an offset into the FADJ input. (Using the DAC and the IIN pin for coarse frequency control allows the FADJ pin to have very fine control with reasonably fast response to switch changes.)

A 50MHz, 50Ω lowpass filter in the output allows passage of 16MHz square waves and triangle waves with reasonable fidelity, while stopping high-frequency noise generated by the ÷N circuit.

Chip Topography

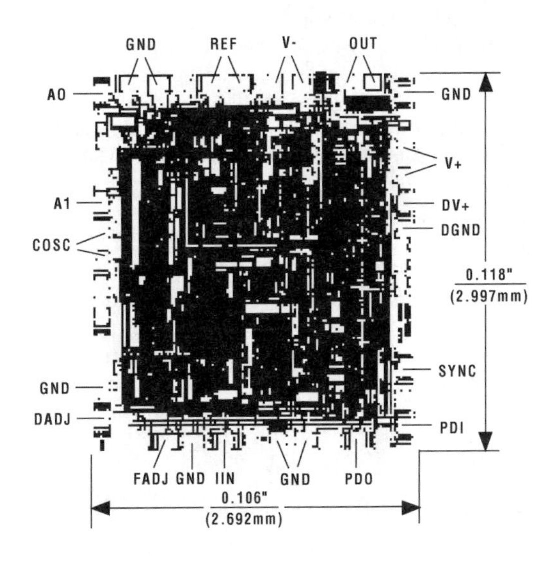

TRANSISTOR COUNT: 855
SUBSTRATE CONNECTED TO GND

19-0129; Rev. 3; 7/94

Single/Dual/Quad High-Speed, Ultra Low-Power, Single-Supply TTL Comparators

MAX907/MAX908/MAX909

General Description

The MAX907/MAX908/MAX909 dual, quad, and single high-speed, ultra low-power voltage comparators are designed for use in systems powered from a single +5V supply; the MAX909 also accepts dual ±5V supplies. Their 40ns propagation delay (with 5mV input overdrive) is achieved with a power consumption of only 3.5mW per comparator. The wide input common-mode range extends from 200mV below ground (below the negative supply rail for the MAX909) to within 1.5V of the positive supply rail.

Because they are micropower, high-speed comparators that operate from a single +5V supply and include built-in hysteresis, these devices replace a variety of older comparators in a wide range of applications.

MAX907/MAX908/MAX909 outputs are TTL compatible, requiring no external pull-up circuitry. All inputs and outputs can be continuously shorted to either supply rail without damage. These easy-to-use comparators incorporate internal hysteresis to ensure clean output switching even when the devices are driven by a slow-moving input signal.

The MAX909 features complementary outputs and an output latch. A separate supply pin for extending the analog input range down to -5V is also provided.

The dual MAX907 and single MAX909 are available in 8-pin DIP and small-outline packages, and the quad MAX908 is available in 14-pin DIP and small-outline packages. These comparators are ideal for single +5V-supply applications that require the combination of high speed, precision, and ultra-low power dissipation.

Applications

- Battery-Powered Systems
- High-Speed A/D Converters
- High-Speed V/F Converters
- Line Receivers
- Threshold Detectors/Discriminators
- High-Speed Sampling Circuits
- Zero Crossing Detectors

Features

- ♦ **40ns Propagation Delay**
- ♦ **700μA (3.5mW) Supply Current per Comparator**
- ♦ **Single 4.5V to 5.5V Supply Operation (or ±5V, MAX909 only)**
- ♦ **Wide Input Range Includes Ground (or -5V, MAX909 only)**
- ♦ **Low, 500μV Offset Voltage**
- ♦ **Internal Hysteresis Provides Clean Switching**
- ♦ **TTL-Compatible Outputs (Complementary on MAX909)**
- ♦ **Input and Output Short-Circuit Protection**
- ♦ **Internal Latch (MAX909 only)**

Ordering Information

PART	TEMP. RANGE	PIN-PACKAGE
MAX907CPA	0°C to +70°C	8 Plastic DIP
MAX907CSA	0°C to +70°C	8 SO
MAX907C/D	0°C to +70°C	Dice*
MAX907EPA	-40°C to +85°C	8 Plastic DIP
MAX907ESA	-40°C to +85°C	8 SO
MAX907MJA	-55°C to +125°C	8 CERDIP

Ordering Information continued on last page.

* *Dice are specified at +25°C, DC parameters only.*

Pin Configurations

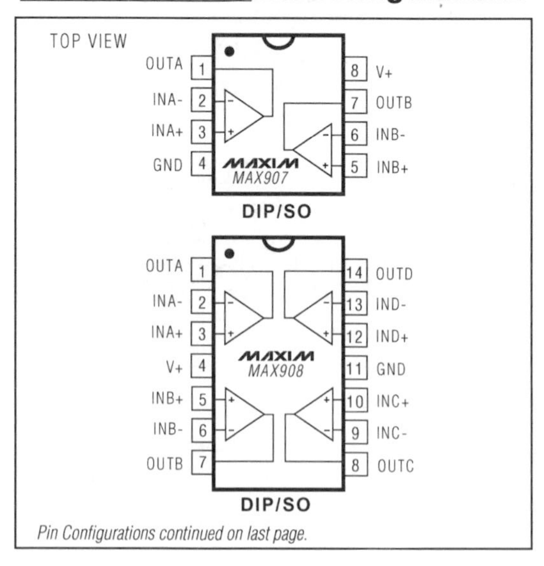

Pin Configurations continued on last page.

MAX907/MAX908/MAX909

Single/Dual/Quad High-Speed, Ultra Low-Power, Single-Supply TTL Comparators

ABSOLUTE MAXIMUM RATINGS

Positive Supply Voltage (V+ to GND) .... +7V
Negative Supply Voltage (V- to GND, MAX909 only) .... -7V
Differential Input Voltage
MAX907/MAX908 .... -0.3V to (V+ + 0.3V)
MAX909 .... (V- - 0.3V) to (V+ + 0.3V)
Common-Mode Input Voltage
MAX907/MAX908 .... -0.3V to (V+ + 0.3V)
MAX909 .... (V- - 0.3V) to (V+ + 0.3V)
Latch Input Voltage (MAX909 only) .... -0.3V to (V+ + 0.3V)
Input/Output Short-Circuit Duration to V+ or GND ... Continuous

Continuous Power Dissipation (T_A = +70°C)
8-Pin Plastic DIP (derate 9.09mW/°C above +70°C) ... 727mW
8-Pin SO (derate 5.88mW/°C above +70°C) 471mW
8-Pin CERDIP (derate 8.00mW/°C above +70°C) 640mW
14-Pin Plastic DIP (derate 10.00mW/°C above +70°C) ... 800mW
14-Pin SO (derate 8.33mW/°C above +70°C) 667mW
14-Pin CERDIP (derate 9.09mW/°C above +70°C) 727mW
Operating Temperature Ranges:
MAX90_C_ _ .... 0°C to +70°C
MAX90_E_ _ .... -40°C to +85°C
MAX90_MJ_ .... -55°C to +125°C
Storage Temperature Range .... -65°C to +160°C
Lead Temperature (soldering, 10sec) .... +300°C

Stresses beyond those listed under "Absolute Maximum Ratings" may cause permanent damage to the device. These are stress ratings only, and functional operation of the device at these or any other conditions beyond those indicated in the operational sections of the specifications is not implied. Exposure to absolute maximum rating conditions for extended periods may affect device reliability.

ELECTRICAL CHARACTERISTICS

(V+ = 5V, T_A = +25°C; MAX909 only: V- = 0V, V_{LATCH} = 0V; unless otherwise noted.)

PARAMETER	SYMBOL	CONDITIONS		MIN	TYP	MAX	UNITS
Positive Trip Point	V_{TRIP+}	(Note 1)			2	4	mV
Negative Trip Point	V_{TRIP-}	(Note 1)			-2	-4	mV
Input Offset Voltage	V_{OS}	(Note 2)			0.5	2.0	mV
Input Bias Current	I_B	V_{CM} = 0V, V_{IN} = V_{OS}			100	300	nA
Input Offset Current	I_{OS}	V_{CM} = 0V, V_{IN} = V_{OS}			25	50	nA
Input Voltage Range	V_{CMR}	(Notes 3, 4)	MAX907/908/909	-0.2		V+ - 1.5	V
			MAX909 only: V- = -5V	-5.2		V+ - 1.5	
Common-Mode Rejection Ratio	CMRR	(Notes 4, 5)			50	100	μV/V
Power-Supply Rejection Ratio	PSRR	(Notes 4, 6)			50	100	μV/V
Output High Voltage	V_{OH}	I_{SOURCE} = 100μA		3.0	3.5		V
Output Low Voltage	V_{OL}	I_{SINK} = 3.2mA			0.3	0.4	V
		I_{SINK} = 8mA			0.4		
Positive Supply Current per Comparator	I+	(Note 7)	MAX907/MAX908		0.7	1.0	mA
			MAX909		1.2	1.8	
Negative Supply Current	I-	MAX909 only: V- = -5V			60	100	μA
Power Dissipation per Comparator	PD	(Note 8)	MAX907/MAX908		3.5	5.5	mW
			MAX909		6	10	
Output Rise Time	t_r	V_{OUT} = 0.4V to 2.4V, C_L = 10pF			12		ns
Output Fall Time	t_f	V_{OUT} = 2.4V to 0.4V, C_L = 10pF			6		ns

Single/Dual/Quad High-Speed, Ultra Low-Power, Single-Supply TTL Comparators

ELECTRICAL CHARACTERISTICS (continued)

($V+ = 5V$, $T_A = +25°C$; MAX909 only: $V- = 0V$, $V_{LATCH} = 0V$; unless otherwise noted.)

PARAMETER	SYMBOL	CONDITIONS	MIN	TYP	MAX	UNITS
Propagation Delay	t_{PD+}, t_{PD-}	$V_{IN} = 100mV$, $V_{OD} = 5mV$, (Note 9)		40	50	ns
Differential Propagation Delay	Δt_{PD}	$V_{IN} = 100mV$, $V_{OD} = 5mV$, (Note 10)		1		ns
Propagation Delay Skew	$t_{PD}skew$	MAX909 only: $V_{IN} = 100mV$, $V_{OD} = 5mV$, (Note 11)		2		ns
Latch Input Voltage High	V_{IH}	(Note 12)	2.0			V
Latch Input Voltage Low	V_{IL}	(Note 12)			0.8	V
Latch Input Current	I_{IH}, I_{IL}	(Note 12)			20	µA
Latch Setup Time	t_s	(Note 12)		2		ns
Latch Hold Time	t_h	(Note 12)		2		ns

ELECTRICAL CHARACTERISTICS

($V+ = 5V$, $T_A = T_{MIN}$ to T_{MAX}; MAX909 only: $V- = 0V$, $V_{LATCH} = 0V$; unless otherwise noted.)

PARAMETER	SYMBOL	CONDITIONS		MIN	TYP	MAX	UNITS
Positive Trip Point	V_{TRIP+}	(Note 1)			2	5	mV
Negative Trip Point	V_{TRIP-}	(Note 1)			-2	-5	mV
Input Offset Voltage	V_{OS}	(Note 2)			1	3	mV
Input Bias Current	I_B	$V_{CM} = 0V$, $V_{IN} = V_{OS}$			200	500	nA
Input Offset Current	I_{OS}	$V_{CM} = 0V$, $V_{IN} = V_{OS}$			50	100	nA
Input Voltage Range	V_{CMR}	C/E temp. ranges (Notes 3, 4)	MAX907/908/909	-0.2		V+ - 1.5	V
			MAX909 only, V- = -5V	-5.2		V+ - 1.5	
		M temp. range (Notes 3, 4)	MAX907/908/909	-0.1		V+ - 1.5	
			MAX909 only, V- = -5V	-5.1		V+ - 1.5	
Common-Mode Rejection Ratio	CMRR	(Notes 4, 5)			75	200	µV/V
Power-Supply Rejection Ratio	PSRR	(Notes 4, 6)			75	200	µV/V
Output High Voltage	V_{OH}	$I_{SOURCE} = 100µA$		2.8	3.5		V
Output Low Voltage	V_{OL}	$I_{SINK} = 3.2mA$			0.3	0.4	V
		$I_{SINK} = 8mA$			0.4		
Positive Supply Current per Comparator	I+	(Note 7)	MAX907/MAX908		0.8	1.2	mA
			MAX909		1.2	2.0	
Negative Supply Current	I-	MAX909 only: V- = -5V			100	200	µA
Power Dissipation per Comparator	PD	(Note 8)	MAX907/MAX908		4	7	mW
			MAX909		6	11	

Single/Dual/Quad High-Speed, Ultra Low-Power, Single-Supply TTL Comparators

ELECTRICAL CHARACTERISTICS (continued)

($V+ = 5V$, $T_A = T_{MIN}$ to T_{MAX}; MAX909 only: $V- = 0V$, $V_{LATCH} = 0V$; unless otherwise noted.)

PARAMETER	SYMBOL	CONDITIONS	MIN	TYP	MAX	UNITS
Propagation Delay	t_{PD+}, t_{PD-}	$V_{IN} = 100mV$, $V_{OD} = 5mV$ (Note 9)		45	70	ns
Differential Propagation Delay	Δt_{PD}	$V_{IN} = 100mV$, $V_{OD} = 5mV$ (Note 10)		2		ns
Propagation Delay Skew	$t_{PD}skew$	MAX909 only: $V_{IN} = 100mV$, $V_{OD} = 5mV$ (Note 11)		4		ns
Latch Input Voltage High	V_{IH}	(Note 12)	2.0			V
Latch Input Voltage Low	V_{IL}	(Note 12)			0.8	V
Latch Input Current	I_{IH}, I_{IL}	(Note 12)			20	µA
Latch Setup Time	t_s	(Note 12)		4		ns
Latch Hold Time	t_h	(Note 12)		4		ns

Note 1: Trip Point is defined as the input voltage required to make the comparator output change state. The difference between upper (V_{TRIP+}) and lower (V_{TRIP-}) trip points is equal to the width of the input-referred hysteresis zone (V_{HYST}). Specified for an input common-mode voltage (V_{CM}) of 0V. See Figure 1.

Note 2: Input Offset Voltage is defined as the center of the input-referred hysteresis zone. Specified for $V_{CM} = 0V$. See Figure 1.

Note 3: Inferred from the CMRR test. Note that a correct logic result is obtained at the output, provided that at least one input is within the V_{CMR} limits. Note also that either or both inputs can be driven to the upper or lower absolute maximum limit without damage to the part.

Note 4: Tested with $V+ = 5.5V$ (and $V- = 0V$ for MAX909). MAX909 also tested over the full analog input range (i.e., with $V- = -5.5V$).

Note 5: Tested over the full input voltage range (V_{CMR}).

Note 6: Specified over the full tolerance of operating supply voltage: MAX907/MAX908 tested with $4.5V < V+ < 5.5V$. MAX909 tested with $4.5V < V+ < 5.5V$ and with $-5.5V < V- < 0V$.

Note 7: Positive Supply Current specified with the worst-case condition of all outputs at logic low (MAX907/MAX908), and with $V+ = 5.5V$.

Note 8: Typical power specified with $V+ = 5V$; maximum with $V+ = 5.5V$ (and with $V- = -5.5V$ for MAX909).

Note 9: Due to difficulties in measuring propagation delay with 5mV of overdrive in automatic test equipment, the MAX907/MAX908/MAX909 are sample tested to 0.1% AQL with 100mV input overdrive. Correlation tests show that the specification can be guaranteed if all other DC parameters are within the specified limits. V_{OS} must be added to the overdrive voltage for low values of overdrive.

Note 10: Differential Propagation Delay is specified as the difference between any two channels in the MAX907/MAX908 (both outputs making either a low-to-high or a high-to-low transition).

Note 11: Propagation Delay Skew is specified as the difference between any single channel's output low-to-high transition (t_{PD+}) and high-to-low transition (t_{PD-}), and also between the QOUT and $\overline{Q}$OUT transition on the MAX909.

Note 12: Latch specifications apply to MAX909 only. See Figure 2.

Single/Dual/Quad High-Speed, Ultra Low-Power, Single-Supply TTL Comparators

Typical Operating Characteristics

(V+ = 5V, T_A = +25°C, unless otherwise noted.)

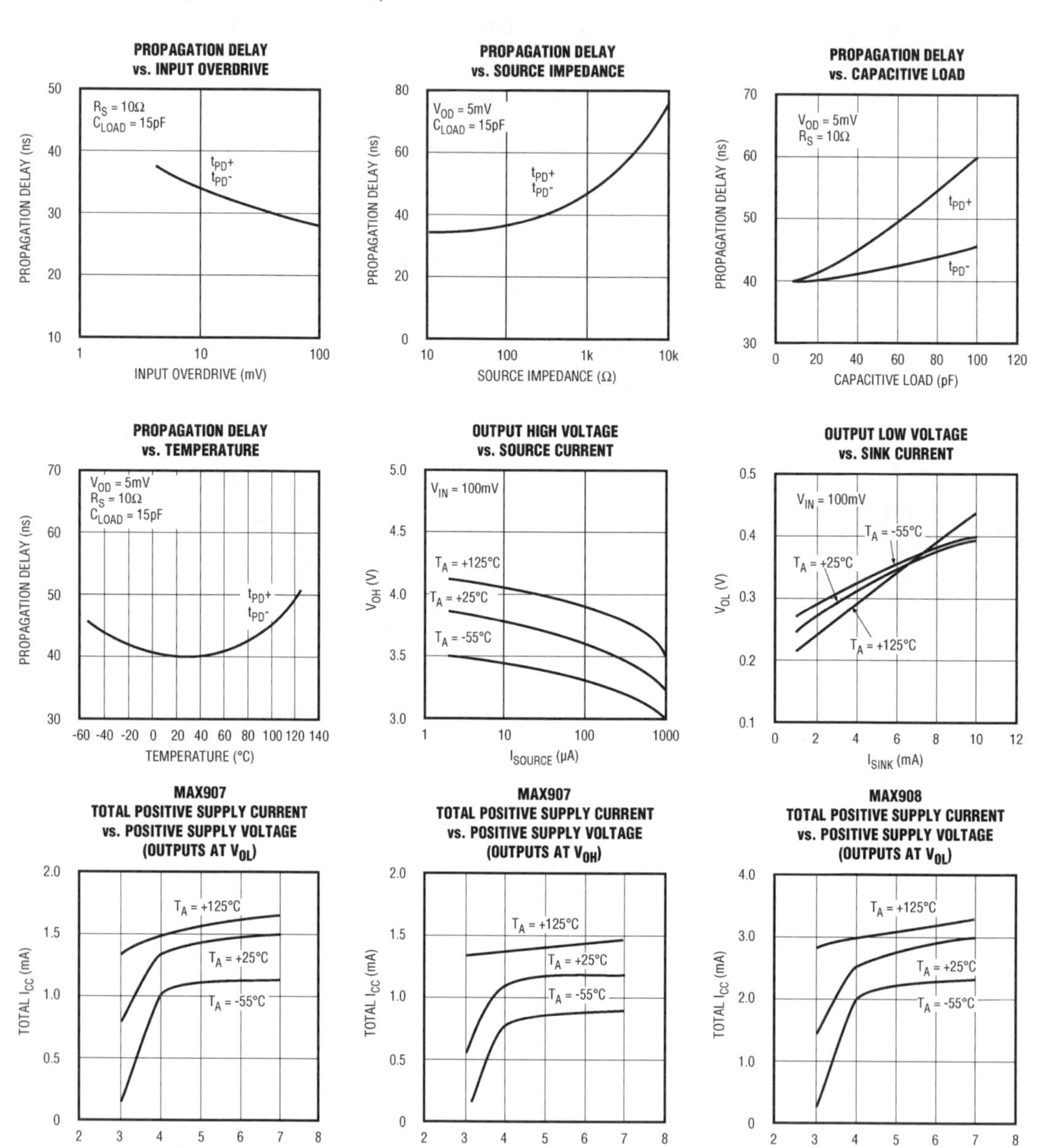

Single/Dual/Quad High-Speed, Ultra Low-Power, Single-Supply TTL Comparators

MAX907/MAX908/MAX909

Typical Operating Characteristics (continued)

(V+ = 5V, T_A = +25°C, unless otherwise noted.)

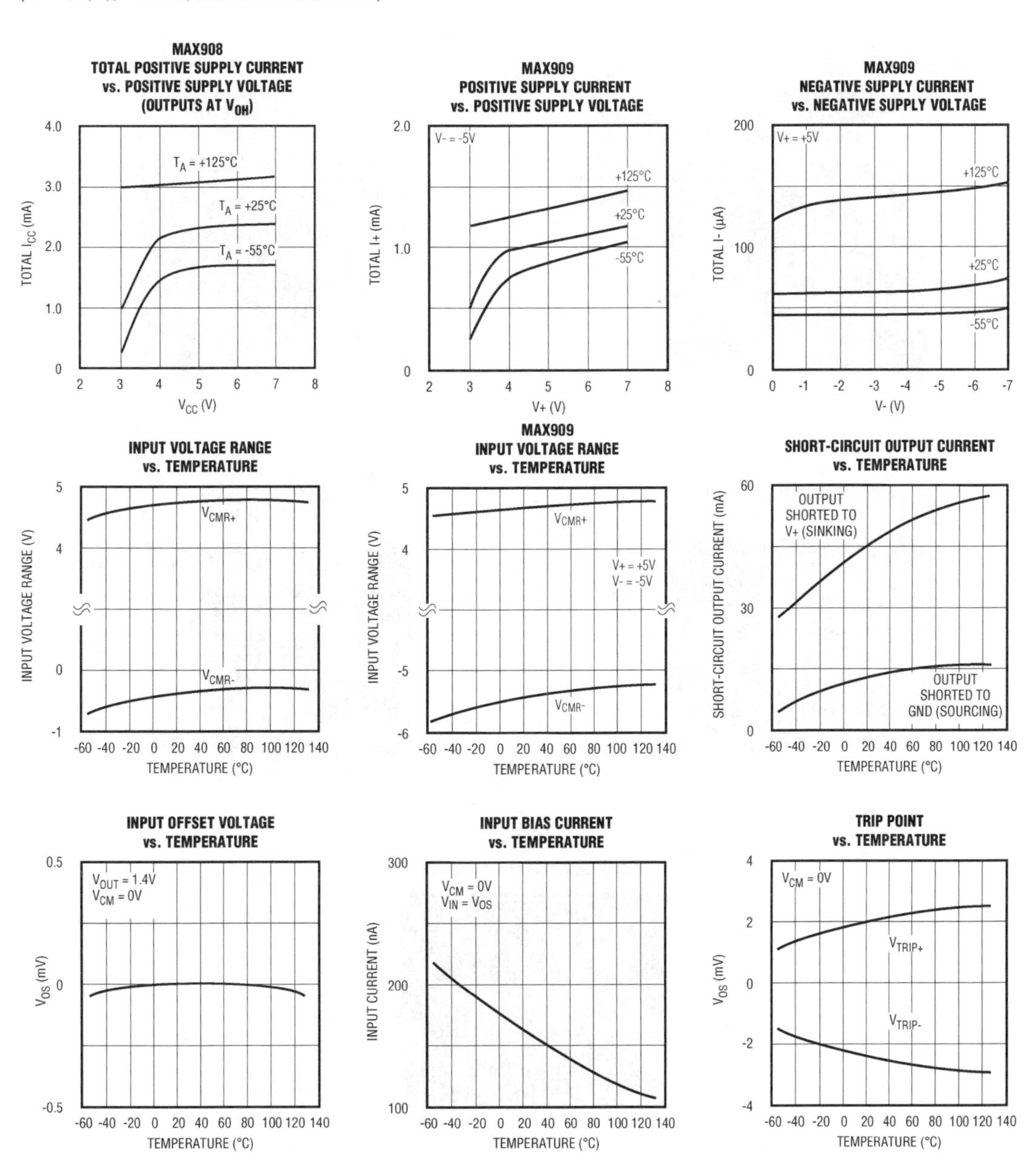

Single/Dual/Quad High-Speed, Ultra Low-Power, Single-Supply TTL Comparators

Typical Operating Characteristics (continued)

(V+ = 5V, T_A = +25°C, unless otherwise noted.)

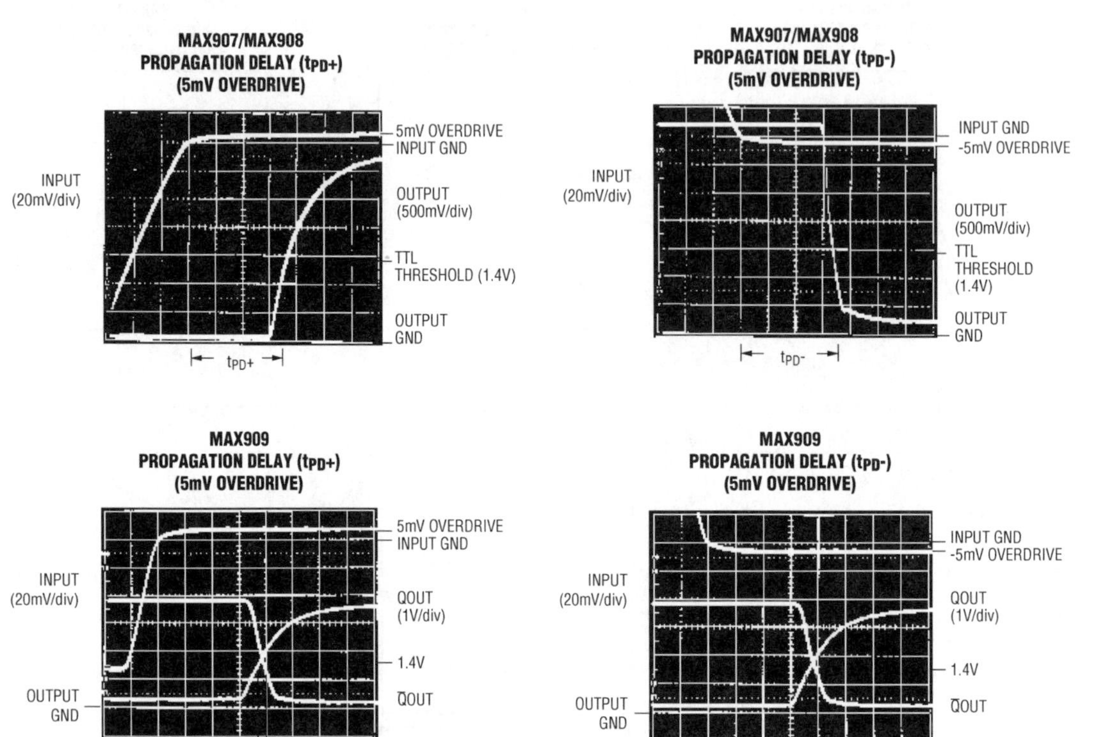

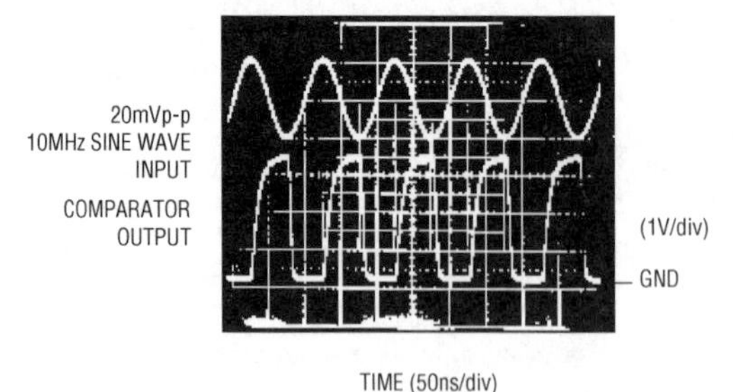

Single/Dual/Quad High-Speed, Ultra Low-Power, Single-Supply TTL Comparators

Pin Description

PIN			NAME	FUNCTION
MAX907	MAX908	MAX909		
1	1		OUTA	Comparator A Output
2	2		INA-	Comparator A Inverting Input
3	3		INA+	Comparator A Noninverting Input
8	4	1	V+	Positive Supply
5	5		INB+	Comparator B Noninverting Input
6	6		INB-	Comparator B Inverting Input
7	7		OUTB	Comparator B Output
	8		OUTC	Comparator C Output
	9		INC-	Comparator C Inverting Input
	10		INC+	Comparator C Noninverting Input
4	11	6	GND	Ground
	12		IND+	Comparator D Noninverting Input
	13		IND-	Comparator D Inverting Input
	14		OUTD	Comparator D Output
		2	IN+	Noninverting Input
		3	IN-	Inverting Input
		4	V-	Negative Supply or Ground
		5	LE	The latch is transparent when LE is low. The comparator output is stored when LE is high.
		7	QOUT	Comparator Output
		8	$\overline{QOUT}$	Inverted Comparator Output

Detailed Description

Timing

Noise or undesired parasitic AC feedback cause most high-speed comparators to oscillate in the linear region (i.e., when the voltage on one input is at or near the voltage on the other input). The MAX907/MAX908/MAX909 eliminate this problem by incorporating internal hysteresis. When the two comparator input voltages are equal, hysteresis effectively causes one comparator input voltage to move quickly past the other, thus taking the input out of the region where oscillation occurs. Standard comparators require that hysteresis be added through the use of external resistors. The MAX907/MAX908/MAX909's fixed internal hysteresis eliminates these resistors (and the equations required to determine appropriate values).

Adding hysteresis to a comparator creates two trip points: one for the input voltage rising and one for the input voltage falling (Figure 1). The difference between these two input-referred trip points is the hysteresis.

Figure 1 illustrates the case where IN- is fixed and IN+ is varied. If the inputs were reversed, the figure would look the same, except the output would be inverted.

The MAX909 includes an internal latch, allowing the result of a comparison to be stored. If LE is low, the latch is transparent (i.e., the comparator operates as though the latch is not present). The state of the comparator output is stored when LE is high. See Figure 2.

Note that the MAX909 can be operated with V- connected to ground or to a negative supply voltage. The MAX909's input range extends from (V- - 0.2V) to (V+ - 1.5V).

Single/Dual/Quad High-Speed, Ultra Low-Power, Single-Supply TTL Comparators

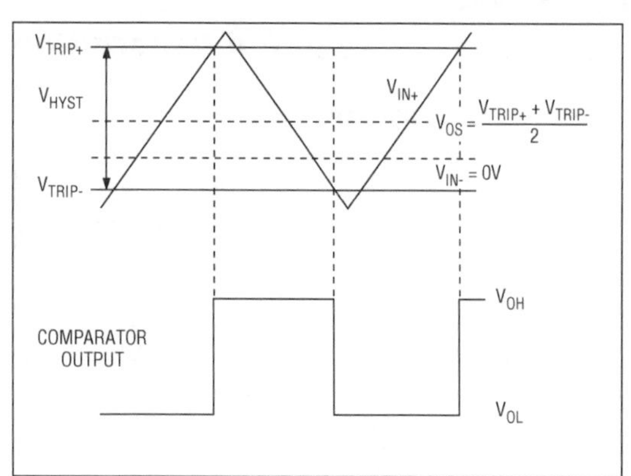

Figure 1. Input and Output Waveforms, Noninverting Input Varied

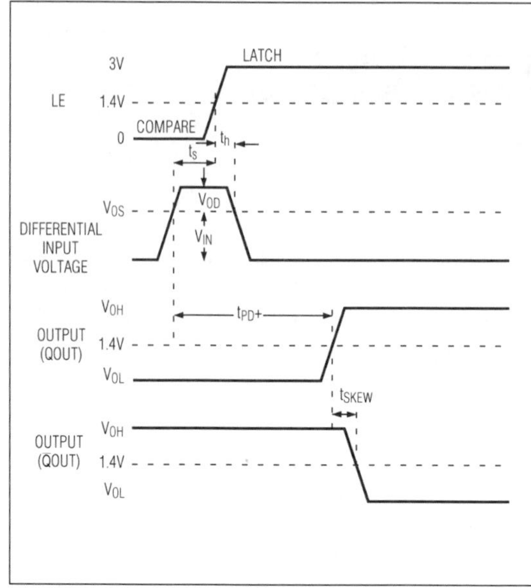

Figure 2. MAX909 Timing Diagram

Applications Information

Circuit Layout

Because of the MAX907/MAX908/MAX909's high gain bandwidth, special precautions must be taken to realize the full high-speed capability. A printed circuit board with a good, low-inductance ground plane is mandatory. Place the decoupling capacitor (a 0.1μF ceramic capacitor is a good choice) as close to V+ as possible. Pay close attention to the decoupling capacitor's bandwidth, keeping leads short. Short lead lengths on the inputs and outputs are also essential to avoid unwanted parasitic feedback around the comparators. Solder the device directly to the printed circuit board instead of using a socket.

Overdriving the Inputs

The inputs to the MAX907/MAX908/MAX909 may be driven beyond the voltage limits given in the *Absolute Maximum Ratings*, as long as the current flowing into the device is limited to 25mA. However, if the inputs are overdriven, the output may be inverted. The addition of an external diode prevents this inversion by limiting the input voltage to 200mV to 300mV below ground (see Figure 3).

Battery-Operated Infrared Data Link

Figure 4's circuit allows reception of infrared data. The MAX403 converts the photodiode current to a voltage, and the MAX907 determines whether the amplifier output is high enough to be called a "1". The current consumption of this circuit is minimal: The MAX403 and MAX907 require typically 250μA and 700μA, respectively.

Single/Dual/Quad High-Speed, Ultra Low-Power, Single-Supply TTL Comparators

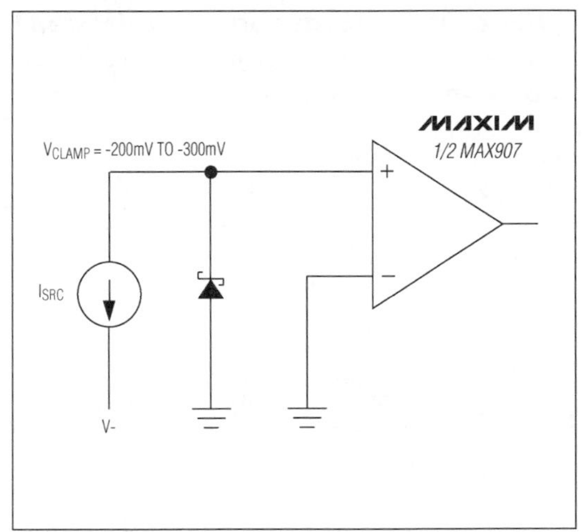

Figure 3. Schottky Clamp for Input Driven Below Ground

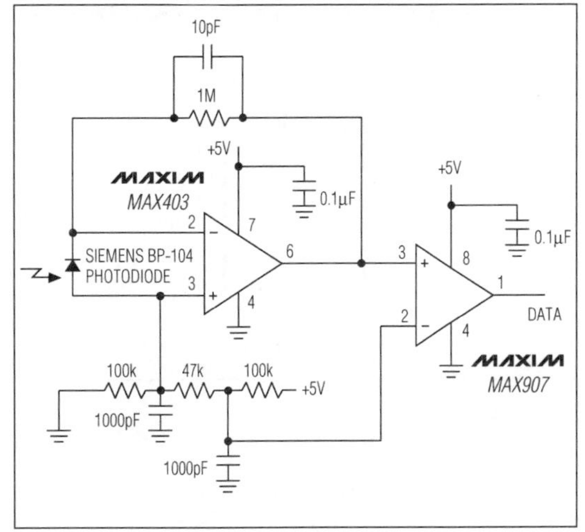

Figure 4. Battery-Operated Infrared Data Link Consumes Only 1mA

Single/Dual/Quad High-Speed, Ultra Low-Power, Single-Supply TTL Comparators

Pin Configurations (continued)

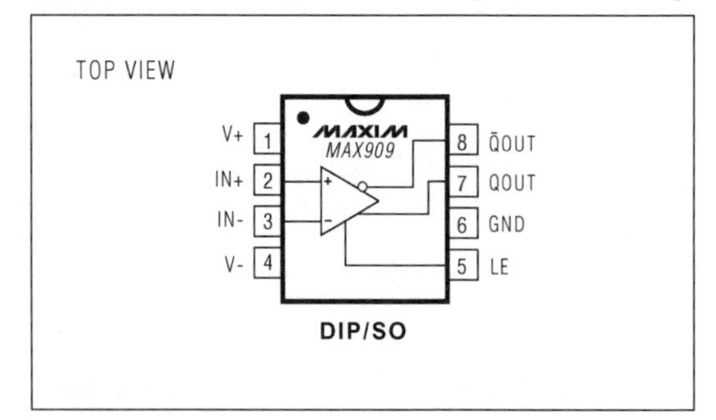

Ordering Information (continued)

PART	TEMP. RANGE	PIN-PACKAGE
MAX908CPD	0°C to +70°C	14 Plastic DIP
MAX908CSD	0°C to +70°C	14 SO
MAX908EPD	-40°C to +85°C	14 Plastic DIP
MAX908ESD	-40°C to +85°C	14 SO
MAX908MJD	-55°C to +125°C	14 CERDIP
MAX909CPA	0°C to +70°C	8 Plastic DIP
MAX909CSA	0°C to +70°C	8 SO
MAX909C/D	0°C to +70°C	Dice*
MAX909EPA	-40°C to +85°C	8 Plastic DIP
MAX909ESA	-40°C to +85°C	8 SO
MAX909MJA	-55°C to +125°C	8 CERDIP

** Dice are specified at +25°C, DC parameters only.*

Single/Dual/Quad High-Speed, Ultra Low-Power, Single-Supply TTL Comparators

Chip Topographies

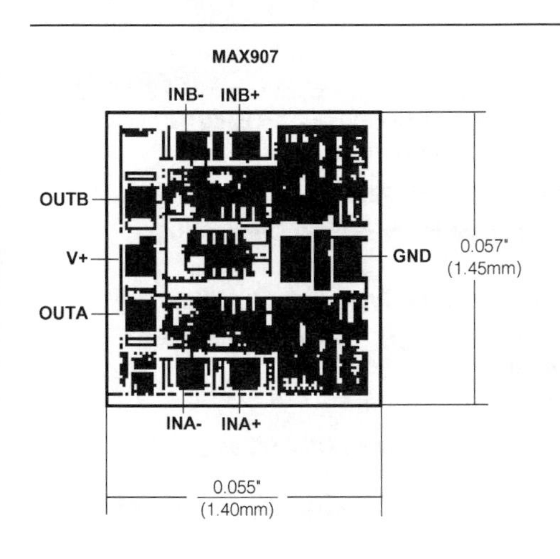

MAX909

V+

IN+

IN-

V-

$\overline{Q}$OUT

QOUT

GND

LE

0.051"
(1.30mm)

0.050"
(1.27mm)

TRANSISTOR COUNT: MAX907: 180
MAX908: 360

SUBSTRATE CONNECTED TO GND.

TRANSISTOR COUNT: 95;

SUBSTRATE CONNECTED TO V-.

19-0481; Rev 2; 11/92

MAXIM

General Purpose Timers

ICM7555/7556

General Description

The Maxim ICM7555 and ICM7556 are respectively single and dual general purpose RC timers capable of generating accurate time delays or frequencies. The primary feature is an extremely low supply current, making this device ideal for battery-powered systems. Additional features include low THRESHOLD, $\overline{\text{TRIGGER}}$, and $\overline{\text{RESET}}$ currents, a wide operating supply voltage range, and improved performance at high frequencies.

These CMOS low-power devices offer significant performance advantages over the standard 555 and 556 bipolar timers. Low-power consumption, combined with the virtually non-existent current spike during output transitions, make these timers the optimal solution in many applications.

Applications

- Pulse Generator
- Precision Timing
- Time Delay Generation
- Pulse Width Modulation
- Pulse Position Modulation
- Sequential Timing
- Missing Pulse Detector

Pin Configuration

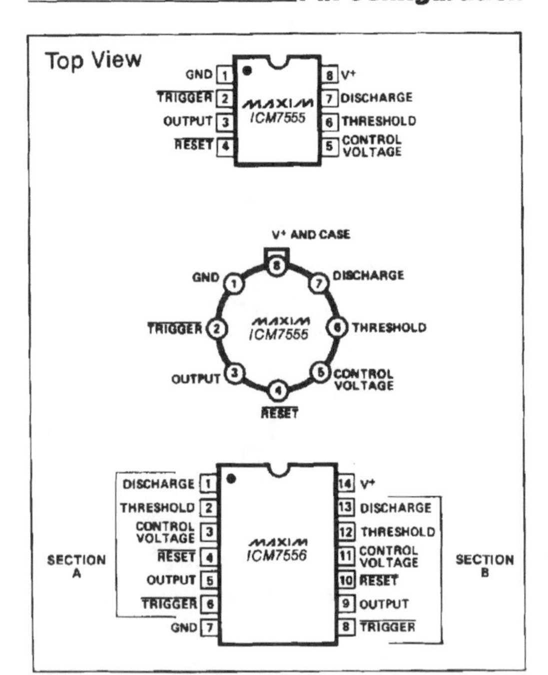

Features

- ♦ Improved 2nd Source! (See 3rd page for "Maxim Advantage™").
- ♦ Wide Supply Voltage Range: 2-18V
- ♦ No Crowbarring of Supply During Output Transition
- ♦ Adjustable Duty Cycle
- ♦ Low THRESHOLD, TRIGGER and RESET Curents
- ♦ TTL Compatible
- ♦ Monolithic, Low Power CMOS Design

Ordering Information

PART	TEMP. RANGE	PACKAGE
ICM7555IPA	-20°C to +85°C	8 Lead Plastic DIP
ICM7555IJA	-20°C to +85°C	8 Lead CERDIP
ICM7555ITV	-20°C to +85°C	TO-99 Can
ICM7555MJA	-55°C to +125°C	8 Lead CERDIP
ICM7555MTV	-55°C to +125°C	TO-99 Can
ICM7555ISA	-20°C to +85°C	8 Lead Small Outline
ICM7555/D	0°C to +70°C	Dice
ICM7556IPD	-20°C to +85°C	14 Lead Plastic DIP
ICM7556MJD	-55°C to +125°C	14 Lead CERDIP
ICM7556ISD	-20°C to +85°C	14 Lead Small Outline
ICM7556/D	0°C to +70°C	Dice

Typical Operating Circuit

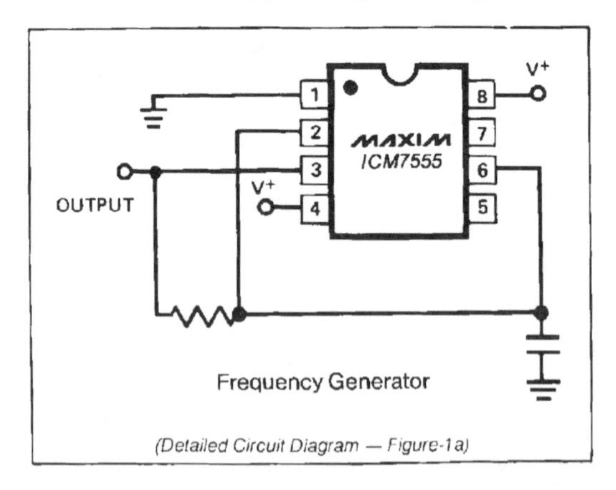

Frequency Generator

(Detailed Circuit Diagram — Figure-1a)

The "Maxim Advantage™" signifies an upgraded quality level. At no additional cost we offer a second-source device that is subject to the following: guaranteed performance over temperature along with tighter test specifications on many key parameters; and device enhancements, when needed, that result in improved performance without changing the functionality.

ICM7555/7556

General Purpose Timers

ABSOLUTE MAXIMUM RATINGS (Note 1)

Supply Voltage . . . . +18 Volts
Input Voltage TRIGGER
Control Voltage THRESHOLD <V+ +0.3V to ≥ −0.3V
RESET
Output Current 100mA
Power Dissipation[2] ICM7556 . . . . 300mW
ICM7555 . . . . 200mW
Operating Temperature Range
ICM7555IJA (Maxim) . . . . −20°C to +85°C
ICM7555ISA (Maxim) . . . . −20°C to +85°C
ICM7555IPA . . . . −20°C to +85°C
ICM7555ITV . . . . −20°C to +85°C
ICM7556IPD . . . . −20°C to +85°C
ICM7555MTV . . . . −55°C to +125°C
ICM7556MJD . . . . −55°C to +125°C
Storage Temperature . . . . −65°C to +150°C
Lead Temperature (Soldering 60 Seconds) . . . . +300°C

Stresses above those listed under Absolute Maximum Ratings may cause permanent damage to the device. These are stress ratings only and functional operation of the device at these or any other conditions above those indicated in the operational sections of the specifications is not implied. Exposure to absolute maximum rating conditions for extended periods may affect device reliability.

ELECTRICAL CHARACTERISTICS

($V^+ = +2$ to $+15$ volts; $T_A = 25°C$, Unless Noted)

PARAMETER	SYMBOL	TEST CONDITIONS	VALUE			UNITS
			MIN	TYP	MAX	
Supply Voltage	V^+	$-20°C \le T_A \le +70°C$	2		18	V
		$-55°C \le T_A \le +125°C$	3		16	V
Supply Current [3]	I^+	ICM7555 $V^+ = 2V$		60	200	μA
		$V^+ = 18V$		120	300	μA
		ICM7556 $V^+ = 2V$		120	400	μA
		$V^+ = 18V$		240	600	μA
Timing Error		$R_A, R_B = 1k$ to $100k$, $5V \le V^+ \le 15V$, $C = 0.1\mu F$				
Initial Accuracy		Note 4		2.0	5.0	%
Drift with Temperature		Note 4 $V^+ = 5V$		50		ppm/°C
		$V^+ = 10V$		75		
		$V^+ = 15V$		100		
Drift with Supply Voltage		$V^+ = 5V$		1.0	3.0	%/V
Threshold Voltage	V_{TH}	$V^+ = 5V$	0.63	0.66	0.67	V^+
Trigger Voltage	V_{TRIG}	$V^+ = 5V$	0.29	0.33	0.34	V^+
Trigger Current	I_{TRIG}	$V^+ = 18V$		50		pA
		$V^+ = 5V$		10		pA
		$V^+ = 2V$		1		pA
Threshold Current	I_{TH}	$V^+ = 18V$		50		pA
		$V^+ = 5V$		10		pA
		$V^+ = 2V$		1		pA
Reset Current	I_{RST}	V_{RESET} = Ground $V^+ = 18V$		100		pA
		$V^+ = 5V$		20		pA
		$V^+ = 2V$		2		pA
Reset Voltage	V_{RST}	$V^+ = 18V$	0.4	0.7	1.0	V
		$V^+ = 2V$	0.4	0.7	1.0	V
Control Voltage Lead	V_{CV}	$V^+ = 5V$	0.62	0.66	0.67	V^+
Output Voltage Drop	V_O	Output Lo $V^+ = 18V$ $I_{SINK} = 3.2mA$		0.1	0.4	V
		$V^+ = 5V$ $I_{SINK} = 3.2mA$		0.15	0.4	V
		Output Hi $V^+ = 18V$ $I_{SOURCE} = 1.0mA$	17.25	17.8		V
		$V^+ = 5V$ $I_{SOURCE} = 1.0mA$	4.0	4.5		V
Rise Time of Output	t_r	$R_L = 10M\Omega$ $C_L = 10pF$ $V^+ = 5V$	35	40	75	ns
Fall Time of Output	t_f	$R_L = 10M\Omega$ $C_L = 10pF$ $V^+ = 5V$	35	40	75	ns
Guaranteed Max Osc Freq	f_{max}	Astable Operation	*500			kHz

Note 1: Due to the SCR structure inherent in the CMOS process used to fabricate these devices, connecting any terminal to a voltage greater than $V^+ + 0.3V$ or less than $V^- - 0.3V$ may cause destructive latchup. For this reason it is recommended that no inputs from external sources not operating from the same power supply be applied to the device before its power supply is established. In multiple systems, the supply of the ICM7555/6 must be turned on first.

Note 2: Junction temperatures should not exceed 135°C and the power dissipation must be limited to 20mW at 125°C. Below 125°C power dissipation may be increased to 300mW at 25°C. Derating factor is approximately 3mW/°C (7556) or 2mW/°C (7555).

Note 3: The supply current value is essentially independent of the TRIGGER, THRESHOLD and RESET voltages.

Note 4: Parameter is not 100% tested. Majority of all units meet this specification.

The electrical characteristics above are a reproduction of a portion of Intersil's copyrighted (1983/1984) data book. This information does not constitute any representation by Maxim that Intersil's products will perform in accordance with these specifications. The "Electrical Characteristics Table" along with the descriptive excerpts from the original manufacturer's data sheet have been included in this data sheet solely for comparative purposes.

MAXIM
ADVANTAGE™
General Purpose Timers

ICM7555/7556

- **Lower Supply Current**
- **Increased Output Source Current**
- **Guaranteed THRESHOLD, $\overline{\text{TRIGGER}}$ and $\overline{\text{RESET}}$ Input Currents**
- **Guaranteed Discharge Output Voltage**
- **Supply Current Guaranteed Over Temperature**
- **Significantly Improved ESD Protection (Note 6)**
- **Maxim Quality and Reliability**

ABSOLUTE MAXIMUM RATINGS
This device conforms to the Absolute Maximum Ratings on adjacent page.

ELECTRICAL CHARACTERISTICS
Specifications below satisfy or exceed all "tested" parameters on adjacent page.

($V^+ = +2$ to $+15$ volts; $T_A = 25°C$, unless noted.)

PARAMETER	SYMBOL	TEST CONDITIONS	MIN	TYP	MAX	UNITS
Supply Voltage	V^+	$-20°C \le T_A \le +85°C$ $-55°C \le T_A \le +125°C$	2 3		16.5 16	V V
Supply Current (Note 3)	I^+	**ICM 7555** $V^+ = 2\text{-}16.5V$; $T_A = +25°C$ $V^+ = 5V$; $T_A = +25°C$ $V^+ = 5V$; $-20°C \le T_A \le +85°C$ $V^+ = 5V$; $-55°C \le T_A \le +125°C$		**30**	**250** **120** **250** **300**	**μA** **μA** **μA** **μA**
		ICM 7556 $V^+ = 2\text{-}16.5V$; $T_A = +25°C$ $V^+ = 5V$; $T_A = +25°C$ $V^+ = 5V$; $-20°C \le T_A \le +85°C$ $V^+ = 5V$; $-55°C \le T_A \le +125°C$		**60**	**500** **240** **500** **600**	**μA** **μA** **μA** **μA**
Timing Error (Note 4)		Circuit of figure 1(b); $R_A = R_B = 100k\Omega$, $C = 0.1\mu F$, $V^+ = 5V$				
Initial Accuracy (Note 5)				2.0	5.0	%
Drift with Temperature		$V^+ = 5V$ $V^+ = 10V$ $V^+ = 15V$		50 75 100		ppm/°C ppm/°C ppm/°C
Drift with Supply Voltage		$V^+ = 5V$		1.0	3.0	%/V
Threshold Voltage	V_{TH}	$V^+ = 5V$	0.63	0.66	0.67	V^+
Trigger Voltage	V_{TRIG}	$V^- = 5V$	0.29	0.33	0.34	V^+
Trigger Current	I_{TRIG}	$V^+ = 16.5V$ $V^+ = 5V$ $V^+ = 2V$		**50** 10 1		**pA** pA pA
Threshold Current	I_{TH}	$V^+ = 16.5V$ $V^+ = 5V$ $V^+ = 2V$		**50** 10 1		**pA** pA pA
Reset Current	I_{RST}	V_{RESET} = Ground $V^+ = 16.5V$ $V^+ = 5V$ $V^+ = 2V$		**100** 20 2		**pA** pA pA
Reset Voltage	V_{RST}	$V^+ = 16.5V$ $V^+ = 2V$	0.4 0.4	0.7 0.7	1.2 1.2	V V
Control Voltage	V_{CV}	$V^+ = 5V$	0.62	0.66	0.67	V^+
Output Voltage Drop	V_O	Output Lo $V^+ = 16.5V$ $I_{SINK} = 3.2mA$ $V^+ = 5V$ $I_{SINK} = 3.2mA$ **Output Hi** $V^+ = 16.5V$ $I_{SOURCE} = 2.0mA$ $V^+ = 5V$ $I_{SOURCE} = 2.0mA$	 15.75 4.0	0.1 0.15 16.25 4.5	0.4 0.4	V V V V
Discharge Output Voltage	V_{DIS}	$V^+ = 5V$, $I_{DIS} = 3.2mA$		**0.1**	**0.4**	**V**
Rise Time of Output (Note 4)	t_r	$R_L = 10M\Omega$ $C_L = 10pF$ $V^+ = 5V$	35	40	75	ns
Fall Time of Output (Note 4)	t_f	$R_L = 10M\Omega$ $C_L = 10pF$ $V^+ = 5V$	35	40	75	ns
Guaranteed Max Osc. Freq. (Note 4)	f_{max}	Astable Operation	500			kHz

Note 1: Due to the SCR structure inherent in the CMOS process used to fabricate these devices, connecting any terminal to a voltage greater than $V^+ +0.3V$ or less than $V^- -0.3V$ may cause destructive latchup. For this reason it is recommended that no inputs from external sources not operating from the same power supply be applied to the device before its power supply is established. In multiple systems, the supply of the ICM7555/6 must be turned on first.

Note 2: Junction temperatures should not exceed 135°C and the power dissipation must be limited to 20mW at 125°C. Below 125°C power dissipation may be increased to 300mW at 25°C. Derating factor is approximately 3mW/°C (7556) or 2mW/°C (7555).

Note 3: The supply current value is essentially independent of the $\overline{\text{TRIGGER}}$, THRESHOLD AND $\overline{\text{RESET}}$ voltages.

Note 4: Parameter is not 100% tested. Majority of all units meet this specification.

Note 5: Deviation from $f = 1.48/(R_A + 2R_B)C$, $V^+ = 5V$.

Note 6: All pins are designed to withstand electrostatic discharge (ESD) levels in excess of 2000V. (Mil Std 883B, Method 3015.1 Test Circuit.)

General Purpose Timers

Typical Operating Characteristics

DISCHARGE OUTPUT CURRENT AS A FUNCTION OF DISCHARGE OUTPUT VOLTAGE

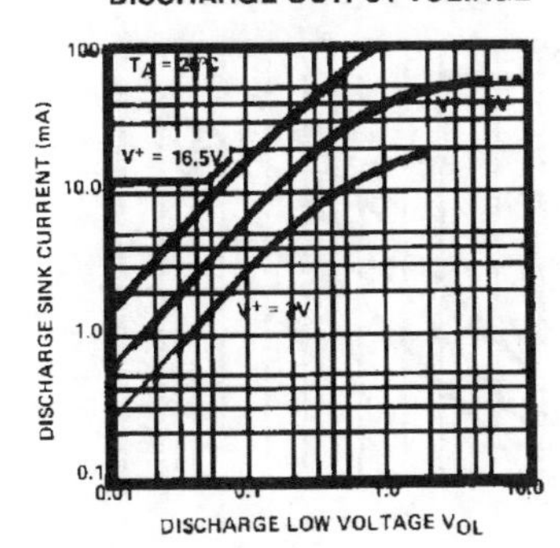

FREE RUNNING FREQUENCY AS A FUNCTION OF R_A, R_B AND C

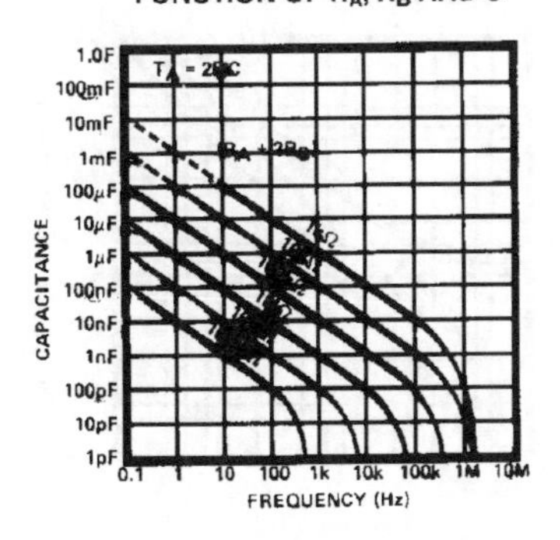

NORMALIZED FREQUENCY STABILITY IN THE ASTABLE MODE AS A FUNCTION OF SUPPLY VOLTAGE

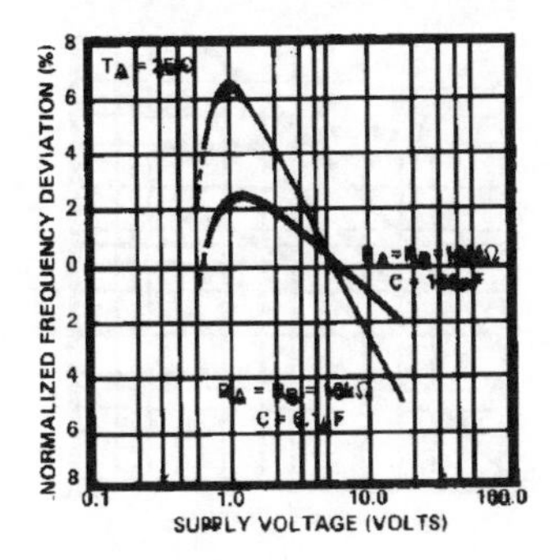

PROPAGATION DELAY AS A FUNCTION OF VOLTAGE LEVEL OF TRIGGER PULSE

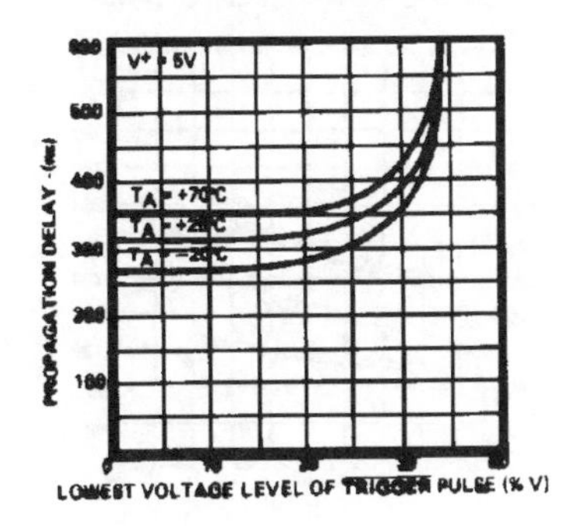

NORMALIZED FREQUENCY STABILITY IN THE ASTABLE MODE AS A FUNCTION OF TEMPERATURE

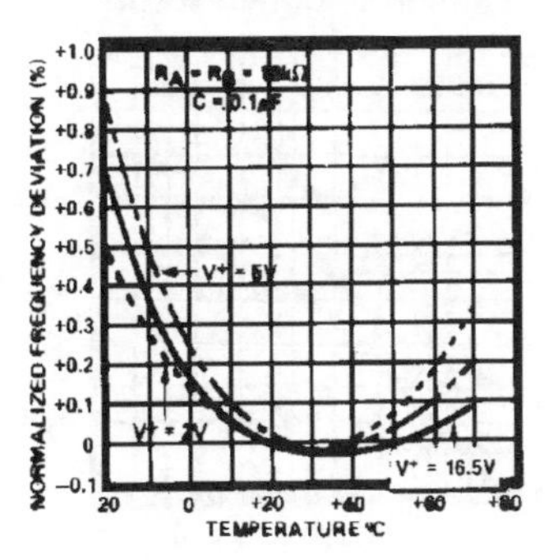

TIME DELAY IN THE MONOSTABLE MODE AS A FUNCTION OF R_A AND C

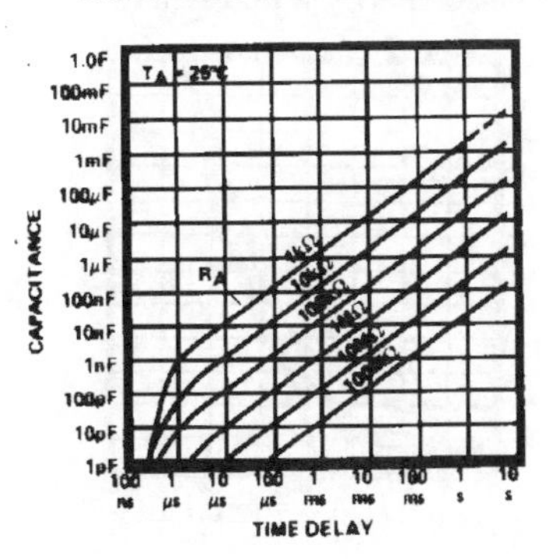

General Purpose Timers

Typical Operating Characteristics

SUPPLY CURRENT AS A FUNCTION OF SUPPLY VOLTAGE

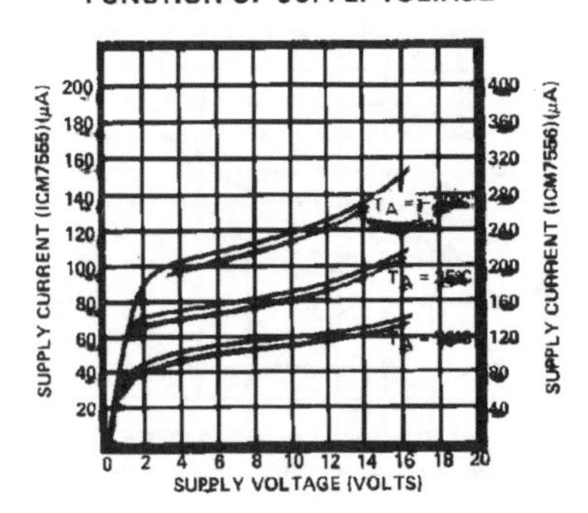

OUTPUT SINK CURRENT AS A FUNCTION OF OUTPUT VOLTAGE

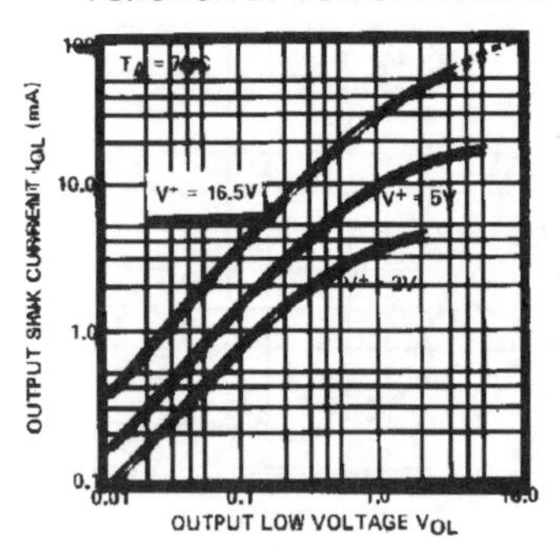

MINIMUM PULSE WIDTH REQUIRED FOR TRIGGERING

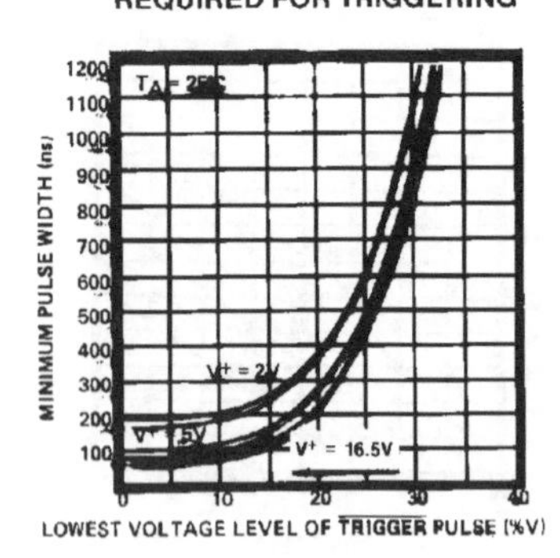

OUTPUT SINK CURRENT AS A FUNCTION OF OUTPUT VOLTAGE

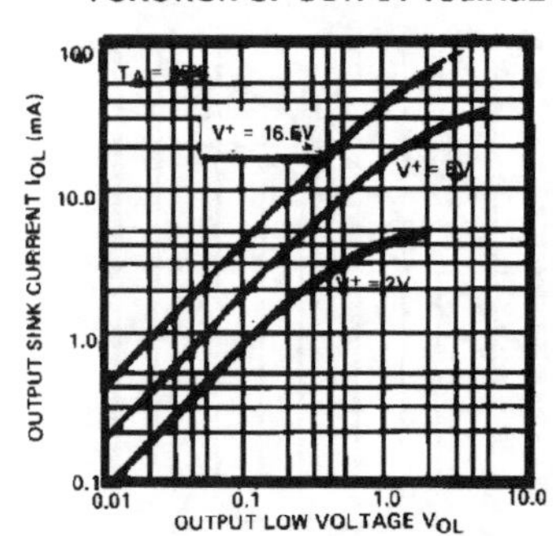

OUTPUT SOURCE CURRENT AS A FUNCTION OF OUTPUT VOLTAGE

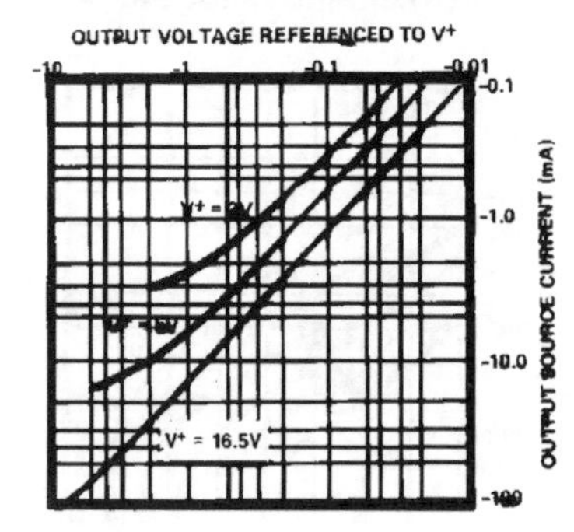

OUTPUT SINK CURRENT AS A FUNCTION OF OUTPUT VOLTAGE

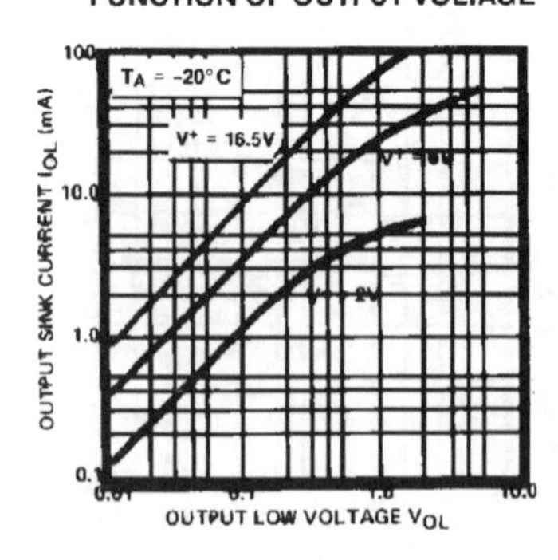

General Purpose Timers

Detailed Description

Both the ICM7555 timer and the ICM7556 dual timer can be configured for either astable or monostable operation. In the astable mode the free running frequency and the duty cycle are controlled by two external resistors and one capacitor. Similarly, the pulse width in the monostable mode is precisely controlled by one external resistor and capacitor.

The external component count is decreased when replacing a bipolar timer with the ICM7555 or ICM7556. The bipolar devices produce large crowbar currents in the output driver. To compensate for this spike, a capacitor is used to decouple the power supply lines. The CMOS timers produce supply spikes of only 2-3mA vs. 300-400mA (Bipolar), therefore supply decoupling is typically not needed. This current spike comparison is illustrated in Figure 3. Another component is eliminated at the control voltage pin. These CMOS timers, due to the high impedance inputs of the comparators, do not require decoupling capacitors on the control voltage pin.

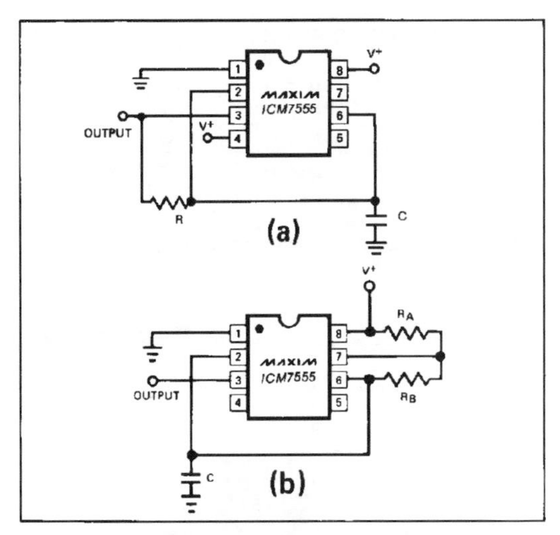

Figure 1. Maxim ICM7555 used in two different astable configurations.

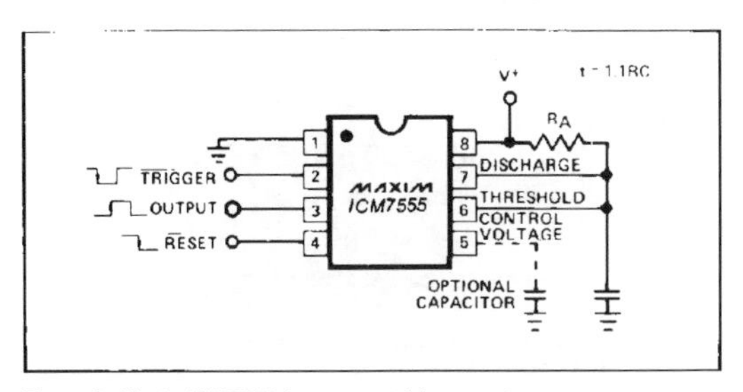

Figure 2. Maxim ICM7555 in a monostable operation.

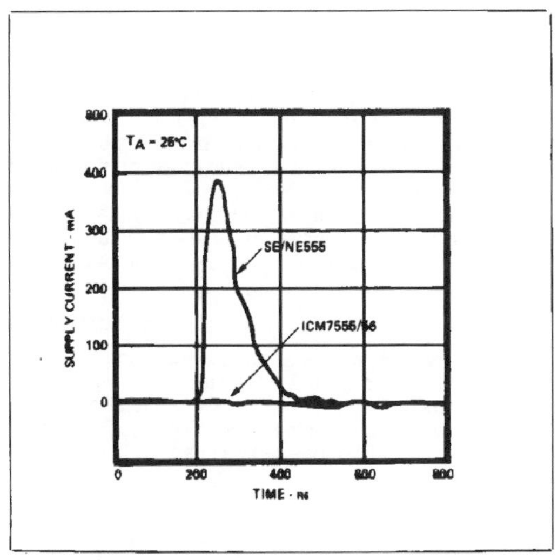

Figure 3. Supply current transient compared with a standard bipolar 555 during an output transition.

Applications Information

Astable Operation

We recommend either of the two astable circuit configurations illustrated in Figure 1. The circuit in (1a) provides a 50% duty cycle output using one timing resistor and capacitor. The oscillator waveform across the capacitor is symmetrical and triangular, swinging from ⅓ to ⅔ of the supply voltage. The frequency generated is defined by:

$$f = \frac{1}{1.4\,RC}$$

The circuit in (1b) provides a means of varying the duty cycle of the oscillator. The frequency is defined by:

$$f = \frac{1.46}{(R_A + 2R_B)C}$$

The duty cycle is:

$$D = \frac{R_B}{R_A + 2R_B}$$

Monostable Operation

The circuit diagram in Figure 2 illustrates monostable operation. In this mode the timer acts as a one shot. Initially the external capacitor is held discharged by the discharge output. Upon application of a negative $\overline{\text{TRIGGER}}$ pulse to pin 2, the capacitor begins to charge exponentially through R_A. The device resets after the voltage across the capacitor reaches $\frac{2}{3}(V^+)$.

$$t_{output} = -\ln(1/3)R_AC = 1.1\,R_AC$$

General Purpose Timers

Reset

The reset function is significantly improved over the standard bipolar 555 and 556 in that it controls only the internal flip-flop, which in turn simultaneously controls the state of the Output and Discharge pins. This avoids the multiple threshold problems sometimes encountered with slow-falling edges of the bipolar devices. This input is designed to have essentially the same trip voltage as the standard bipolar devices (0.6 to 0.7V). At all supply voltages this input maintains an extremely high impedance.

Control Voltage

The control voltage regulates the two trip voltages for the THRESHOLD and TRIGGER internal comparators. This pin can be used for frequency modulation in the astable mode. By varying the applied voltage to the control voltage pin, delay times can be changed in the monostable mode.

Power Supply Considerations

Since the TRIGGER, THRESHOLD and Discharge leakage currents are very low, high impedance timing components may be used, keeping total system supply current at a minimum.

Output Drive Capability

The CMOS output stage is capable of driving most logic families including CMOS and TTL. The ICM7555 and ICM7556 will drive at least two standard TTL loads at a supply voltage of 4.5V or greater. When driving CMOS, the output swing at all supply voltage levels will equal the supply voltage.

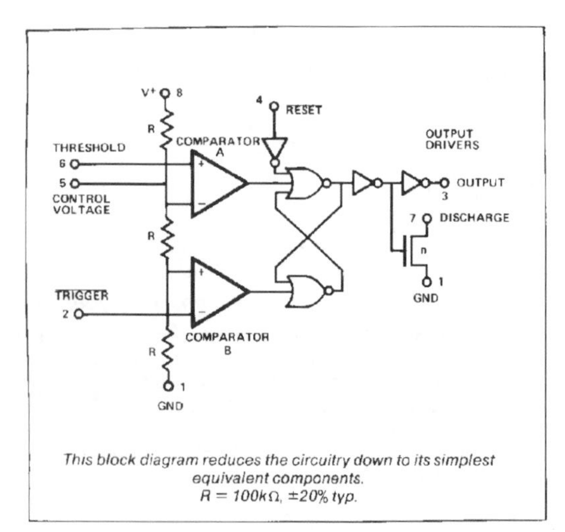

This block diagram reduces the circuitry down to its simplest equivalent components.
R = 100kΩ, ±20% typ.

Figure 4. Block diagram of ICM7555.

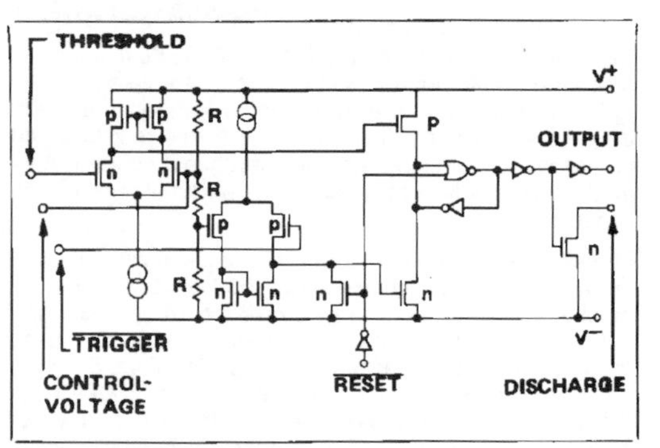

Figure 5. Equivalent circuit.

Function Table

RESET	TRIGGER VOLTAGE†	THRESHOLD VOLTAGE†	OUTPUT	DISCHARGE SWITCH
Low	Irrelevant	Irrelevant	Low	On
High	$< \frac{1}{3} V^+$	Irrelevant	High	Off
High	$> \frac{1}{3} V^+$	$> \frac{2}{3} V^+$	Low	On
High	$> \frac{1}{3} V^+$	$< \frac{2}{3} V^+$	As previously established	

†Voltages levels shown are nominal.

NOTE: RESET will dominate all other inputs. TRIGGER will dominate over THRESHOLD.

Chip Topographies

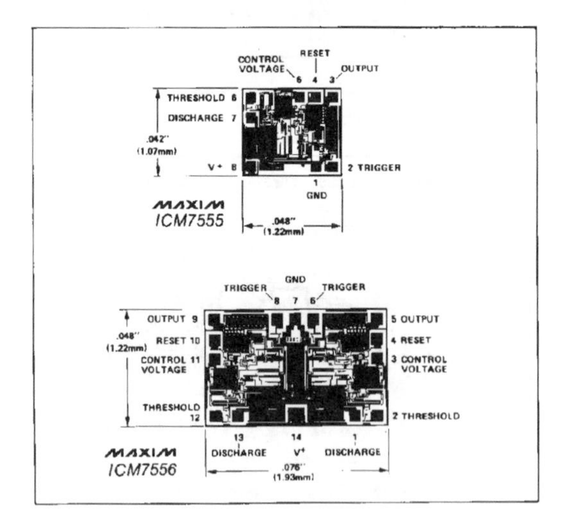

General Purpose Timers

Package Information

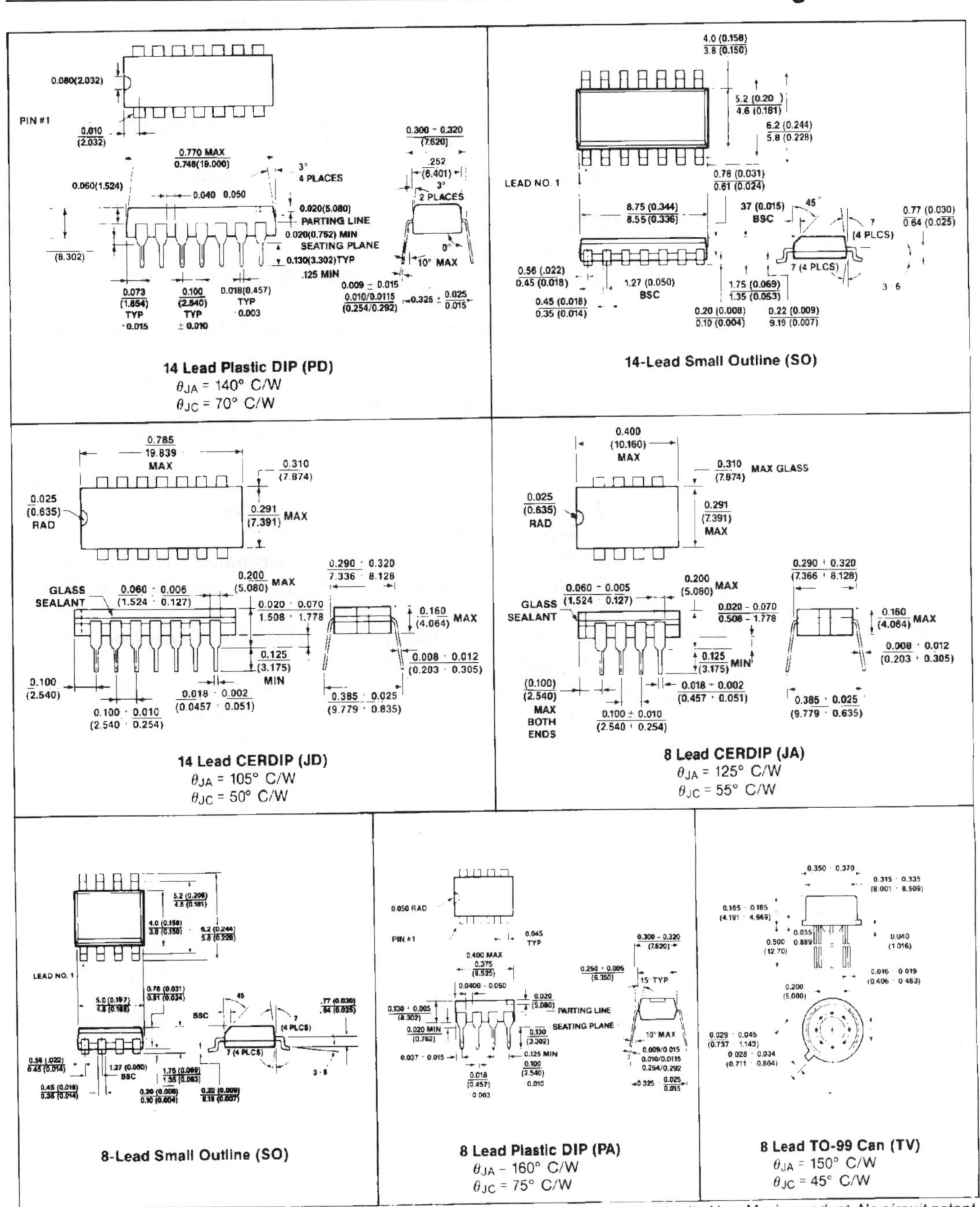

14 Lead Plastic DIP (PD)
θ_{JA} = 140° C/W
θ_{JC} = 70° C/W

14-Lead Small Outline (SO)

14 Lead CERDIP (JD)
θ_{JA} = 105° C/W
θ_{JC} = 50° C/W

8 Lead CERDIP (JA)
θ_{JA} = 125° C/W
θ_{JC} = 55° C/W

8-Lead Small Outline (SO)

8 Lead Plastic DIP (PA)
θ_{JA} = 160° C/W
θ_{JC} = 75° C/W

8 Lead TO-99 Can (TV)
θ_{JA} = 150° C/W
θ_{JC} = 45° C/W

Bibliography

[1] Terrell, *Op Amps: Design, Application and Troubleshooting*, 1996 (Butterworth-Heinemann) ISBN 0–7506–9702–4

[2] Graeme, *Optimizing Op Amp Performance*, 1997 (McGraw-Hill) ISBN 0–07–024522–3

[3] Dostal, *Operational Amplifiers*, 1993 (Butterworth-Heinemann) ISBN 0–7506–9317–7

[4] Texas Instruments, *Linear Design Seminar Slide Book*, 1992 (Texas Instruments) SLYZE01.

[5] Texas Instruments, *Linear Mixed Signal Design Seminar Reference Book*, 1994 (Texas Instruments) SLY6E03.

[6] National Semiconductor, *1999 Analog Seminar Reference Book*, 1999 (National Semiconductor) Literature Number 570141–004.

[7] Various Application Notes from Texas Instruments, National Semiconductors, Maxim, Analog Devices

[8] Various magazine articles from *Electronics World, Electronic Engineering,* EDN,

[9] Horowitz and Hill, *The Art of Electronics*, 1989 (Cambridge University Press) ISBN 0–521–37095–7

[10] Graf, *Amplifier Circuits*, 1997 (Butterworth-Heinemann) ISBN 0–7506–9877–2

[11] Hickman, *Electronic Circuits, Systems and Standards*, 1991 (Butterworth-Heinemann) ISBN 0–7506–0068–3

[12] Hickman and Travis, *EDN Designers Companion*, 1994 (Butterworth-Heinemann) ISBN 0–7506–1721–7

[13] Savant, Roden and Carpenter, *Electronic design: Circuits and Systems*, 1991 (Benjamin Cummings) ISBN 0–8053–0285–9

[14] Winder, *Analog and Digital Filter Design*, 2002 (Newnes) ISBN 0–7506–7547–0.

Index